Essentials of Human Physiology

Essentials of Human Physiology

EDITED BY

GORDON ROSS, M.D.
Departments of Medicine and Physiology
University of California, Los Angeles,
School of Medicine

YEAR BOOK MEDICAL PUBLISHERS, INC.
Chicago • London

Library of Congress Catalog Card Number: 77-081527

International Standard Book Number: 0-8151-7438-1

Contributors

EUGENE D. JACOBSON, M.D., Professor and Associate Dean, University of Cincinnati Medical Center, Cincinnati, Ohio

WILFRIED F. H. M. MOMMAERTS, M.D., Ph.D., Professor and Chairman, Department of Physiology, University of California, Los Angeles, School of Medicine

BRIAN W. PAYTON, M.D., Professor of Physiology, Director of Medical Audio-Visual Services, Memorial University of Newfoundland, St. Johns, Newfoundland, Canada

GORDON ROSS, M.D., Professor, Departments of Medicine and Physiology, University of California, Los Angeles, School of Medicine

CLARK T. SAWIN, M.D., Chief, Endocrine-Diabetes Section, V.A. Hospital, Boston, Associate Professor of Medicine, Tufts University School of Medicine, Boston, Massachusetts

RAYMOND G. SCHULTZE, M.D., Associate Professor and Vice-Chairman, Department of Medicine, University of California, Los Angeles, School of Medicine

BRIAN J. WHIPP, Ph.D., Associate Professor, Department of Physiology, University of California, Los Angeles, School of Medicine

ERNEST M. WRIGHT, Ph.D., Professor, Department of Physiology, University of California, Los Angeles, School of Medicine

Preface

DURING THE LAST DECADE important changes have occurred in medical and dental school curricula. The total period of formal instruction has decreased and in many schools this has been accomplished at the expense of the basic sciences. There has also been increasing stress on interdisciplinary courses, electives and self-study. At the same time the expansion of knowledge has continued unabated and most schools rightly continue to demand a thorough understanding of the basic sciences as the cornerstone of medical education and a necessary prerequisite of professional practice. Thus today's students are being compelled to learn more basic science in a shorter time.

Discussions with medical and dental students have indicated dissatisfaction with many of the current physiology texts. Some were thought to overwhelm the student with a mass of detail, while others were considered to have such a terse, telegraphic style as to make them difficult to read. A common complaint was that students felt unable to discern the relationship between the material presented and their career goals.

The present work is an attempt to present the essential facts and concepts of medical physiology in a relatively short text without sacrificing readability. In order to achieve this, some aspects of physiology such as historical development, experimental methods and comparative physiology have received scant attention. To compensate for the brevity of the text a short list of recommended reviews and books has been provided at the end of each section. Original papers are only rarely referred to as we have found that most medical and dental students do not read them and those individuals with special interests readily find their own way in the literature when guided by appropriate reviews.

A major goal of the book is to emphasize the way in which knowledge of human physiology aids in the understanding of clinical medicine and vice versa. Abnormalities of structure and function frequently occur in nature in a way that cannot be duplicated in the laboratory and that illumine the mechanisms underlying normal function. Conversely, many diseases are better understood when considered in terms of disordered physiology. We have tried to make the student aware of these interrelations in all sections of the book, which therefore has a greater clinical emphasis than usual. The clinical examples are intended to arouse interest, to alert the student to look for more extended discussions in other parts of his curriculum and to enhance his understanding of the normal. They are not intended to usurp pathophysiology.

The book has been written primarily for medical and dental students, but we believe it will also appeal to those physiology, bioengineering and nursing students who need a brief text on organ physiology to supplement their other studies.

I am grateful indeed to many UCLA colleagues who not only provided helpful material but read and criticized portions of the manuscript. It is also a pleasure to acknowledge the contributions of Gwynne Gloege, who drew many of the illustrations, and of Ethel Mason, who typed much of the manuscript. I am indebted to C. Joseph Frank, senior medical editor of Year Book Medical Publishers for his cordial collaboration, which made the task of producing the book so much easier.

GORDON ROSS

Table of Contents

1

General Physiology

ERNEST M. WRIGHT, Ph.D.

A PREREQUISITE for the study of human physiology is a survey of the general properties of cells and tissues. Cells are the smallest units of the body that are able to carry out and control the fundamental processes of life—energy production, synthesis of simple and complex molecules and reproduction. In this chapter one particular aspect of cell physiology will be considered, namely, the physiology of cell membranes. Our immediate purpose is to explain how cell membranes control the volume and composition of cells. There are striking differences between the aqueous protein gel of the cell interior and the interstitial fluid that surrounds most cells of the body. For example, the sodium concentration within the cell is only one 20th of that in the extracellular fluid, whereas cell potassium concentration is about 30 times higher than the external value. It is necessary to consider the origin of these concentration differences because they have an extremely powerful influence on (1) metabolic activity of the cell, (2) regulation of cell size and (3) generation of electric potential gradients across the cell membrane. Furthermore, the same fundamental transport processes responsible for regulation of the composition of cells also account for the functions of the major organs of the body, e.g., absorption of food by the intestine, excretion of waste by the kidney, transmission of information by the nervous system and circulation of blood by the heart.

CELL STRUCTURE

Mammalian cells vary widely in shape and size, but almost all are less than 20 μm in diameter. (One micrometer is equal to 1×10^{-6} meter.) Their structure also varies considerably and this is related to the special functions they perform. A representative cell is shown in Figure 1–1. Cells are contained within a limiting membrane, and they consist of various organelles suspended in cytoplasm. The most prominent organelle is the nucleus (N), and this contains genetic material and enzymes used in synthesis of the nucleic acids required for cell reproduction. Cells without a nucleus, e.g., human red blood cells, are unable to reproduce themselves. The only structure usually visible

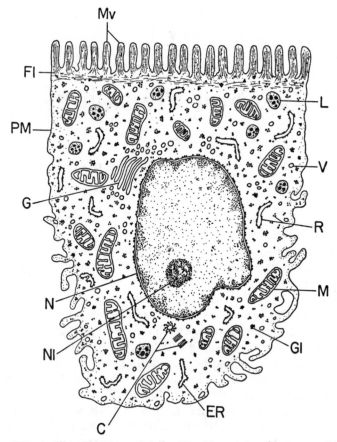

Fig 1–1.—"Typical" cell. Mv = microvilli; Fl = microfilaments; PM = plasma membrane; G = Golgi bodies; N = nucleus; Nl = nucleolus; C = centrioles; ER = endoplasmic reticulum; Gl = glycogen; M = mitochondria; R = ribosomes; V = vacuole; L = lysosomes. (See text.)

within the nucleus is the nucleolus *(Nl),* which is concerned with production of ribonucleic acid (RNA). The nucleus is separated from the cytoplasm by the nuclear membrane.

Mitochondria *(M)* are the primary source of energy in the cell, and all metabolic pathways are linked directly or indirectly to these organelles. Using molecular oxygen, mitochondria oxidize metabolites such as glucose to carbon dioxide and water, and much of the free energy liberated is conserved by synthesis of adenosine triphosphate (ATP). This high energy compound is used to carry out active transport, muscle contraction and biosynthesis. Consequently, mitochondria are found concentrated in regions of the cell where there is a high demand for ATP, e.g., close to plasma membranes specialized for ion transport and to the internal membranes that synthesize proteins.

The outer membrane of the mitochondrion is contiguous with the endoplasmic reticulum *(ER),* a system of continuous intracellular membranes within the cytoplasm. The rough ER, internal membranes with dense granules called ribosomes *(R)*, is concerned with protein synthesis, whereas the smooth ER is involved in glycogen and lipid production. Ribosomes are also found freely distributed throughout the cytoplasm and on the external surface of the nuclear membrane.

Proteins synthesized on the rough ER are transported to the Golgi bodies *(G)*, where they are either packaged into secretory granules for export or are incorporated into intracellular organelles, e.g., hydrolytic enzymes are packaged into lysosomes *(L)*, which are responsible for digestion of ingested proteins or effete cellular constituents.

Each cell contains a pair of centrioles *(C)*, which are thought to play a part in synthesis of cilia and microtubules. Cilia are motile hairlike structures that project from the surface of cells such as those in the respiratory epithelium and oviduct. Fluid and particles are moved over these surfaces by ciliary mo-

Fig 1–2.—Distribution of the major ions between intra- and extracellular fluids. X^- represents the balance of the anions in the two fluids (mainly proteins in the cytoplasm).

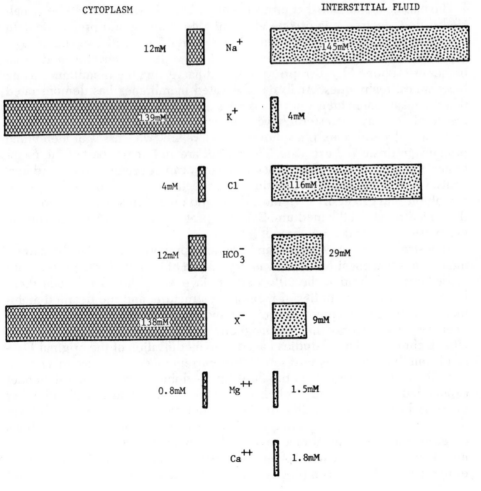

CYTOPLASM   INTERSTITIAL FLUID

Cytoplasm	Ion	Interstitial Fluid
12mM	Na^+	145mM
139mM	K^+	4mM
4mM	Cl^-	116mM
12mM	HCO_3^-	29mM
138mM	X^-	9mM
0.8mM	Mg^{++}	1.5mM
	Ca^{++}	1.8mM

tion. Microtubules and microfilaments *(Fl)* occur throughout the cytoplasm and provide structural support and a basis for cell motility.

Numerous granules and droplets such as zymogen, pigment granules, lipid droplets and glycogen are suspended in the cytoplasm. Metabolic aspects of cell organelles and inclusions are covered in depth in standard biochemical texts and will not be considered further here.

Cytoplasm is an aqueous protein gel composed of 80% water; 14% protein; 2% lipid; and 4% organic and inorganic ions, nucleic acids and polysaccharides. The main cation is potassium and the major anions are chiefly metabolic intermediates and polyvalent proteins. This contrasts strikingly with extracellular fluids, which contain little protein and in which the major ions are sodium, chloride and bicarbonate (Fig 1–2). Later in this chapter we shall discuss the origin of the differences in ionic composition between intra- and extracellular fluid compartments.

CELL MEMBRANES

The plasma membrane that forms the outer boundary of cells can be visualized in the electron microscope after suitable staining and preparation. In cross section it appears as two dense lines, each 2.5 nm wide, separated by a gap 2.5 nm wide. (One nanometer is equal to 1×10^{-9} meter.) Purified membranes are obtained by disrupting cells and harvesting the membranes using biochemical techniques. Analysis of isolated membranes has demonstrated that the major constituents of membranes are lipids and proteins. The lipids are sterols, chiefly cholesterol, glycerides and long chain fatty acids; the most important glycerides are phospholipids. There are two classes of membrane proteins, intrinsic and extrinsic. The former are an integral part of the membrane, e.g., glycophorin of red blood cells; they can be removed only by detergents, whereas the extrinsic proteins, e.g., spectrin of red blood cells, are water soluble and can easily be removed by mild procedures such as increasing the ionic strength of the medium. Extrinsic proteins are usually located on the cytoplasmic side of the membrane.

Almost 50 years ago Davson and Danielli proposed a model for natural membranes that consisted of a bimolecular film of lipid with the proteins adsorbed onto the lipid at the oil/water interface. Over the past decade there have been extensive studies of membrane structure, and the detailed architecture of the arrangement of proteins and lipids in intact membranes has been revealed by x-ray, nuclear magnetic resonance, fluorescence and spin label techniques. These studies have led to modification of the original Davson-Danielli model. The most satisfactory current model that accounts for the experimental observations is the fluid lipid–globular protein mosaic model expounded by Singer and Nicholson (Fig 1–3). In this model the globular proteins float like icebergs in a lipid sea. The lipids are arranged in a bilayer with their ionic head groups in direct contact with the aqueous salt solutions on each side of the membrane, and with their hydrocarbon tails aligned to minimize their contact with water. Globular proteins, the intrinsic proteins, either span the bilayer completely or are restricted to one or other of the lipid

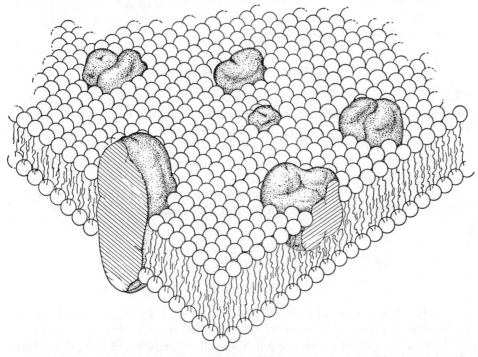

Fig 1–3.—Representation of three-dimensional organization of plasma membranes. A bimolecular film of phospholipids forms the matrix of the membrane, and globular proteins are embedded in the lipid core. Some proteins span the membrane; others are embedded in one of the lipid monolayers. (Courtesy of Singer, S. J., and Nicolson, G. L.: The fluid mosaic model of the structure of cell membranes, Science 175:720, 1972. Copyright 1972 by the American Association for the Advancement of Science.)

monolayers. In either case the hydrophobic regions of the proteins are embedded in the hydrocarbon core of the lipid bilayer, and the more polar regions are exposed to the aqueous solutions at one or both sides of the membrane.

This membrane model is consistent with the known chemical and physical properties of biologic membranes, e.g., the rates of permeation of neutral molecules (see below) and their electric capacitance and resistance. It is now possible to synthesize artificial bilayer lipid membranes, and these have provided powerful models for biologic membranes. In both biologic membranes and artificial bimolecular lipid membranes, the capacitance is about 0.5 microfarad cm^{-2} and the electric resistance is greater than 5,000 ohms cm^2. At this juncture it should be noted that the lipid composition of plasma membranes greatly influences the properties of membranes. For example, increasing the concentration of cholesterol in blood increases the amount of cholesterol in cell membranes, and this in turn lowers membrane permeability to solutes such as urea and lowers the activity of membrane-bound enzymes such as sodium- and potassium-dependent ATPase. This enzyme is an intrinsic membrane protein that controls the transport of sodium and potassium across plasma membranes and so, as the cholesterol in the blood increases, more enzyme is required to transport the same amount of sodium and potassium across a given cell membrane.

MEMBRANE TRANSPORT

To explain the difference in composition between the cytoplasm and extracellular fluids, the entrance of metabolic substrates into cells and the exit of waste products, it is necessary to consider transport processes. These range from the relatively simple process of diffusion to extremely complex mechanisms that require the presence of special molecules within cell membranes and are linked closely to cellular metabolism. First we shall discuss the transport of molecules, e.g., water, glucose and amino acids, and then the transport of inorganic ions such as sodium, potassium and chloride.

NONELECTROLYTES

Diffusion

Diffusion is the process by which molecules distribute themselves throughout the whole space available to them. In a solution molecules are distributed evenly as the result of rapid, random thermal movements of both the solute and solvent and frequent elastic collisions among all molecules. Although it is impossible to predict the movement of the individual molecules, it is relatively simple to describe the net movement of molecules from a region of high concentration to a region of lower concentration. The rate of transport of molecules down a concentration gradient is given by Fick's law, which states that the rate of diffusion (dn/dt, moles sec^{-1}) is directly proportional to the concentration gradient (dc/dx, moles cm^{-3} cm^{-1}), i.e.,

$$dn/dt = J = - DA(dc/dx) \qquad (1)$$

where the proportionality constant (D) is the diffusion coefficient and A the area of the plane of solution through which diffusion occurs. The negative sign indicates that diffusion occurs in the direction of decreasing concentration. When the concentration gradient is 1 mole cm^{-3} cm^{-1} and the area of the plane is 1 sq cm, the diffusion coefficient is the number of moles diffusing down the concentration gradient in one second. The actual value of the diffusion coefficient, normally quoted in units of sq cm sec^{-1}, depends on the size of the molecule and viscosity of the solution through which diffusion occurs—the larger the molecule and the greater the viscosity of the solution the lower the diffusion coefficient. The self-diffusion coefficient of water in water is 2.4×10^{-5} sq cm sec^{-1}, and in water the sucrose and hemoglobin diffusion coefficients are 5.2×10^{-6} and 6×10^{-7} sq cm sec^{-1}, respectively, at 25 C.

An important concept in cell biology is that the net transport of a molecule is the difference between two unidirectional fluxes. This is shown diagrammat-

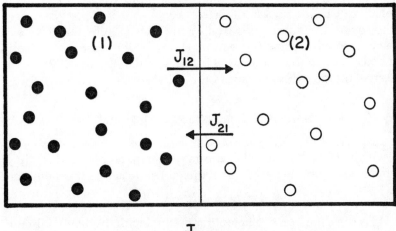

Fig 1-4.—Net transport of a molecule (J_{net}) is sum of two unidirectional fluxes: J_{12} (compartment *1* to *2*) and J_{21} (compartment *2* to *1*).

ically in Figure 1-4. The unidirectional diffusional flux from compartment 2 to compartment 1 (J_{21}) is proportional to the concentration of the solute in compartment 2, whereas the unidirectional diffusional flux from 1 to 2 (J_{12}) is proportional to the concentration in compartment 1. The net flux from compartment 2 to 1 (J_{net}) is the difference between the two unidirectional fluxes, i.e., $J_{net} = J_{21} - J_{12}$. The net flux is zero when the concentration of the solute is the same in both compartments, i.e., $J_{net} = O = J_{21} - J_{12}$. Normally only the net flux can be observed, but with the use of radioactive isotopes it is possible to measure J_{net}, J_{21} and J_{12} simultaneously. This concept is important when we consider the transport in and out of cells, where the net rate of uptake of each solute into the cell depends upon the sum of all transport processes occurring in the membrane, e.g., diffusion, facilitated diffusion and active transport (see Fig 1-11).

Although the actual solution of the Fick equation is complex for diffusion in any real system, an extremely important generalization can be made, i.e., that diffusion is an effective transport process only over short distances. In fact, the efficiency of diffusion is inversely proportional to the square of the diffusion distance. For example, the time for equilibration by diffusion over distances of a few microns is in the order of seconds, but it increases to days for diffusion over a few centimeters.

This simple physical fact about diffusion has remarkable consequences in biology. First, it imposes serious constraints on the size of individual cells because the rate of metabolism in large cells would be severely limited by diffusion of oxygen and substrates from the membrane to the interior of the cell. In practice, mammalian cells with their high rates of metabolism are less than 20 μm in diameter. Second, multicellular organisms require sophisticated mechanisms to permit rapid, effective distribution of substances, material and information from one cell to another. In other words, mammals require huge, efficient cardiovascular, digestive, renal and nervous systems to overcome the severe limitations imposed by the laws of diffusion. Two examples are sufficient to emphasize the point: (1) no metabolically active cell in the human body is more than 20 μ away from a capillary bed and (2) the capillary surface area in 1 gm of brain tissue is about 250 sq cm.

Diffusion across Cell Membranes

Fick's law can be applied to the diffusion of molecules across cell membranes where it can be seen that the rate of transport (J) is directly proportional to the concentration gradient of the molecule across the membrane, i.e.,

$$J = D_m A_m (C_o^m - C_i^m)/x_m \tag{2}$$

where D_m is the diffusion coefficient for the molecule in the membrane, A_m is the area of the membrane, x_m is the membrane thickness and C_o^m and C_i^m are the concentration of the molecule just within the membrane on the outside and inside of the cell, respectively (see eq. 1).

The values of D_m are two to three orders of magnitude lower than those for diffusion in water owing to the fact that the viscosities of membranes are 100–1,000 times greater than that of water.

Concentration of the solutes just within the membrane, $C_o{}^m$ and $C_i{}^m$, is related to concentrations of solutes in the cytoplasm and extracellular fluids by the partition coefficient (K_m) where $K_m = C^m/C^{water}$. Partition coefficients are determined by measuring the ratio of the solute concentrations in the membrane and in water when the solutes are distributed at equilibrium between extracellular fluids and the membrane.

Using this relationship between the concentration of a solute in a membrane and in water, the rate of diffusion across a cell membrane can be rewritten in terms of the solute concentrations in the intra- and extracellular fluids, if diffusion through the membrane is rate limiting, i.e., eq. 2 becomes

$$J = D_m A_m K_m (C_o - C_i)/x_m \tag{3}$$

i.e., the rate of diffusion of a molecule into a cell is directly proportional to the concentration gradient between the outside and inside of the cell. The proportionality coefficient is commonly referred to as the *permeability coefficient* (P), and from eq. 3 it can be deduced that $P = D_m K_m/x_m$ and the units used for P are cm sec^{-1}, i.e.,

$$J = A_m P(C_o - C_i) \tag{4}$$

An overwhelming majority of the neutral molecules that permeate across biologic membranes behave as predicted by eq. 4, i.e., the rate of transport across plasma membranes is directly proportional to the concentration gradient. Transport that can be described in terms of the known forces across membranes is referred to as *passive transport*, and the particular process described by eq. 4 is called the solubility/diffusion mechanism.

Factors Influencing the Magnitude of P

Permeability coefficients range from as low as 1×10^{-10} cm sec^{-1} for sucrose in red blood cells up to 1×10^{-2} cm sec^{-1} for water in red cells. Furthermore, the permeability coefficient for any given molecule may vary by orders of magnitude from one membrane to another. For example, the water permeability of human red blood cells is about 100 times greater than that of frog eggs. This large range in P is controlled by three variables: membrane thickness (x_m), membrane diffusion coefficients (D_m) and membrane partition coefficients (K_m). For any given membrane the spread in Ps for a variety of solutes, up to eight orders of magnitude, is due mainly to variations in the partition coefficients. Diffusion coefficients vary by less than a factor of 10, and x_m is constant for any given cell. An experimentally determined relationship between P and K is illustrated in Figure 1–5.

Membrane partition coefficients are controlled by the same molecular forces that control partition of molecules between water and bulk lipid solvents such as olive oil and ether. The most important force is hydrogen bonding, and the greater the number and strength of hydrogen bonds that a solute can make with water the lower the partition and permeability coefficients. A hydrogen bond is the bridge formed by hydrogen between two electronegative atoms acting as proton acceptors and donors, respectively, e.g.,

$$\underset{\text{water}}{\text{H}\text{———}\text{O}\text{———}\text{H}} \underset{\text{H bond}}{————} \underset{\text{urea}}{\text{NH(CH}_3)_2}$$

The greater the hydrogen bonding between a molecule and water the more energy is required to tear the molecule from water and insert it into a lipid environment. The most effective groups in promoting hydrogen bonding are amides (R—C—NH$_2$) > amines (R—NH$_2$) > hydroxyl (R—OH) ~ carboxylic
$$\overset{\|}{\text{O}}$$

acids (R—COOH) > ethers (R—O—R) ~ esters (R—C—OR). Addition of
$$\overset{\|}{\text{O}}$$

one —OH group to a molecule can reduce partition and permeability coefficients by over two orders of magnitude.

A second important factor is the number of methylene (—CH$_2$) groups on the molecule. Each additional —CH$_2$ group can increase permeability and partition coefficients by up to 10-fold, e.g., permeability of the brush border membrane of the intestinal epithelium to valeric acid (CH$_3$CH$_2$CH$_2$CH$_2$COOH) is about four times greater than to butyric acid (CH$_3$CH$_2$CH$_2$COOH). The origin of the effect of methylene groups is most complex, but it is due in part to entropy effects that "push" hydrocarbons out of water and "pull" them into membranes.

Finally, differences in Ps and Ks occur among molecular isomers, i.e., compounds with the same number of methylene and polar groups but with different conformations, and these are explained mainly by intramolecular hydrogen bonding and inductive effects. For example, 1,4-butanediol is less

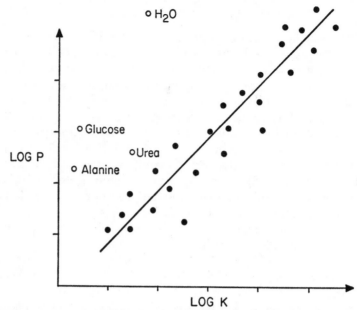

Fig 1–5.—Experimental relationship between permeability coefficient (*P*) and partition coefficient (*K*). For the majority of molecules, *P* is directly proportional to *K*. but for a few molecules, such as water, urea, L-alanine and D-glucose, permeabilities are much higher than predicted.

permeable than 1,2-butanediol owing to the fact that intramolecular H bonding occurs between the groups in the 1 and 2 positions. Intramolecular H bonding reduces the ability of a molecule to form intermolecular H bonds with water. On the other hand, inductive effects explain the increase in permeability of amides when they are halogenated. The halogens cause electron withdrawal from the amide N and O and reduce the strength of the H bonds between the amide group and water. Nitro groups are also electron withdrawing, but alkyl groups are electron releasing.

An understanding of the relations between the structure and conformation of molecules and their rates of permeation across membranes is most important in pharmacology and therapeutics. For example, rates of absorption of drugs from the gastrointestinal tract can be either enhanced by the addition of methylene groups or halogenation of the drug, or retarded by the substitution of —OH groups for H atoms.

In general, the permeability of one cell membrane to nonelectrolytes is very similar to the permeability of any other membrane. The major difference lies in the absolute magnitude of the actual permeability coefficients and the difference in permeability between a pair of molecules. For example, in the gallbladder the permeability of 1,2-propanediol is only two times greater than the permeability of glycerol (propanetriol), but in the urinary bladder the permeability of the diol is five times greater than that of the triol. This is due to differences in the composition and organization of the membrane lipids in the two membranes, i.e., difference in both D_m and K_m.

PERMEATION THROUGH PORES

Although most molecules obey the general relationship between P and K, there are several notable exceptions, which include water, urea, D-glucose and amino acids (see Fig 1–5). All these have higher Ps than would be predicted from the general pattern; the permeability of water is frequently 1,000 times higher than expected. All small molecules, e.g., water, urea, ethylene glycol and acetamide, are highly permeable, and in this case the dominant force controlling permeation is of molecular size – the larger the molecule the lower the permeability ($P \propto$ molecular volume^{-3}). This size effect is only seen for molecules smaller than glucose (0.42 nm radius), and this has led to the hypothesis that a few small aqueous channels cross the membrane, i.e., small molecules pass through membranes via small pores rather than through membrane lipids. If the pores have a radius of 0.4 nm, it is estimated that only 0.02% of the cell membrane area is occupied by pores.

The pore hypothesis is highly controversial, and at least two other explanations for the high permeability of small molecules have been proposed: (1) small molecules cross membranes via carriers (see below), and (2) small molecules can pass more readily between the highly ordered hydrocarbon tails of the membrane lipids than can larger molecules. According to this latter hypothesis, the spaces between the hydrocarbon tails of the lipid (see Fig 1–3), are just large enough for small molecules to squeeze through, but larger molecules have to distort the quasi-crystalline array of membrane lipids to enter the membrane and exit from the other side. Consequently, more energy is needed for the larger molecules than for small molecules to diffuse through the membrane.

WATER PERMEABILITY AND VOLUME FLOW

Implicit in the above discussion of water permeability is the fact that water can be treated as any other molecule, i.e., transport of water across cell membranes is determined by the water concentration gradient and P_{H_2O}. Diffusional water permeability, P_d, can be measured by the use of isotopically labeled water (THO or D_2O) and ranges from 1×10^{-6} to 1×10^{-2} cm sec^{-1} in various cells.

Water also passes across membranes in response to hydrostatic (ΔP^*) and osmotic ($\Delta \pi$) pressure gradients. In practice, this transport of water is recorded as a volume flow, which can be converted into moles by multiplying by the partial molar volume of water ($\overline{V}_m = 0.0181$, mole^{-1}). The flow of water is directly proportional to the hydrostatic pressure gradient, flow = $L_p A_m \Delta P^*$, or to the osmotic pressure gradient, flow = $L_p A_m \Delta \pi = L_p A_m RT \Delta C$, where L_p is the hydraulic coefficient. The usual units of L_p are cm sec^{-1} atm^{-1}, and in both cases water flows down the water chemical potential gradient.

The hydraulic water permeability of a membrane can be compared with the diffusional permeability by converting L_p into the same units as P_d, i.e.,

$$P_{os} = L_p \times \frac{RT}{V_m} = L_p \times 1,413 \text{ cm sec}^{-1}$$

P calculated in this manner is referred to as P_{os}. In most plasma membranes P_{os} is fourfold greater than P_d, but in bilayer lipid membranes $P_{os} = P_d$: this has been taken as evidence for the presence of pores in biologic membranes (see above). The argument is that $P_d = P_{os}$ if water permeates across the membrane solely by a solubility diffusion mechanism, but, if there are pores in the membrane, $P_{os} > P_d$ because bulk flow through pores in response to pressure gradients should be greater than diffusional flow through pores. As outlined above, the subject of pores in cell membranes is controversial.

Osmotic pressure is defined as the hydrostatic pressure required to stop volume flow across a membrane when the membrane separates two solutions of different composition. When there is only a single solute and the membrane is permeable only to water, the osmotic pressure is given by the van't Hoff relationship, i.e., $\Delta\pi = RT\Delta C$, where R is the gas constant, T the absolute temperature and ΔC the concentration gradient of the solute across the membrane. A concentration gradient of 10 mM sucrose produces a water flow across a semipermeable membrane that is balanced by a hydrostatic pressure of 0.22 atm, whereas a 10 mM NaCl gradient requires about 0.44 atm. This is because NaCl is a strong electrolyte that dissociates in solutions to Na^+ and Cl^-, i.e., there are two osmotically active solutes per molecule of NaCl.

When a membrane is permeable to *both* the solute and water, the osmotic pressure developed by a concentration gradient is less than predicted by the van't Hoff relationship. For example, when the membrane is equally permeable to water and the solute, there is virtually no volume flow in response to a solute concentration gradient owing to the fact that the volume flow of solute down the solute concentration gradient is about equal and opposite to the volume flow of water down the water "concentration gradient."

In animal cells hydrostatic pressure gradients are less important driving forces for water flows across membranes than are osmotic gradients. The reasons for this are that only very small hydrostatic pressures normally exist across cell membranes, and small concentration gradients generate relatively large osmotic pressure gradients, e.g., 10 mM concentration gradient of an impermeable nonelectrolyte produces an osmotic gradient of 0.22 atm. Nevertheless, hydrostatic pressures play important roles in fluid exchanges across capillaries and in glomerular filtration by the kidney (see Chapter 4, section on the microcirculation; Chapter 6, renal blood flow).

PERMEATION AND SURFACE AREA

The rate of passive permeation of molecules across a membrane is directly proportional to the membrane area (eq. 4). Thus all that is necessary to increase the rate of transport between two compartments of the body is to increase the area of the membrane separating the compartments. A prime example of this is in the lung. The total surface area of the lung in man is 75 sq m, about half the area of a tennis court, and at rest only a fraction of the total area is used for gas exchange. However, as the demand for oxygen by the body goes up during exercise, the functional area of the lung increases. In single cells, specialization of the cell surfaces by the development of microvilli (see

Fig 1-1) may serve the same purpose. Particularly well-developed microvilli are seen in the epithelial cells of the small intestine and kidney.

PERMEATION OF GLUCOSE AND AMINO ACIDS

Very polar molecules such as glucose and amino acids are essential for cell metabolism and it is necessary that they enter cells very rapidly. Special mechanisms have evolved to allow these nutrients to permeate across cell membranes. Two types of processes have evolved: facilitated diffusion and sodium co-transport.

FACILITATED DIFFUSION.—In this process, as in simple diffusion, the driving force for permeation is the solute concentration gradient across the cell membrane. If the concentration outside the cell is higher than in the cytoplasm, there is net transport of the substrate into the cell; there is net transport out of the cell if the gradient is reversed. A good example of this mechanism is transport of glucose into red blood cells. The following features distinguish facilitated diffusion of glucose in the red cell from the simple solubility-diffusion process previously discussed. (1) The rate of glucose entry is greater than predicted from the glucose partition coefficient (see Fig 1-5), i.e., sugar enters much faster than expected for a simple solubility diffusion mechanism. (2) The rate of glucose entry as a function of glucose concentration deviates from the diffusion law (Fig 1-6). At low glucose concentrations the rate of entry is high and the rate increases with concentration, but at higher concentrations transport becomes saturated, i.e., a concentration is reached above which there is no further increase in the rate of transport. Glucose transport can formally be described by Michaelis-Menten kinetics with a maximum rate of transport (V_{max}) of 500 μM/ml of cells per minute and a K_m of 5 mM, respectively. The K_m is simply the concentration of glucose that produces 50% of the maximal rate of transport, but it is taken as an index of the affinity of the transport system for glucose. The rate of glucose entry at any given glucose concentration (C) is given by $CV_{max}/(K_m + C)$. (3) Glucose entry is a stereospecific process. Although D-glucose, the biologic isomer, rapidly enters the red

Fig 1-6.—Rate of transport as a function of concentration for simple diffusion and carrier-mediated transport.

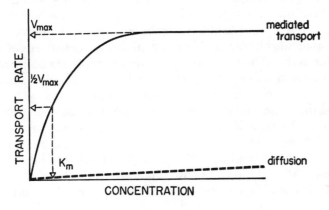

cell, the optical isomer L-glucose does not. The K_m for L-glucose transport is more than 600 times greater than is the D-glucose K_m. To obtain a rate of L-glucose transport equal to half the maximal D-glucose, the rate obtained at 5 mM D-glucose, the L-glucose concentration has to be raised to more than 3 M. (4) Other sugars are also transported by the D-glucose system. These include D-mannose (K_m 20 mM), D-galactose (K_m 30 mM), D-xylose (K_m 60 mM) and L-arabinose (K_m 100 mM). All are hexoses or pentoses with a pyranose (6-membered) ring in the same chair conformation as D-glucose (C1 conformation), and a number of —OH groups in the equatorial plane of the ring. In the red cell there is a good correlation between the number of —OH groups in the equatorial plane and the K_m, the sugar with the greater number of —OHs in this plane having the lower K_m. Specificity of sugar transport is explained by a three-dimensional "lock-key" relationship between the transport site and the sugar. (5) Transported sugars show mutual competition. This is a special case of (2), where competition may be considered to be between identical, as opposed to similar, molecules. An example of the competition effect is that galactose entry into the red cell can be strongly blocked by increasing the glucose concentration in plasma. The kinetics of the inhibition are of the true competitive type in that D-glucose causes an apparent increase in the galactose K_m with no concurrent effect on the maximal rate. (6) The rate of facilitated transport is markedly inhibited by specific poisons. Many of these reagents, e.g., $HgCl_2$ and dinitrofluorobenzene, react with proteins in the cell membrane. On the other hand, nonspecific metabolic poisons, e.g., cyanide and dinitrophenol, have no influence on facilitated transport. This is consistent with the observation that the direction of sugar transport across the red cell membrane is always in the direction of the glucose concentration gradient, i.e., no input of metabolic energy is required. An important feature of facilitated transport is that there is no chemical change in the sugar molecules during transport, i.e., D-glucose is transported across the membrane and appears in the cytoplasm as D-glucose.

A carrier transport model (Fig 1–7) is frequently used to explain facilitated sugar transport across cell membranes. The basic tenet of this hypothesis is that a small number of "carrier" molecules are present in the cell membrane to ferry the sugar molecules across the membrane. These carriers are thought to reversibly bind the sugar at one side of the membrane, shuttle it across to the other side of the membrane and then dissociate to deliver the sugar into the solution on the other side of the membrane. The cyclic movements of specific carrier molecules in the membrane explain all the features of sugar transport across the red cell membrane. Currently attempts are being made to isolate sugar carriers from cell membranes and, so far, specific sugar transport proteins have been isolated from red blood cell membranes.

Facilitated diffusion systems have been found for sugars in erythrocytes, leukocytes, muscle, fat cells, the choroid plexus and the blood capillaries of the brain. The hormone insulin increases the rate of sugar transport into muscle and adipose tissue, and in diabetes mellitus, where there is an insulin deficiency, the elevated blood glucose is largely accounted for by reduction in the rate of glucose uptake into cells. Facilitated amino acid transport has been

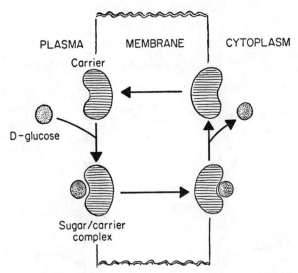

Fig 1–7.—Simple carrier model for sugar transport across red cell membrane.

described in erythrocytes, leukocytes and the basal-lateral membranes of the intestinal epithelium.

SODIUM CO-TRANSPORT.—This transport process, which is more complex than the facilitated diffusion process just described, is widely distributed throughout the cells and tissue of the body. Perhaps the best example is the transport of sugars across the brush border membrane of the epithelial cells lining the small intestine. There are many similarities between this process and facilitated diffusion: (1) The rate of D-glucose transport is higher than predicted from its partition coefficient (see Fig 1–5). (2) The rate of sugar accumulation is a saturable process (see Fig 1–6). (3) Sugar transport is a stereospecific process; L-glucose is not accumulated by the intestinal epithelium. (4) Several sugars share the transport system. However, the process of sodium co-transport is more highly specific in that hexoses, and only hexoses with an —OH group in the equatorial configuration on carbon number two, i.e.,

$$
\begin{array}{c}
CH_2OH \\
| \\
C\text{—}O \\
\diagup \qquad \diagdown \\
C \qquad\qquad C \\
\diagdown \qquad \diagup \\
C\text{—}C \\
| \\
OH
\end{array}
$$

are transported to any significant extent. According to this scheme, 2 deoxyglucose, which is lacking a —OH group on carbon number 2, is not accumulated by the intestine. (5) All the sugars that are accumulated by the intestinal

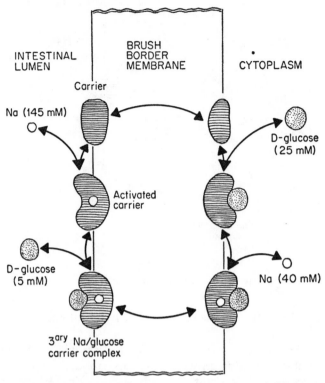

INTESTINAL LUMEN   BRUSH BORDER MEMBRANE   • CYTOPLASM

Carrier

Na (145 mM)

D-glucose (25 mM)

Activated carrier

D-glucose (5 mM)

Na (40 mM)

3^{ary} Na/glucose carrier complex

Fig 1−8.—Simple carrier model for co-transport of glucose and sodium across the brush border membrane of the intestinal epithelium.

epithelial cell exhibit mutual competition. (6) There is no measurable change in the structure of the sugar molecule during transfer across the membrane.

There are several major differences between facilitated diffusion and sodium co-transport. The most important are as follows: (1) The co-transport process is more specific than facilitated diffusion (see 4 above). (2) Sugars are accumulated within the epithelium above the concentration in the intestinal lumen. The intracellular concentration of 3−O−methyl glucose, a poorly metabolized sugar, is often 10 times higher than the external concentration. Stated another way, sugars are transported into the intestinal cell *against* their concentration gradient. (3) The accumulation of sugars depends on cell metabolism, whereas entry into the red cell does not. General metabolic poisons, e.g., cyanide, eliminate uphill accumulation of sugars in the intestine. Ouabain, a potent inhibitor of membrane-bound sodium- and potassium-dependent ATPase (Na/K ATPase), also blocks sugar transport. (4) Uphill transport of sugars requires the presence of sodium in the lumen of the gut. Replacing sodium with lithium, potassium or any other solute severely reduces the transport of sugar across the brush border membrane. Lowering the sodium concentration lowers the affinity of the transport carrier for sugars. (5) Sugars stimulate the transport of sodium across the brush border membrane into the intestinal epithelium. At normal sodium concentrations one sodium ion is transported into the cell with each sugar molecule.

These and other observations have led to the model shown in Figure 1–8. Sugars and sodium enter the epithelial cell from the intestinal lumen via a common carrier. The rate of binding of sugar to the carrier at either surface of the membrane depends on sugar and sodium concentrations. The higher the sodium concentration the higher the affinity of the carrier for sugar. The external sodium concentration is high compared to that in the intestinal cell, and so the concentration of the carrier-sugar-sodium complex at the outer surface of the brush border membrane is higher than at the inner surface of the membrane. Consequently there is a net transfer of the tertiary carrier complex from the outer to the inner surface where it dissociates and delivers sodium and sugar to the cell interior. Net transport into the cell occurs until the sugar and sodium concentrations within the cell reach the level at which the rate of association of the sodium-sugar-carrier complex is equal to the rate of dissociation. Normally, the internal sodium concentration is maintained at a low level by a Na/K exchange pump (see below), and the rates of association and dissociation of the carrier become equal only when the sugar concentration in the cell rises above the external concentration. If the sodium pump is inhibited, either by cyanide or ouabain, intracellular sodium concentration approaches the external concentration, and sugars are no longer accumulated within the cell. Until recently it was claimed that sugar accumulation within the intestinal epithelium was an active process, but it is now clear that the uphill transport of sugars is due to coupling to the transport of sodium down its electrochemical potential gradient (see below). The co-transport of sugars and sodium also has been observed in the renal proximal tubule and gallbladder. Amino acids are transported into a wide variety of cells by a similar process, but in this case there appear to be at least four separate carriers for neutral amino acids, dibasic amino acids, imino acids and dicarboxylic acids. These transport carriers are found in the intestine, renal proximal tubule, choroid plexus, leukocytes, ascites tumor cells and brain cells. Hereditary disorders of sugar and amino acid co-transport systems have been well documented, e.g., glucose-galactose malabsorption by the intestine, Hartnup's disease (a disorder of neutral amino acid absorption in the intestine) and cystinuria. These are all rare autosomal recessive diseases. Furthermore, in the intestine the coupling between sodium and nonelectrolyte transport across the brush border membrane provides the rationale for treatment of cholera with orally given, balanced salt solutions containing sugars and amino acids.

Finally, it should be pointed out that other molecules are probably transported into cells by similar sodium-dependent carrier systems. These include transport of choline and para-aminohippuric acid by the kidney and choroid plexus, and transport of catecholamines by cells in the nervous system.

ION TRANSPORT

ELECTROLYTES

Consideration of ion transport is more complex than the transport of neutral molecules, since ions are charged, and there usually is an electric potential difference across membranes. The rate and direction of ion transport across a cell membrane depend upon the valence of the ion, the magnitude of the membrane potential (E_m), the ion concentration gradient and the permeability of the membrane. The major permeating ions are monovalent cations and anions, and the membrane potential ranges from −50 mv in liver cells up to −90 mv in muscle—the inside of cells are negative with respect to the external fluids—and there are concentration gradients of Na, K, Cl and HCO_3 across cell membranes (see Fig 1–2). The potential difference (PD) across cell membranes tends to drive Na and K into, and Cl and HCO_3 from, the cell, whereas the concentration gradients tend to drive Na, Cl and HCO_3 into, and K out of, the cell.

In solutions it should be recalled that voltage gradients produce flow of current according to Ohm's law

$$I = G(V_2 - V_1) \tag{5}$$

where I is the current in amperes, V_2-V_1 is the voltage gradient and G is the conductance of the solution ($G = 1/R$, where R is the resistance of the solution in ohms). Current in solution is carried by ions, not electrons, and the number of ions flowing per second is obtained by dividing the current by the Faraday constant. The fraction of current carried by a particular species of ion is called the transport number. In NaCl solutions the Na and Cl transport numbers are 0.4 and 0.6, respectively, i.e., Na carried 40% of current in NaCl solutions. In membranes ion transport numbers range from 0 to 1.

The conductance of plasma membranes is about eight orders of magnitude lower than a film of plasma of comparable thickness (10 nm), and this is consistent with the lipid composition of membranes. As a result, ion fluxes across membranes in response to electric (and concentration) gradients are small compared to those observed in aqueous solution.

In general, the rate of ion diffusion (J) in solution is described by the relation

$$J = -DA[dc/dx + (zCF/RT)dV/dx] \tag{6}$$

where D is the ion diffusion coefficient, A the area, dc/dx the concentration gradient, dV/dx the electric potential gradient, R the gas constant, T the absolute temperature, z the ion valence, F the Faraday constant and C the ion concentration. In the absence of electric gradients, ions diffuse down their concentration gradients, as described for neutral molecules (see eq. 1), and in the absence of concentration gradients, ions move according to the magnitude and polarity of the membrane PD.

Using this equation (Nernst-Planck equation) as a starting point, it is possible to derive relationships between the membrane potential (E_m), the concentration of ions on each side of the membrane and the flux of ions across the membrane. Using an approach very similar to that employed for nonelectrolytes, a useful equation for the net flux across a membrane J is obtained, i.e.,

$$J = E_m(F/RT)AP \frac{(C_i - C_o e^{-E_mF/RT})}{(1 - e^{-E_mF/RT})} \tag{7}$$

where $P = K_m D_m / x$ (see eq. 4), and C_i and C_o are the ion concentrations on the inside and outside of the cell, respectively. This equation has been used to obtain Ps for a variety of biologic membranes, e.g., in frog muscle P_K and P_{Na} were estimated from radioactive tracer fluxes and measurements of E_m to be 6×10^{-7} and 8×10^{-9} cm sec^{-1}, respectively. A general pattern emerges from studies such as these in that cell membranes are usually more permeable to cations than to anions and that potassium is more permeable than sodium.

MECHANISMS OF ION PERMEATION

Because of technical problems, little is known about the mechanisms by which ions permeate across biologic membranes, and so artificial membranes, e.g., lipid bilayer membranes, have been used extensively as model systems. It is unlikely that simple solubility mechanisms operate because (1) the resistance of model lipid bilayer membranes ($\sim 1 \times 10^8$ ohms cm^2) is orders of magnitude higher than cell members ($\sim 5 \times 10^3$ ohms cm^2) and (2) these model lipid membranes do not exhibit any significant degree of ion selectivity, e.g., in phosphatidylcholine bilayer membranes $P_K = 3.4 \times 10^{-12}$ cm sec^{-1}; $P_{Na} = 1 \times 10^{-12}$ cm sec^{-1}; and $P_{Cl} = 1 \times 10^{-11}$ cm sec^{-1}. Two models are frequently used to explain ion permeation across cell membranes. The first is a mobile carrier model (see Fig 1–7) in which a carrier molecule in the membrane shuttles ions across the barrier. The macrocyclic antibiotic valinomycin behaves as a mobile potassium carrier when added to artificial lipid bilayer membranes. The second is the pore model, where ions permeate across the membrane via small aqueous channels or pores. The antibiotics gramicidin and nystatin form channels for monovalent ions in artificial bilayer membranes. Available information from biologic systems suggests that Na and K permeate across nerve and muscle membranes via channels (see Chapter 2).

ION SELECTIVITY

As indicated above, cell membranes are able to distinguish between ions, e.g., in muscle and nerve during rest the K permeability is about 100 times greater than Na permeability. Eisenman has formulated a theory to account for both the magnitude and sequence of the ion selectivity patterns observed in both biologic and artificial systems. In the case of the five alkali metal cations, Li, Na, K, Rb and Cs, only 11 of the possible 120 permutated selectivity sequences are actually observed in nature, and Eisenman was able to correctly predict the observed sequences and the quantitative relationships between them. He reasoned that selectivity sequences are related to differences among cations in the free energy of transferring the ion from water to the membrane sites controlling permeation. The free energy of transfer is governed by the difference between ion hydration energy and the free energy of ion interaction with the membrane sites. In the simplest case of negatively charged membrane sites, the free energy of interaction is dominated by electrostatic attractive forces, and so for a given ion the free energy of interaction with the site depends upon the charge on the site—the greater the charge the greater the electrostatic interaction and therefore the larger the binding energy. In the case of very strong sites, i.e., sites with high negative charge, the free energy of ion transfer from water to membrane is controlled entirely by ion binding at the site. The smallest cation is preferred because of the greater attractive coulombic forces between the ion and the site; the binding energy is inversely proportional to the square of the distance between the center of the ion and the site, and the permeation sequence is Li > Na > K > Rb > Cs. In nature, this selectivity sequence is observed for the "sodium channel" in nerve, and this implies that there are strongly charged negative sites lining the Na channel.

At the opposite end of the spectrum where the membrane charge is very weak, the free energy of ion transfer between water and membrane is controlled by the strength of the attractive forces between the ion and water, i.e., the ion hydration energies. In this case the permeation sequence is Cs > Rb > K > Na > Li, where the largest ion is most permeating because it takes less energy to tear it free from water than the smaller ions such as Li and Na.

As the strength of the membrane charge varies from strong to weak, nine intermediate sequences are observed, e.g., the sequence K > Rb > Na > Cs > Li was observed for permeation through the "potassium channel" in nerve.

The 11 sequences are as follows:

I. Cs > Rb > K > Na > Li
II. Rb > Cs > K > Na > Li
III. Rb > K > Cs > Na > Li
IV. K > Rb > Cs > Na > Li
V. K > Rb > Na > Cs > Li
VI. K > Na > Rb > Cs > Li
VII. Na > K > Rb > Cs > Li
VIII. Na > K > Rb > Li > Cs
IX. Na > K > Li > Rb > Cs
X. Na > Li > K > Rb > Cs
XI. Li > Na > K > Rb > Cs

In biologic membranes the negatively charged sites are probably either carboxyl ($-COO^-$) or phosphate groups ($-O-\overset{\overset{\displaystyle O}{\|}}{\underset{\underset{\displaystyle OH}{|}}{P}}-O^-$) with a pK_a between 4 and 5. Protonation of the sites at low pH values produces predictable changes in selectivity, e.g., lowering the pH changes the ion selectivity of gallbladder epithelium from sequence IV to II. Currently, work is in progress to identify the sites controlling permeation across membranes.

Eisenman's theory can be extended to account for the selectivity of alkaline earth cations and for the halide and monovalent polyatomic anions. The above observations, although apparently esoteric, are vital to the understanding of such processes as electric excitation in nerve; calcium accumulation by muscle sarcoplasmic reticulum, which governs the force of contraction of heart and skeletal muscle; and accumulation of iodide in the thyroid gland.

MEMBRANE POTENTIALS

The diffusion of ions across membranes produces electric potentials. In the simplest case where the membrane is only permeable to one ion, the size of the diffusion potential (E) is given by the Nernst equation:

$$E = V_2 - V_1 = (60/z)\log_{10} C_1/C_2 \tag{8}$$

where z is the valence of the ion; C_1 and C_2 are the ion concentrations, more accurately ion activities, on each side of the membrane; and E is the potential (in millivolts) of side 2 with respect to side 1. This expression is derived from eq. 6, the assumption being that the net transport of the ion across the membrane J is zero.

As an example, let us consider a membrane that separates a 150 mM NaCl solution (side 1) from a 15 mM NaCl solution (side 2). If the membrane is permeable only to sodium, the diffusion potential is

$$E = E_{Na} = (60/+1)\log_{10} 150/15 = +60\ mv$$

with the dilute salt solution being electrically positive with respect to the concentrated solution. This is the value of the PD at equilibrium, where there is no net transport of Na across the membrane; the diffusion of Na down the concentration gradient from side 1 to 2 is balanced by an equal and opposite Na flux driven by the diffusion PD, i.e.,

$$J_{Na(1\rightarrow2)} + J_{Na(2\rightarrow1)} = P_{Na}\Delta C_{Na} + G_{Na} E_{Na} = 0$$

On the other hand, if the membrane is permeable only to Cl ions, the chloride equilibrium PD is

$$E_{Cl} = (60/-1)\log_{10} 150/15 = -60\ mv$$

with the dilute salt solution negative. This shows that the equilibrium PD across a membrane separating a 10-fold concentration gradient of NaCl can either be +60 or −60 depending on the permeability of the membrane to the two ions.

The Nernst equation can be used to calculate the PD that would need to exist across a membrane to account for an observed ion concentration gradient, and comparison of the ion equilibrium potentials, E_{Na}, E_K and E_{Cl}, with the actual membrane potential enables us to conclude whether or not ions are distributed across a membrane simply as a result of the difference in PD. This approach will be used later when we consider the distribution of ions across cell membranes.

If the membrane is permeable to both Na and Cl ions, a more complex expression is required to relate the magnitude of the diffusion potential (sum of the Na and Cl diffusion potentials) and the ion gradients. Obviously with a simple membrane permeable to both Na and Cl ions, equilibrium will only be attained when there is no electric or concentration gradients across the membrane: if the membrane is sodium selective and the membrane is slightly permeable to chloride, the Na diffusion potential will enhance the flux of Cl from the concentrated to the dilute solution and partially short-circuit the Na diffusion potential; thus, the PD never becomes large enough to balance the diffusion of Na down the concentration gradient.

Under the condition that the ion concentrations on each side of the membrane are maintained constant, e.g., if the volumes of the solutions on each side of the membrane are infinitely large and the flux of ions across the membrane is small, the diffusion PD will be fairly constant with time. In the absence of net current flow across the membranes, i.e., $I_{Na} + I_K + I_{Cl} = 0$, flux eq. 7 can be solved for the membrane PD:

$$E_m = 60 \log_{10} \frac{P_K[K]_o + P_{Na}[Na]_o + P_{Cl}[Cl]_i}{P_K[K]_i + P_{Na}[Na]_i + P_{Cl}[Cl]_o} \tag{9}$$

It should be noted that this applies only to monovalent ions; a much more complex relation is required for mixtures of monovalent and divalent ions. The equation, first derived by Hodgkin and Katz and known as the constant field equation because of assumptions made in the derivation, describes the value of the membrane PD (E_m) in terms of the concentrations (more correctly ion activities) on each side of the membrane and the ion permeability coefficients. In many cells the membrane potential predicted by this equation is in close agreement with that actually recorded by microelectrode techniques. For example, in frog skeletal muscle, the diffusion potential predicted by the constant field equation from estimates of the sodium and potassium concentrations and permeability coefficient is

$$E_m = 60 \log_{10} \frac{P_K[K]_o + P_{Na}[Na]_o}{P_K[K]_i + P_{Na}[Na]_i}$$

$$= 60 \log_{10} \frac{6 \times 10^{-7} \times 2.5 + 8 \times 10^{-9} \times 120}{6 \times 10^{-7} \times 140 + 8 \times 10^{-9} \times 9}$$

$$= -92 \text{ mv}$$

and the value recorded by microelectrodes is −90 mv. In this cell it is concluded that the resting membrane potential is due to asymmetric distribution of Na and K across the cell membrane.

The constant field equation may also be used to obtain estimates of the relative ion permeability coefficients, e.g., P_{Cl}/P_K and P_{Na}/P_K, by electric methods when it is impractical to obtain permeability coefficients by more direct techniques. Membrane potentials are recorded as a function of the ionic composition of the intra- and extracellular compartments, and the equation is solved for the relative permeability coefficients. It may be readily shown using eq. 9 for frog muscle, given $E_m = -92$ mv and the ion concentrations, that K is 77 times more permeant than Na.

A different description of ion fluxes and membrane potentials often used by electrophysiologists is the electric analogue or equivalent circuit approach (see Chapter 2). In this the flow of ions is related to the amount of current each ion carries through the membrane, i.e., I_K, I_{Na} and I_{Cl}. The current flows are given by

$$I_{Na} = G_{Na}(E_{Na} - E_m) \qquad (10)$$
$$I_K = G_K(E_K - E_m) \qquad (11)$$
$$I_{Cl} = G_{Cl}(E_{Cl} - E_m) \qquad (12)$$

where E_m is the membrane PD; E_{Na}, E_K and E_{Cl} are the ion equilibrium potentials (Nernst equilibrium potentials); and G_K, G_{Na} and G_{Cl} are the partial ionic conductances (G_m, the total membrane conductance, $1/R_m$, is the sum of the partial ionic conductances).

The terms permeability (P) and conductance (G) are often used interchangeably, as changes in the partial ionic conductances are often associated with changes in ionic permeability. However, it should be noted for any given ion X that

$$G_X \propto [X] P_X \qquad (13)$$

i.e., changes in G_X may occur as a result of changes in the concentration of X without changes in P_X.

DONNAN EQUILIBRIUM

It should be recalled from Figure 1–2 that the major anions within cells are large organic anions such as metabolic intermediates and protein. This fact has important implications in biology. First, let us consider a simple cell of constant volume that contains KCl and 20 mM KX, where X^- is a monovalent protein (Fig 1–9). When this cell is suspended in a solution of 155 mM KCl, there are concentration gradients across the cell membrane. If the

Fig 1–9.—Gibbs-Donnan distribution of ions across a permeable cell membrane. The *rigid cell membrane*, which maintains the cell volume constant, is permeable to K^+ and Cl^-, but not X^- (monovalent protein).

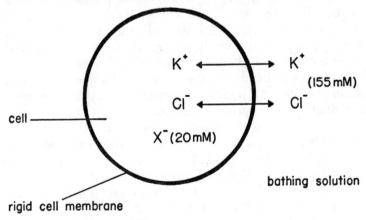

membrane is freely permeable to K^+, Cl^- and X^-, all three ions will diffuse across the membrane until equilibrium is reached; this occurs when the composition of the internal and external compartments is identical and there is no PD across the membrane. However, if the membrane is only permeable to K^+ and Cl^- and not to X^-, there are concentration gradients of K^+, Cl^- and X^- across the membrane at equilibrium. The Nernst equation describes the relationship between membrane PD and ion concentration gradients at equilibrium, i.e.,

$$E_m = E_K = 60 \log_{10} K_o/K_i$$
$$E_{Cl} = 60 \log_{10} Cl_i/Cl_o$$

where o and i refer to the solutions on the outside and inside of the cell, respectively. It follows that $K_o/K_i = Cl_i/Cl_o$, and this ratio (r) is defined as the Donnan ratio. The following relations must also hold:

$$K_o = Cl_o = 155 \text{ mM} \tag{14}$$
$$K_i = Cl_i + 20 \text{ mM } X_i \tag{15}$$

It follows that dividing eq. 15 by 14

$$K_i/K_o = Cl_i/Cl_o + X/Cl_o \tag{16}$$

and substituting for r gives

$$1/r = r + X/Cl_o \tag{17}$$

Rearranging and solving for r, it may be shown that when the concentration of X^- is small relative to Cl^-_o, as is true in our example, the following simple approximation holds:

$$r \sim l - X/2 \ Cl_o \tag{18}$$

So from a knowledge of the intracellular concentration of the impermeable anion, we can estimate the Donnan ratio ($r \sim 0.94$) and the magnitude of the membrane potential ($E = 60 \log_{10} K_o/K_i \sim 60 \log_{10} 0.94 \sim -2$ mv), and from the Nernst equation we can estimate the intracellular K and Cl concentrations at equilibrium ($K_i \sim 167$ mM, and $Cl_i \sim 143$ mM). In general, when a cell contains impermeable anions, we can conclude (1) that there will be an electric PD across the cell membrane, with the cell interior negative, (2) that intracellular concentration of the freely permeable anions will be less than external concentrations and (3) that intracellular concentration of the freely permeable cations will be greater than extracellular concentrations. The size of this membrane potential and these concentration gradients are all related to the concentration of impermeable anions.

A final point to be made about the Donnan equilibrium is that the osmotic pressure of the intracellular fluid will be greater than the extracellular fluid: in our example, $K_i + Cl_i + X_i \sim 167 + 143 + 20 \sim 330$ mM compared with $K_o + Cl_o = 155 + 155 = 310$ mM. Consequently, there will be an osmotic flow of water into the cell until the hydrostatic pressure within the cell balances the osmotic flow. In our example, the osmotic pressure developed will be approximately half an atmosphere ($\pi = RT\Delta C = 0.51$ atm).

How applicable is the Donnan equilibrium to real cells? Certainly, large impermeable anions do exist in the cytoplasm (see Fig 1-2), and these include proteins, amino acids and metabolic intermediates. This would imply that freely permeable ions should be distributed according to the Donnan relation and that cells should develop large hydrostatic pressures at equilibrium. Extensive studies with frog skeletal muscle have shown that K and Cl ions are distributed between the intra- and extracellular fluids, as expected from the Donnan effect, but Na is not. As we shall see, it is a common feature of all cells that intracellular Na is *not* at equilibrium, and that it is regulated by a Na pump in the cell membrane. Estimates suggest that hydrostatic pressures of over 1,000 mm Hg are required to balance the Donnan osmotic effect, but animal cells, unlike plant cells, are unable to maintain large pressure gradients. Animal cell membranes are not rigid, and cells change their volume rapidly in response to changes in osmotic pressure. Normally, cell volume is regulated by the Na pump, which controls both intracellular Na concentration and intracellular osmotic pressure. Inhibition of the pump causes a cell to gain Na and swell; e.g., cooling liver cells to 2 C reversibly increases their water content by 50% in 1 hour, and the gain in water is accounted for by the increase in cell Na and Cl. In terms of the Donnan equilibrium, the Na pump can be regarded as equivalent to a rigid cell membrane that is impermeable to Na ions.

Although the Donnan equilibrium is not strictly applicable to cells, the system does apply to the exchange of water and solutes across blood capillaries (see Chapter 4). The protein concentration of blood is about 1 mM, the interstitial fluid is virtually protein free and the capillary wall is impermeable to large proteins. Although proteins generate an osmotic flow of water into the capillary from the interstitial fluid by virtue of their concentration difference and impermeability, there is an additional effect owing to the fact that plasma proteins behave as polyvalent charged anions. The Donnan effect increases the osmotic pressure of blood by about 0.5 milliosmole (mOsm). The overall effect of proteins in the capillaries is often referred to as colloid osmotic pressure, and it is an important concept in the fluid balance of the body.

Ion Distribution across Cell Membranes

Intracellular ion composition is very different from the composition of the fluids bathing the cells (see Fig 1-2). In particular, intracellular anions are mostly large impermeable organic anions, intracellular Na and Cl concentrations are low, and intracellular K concentration is high with respect to extracellular fluid. The first question to resolve is whether or not the internal concentration of permeable ions is accounted for by the membrane potential. The Nernst equilibrium potentials for Na, Cl and K are

$$E_{Na} = (60/+1) \log Na_o/Na_i = 60 \log \frac{145}{12} = +65 \text{ mv}$$

$$E_{Cl} = (60/-1) \log Cl_o/Cl_i = -60 \log \frac{116}{4} = -88 \text{ mv}$$

$$E_K = (60/+1) \log K_o/K_i = 60 \log \frac{4}{139} = -92 \text{ mv}$$

Comparison of these equilibrium potentials with the membrane PD, which is −90 mv, shows that intracellular K and Cl concentrations are close to the expected equilibrium values. On the other hand, Na is not at equilibrium; the membrane PD would have to be +65 mv to account for the low intracellular concentration.

To maintain the low intracellular Na concentration in face of the leak of Na into the cell, Na must be pumped back out of the cell, and between 20% and 45% of the total cellular energy is needed to pump the Na out against its electrochemical potential gradient.

ACTIVE SODIUM TRANSPORT

Transport of Na out of the cell against the electrochemical potential is referred to as active Na transport. Before discussing the mechanism of Na transport, it should first be recalled that the net transport of an ion or molecule across a membrane is composed of two unidirectional fluxes, i.e., $J_{net} = J_{in} - J_{out}$. In the steady state where there is no change in the internal Na concentration, $J_{in} = J_{out}$. The Na outflux is composed of two components—active and passive fluxes—but normally the active flux is much larger than passive flux. Active transport of Na out of cells has been carefully analyzed in nerve, muscle and red blood cells, where the following have been established: (1) The rate of active flux is a saturable function of the internal Na concentration. The K_m is about 20 mM, which is close to the normal internal Na concentration. (2) Active Na transport only occurs in the presence of K in the external solution. Potassium is simultaneously transported into the cell, and this component of the influx saturates as external K concentration increases. The potassium K_m is about 2 mM. The saturable component of the K influx also requires the presence of Na inside the cell, and this suggests that the Na and K fluxes are coupled. Three Na ions are pumped out for every two K ions pumped in by this mechanism. (3) Transport of Na and K requires the presence of Mg-ATP inside the cell. One molecule of ATP is hydrolyzed for every three Na ions pumped out. (4) Cardiac glycosides, e.g., ouabain, present in the external solution are potent inhibitors of the linked Na/K fluxes. The ouabain K_i is about 1×10^{-7} M, and the effect of ouabain can be antagonized to some extent by increasing the external K concentration.

These observations have been included in the model illustrated in Figure 1–10. The pump is shown as a cyclic carrier that transports three Na ions out and two K ions in for each cycle. The carrier is energized by hydrolysis of ATP at the inner surface of the membrane, leading to release of ADP and inorganic phosphate (P_i) within the cell. Ouabain inhibits the pump at the outer surface of the membrane, and it has been estimated from the amount of ouabain bound to various cells that there are between 10^3 and 10^6 pumps per cell.

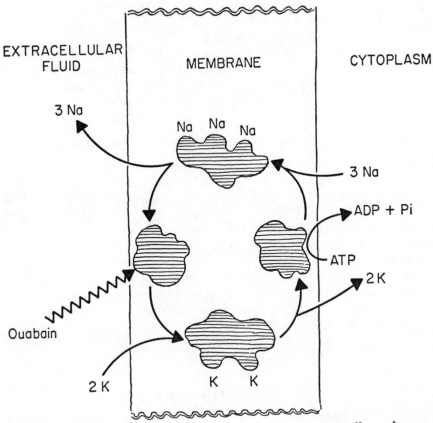

Fig 1–10.— Simple carrier model for Na/K exchange pump in cell membranes.

NA/K TRANSPORT AND NA/K ATPASE

In 1957 Skou made the important discovery of a membrane-bound enzyme that hydrolyzed ATP to ADP and P_i in the presence of Na and Mg. It is now well established that this enzyme is intimately linked to the coupled active transport of Na and K across cell membranes. Major observations supporting this conclusion are as follows: (1) The enzyme requires the simultaneous presence of Na ions inside the cell and K outside the cell to hydrolyze ATP. The Na and potassium K_m of the enzyme are close to the transport K_m. (2) Mg-ATP is required inside the cell—ATP on the outside of cells is not hydrolyzed. In addition there was a good correlation between rate of ion transport and ATP hydrolysis. (3) Cardiac glycosides, e.g., ouabain, in the external fluids inhibit both ion pumping and ATP hydrolysis. The K_i for inhibition of the enzyme is in the same range as the K_i for inhibition of transport; (4) There is a good correlation between distribution of the enzyme and Na/K pumps in the animal kingdom.

Recently much progress has been made in the attempt to solubilize, purify and identify the enzyme. The "purified" enzyme consists of two proteins, which have molecular weights of 95,000 and 55,000 daltons, respectively. The larger protein has binding sites for Na, ATP, K and ouabain; the smaller, a glycoprotein, has no known function. Reports have demonstrated that the purified enzyme can be incorporated into artificial lipid bilayer membranes and that these membranes are capable of transporting Na and K in the presence of ATP. This strongly suggests that the enzyme and the pump are one and the same molecule.

SUMMARY OF ION FLUXES ACROSS CELL MEMBRANES

Normally the intracellular ionic composition is constant and the net flux of each ion across the cell membrane is zero. The unidirectional Na and K fluxes across the membrane of an idealized cell are summarized in Figure 1–11. The membrane PD is −90 mv, cell interior negative; intracellular Na and K concentrations are 12 mM and 139 mM, respectively, and the interstitial fluid Na and K concentrations are 145 mM and 3 mM, respectively. The P_K is 5×10^{-7} cm/sec and P_{Na} is 5×10^{-9} cm/sec. By means of the Nernst equation, E_K is estimated to be −100 mv, i.e., close but not identical to E_m, and E_{Na} is +65 mv. The driving force for Na entry into the cell is therefore equivalent to +155 mv ($E_{Na} - E_m = 65$ mv + 90 mv). The unidirectional influx of Na is obtained from equation 7, assuming $Na_i = 0$, i.e.,

$$J_{in} = E_m(F/RT)A\,P_{Na}\frac{(Na)_o e^{-E_m F/RT}}{1 - e^{-E_m F/RT}}$$

$$= 0.9 \times 10^{-8} \text{ moles/sq cm/hour}$$

The unidirectional passive outflux of Na is estimated from equation 7, assuming that $Na_o = 0$, i.e.,

$$J_{out} = E_m(F/RT)A\,P_{Na}\frac{(Na)_i}{(1 - \exp^{-E_m F/RT})}$$

$$= 1.8 \times 10^{-11} \text{ moles/sq cm/hour}$$

which is negligible compared to the unidirectional influx. This reveals that the net passive flux of Na into the cell is 0.9×10^{-8} moles/sq cm/hour. Similarly, the passive K fluxes are 2×10^{-8} moles /sq cm/hour for the unidirectional influx, 2.6×10^{-8} moles/sq cm/hour for the unidirectional outflux, i.e., there is a net passive flux of K out of the cell of 0.6×10^{-8} moles/sq cm/hour. Despite the fact that the E_K (−100 mv) is close to E_m(−90 mv), this relatively large net passive flux is due to the high membrane permeability to K.

The net passive K efflux from the cell and the net passive Na influx into the cell are balanced by the activity of the Na/K exchange pump, and this explains why (1) intracellular Na concentration is maintained at such a low value in the face of the leak of Na into the cell down the chemical and electric gradients; (2) intracellular K concentration is slightly higher than expected from the membrane PD (this difference is small because any increase in internal K concentration results in an immediate significant increase in passive efflux of K from the cell); and (3) the cell volume is fairly constant. Intracellular Na, K

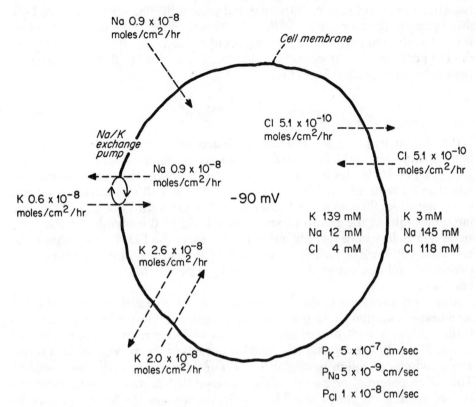

Na 0.9×10^{-8} moles/cm^2/hr

Cell membrane

Cl 5.1×10^{-10} moles/cm^2/hr

Na/K exchange pump

Cl 5.1×10^{-10} moles/cm^2/hr

Na 0.9×10^{-8} moles/cm^2/hr

K 0.6×10^{-8} moles/cm^2/hr

−90 mV

K 139 mM   K 3 mM
Na 12 mM   Na 145 mM
Cl 4 mM   Cl 118 mM

K 2.6×10^{-8} moles/cm^2/hr

K 2.0×10^{-8} moles/cm^2/hr

P_K 5×10^{-7} cm/sec
P_{Na} 5×10^{-9} cm/sec
P_{Cl} 1×10^{-8} cm/sec

Fig 1–11. — Summary of unidirectional fluxes across a "typical" cell membrane.

and Cl and the cell volume are determined by the number of pump sites, pump turnover time, pump coupling ratio and passive permeability of the cell membrane to ions. Consequently, the cell volume will increase following inhibition of the pump, a decrease in pump coupling ratio or an increase in permeability of the membrane to Na. Swelling occurs when cells are incubated in the cold or in the presence of metabolic inhibitors.

The membrane is relatively permeable to Cl, $P_{Cl} = 1 \times 10^{-8}$ cm/sec and in the absence of anion pumps it is distributed according to the membrane PD. The unidirectional influx 5.1×10^{-10} moles/sq cm/hour, is identical to the unidirectional influx in the steady state.

CELL MEMBRANE POTENTIAL

To a first approximation the membrane PD is that expected from the K and Cl equilibrium potentials, i.e., $E_m \sim E_K \sim E_{Cl}$, but, as indicated above, a more accurate description of the PD is given by the constant field equation (eq. 9), which takes into account the Na concentration gradient and relative permeabilities of Na, K and Cl, i.e.,

$$E_m = 60 \log \frac{P_{Na}[Na]_o + P_K[K]_o + P_{Cl}[Cl]_i}{P_{Na}[Na]_i + P_K[K]_i + P_{Cl}[Cl]_o}$$

Inserting the values of concentrations and permeabilities used in Figure 1–9, this equation gives a membrane PD of −90 mv, which is the measured value. It is interesting to note that at high values of K_o, i.e., $K_o > 10$ mM, the value of the PD predicted by the constant field equation is close to that calculated using the Nernst equation for K, i.e.,

$$E_m = 60/+1 \, \log_{10} \frac{K_o}{K_i}$$

It might be expected that the Na/K exchange pump contributes to the membrane PD in view of the fact that three Na ions are pumped out for every two K ions pumped in. Under normal steady state conditions, the pump contributes less than 3 mv to the total membrane PD, i.e., the cell membrane is slightly more negative than expected from the constant field equation. The most important role of the pump in setting the level of the membrane PD is in smooth muscle where it contributes 15–20 mv to the PD. In conclusion, the membrane PD arises largely as a result of the concentration gradients between the cell and external solutions and the permeability of the cell membrane to ions.

An important feature of excitable cells, such as nerve and muscle, is that the membrane potential undergoes rapid changes, suddenly changing from the resting potential (−60 to −90 mv) to +30 mv and then returning to its original value. These PD changes, called action potentials, occur within 2–3 msec, and hundreds of action potentials can occur in excitable cells every second. The action potential originates in the unequal distribution of ions across the cell membrane and the permeability of the membrane. We have seen that the membrane potential at rest approximates a K equilibrium potential but, during an action potential, there is a large transient increase in permeability of the membrane to Na ions and in potential swings close to that expected for the Na equilibrium potential. The increased permeability only lasts about a millisecond, and the amount of Na that actually enters the cell is small. However, during a rapid train of action potentials, a significant amount of Na enters the cell and, unless this is pumped back out, this leads to changes in the ion gradients and membrane potential. This important extension of the biophysics of cell membranes will be discussed in later chapters on the electrophysiology of nerve and muscle.

Summary

The origin of the differences in ionic composition between the cytoplasm and interstitial fluid is most complex. Contributing factors include (1) the concentration of impermeable anions within the cell, (2) permeability of the cell membrane to anions and cations, (3) kinetics of the Na/K exchange pump and (4) cell metabolism, which produces impermeable organic anions and ATP to energize Na/K active transport.

It is instructive to consider briefly the effect of a disturbance of the steady state in the cell, e.g., a rapid permanent increase in permeability of the cell membrane to Na ions. Initially this leads to an increase in the leakage of Na into the cell down the steep electrochemical potential gradient. This in turn causes a decrease in the membrane PD (toward 0), which can be predicted from the constant field equation. The decrease in PD is followed by an increase in the net flux of K out of the cell and a net Cl flux into the cell, the magnitude of the fluxes depending on the change in PD and permeability coefficients. Finally, the change in the fluxes may cause a transient increase in osmotic pressure of the cytoplasm, which in turn causes water flow into the cell to produce a slight increase in volume.

An important factor in bringing the cell to a new steady state is the increase in pump activity related to the increase in internal Na concentration. The rate of sodium pumping is a steep function of the internal Na concentration (rate $\propto$ $[Na]^3$), and small increases in Na_i cause a large increase in pumping. This is associated with an increase in the utilization of ATP, synthesis of ATP and rise in oxygen consumption. The higher pump rate directly leads to a small increase in the membrane PD owing to the electrogenic nature of the pump, and this in turn partially compensates for the increase in P_{Na}. In sequence, this produces a decrease in the net passive efflux of K and net passive influx of Cl. Eventually the cell reaches a new steady state where the membrane PD is slightly reduced, internal Na and Cl concentrations are slightly elevated, cell volume is a little larger, internal K concentration is slightly lower, the pump is working a little faster and ATP is being synthesized and utilized at a greater rate.

TRANSPORT ACROSS EPITHELIA

The transport of solutes and water across epithelial membranes is the most important function of many organs in the body, e.g., digestion and absorption of food by the gastrointestinal tract and formation of urine by the kidneys. Epithelia are continuous sheets of cells that line the surfaces of organs and body cavities and, in principle, the mechanisms of transport of material across these cells are identical to those already discussed. However, in some epithelia, e.g., muscle capillaries, solute can bypass the cells completely and permeate from one side of the tissue to the other between the cells. The rate of permeation depends on the nature of the junctions between cells and the dimensions of the lateral intercellular spaces (Fig 1–12). In many epithelia, e.g., intestine and renal tubule, the junctions are formed by the plasma membranes of adjacent cells coming into direct contact to make a complete seal around the entire circumference of the lateral borders of the cell. These belt-like structures are referred to as *zonae occludentes* or "tight junctions." However, in other epithelia, e.g., in muscle capillaries and brain ependyma, the plasma membranes do not come into direct contact and do not form seals around the entire cell. Consequently there may be aqueous channels or pores between the epithelial cells. In muscle capillaries these have a diameter of 3–5 nm and occupy about 0.2% of the total membrane area, resulting in unexpected high fluxes of polar molecules such as sucrose, inulin and hemoglobin across the capillary wall. Even in epithelia with continuous occluding junctions, there is a wide spread in the rates of permeation through the junctions.

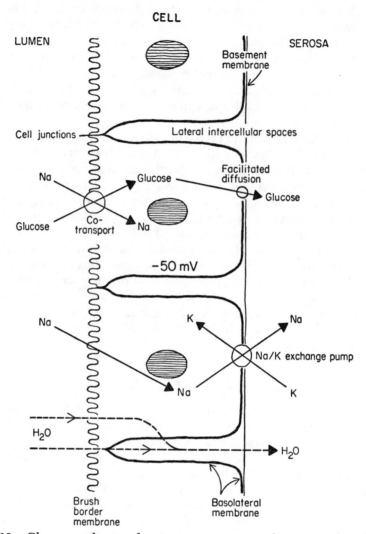

Fig 1–12. — Glucose, sodium and water transport across the intestinal epithelium.

For instance in the gallbladder, small intestine, proximal renal tubule and choroid plexus, most of the passive Na and Cl fluxes proceed via the extracellular route, but in tight epithelia, e.g., the urinary bladder and renal distal tubule, there is no significant permeation between the cells. The cause of this variation between epithelia is not clear.

When molecules permeate across epithelial membranes exclusively via the cellular route, the permeability coefficient for the epithelium, P_e, is related to the permeability coefficients of the two plasma membranes, $P_m{}^1$ and $P_m{}^2$, by the expression for two series resistance barriers,

$$1/P_e = 1/P_m{}^1 + 1/P_m{}^2$$

i.e., the overall permeability coefficient is determined by the properties of the two faces of the epithelium. The factors controlling the magnitude of $P_m{}^1$ and $P_m{}^2$ are identical to those discussed above for single cell membranes. Irrespective of the actual route of permeation across the epithelium, the rate and direction of passive permeation are determined by concentration and electric gradients across the whole epithelium.

ACTIVE TRANSPORT ACROSS EPITHELIA

It is well known that a net transport of solutes and water can occur across some epithelia in the absence of external driving forces, and in some cases transport can actually occur uphill against the concentration and electric gradients. Reid in 1892 was one of the first to demonstrate that water was transported across the intestine and frog skin in the absence of both osmotic and hydrostatic pressure gradients and that this was a property of the living tissues. These observations have been confirmed and extended to the net transport of ions, sugars and amino acids in other epithelia such as the renal tubule, stomach, gallbladder, pancreas and choroid plexus.

The mechanisms involved in these "active" transport processes can best be clarified by briefly considering the transport of glucose, sodium and water across the intestinal epithelium. A simple scheme showing the mechanisms involved is illustrated in Figure 1–12.

Glucose is accumulated in the epithelium across the brush border (see Fig 1–12) membrane by the co-transport process described earlier. Subsequently there is a net downhill transport of the sugar to the serosal side of the epithelium across the basolateral cell membrane. Quite recently it has been established that sugar transport across the serosal surface of the cell occurs by facilitated diffusion, i.e., the same mechanism as described above for the entry of glucose into the red blood cell. Thus, the actual rate of glucose transport across the intestine depends on the kinetics of the sodium/glucose carrier in the brush border membrane and the kinetics of the facilitated glucose carrier in the basolateral cell membrane. Glucose is absorbed into the blood across the intestinal epithelium as long as the glucose concentration in the cell is higher than in the blood. The process depends on cellular metabolism to the extent that the continued operation of the Na/K exchange pump is necessary to maintain the sodium concentration gradient across the brush border membrane. Both the intracellular accumulation of glucose and uphill transport of glucose from the lumen of the intestine to the blood cease in the presence of metabolic inhibitors.

Sodium ions can enter the cell down the concentration and electric gradients from either the intestinal lumen or the blood but, because of the higher permeability of the brush border membrane, more enters from the lumen (see Fig 1–12). In addition, sodium is transported into the cell across the brush border membrane with amino acids and sugars. In the steady state all the sodium that leaks into the cell is pumped out into the serosal fluid by Na/K pumps on the basolateral cell membrane. The exclusive distribution of the pumps on the serosal side of the cell has been established by (1) localization of radioactively labeled ouabain by autoradiography, and (2) isolation and purification of brush border and basolateral membranes of the intestinal epithelium by biochemical techniques. As far as is known, the properties of the Na/K exchange pump and the Na/K ATPases of the basolateral membranes of the intestinal epithelium are similar to those in muscle, nerve and red blood cells. This two-stage model for sodium transport originates from the work by the Danish physiologist Ussing on sodium transport across frog skin, and it also accounts for the net transport of sodium across other epithelial membranes such as the renal tubule, gallbladder and choroid plexus. In epithelia in which the permeability of the extracellular route is very low, e.g., frog skin and urinary bladder, sodium can be transported against concentration gradients as high as 100 to 1.

The rate of net transport of water is directly proportional to the rate of sodium transport across the tissue. Furthermore, the proportionality constant is such that 300–400 water molecules are absorbed for each sodium ion, i.e., the fluid transported has the same osmolarity as plasma. Current concepts of water transport across epithelia stem from Diamond's work on the gallbladder, and his model is applicable to the intestine. The major feature of this model is that the sodium pumps are localized along the lateral borders of the epithelial cells, and that salt is actively transported into the spaces between the cells (see Fig 1–12). This increases the salt concentration in the spaces, and the resulting increase in osmotic pressure draws fluid into the spaces from the cells and from the intestinal lumen. The subsequent increase in pressure in the spaces forces the fluid across the basement membrane and into the blood. The rate and composition of the fluid transported out of the epithelium depend (1) on the rate of solute pumping into the lateral spaces, (2) on the hydraulic water permeability of the epithelium and (3) on the geometry of the lateral intercellular spaces. Using measured values of these parameters, it can be shown theoretically that the model accounts for the rate and composition of fluid transport across epithelia.

In conclusion, it bears emphasis that the Na/K exchange pump is one of the most fundamental processes in physiology. This pump is involved directly or indirectly in almost all processes in the body, and the energy consumed by the pump represents a major proportion of the total energy consumption by the body.

BIBLIOGRAPHY

Davson, H.: *A Textbook of General Physiology* (4th ed.; London: Churchill Livingstone, 1970).
Dowben, R. M.: *General Physiology: A Molecular Approach* (New York: Harper & Row, 1970).
Weissmann, G., and Clairborne, R.: *Cell Membranes: Biochemistry, Cell Biology and Pathology* (New York: H. P. Publishing Co., 1975).

2

Excitation and Conduction

WILFRIED F. H. M. MOMMAERTS, Ph.D.

NERVES

IN THE PREVIOUS CHAPTER it was shown that cells maintain a set of ionic concentration differences between their interiors and the extracellular medium. It was further indicated that permeability of the cell membrane for different ionic species varies considerably. These variations in ionic permeability were shown to provide the basis for the cell membrane potential to which each ion contributes in consequence of its concentration gradient, its charge and sign and its permeability coefficient. Most bioelectric phenomena to be encountered in this chapter can be explained by these factors with respect to K^+, Na^+ and Cl^- and can be described by the constant-field or Goldman equation:

$$E = \frac{RT}{F} \ln \frac{P_K[K^+]_o + P_{Na}[Na^+]_o + P_{Cl}[Cl^-]_i}{P_K[K^+]_i + P_{Na}[Na^+]_i + P_{Cl}[Cl^-]_o} \tag{1}$$

RESTING POTENTIAL

In resting cells, the values of P_K and P_{Cl} are often similar, and the ratios of the inside and outside concentrations of K^+ and Cl^- operate in the same sense and have similar magnitudes. Thus, the resting potential, usually 60–100 mv (inside negative), is often described as the equilibrium potential for K^+, or as something intermediary between the K^+ and Cl^- distribution potentials. On the other hand, Na^+, which is present in much higher concentrations outside cells than inside, would strongly tend to move inward on account of both the concentration gradient and transmembrane potential. The low resting intracellular Na^+ concentration is maintained, however, by the low value of P_{Na} together with the action of the Na^+ pump: the cell is constantly baling out the Na^+ that enters through a small leak.

The development of capillary microelectrodes has allowed direct experimental study and manipulation of intracellular potentials, first in giant nerve fibers such as the squid axon, which measures up to several tenths of a milli-

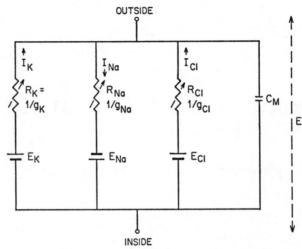

Fig 2–1.— Diagram of electric properties of a cell membrane. (From Hodgkin, A. L., and Huxley, A. F.: J. Physiol. [Lond.] 117:500, 1952. Used by permission.)

meter in diameter, and then in muscle fibers, large neurons and increasingly in well-nigh all but the smallest cells.

Figure 2–1 is a diagram of the meaning of equation 1. Transmembrane potential E is measurable with a microelectrode placed inside the cell and a reference electrode placed extracellularly. It results from the concentration potentials E_K, E_{Na} and E_{Cl}, determined by the concentration differences for these ions, and the resistances R_{sub} offered by the membrane against the currents I_{sub} ascribed to each ion. Instead of R, we often use its inverse, the conductances g_{sub} that can be related more directly to the permeability constants P_{sub} in eq. 1. The membrane also is shown to have a capacitance C_M, which results

TABLE 2–1.—APPROXIMATE OR POSSIBLE VALUES OF INTRA-
CELLULAR (C_1) AND EXTRACELLULAR (C_2) CONCENTRATIONS
OF MAJOR ION SPECIES AND EQUILIBRIUM POTENTIALS*

TISSUE, AND TYPICAL OBSERVED RESTING POTENTIAL	ION	C_1	C_2	EQUILIBRIUM POTENTIAL
Frog muscle −98 mv	Na+	9.0–13.0	120.0	+55
	K+	140.0	2.5	−101
	Cl−	3.5	120.0	−86
Mammalian spinal cord†	Na+	15.0	150.0	+60
motor neuron	K+	150.0	5.5	−90
	Cl−	9.0	125.0	−70
Mammalian brain†	Na+	46.0	130.0	+27.5
	K+	135.0	5.2	−87
	Cl−	27.5	110.0	−37
Squid axon −60 mv	Na+	50.0	460.0	+55
	K+	400.0	20.0	−75
	Cl−	40–100.0	560.0	−50

*Equilibrium potentials calculated for each ion species as if it were to dominate eq. 1, which then simplifies to $E = 58 \log C_1/C_2$.
†Values for mammalian CNS estimated under simplifying assumptions by Eccles and McLennan, respectively (values for frog and squid axons well known).

from separation of the conducting phases on either side by the dielectric membrane. The capacitance is the main determinant of the *time constant,* to be described later.

Inspection of the numerical examples of Table 2–1 illustrates the validity of the above view of the membrane potential. The calculated equilibrium potentials for both K^+ and Cl^- are close to the observed resting potential, so that both can be said to contribute to its maintenance, though there are situations in which K^+ seems more determinant. A test for the nature of the resting potential is to measure its dependence on the external K^+ concentration, experimental variation of which would then change the potential in accordance with equation 1 written for K^+ only. This it does (Fig 2–2), except at very low $[K^+]_o$ into which other factors enter, but where the approximation is improved by taking into consideration the sodium concentration gradient across the membrane and the membrane permeability to sodium. Depolarization by increased external K^+ is an often used experimental artifice to alter membrane potential. Alternatively, the internal ionic concentrations can be manipulated. Much has been learned by electrophoretically injecting ions through the intracellular microelectrode; thus, the Cl^-_i and Na^+_i could be increased, although not the K^+_i, because this is already predominant, and the injected current, although entering as K^+, also leaves the cell in that way.

Fig 2–2.—Measured cellular potential (ordinate) of frog muscle fiber determined during experimental variation of external K^+ concentration (abscissa, plotted logarithmically). Internal K^+ concentration is 140 mM. In the higher range of $[K^+]_o$, the points fall on the straight line, which expresses eq. 1 written for the K^+ term only, i.e.,

$$E = -58 \log \frac{K^+_o}{140}$$

At low K^+_o the measured values become too positive but can now be approximated by the formula

$$E = -58 \log \frac{[K^+]_o + 0.01\,[Na^+]_o}{140}$$

which expresses a contribution by the Na^+ ions commensurate with a low P_{Na}. (After Hodgkin, A. L., and Horowicz, P.: J. Physiol. [Lond.] 145:405, 1959.)

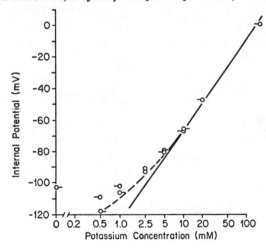

The mechanism of the maintenance of the differential distribution of ions has been advantageously studied in the giant axon of the squid, perfusing the inside and outside with solutions containing radioactive markers. With intact metabolism, there was a continuous extrusion of Na^+ against its concentration gradient. This extrusion, much reduced in a K^+-free external medium, suggested the existence of a pump mechanism that exchanged Na^+ for K^+. When metabolism was stopped, the extrusion rate dropped drastically but could be restored by injection of substances such as phosphoryl-arginine or -pyruvate that regenerate ATP. The concept of an ATP-driven transport mechanism has been found widely applicable in active tissues (Chapter 1).

The meaning of the differential ionic distribution and conjugate potential in vegetative cells is not in all respects clear. It is plausible that many enzymes more closely approach optimal activity in the cytosol, where K^+ and Mg^{2+} predominate, than in a Na^+ and Ca^{2+} medium; but other enzymes might well have developed instead had there been no ionic segregation. However, once we accept that segregation has taken place, the possibility of a property of tremendous consequences is created. We shall describe this as *excitability*, and it is *a property upon which any form of active or animal life depends,* and is *the basis of all behavioral and mental activity*. This section will discuss mainly the excitability of peripheral nerves, but similar considerations apply to muscles, sensory cells and glands, be they exocrine or endocrine.

STRUCTURE OF NERVES

Nerves consist of bundles of nerve fibers or axons enmeshed in and surrounded by connective tissue. Each axon is a greatly elongated process of a *cell body,* located in the brain stem or in an autonomic ganglion. Each axon is enveloped by satellite cells called Schwann's cells, the function of which is largely unknown in most cases, although it has been suggested that they may provide some type of metabolic support for the axon. During development Schwann's cells of some nerve fibers rotate around and around the axon. The resulting numerous turns of compressed cytoplasm and cell membranes form an insulating sheath termed *myelin*. This sheath is interrupted every few millimeters by a zone about 1 μ long, termed the *node of Ranvier,* where the axon membrane is exposed. Thus, we can distinguish two types of axon, myelinated and unmyelinated (Fig 2–3).

NERVE EXCITATION

The excitation process is a sudden *depolarization* or diminution of the negative resting potential. Not infrequently an *overshoot* occurs, i.e., the membrane potential actually becomes transiently positive. The membrane then returns to its resting state by a process termed *repolarization*. These electric events constitute an *action potential,* colloquially known as a *spike*. The principal mechanism in nerve and skeletal muscle is a drastic increase of the permeability and conductance for the Na^+ ion. With P_{Na} becoming predominant in eq. 1, the membrane potential now approaches the equilibrium potential

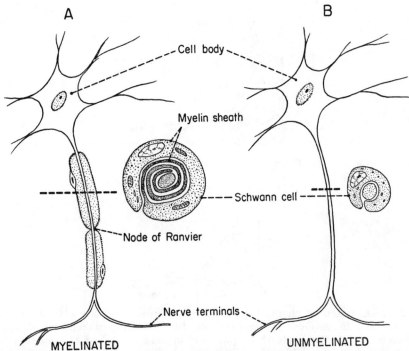

A   B

Cell body

Myelin sheath

Schwann cell

Node of Ranvier

Nerve terminals

MYELINATED   UNMYELINATED

Fig 2–3.—Diagram of neurons with myelinated axon (**A**) and nonmyelinated axon (**B**). Cross sections of axons at levels indicated are also shown. (After Katz, B.: *Nerve, Muscle and Synapse* [New York: McGraw-Hill Book Co., 1966].)

for Na+, which it may or may not reach depending on the magnitude and time courses of the conductance changes for sodium and the other permeable ions.

The nature of the phenomenon of excitation is best introduced by reference to a classic study by Hodgkin on a crab nerve axon, done before the use of intracellular microelectrodes. The essential details are found in Figure 2–4 and its legend. The starting potential indicated by a small external electrode is taken as zero and depolarizations are plotted upward. The responses shown were recorded upon applying brief shocks (less than 0.1 msec) of either sign. When anodic shocks are applied, the resulting increases in membrane potential (hyperpolarizations) come and go with a characteristic time course and extend over a certain distance. These are determined by the same electric laws that govern the behavior of incompletely insulated, "leaky" cables and are described by a *time constant* τ and a *length constant* λ. These are, respectively, the time after which an imposed voltage has reached the value $1/e$ of its completion, and the distance over which it has declined to $1/e$ of its original value. If r_m is the membrane resistance, r_i the internal resistance and c_m the membrane capacitance, all expressed per centimeter of fiber length, then $\tau = r_m c_m$ and $\lambda = r_m/r_i$. Such passive spreads of imposed potentials are called *electrotonic*. When cathodic, i.e., depolarizing, shocks of small magnitude are applied, the induced potential changes behave similarly, but at increasing intensities of shock, the potential changes become greater and last longer than

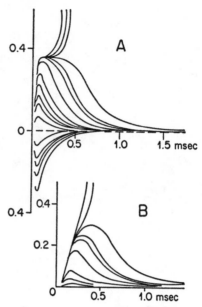

Fig 2–4.—A, potential changes detected by external electrode placed on a large crab nerve fiber in response to shocks of very brief duration (0.06 msec), all applied at the same site. The ordinate is the potential change expressed as a fraction of the peak value of the conducted spike (about 40 mv). The upgoing curves are responses to depolarizing shocks. The curves were elicited by shocks of 1.0 (upper six curves), 0.96, 0.85, 0.71, 0.57, 0.43, 0.21, −0.21, −0.43, −0.57, −0.71, −1.0 times the value of the standard threshold. B, local responses produced by shocks of 1.0 (upper five curves), 0.96, 0.85, 0.71, 0.57 times the standard threshold and obtained from A by subtracting the hyperpolarizing curves from the corresponding depolarizing curves. Note that at 0.57 threshold the local response is virtually undetectable, whereas at 1.0 threshold the local response is large and it is a matter of chance whether and when a propagated spike develops. (From Hodgkin, A. L.: Proc. R. Soc. Lond. (Biol.) 126:87, 1938. Used by permission.)

their hyperpolarizing counterparts, as if some active component were added.

This difference, called the *local response*, is shown in Figure 2–4, B, where its magnitude is plotted as a function of the stimulus strength. The passive response occurs whether or not the nerve is excitable, but the active local response is characteristic for depolarization of an excitable tissue and leads to an excitatory event when it reaches a certain value, the *threshold*. Accordingly, Figure 2–4, B, shows that the local response comes and goes when the applied stimuli are below threshold, that very close to threshold the membrane becomes unstable and may react either way and that just above threshold the potential veers drastically and changes into a conducted spike or action potential. This concept of a subthreshold "local response," which when reaching threshold becomes a conducted excitatory event, is of fundamental importance and will be encountered in a number of situations. It is characteristic that, once the exciting threshold has been reached, the magnitude of the resulting conducted spike is determined only by local conditions and not by

the intensity of the applied stimulus. Thus a conducted impulse is said to be
all-or-none.

It is to be noted that from a site of initiation the excitatory process is transmitted equally in both directions. In functional situations, however, the initiation of events is so organized that there is a natural or *orthodromic* direction in
which impulses normally progress.

Generation of a spike potential from a local response, i.e., the occurrence of
an all-or-none conducted excitatory event, is subject to some interesting rules,
explained in Figure 2–5. The presentation is schematic, applying equally
whether the stimulus is administered by an extra- or intracellular electrode or,
for that matter, by stimulus modes other than electric ones. Figure 2–5
merely shows that the impulse is switched on as a sudden step, and implies
that this sets up a local response in some proportion to the stimulus intensity.
Two ways of presenting the concepts exist. Either the step-function is
switched on and the time interval to excitation is measured, i.e., *the utilization time of the stimulus* or *the latency time of the spike appearance* is measured, or square pulses of a set intensity are switched on for various times on a
trial basis, it being determined which conjugate values of *strength and duration* just suffice to elicit an excitatory event. Such measurements fall on the so-
called *strength-duration curve* (Fig 2–6), which shows (1) that a weaker stimulus needs a correspondingly longer time to be effective, but (2) that below an
absolute threshold intensity, the *rheobase,* excitation never ensues. The utili-

Fig 2–5.—Diagram of the concepts underlying strength-duration or strength-
latency relations. A square pulse of electric stimulation is applied. In series **A** this is
administered for an indefinite time (**insert,** top), and the interval before an action
potential occurs is measured. This is called the utilization time (of the stimulus) or
the latency time (of the response). The time becomes shorter the greater the stimulus
intensity. Series **B** and the lower part of the **insert** show how a rectangular pulse of
limited duration can be applied, equivalent to the utilization time at a given intensity. With such test pulses, we determine whether or not a response occurs.

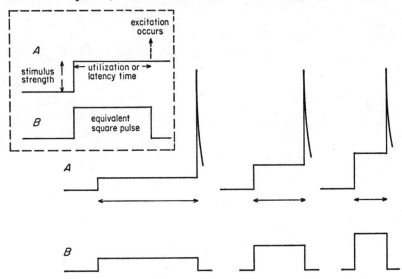

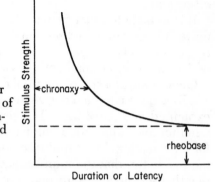

Fig 2–6.—Schematic strength-duration or strength-latency curve as obtained by either of the routes described in Figure 2–5. The concepts of *rheobase* and *chronaxy* are explained in the text.

zation time needed at a stimulus intensity twice the rheobase is called the *chronaxy,* and this is some measure of the reaction speed of a tissue's excitability. For example, the chronaxy for muscle is decidedly longer than for nerve. This is used clinically to judge the progress of regeneration of an injured nerve. It is possible to stimulate a muscle transdermally with electric pulses. With the innervation intact, we really stimulate the nerve branches rather than the muscle and the chronaxy is correspondingly brief (Fig 2–7),

Fig 2–7.—Examples of strength-duration curves obtained by transdermal stimulation of a muscle in the normal state (*N*) or in the denervated condition (*A*). In the latter case, the chronaxy of 4 msec is that of muscle, whereas the control at a chronaxy of 0.08 msec is that of nerve. During healing of the innervation, the mixed transitional curves illustrate the re-establishment of muscle innervation. (From Ritchie, A. E.: Brain 76:322, 1944. Used by permission.)

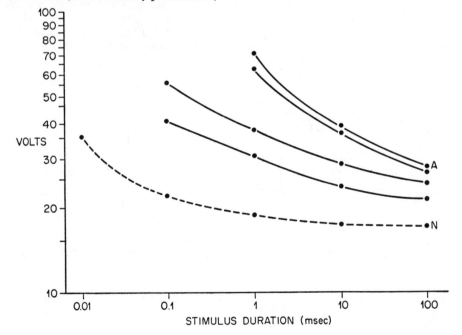

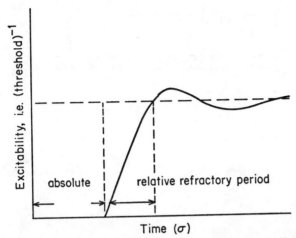

Fig 2–8.—Diagram of refractory period, absolute and relative, immediately follow-ing an excitatory event. The ordinate shows the excitability expressed by the inverse of the threshold. After an absolute refractory period of the order of a millisecond or a fraction thereof, excitability returns and goes through some fluctuations. The *hori-zontal dashed line* represents the normal resting level of excitability.

but with the innervation dead, the chronaxy measured is the muscle's, be-cause now direct stimulation of the muscle has to occur. As nerve regenera-tion occurs, the chronaxy becomes progressively shorter.

One other property of excitation can be mentioned at the present level of discussion. After a stimulus has taken effect, there is a time, the *refractory pe-riod,* during which the tissue cannot be stimulated again. Actually, there is an *absolute refractory period* during which no stimulus, however strong, is effec-tive. This is followed by a *relative refractory period* during which a signif-icantly greater stimulus intensity is required to induce excitation. Subse-quently a few fluctuations of hypo- and hyperexcitability may occur until the normal steady resting level of excitability is restored (Fig 2–8). In nerve and other tissues commonly encountered, these time spans are measured in milli-seconds. If the refractory period of a nerve fiber were to be 1 msec, the fiber in question could transmit at a frequency up to about 1,000 Hz, a figure not quite reached in physiologic stimulations commonly encountered in a number of situations.

Having understood the principle of the local response, we refer to Figure 2–9 to review some matters from this point of view. Two electrodes are ap-plied intracellularly to a large fiber. The first is used to record responses, the second to give small shocks not unlike those illustrated in Figure 2–4 (extracellularly then), but with a longer duration of the pulses. The two elec-trodes are shown to be inserted at the same site, but they can also be placed at several distances. When we apply square wave pulses (lower tracing in each set), we see (see Fig 2–9, B) that the responses follow the applied pulses pro-portionally but for the time constant imposed by the capacitative properties, and that these electrotonic responses die away with distance according to the

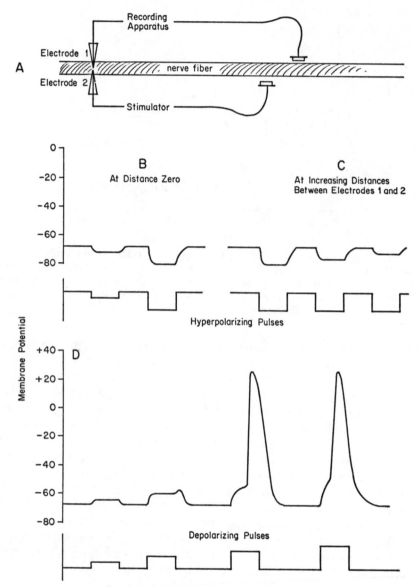

Fig 2–9.—A, nerve fiber with two intracellular microelectrodes for stimulation and for recording. The latter can be placed at the site of stimulation as shown, or at varying distances therefrom, as implied in **C.** The bottom traces in **B, C** and **D** show the applied rectangular pulses with respect to magnitude and polarity. The tracings are to be compared with Figure 2–4 except that now the pulses are meant to be of much longer duration. When hyperpolarizing pulses are applied (**B** and **C**), no active responses occur. When depolarizing stimuli are administered (**D**), an active contribution by the membrane becomes evident (as in Fig 2–4) and turns into a conducted spike; the greater the depolarization the earlier the spike.

space constant (see Fig 2–9, C). In contrast, depolarizing pulses, when approaching threshold, show signs of an active response, which, upon exceeding threshold, turns into a conducted spike, the sooner the greater the stimulus (see Fig 2–9, D).

IONIC BASIS OF ACTION POTENTIAL

We have already seen that the nerve action potential is associated with a dramatic increase in membrane sodium conductance (g_{Na}). The analysis of this change and the changes in other ionic conductances is very complex since they are functions of time and membrane voltage (E_m) and these functions differ for each ion. If we consider only the two most important cations, sodium and potassium, we have four variables—g_K, g_{Na}, E_m and t. The first three variables all change with time and interact so that changing one also changes the others.

Hodgkin, Huxley and Katz devised an ingenious technique, *the voltage clamp,* which enabled them to eliminate one variable (membrane potential, E_m) and to study the others in isolation. An apparatus was designed that enabled a giant axon of the squid to be rapidly depolarized to a new level of membrane potential and held or "clamped" at this potential. Any deviation from the preset potential was sensed electronically, and appropriate circuitry caused current either to be injected or removed from the fiber so as to keep the potential constant. The sudden imposed shift in membrane potential alters the ionic permeability of the membrane. During an unclamped action potential, these permeabilities change with time and membrane potential. However, when the membrane is clamped, they change only as a function of time and in a manner characteristic of the value of the clamped potential. The altered permeabilities produce ionic currents, which are sensed and opposed

Fig 2–10.—Current recorded from a nerve axon when its membrane is depolarized and clamped at 0 mv.

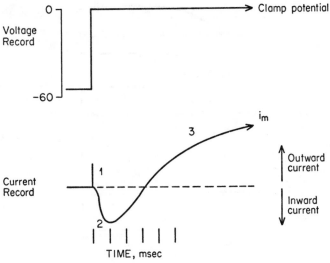

by the feedback circuitry described above. Thus the current injected or withdrawn is an electric measurement of the total membrane current (i_m) and approximately the sum of potassium (i_K) and sodium (i_{Na}) currents. Experiments were devised to separate these currents and to show how each changed with time at the clamped potential. Knowing i_K and i_{Na}, it was a simple matter, using Ohm's law, to calculate membrane conductance changes, since $i_{ion} = g_{ion}$ $(V - E_{ion})$, where V is the clamped voltage and E_{ion} is the equilibrium potential for the ion under consideration. Thus under these experimental conditions, changes in ion flow with time directly reflect changes in membrane conductance. A typical record obtained by the voltage clamp technique is shown in Figure 2–10. When the membrane potential was suddenly altered from −60 mv to 0 mv and clamped, the current record showed three components: (1) a "blip" owing to the discharge of the membrane capacity, (2) a brief surge of inward current and (3) an outward current, which was maintained as long as the voltage remained clamped at its new level. This type of experiment was repeated, each time with the membrane voltage clamped at a different level. As can be seen from Figure 2–11, for potential displacements up to 98 mv, i_m was first inward then outward, but for potential displacements greater than 100 mv (which brings the membrane potential to the equilibrium

Fig 2–11.—Membrane currents recorded when the axon was depolarized and clamped at different voltages. The numbers at the left of each trace indicate the number of millivolts' *displacement* of the clamped voltage from the original membrane potential. The ordinate scale indicates a membrane current of 0.5 ma/sq cm per division. Outward currents are upward, inward currents downward. (After Hodgkin, A. L., and Huxley, A. F.. J. Physiol. [Lond.] 116:449, 1952. Used by permission.)

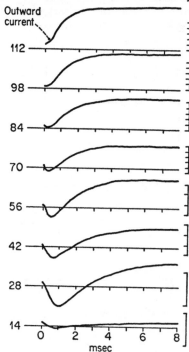

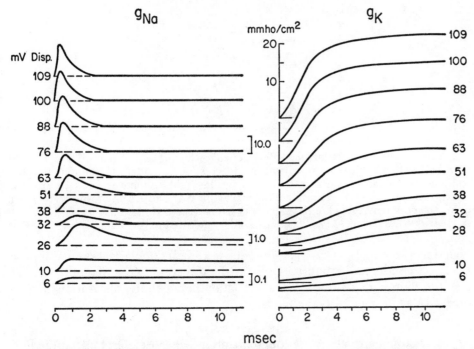

Fig 2–12.–Left, derived curves showing changes in Na⁺ conductance (g_{Na}) when membrane voltage was displaced by amounts indicated, and clamped. The ordinate scale is 10 mmho/sq cm for voltage displacements of 32 mv and above, 1 mmho/sq cm for the 26 mv displacements and 0.1 mmho/sq cm for the remainder. **Right,** derived curves, similarly obtained, showing changes in K⁺ conductance (g_K). The Ordinate scale at the upper left applies to all curves for voltage displacements above 28 mv. The scale for the smaller displacements is increased fourfold. (After Hodgkin, A. L., and Huxley, A. F.: J. Physiol [Lond.] 117:500, 1952. Used by permission.)

potential for sodium E_{Na}), the initial inward current reverses to become an outward current. This reversal would be expected if the early phase of current flow is carried by sodium ions and the membrane potential exceeds E_{Na}. By carrying out similar measurements in solutions of varying Na⁺ and K⁺ concentrations, the i_K and i_{Na} components of the measured membrane current were separated and a family of curves showing the variations of i_{Na} and i_K against time at different levels of membrane potential was obtained. The corresponding g_{Na} and g_K curves were then derived (Fig 2–12). These curves show that large sudden depolarizations cause (1) an initial rapid increase in g_{Na} followed by a somewhat slower decrease in g_{Na} to control levels (sodium inactivation)— the increase in g_{Na} becomes progressively larger as membrane voltage becomes less negative—and (2) a slow increase in g_K (delayed by a few milliseconds). It was therefore possible to show how the Na⁺ and K⁺ conductances would change during an action potential and to reconstruct the waveform of the action potential from the conductance curves (Fig 2–13).

We are now in a position to describe the ionic basis of the membrane response to depolarizing currents. When these are less than threshold, the in-

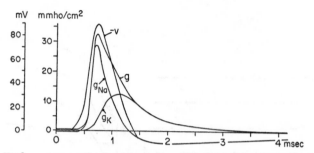

Fig 2–13.—Redrawn record of voltage changes during action potential in a squid axon (left ordinate scale in mv with respect to resting potential). Curve g depicts the conductivity of the membrane, measured by the right ordinate scale. This curve is interpreted in terms of separate conductances for Na^+ and K^+, the former's increase being the dominant aspect of the excitatory event. (From Hodgkin, A. L., and Huxley, A. F.: J. Physiol. [Lond.] 117:500, 1952. Used by permission.)

duced increase in g_{Na} is small and produces a small depolarization that is only passively conducted. Stimuli above threshold increase g_{Na} sufficiently so that a self-reinforcing, positive feedback relation between depolarization and increased Na^+-permeability develops:

$$\text{electrotonic arrival} \longrightarrow \overset{\text{increased } g_{Na^+}}{\underset{\text{depolarization}}{\circlearrowleft}} \longrightarrow \text{electrotonic spread}$$

as a result of which the full excitatory event as well as its propagation occur. The new membrane potential tends toward the equilibrium potential for Na^+, but the subsequent increase in g_K, together with the termination of the state of increased g_{Na} (Na^+-inactivation), causes repolarization.

One point should not be misunderstood. Although the phenomena described are due to drastic alterations in ionic permeabilities, the actual amounts of ions transferred are negligible. From transfers of isotopically labeled K^+ and Na^+ in the squid axon, it emerged that their exits and entries are of the order of 10^{-12} M/sq cm of membrane surface. Small though these amounts are in comparison to analytically detectable concentration changes, they are commensurate with the charge transfer accounting for a potential change of about 100 mv across a condenser gap with a capacitance such as the membrane has. Because 1 M of ion carries approximately 10^5 coulombs of charge, a transfer of 10^{-12} M of ion across a capacitance of about 10^{-6} farad/sq cm would correspond to a change in charge of 0.1 v.

MECHANISM OF CHANGES IN IONIC MEMBRANE CONDUCTANCES

There is considerable evidence that ion conductance changes in nerve occur as the result of potential-dependent conformational changes in aqueous channels located between protein subgroups in the surface membrane. One plausible model suggests that there are two distinct types of channel, the "early" channel and the "late" channel. The early channel carries the early inward current and normally allows only sodium ions to pass (it will also admit lithium but excludes potassium and rubidium). The late channel carries the outward current and normally allows only potassium ions to pass (it will also admit rubidium but excludes sodium and lithium). Support for this model comes from the fact that pharmacologic agents exist that can block sodium and potassium conductance changes independently of each other. Thus the puffer fish poison, tetrodotoxin, specifically blocks the early (sodium) channel and prevents action potential generation in nerve. On the other hand, tetraethylammonium selectively blocks the late channel when suitably applied and thereby prolongs the action potential while having no effect on the early channel.

ACTION POTENTIALS IN OTHER TISSUES

Action potentials vary greatly among tissues. For example, the action potential of cardiac muscle shows a prolonged plateau and may last several hundred milliseconds. The ionic conductance changes responsible for this plateau are complex and are discussed later (section on neuromuscular junction).

Smooth muscle also has much longer action potentials than in nerve, and in many cases the ionic mechanisms are different. Thus, the charge-carrying ion responsible for the initial rapid depolarization or "spike" phase of the action potential in several smooth muscles is calcium rather than sodium. These "calcium spikes" are unaffected by tetrodotoxin but can be inhibited by certain metallic ions such as Mn^{2+} and Cd^{2+} and by specific drugs such as verapamil, which have little effect on "sodium spikes."

CONDUCTION OF NERVE IMPULSE

In earlier paragraphs it was indicated that conduction of an action potential along a nerve is due to electrotonic spread, which successively "ignites" adjacent portions of the membrane. The velocity of conduction in unmyelinated fibers is approximately proportional to the square root of the axon diameter. Thus conduction in the small ($0.2-1.0$ μ in diameter) unmyelinated type C fibers, which subserve pain or innervate smooth muscle or glands, is relatively slow, about 0.5 m per second. Conduction is much faster in myelinated axons and is roughly proportional to the axon diameter itself. It will be recalled that these axons are surrounded by an insulating myelin sheath, which is interrupted at intervals (nodes of Ranvier) where the axon is in conductive contact with the surrounding medium. Owing to its multilayer structure, the resistance of the sheath is some 10,000 times higher than that of typical axonal membranes and its capacitance 1,000-fold lower.

As one illustration of the effect of this insulation, stimulus thresholds vary periodically along the length of a myelinated axon. The highest stimulus is required at the middle of an internode, the lowest at a node. Since the space constant varies with the square root of the membrane resistance, it becomes relatively large in myelinated nerves, where it amounts to several millimeters. This means that an action potential at a node can spread electrotonically

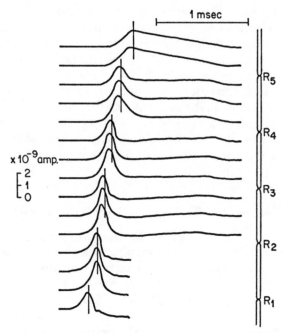

Fig 2–14.—Demonstration of saltatory conduction. A nerve fiber was strung through a very fine hole in the partition between two chambers. The fiber was stimulated at one end, while recording was done from the other chamber. In successive experiments, from the bottom traces upward, the nerve was moved so that the recordings were from increasing distances. Each time a new Ranvier node (R_{1-5}) is passed, the arrival of the spike is shifted by fraction of a millisecond, illustrating saltatory conduction. (After Huxley, A. L., and Stämpfli, R.: J. Physiol. [Lond.] 108:315, 1949. Used by permission.)

to the next node, causing threshold to be exceeded and another action potential to be generated. Thus, in myelinated nerves the excitatory event jumps passively from one Ranvier node to the next and at each node the action potential is regenerated. This type of conductance is termed "saltatory" and the conduction velocity is faster than in nonmyelinated fibers of the same diameter.

The most explicit demonstration of this was given by threading a single isolated myelinated axon through a fine hole in a glass partition between two chambers filled with saline. The fluid space around the fiber was narrow to represent a high enough resistance for measuring the longitudinal current flowing when an action potential passed. The fiber was stimulated at one end and the currents were measured in terms of their appearance after the stimulus artifact as the partition was moved farther and farther from the origin of excitation (Fig 2–14). It is seen that arrival shifts with a finite difference upon passing a Ranvier node, but that arrivals are the same whenever the partition is placed around an internode.

There has been controversy about the precise spacing of myelin in the white matter tracts of the CNS, and the question of the mechanism of conduc-

tion in such tracts does not appear to have been definitively studied. However, there seems to be little reason to doubt that the same mechanism applies, for Tasaki has tested the stimulus sensitivity along a fiber of a spinal
cord tract and found an alternation of sensitive spots and insensitive stretches,
as we would expect; also, the conduction velocities in such tracts are typical
for large myelinated fibers with saltatory conduction.

COMPOUND NERVES

Many of the studies described above have been made on single axons. If we
wish to study whole nerves in animals or man, only external electrodes can be
used and the records represent activity in whole bundles, which consist of
cellular and extracellular phases in parallel. Figure 2–15 shows a comparison
of intra- and extracellular recordings of an action potential in a single axon. It
is to be noted that the "spike" is reversed in the extracellular record. Figure
2–16 shows recordings obtained from injured and uninjured compound
nerves. The electrode on the injured surface appears to have access to the
negative inside of the fiber. The potential recorded is termed the *demarcation*
or *injury potential*. It does not indicate the full value of the resting potential
because the two electrodes are partly shorted by the extracellular space. For
the same reason, the recorded action potential does not allow conclusions
about its actual magnitude, although the shape of the spike is the same as that
detected by fundamental methods. Recordings from an injured nerve show

Fig 2–15.—Comparison between intracellular **(top)** and extracellular **(bottom)**
recording through otherwise comparable sampling electrodes. The intracellular
action potential **(top insert)** is already familiar; the extracellular one **(bottom insert)**
is reversed. It is also smaller because it is short-circuited by extracellular medium.

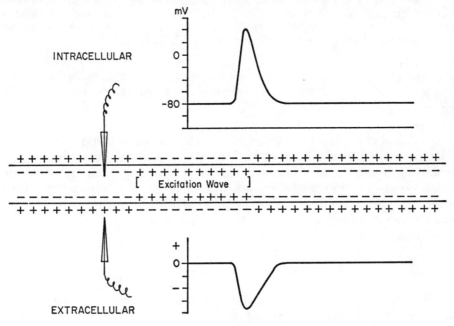

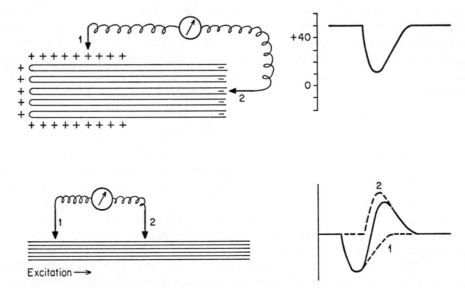

Fig 2–16.—Monophasic (**top**) and biphasic (**bottom**) recordings of action potential in a multifiber nerve. **Top**, the nerve is assumed to be intact at left, but cut at right. Electrode 2, placed at the cut end, would measure internal potential vis-à-vis electrode *1*, but does so only partly because of shorting by the extracellular phase. Electrode 2 records a fixed negative potential. Electrode *1* is positive with respect to *2* in the resting state but becomes progressively less so as it becomes depolarized by a stimulus. As depolarization subsides, the potential at *1* again becomes more positive until the resting state is restored. Thus, a monophasic record is obtained. On the other hand, when both *1* and *2* are applied to uninjured nerve in which excitation is being propagated in the direction shown, biphasic potentials are obtained, since *1* becomes initially negative, then subsequently positive with respect to *2*. The precise pattern obtained depends on the distance between the electrodes in proportion to the length of the excitation wave. **Bottom**, *dashed curves 1* and *2* show the potential variations occurring beneath each electrode as the excitation passes. The *continuous curve* is the actual recording obtained and is the sum of the *dashed curves*.

monophasic potentials but, if the nerve is intact, both electrodes indicate activity, one after the other, and a biphasic signal results (see Fig 2–16). The precise appearance of this signal depends on the geometry of the electrodes and tissues.

In clinical nerve testing, various patterns are obtained when electrodes are placed in the vicinity of nerves. Both stimulation and recording can be achieved through the skin, and conduction velocity readily measured. Marked reduction in speed of conduction may occur in all types of peripheral neuropathy, and its measurement may be useful in following the progress of the disease and in differentiating it from central nervous or primary muscle abnormalities.

INITIATION OF IMPULSES IN VIVO

The emphasis upon experimental generation of excitation by means of electric stimulators should not distract us from the biologic problem of how impulses arise in the functioning intact body. Drawing partly upon the interpretation of cumulated physiologic and behavioral knowledge, and partly upon studies of the behavior of the lowest invertebrate animals with simple nervous systems, we can discern two different though not unrelated mechanisms.

On the one hand, excitatory inputs can arise in *sensory* or *receptor cells* in the broadest sense of the word. A simple case would be a cell responding with generation of a train of impulses when excited by light. In the human body, visual function has become tremendously complex because of the peripheral organization of the receptor organ and the central organization of processing circuits and mechanisms. Still, it remains that chains of excitations and responses can be traced to events starting in such receptor cells.

On the other hand, certain excitatory cells can generate patterns of impulses seemingly spontaneously. It would become a point of sophisticated discussion whether such activity indeed generates "itself" or represents a response to a constant condition of, for instance, pH or oxygen content or ambient temperature. Still, such activity patterns can occur independently of obvious stimulatory happenings. Such systems are called *pacemakers*. The automatic cells of the heart are a non-neural example of this, and neuronal pacemakers have been well studied in lower invertebrates especially.

An interesting additional principle is that both sensory receptors and pacemakers can have their outputs modulated by regulatory nerves. The slowing of the cardiac pacemaker by the vagus nerve and its acceleration by sympathetic innervation are classic phenomena, whereas inhibition of auditory perception by the olivocochlear tract is an example from the realm of higher sensory and cerebral function. Further elaboration of the actual workings of the higher nervous system is left to a later section in this book.

NEUROMUSCULAR JUNCTION

This section deals with the question of how an excitatory process is transferred from a motor nerve axon to an effector organ, skeletal muscle. This is a special case of the general problem of how excitation in the nervous system is transmitted from one neuron to the next, i.e., across synapses. Indeed, the neuromuscular junction has been one of the principal models for comprehension of synaptic transmission whether—a matter of definition—we call it a synapse or not.

ANATOMY OF NEUROMUSCULAR JUNCTION

Motor nerve axons give off a variable number of branches, sometimes a few, sometimes hundreds. Each branch innervates a muscle fiber via a number of expanded terminals, which are separated by 40–50 nm from the adjoining muscle. The region in which muscle and nerve are closely associated is termed the motor *end-plate;* and its area is only a small fraction of the total surface of the muscle cell. However, it is highly specialized (Fig 2–17). The muscle cell membrane is greatly infolded at the end-plate region and this clearly increases the surface interaction between nerve and muscle. The axon terminals are devoid of myelin sheaths, although they are associated with a

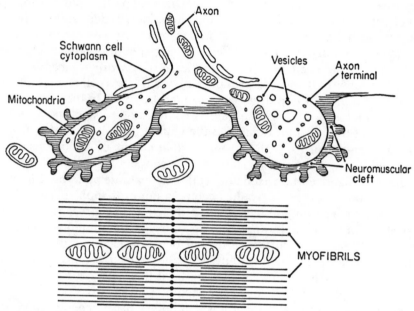

Fig 2–17.—Neuromuscular junction. (From Hubbard, J.: Physiol. Rev. 53:674, 1973. Used by permission.)

thin layer of Schwann cell cytoplasm. The terminals contain numerous vesicles and mitochondria. As a rule, there is only one end-plate per muscle cell and it appears that, once a cell has been innervated by an axon during embryonic development, it usually rejects further contacts. However, polyneural innervation has now been achieved experimentally and may well occur naturally in special cases.

MECHANISM OF NEUROMUSCULAR TRANSMISSION

We can learn a great deal about the mechanism of transmission by studying a classic experiment performed by Kuffler (Fig 2–18). A frog muscle fiber was dissected with its axonal branch attached, and mounted at the surface of a layer of saline covered with oil. The junctional region was then lifted into the oil with a glass hook, which on its inside curvature carried a platinum wire (Fig 2–18, A). The latter, being now isolated from the bathing fluid, serves as an extracellular electrode sampling selectively from the postjunctional region of the muscle fiber. Responses were recorded upon arrival of à nerve impulse. Figure 2–18, B, shows, from below upward, the changes taking place during increasing poisoning with d-tubocurarine, the active principle of curare that blocks transmission at the neuromuscular junction (and which for this reason finds application as a muscle relaxant in surgery). This block is reversible and transmission is restored when the d-tubocurarine is withdrawn. It is seen that, in the partly poisoned junction, the arrival of a nerve impulse gives rise to a postjunctional response, the *end-plate potential* (EPP), which appears after a

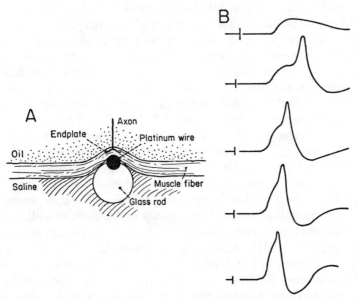

Fig 2–18.—A, classic extracellular demonstration of end-plate potential. (From Kuffler, S. W.: J. Neurophysiol. 5:18, 1942. Used by permission.) B, recordings from end-plate region after exposure to curare. Top traces show end-plate potential. Lower traces show lesser degrees of curarization; end-plate potential is larger and it gives rise to an action potential.

small delay and comes and goes with its own time course. We shall ascribe transmission later to the action of a transmitter substance, released by the axon terminal, which diffuses across the neuromuscular gap in a matter of microseconds, and reacts with a postjunctional receptor with a kinetics determined by the properties of the latter. When curare poisoning is less (see Fig 2–18, B, lower traces), the EPP is somewhat greater and, when it is large enough to exceed threshold, it gives rise to an action potential that leaves the junctional region and changes into the muscle action potential, which is conducted away in both directions.

We might ask whether the transmission from nerve to muscle is not merely a continuation of the nerve's action potential across the boundary. This is not the case. The smallness of the input structure (the axon terminal) focuses attention on the electric mismatch between it and the postsynaptic structure (the muscle cell). Specifically, the muscle cell has a low membrane impedance and its depolarization would require a greater current than the nerve terminal can yield. Instead, transmission is chemical and involves a specific neurotransmitter, acetylcholine.

ROLE OF ACETYLCHOLINE

One of the early outcomes of empirical pharmacology was the discovery that acetylcholine mimics certain actions of the autonomic nervous system such as the slowing of the heart produced by stimulating the vagus nerve. The effects of acetylcholine pass rapidly because it is hydrolyzed by acetylcholine esterase and other cholinesterases, enzymes of widespread occurrence. These enzymes are inhibited by such substances as eserine and prostigmine, which therefore protract or enhance the acetylcholine effects. When Loewi, in 1921, found that perfusion fluid of a frog heart stopped by vagal stimulation would cause arrest of a second heart, especially when eserine was present, the conclusion was drawn that the effect of the vagus on the heart was due to liberation of a "vagus-substance," to wit, acetylcholine. This became a cornerstone in physiology, though the identification of acetylcholine was tentative and only relatively recently confirmed explicitly by chromatography and mass fragmentography. A large part of the classic development was owing to Dale, who found that skeletal neuromuscular transmission was also due to acetylcholine. Modern investigations have much improved the measurement of the amounts of acetylcholine released from motor end-plates, as well as the determination of the minimal amount that, by close application, can cause transmission. These quantities, of the order of femtograms, have come close enough to warrant the conclusion that indeed neuromuscular transmission is due to the axonal release of acetylcholine, its diffusion across the gap and its reaction with a myonal receptor causing the EPP. Because acetylcholine is rapidly hydrolyzed at the end-plate by acetylcholine esterase, the whole event is over in something like a millisecond.

Synthesis and Release of Acetylcholine

Acetylcholine is synthesized by the neuron from acetate, and choline by the enzyme choline acetylase and coenzyme A. The acetate is derived from nerve cell metabolism but the choline originates from outside the nerve cell and is ultimately derived from the plasma. Once formed, acetylcholine is stored in the vesicles of the terminal and is released by the process of *exocytosis* (Fig 2–19). A vesicle containing acetylcholine fuses with the presynaptic terminal cell membrane, discharges its contents and becomes incorporated into the membrane. Vesicles are then re-formed by invagination of other parts of the

Fig 2–19.—Release of acetylcholine by exocytosis. **1,** A vesicle containing acetylcholine is shown within an axon terminal. It moves to the surface of the terminal, and the membranes fuse (**2**). Acetylcholine is then discharged from the vesicle (**3**) and the vesicular membrane becomes incorporated into the terminal membrane. New vesicles are formed by invagination (**4**) and become filled with acetylcholine synthesized in the neuron (**5**).

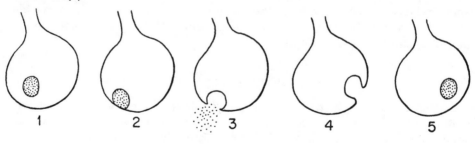

membrane, as shown. The release of acetylcholine is therefore *quantal,* i.e., it is released in packets. Each packet or quantum is equivalent to the acetylcholine content of one vesicle.

MICROPHYSIOLOGY OF THE END-PLATE

The use of intracellular electrodes has revealed details not accessible in the Kuffler experiment described above. When curare is used to lower the EPP below the threshold for action potential generation, nonpropagated transient depolarization occurs at the end-plate and decays exponentially with distance with a space constant of about 1.5 mm (Fig 2–20, A). This response is similar to that met with in nerve (the "local response") and postsynaptically ("excitatory postsynaptic potential"). In the absence of curare, the EPP ex-

Fig 2–20.—Intracellular recording of end-plate potential (EPP). **A,** curarized end-plate. The top curve was obtained in the junctional region, the lower ones by placing the microelectrode successively farther away in 0.5 mm steps, illustrating the space constant of the electrotonic spread (in this case about 1.5 mm). **B,** noncurarized end-plate. The left-most curve was obtained in the junctional region and shows the onset of the EPP, followed by the generation of the spike, which is lower than full magnitude because of the less selective ionic permeability increase in this region. The other curves were obtained farther away, by about 2 mm in the last one, and show how upon longitudinal spread the EPP component vanishes and full spike is generated. Examination of these potentials indicates that conduction velocity in this fiber is about 1 m per second. (From Fatt, P., and Katz, B.: J. Physiol. 115:320, 1951. Used by permission.)

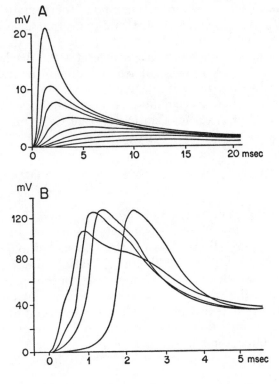

ceeds threshold and becomes a conducted spike. At the junction, the recorded potential shows its composite character, but more distantly it becomes the muscle's action potential pure and simple (see Fig 2–20, B). The EPP is due to an unselective permeability increase for all cations, or at least for K^+ and Na^+, so that the potential becomes clamped toward zero (hence the spike at the end-plate is less than farther out). The muscle action potential itself is primarily due to an increase in Na^+ permeability, much as it is in nerve, and it does show a sizable overshoot.

The change in ionic permeabilities of the postjunctional membrane is owing to activation of specific end-plate receptors by acetylcholine. In the frog the neuromuscular junction is less compact than the mammalian and consists of distinguishable axonal branches running in grooves in the muscle for a sizable part of a millimeter and $10–100$ μ apart. When mapped with a micropipet from which acetylcholine was ejected by current pulses (since acetylcholine is cationic, this requires positive, hence nonstimulatory, pulses), maximal sensitivity was found in the immediate vicinity of the axonal branches, showing that presynaptic inputs and postsynaptic receptors are placed in optimal proximity. It was in these experiments that the true threshold for acetylcholine was most closely approximated, determining that less than 10^{-16} M is needed to mimic a response to an arriving impulse.

Even when the axon is not excited, an electrode placed intracellularly in the muscle at the end-plate region detects spontaneous depolarizations of about 0.5 mv occurring at a rate of a few per second. These are called miniature EPPs or *minepps*. It is believed that they result from random local discharge of vesicles, each containing a quantum of acetylcholine. If several different electrodes are implanted in the end-plate region, it is seen that the minepps are not synchronous. However, the passage of a nerve impulse into the presynaptic nerve terminal greatly accelerates the discharge of vesicles and causes several hundred quanta of acetylcholine to be discharged almost simultaneously into the neuromuscular cleft, thus inducing the EPP. Calcium ions play an essential role in acceleration of acetylcholine release. The mechanism is unknown but is yet another example of Ca^{2+} linking a membrane electric change to a motor or secretory response.

PHARMACOLOGIC ASPECTS OF NEUROMUSCULAR TRANSMISSION

Acetylcholine is a quaternary ammonium compound (Fig 2–21). Its action on postsynaptic muscle membrane is initiated by combination with a proteolipid receptor that normally is almost entirely localized to the end-plate. Because of this receptor localization, this region is far more sensitive to locally applied acetylcholine than is the surrounding muscle membrane. As previously mentioned, the receptor is reversibly blocked by curare derivatives such as tubocurarine. Irreversible block can be produced by α-bungarotoxin, a snake venom constituent, which is firmly bound by the receptor. Biochemical studies of the bungarotoxin-receptor complex suggest that the receptor is a tetramer with a molecular weight of about 320,000. It appears that the quater-

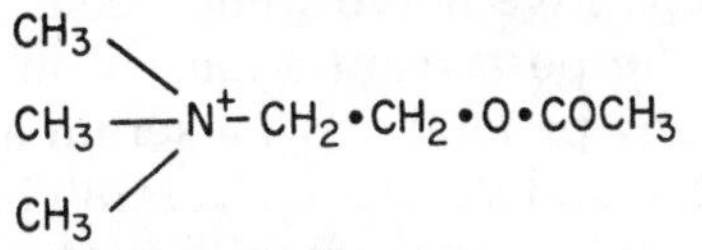

$$CH_3 - N^+ - CH_2 \cdot CH_2 \cdot O \cdot COCH_3$$

with CH_3 and CH_3 groups completing the nitrogen substituents.

Fig 2–21. — Chemical formula of acetylcholine.

nary ammonium component of the acetylcholine molecule forms an ionic link with a cation-binding site on the receptor. This interaction is followed by a conformational change that locally increases Na^+ and K^+ conductance and depolarizes the muscle membrane. The action of acetylcholine is terminated by acetylcholinesterase, which is present in high concentration at the end-plate and by diffusion of acetylcholine away from the receptor. Enzymic hydrolysis of acetylcholine is very rapid and its concentration falls below the threshold required for excitation before the end of the muscle's refractory period. This means that a single nerve impulse can elicit only one twitch. Drugs such as neostigmine and eserine inhibit acetylcholinesterase so that enough acetylcholine remains at the end of the refractory period to elicit another action potential, and these agents may therefore induce repetitive discharges and sustained muscle contractions.

OTHER INTERACTIONS AT NEUROMUSCULAR JUNCTION

The nerve has other effects on muscle apart from merely transmitting action potentials to it. In an innervated muscle the muscle membrane, except at the end-plates, is relatively insensitive to acetylcholine. After denervation the sensitivity of the entire surface membrane rises to almost equal that of the end-plate region. The increased sensitivity appears to be due to increased synthesis of acetylcholine receptors in the surface membrane outside the end-plates. The molecular mechanisms responsible for this remain to be worked out.

The innervation of muscle also strikingly determines some of the biochemical characteristics and the type of contraction of the muscle (e.g., "fast" or "slow," see subsection on specialization of motor units).

Finally, the muscle may exert some influence on nerve and may in part be responsible for initiating the "sprouting" of branches from the injured nerve terminal after a motor axon is cut or damaged. This possibility is speculative but suggests that the muscle may influence its re-innervation after nerve injury.

DISEASE OF NEUROMUSCULAR JUNCTION

Myasthenia gravis.— This uncommon disease of the neuromuscular junction is characterized by weakness especially of the muscles innervated by the cranial nerves. Patients may therefore have drooping eyelids, double vision and difficulty in speaking or swallowing. The end-plates of affected muscle show reduced minepp amplitudes and impaired generation of muscle action potentials during repetitive nerve stimulation. The electron microscope shows degeneration of the postsynaptic muscle folds and widening of the neuromuscular cleft. Many patients have a serum antibody that reacts with the acetylcholine receptor, and a condition resembling myasthenia gravis can be induced in animals by injecting them with antibodies to this receptor. The condition is probably an autoimmune disease, and it is noteworthy that two thirds of myasthenic patients have hyperplasia or neoplasia of the thymus, an organ closely associated with immune mechanisms. These patients may have a pathologic deficiency of effective cholinergic receptors at their end-plates, perhaps due to an antibody, but the picture is far from clear. Striking benefit may follow the administration of inhibitors of acetylcholinesterase, such as neostigmine. These agents raise the local concentration of acetylcholine and permit an enhanced response from the remaining receptors.

MUSCLE

Appreciation of the study of muscle has had ups and downs among medical students and physicians, the negative having been expressed about 50 years ago by a student walking out of a lecture muttering "muscle, muscle, who has ever died of muscle?" Actually, there are several good reasons why the topic deserves full interest. One is that muscle makes up half the body and, to paraphrase the student, "who has ever lived without it?" The tissue serves numerous motile functions, without which life in the human sense would not exist As Setchenow has emphasized, without muscles nothing of what goes on in the brain would be revealed. The body would be a vegetative thermo- and chemostat, perhaps with a mind, but who would know? A second reason is that between bodily rest and activity there can be a several hundred-fold increase in muscle metabolism. Thus the level of muscle activity is a substantial determinant of the adjustments in circulation and respiration that are so important in many disease states as well as in health. And there *are* muscle diseases, sometimes very debilitating and dramatic, which afflict about 0.1% of the population. Finally, muscle physiology has been and still is one of the most advanced scientific frontiers, often setting precedents for other areas. It is the direct basis upon which clinical cardiology rests; after all, the heart is a muscle and the target of major disease entities.

CLASSIFICATION OF MUSCLE

There are three main categories of muscle: smooth, cardiac and skeletal. They differ in microscopic structure, type of activation, location and physiologic properties.

SMOOTH MUSCLE.—This is composed of narrow ($2-4$ μ in diameter) fusiform cells, which do not show the microscopic crossbanding of the other types. It has no structurally defined end-plates and is not under the control of the will, hence it is often termed *involuntary* muscle. It usually surrounds tubes or sacs such as blood vessels, intestines, bronchioles, urinary bladder.

Generally its contraction is slow and it can maintain a given length or tension at low energy cost and without fatigue.

CARDIAC MUSCLE. — This is found *only* in the heart. It shows crossbanding, as does skeletal muscle, but its most characteristic structural feature is that the cells are branched and make complex contacts with each other at *intercalated disks*. These are low-resistance junctions that cause the heart to behave as a *functional syncytium*, i.e., the heart contracts and relaxes as a whole (see Chapter 4). There are no structurally defined end-plates and cardiac muscle is not normally under voluntary control. Neither cardiac nor smooth muscle requires the presence of nerves to initiate or maintain its contractions, which are therefore termed *myogenic*.

SKELETAL MUSCLE. — This is composed of large unbranched cells, which microscopically show characteristic crossbanding. There are well-defined end-plates and the muscle is under voluntary control. Contractions are initiated by nerve impulses and are therefore termed *neurogenic*. When the nerves are cut or diseased, the muscles that they supply are paralyzed and waste rapidly.

This section will concern itself mainly with skeletal muscle, although, where relevant, similarities to and differences from other types will be pointed out.

GENERAL FUNCTION OF MUSCLE

Muscles show only two mechanical responses to stimulation, shortening and force development, and both usually occur together. The action of skeletal muscle is modified by the leverage resulting from its attachment to the skeleton. For example, a small degree of shortening of the hip flexor muscles may produce a large excursion of the foot. Muscles that produce a given movement are termed *agonists;* those that oppose it are called *antagonists*. Thus the principal agonist for flexion of the forearm is the *biceps brachialis* and the principal antagonist the *triceps brachialis*. Most muscle movements occur smoothly as the result of accurately controlled activation of agonists and of inhibition of antagonists. On the other hand, the skeleton can be maintained in a fixed position against considerable force by simultaneous contraction of both agonists and antagonists. These coordinated responses are the function of the nervous system and are discussed in Chapter 9 (section on sensory mechanisms).

Cardiac muscle also clearly exhibits the functions of force development and shortening. Force development is shown by the development of pressure within the heart chambers during contractions; shortening is the mechanism whereby chamber size is reduced and blood is expelled from the heart into the rest of the circulation.

Shortening of smooth muscle is most dramatically seen when cavities are emptied, e.g., expulsion of urine from the bladder or a child from the uterus. Everyday experience demonstrates that considerable force may be developed during these activities.

STRUCTURE OF SKELETAL MUSCLE

Skeletal muscle is composed of numerous cylindric cells called *muscle fibers*, which are about 100 μ in diameter and range in length from 2 mm to 50 cm. The cells are embedded in connective tissue, which supports the muscle fibers and provides pathways for the blood vessels, lymphatics and nerves which are necessary for its function. Close to the area where a muscle is inserted into bone, the connective tissue often becomes very dense and strong to form a *tendon*.

Each muscle cell contains numerous contractile elements, the *myofibrils* (Fig 2–22), as well as nuclei, mitochondria and a well-developed system of fenestrated tubules called the *sarcoplasmic reticulum*, or SR. The cell membrane or *sarcolemmal membrane* sends transverse tubular extensions (T tubules) into the cells, which make intimate contacts with the SR. The T tubules and SR are very important in providing the link between the action potential that excites the cell and its subsequent contraction. This link involves a number of reactions collectively termed *excitation-contraction coupling*, to be discussed in detail later.

Fig 2–22.—Structure of muscle. Whole muscle consists of numerous muscle cells or fibers inserted into a tendon. One fiber is enlarged to show its interior packed with myofibrils. Part of a single myofibril is enlarged still further to show the sarcomere, A and I bands, Z lines and H zone.

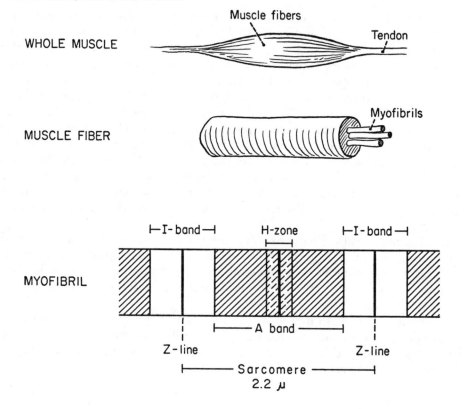

The contractile elements or myofibrils are 0.5–1 μ in diameter, and longitudinal sections show a striated or banded appearance under the microscope. The darker bands are called A bands because they are anisotropic and the lighter bands are termed I bands because they are mainly isotropic. Each I band is bisected by a thin, dark line, the Z line, and the portion of the fibril between successive Z lines is called the *sarcomere*, the basic unit of contraction. The sarcomere length of resting muscles is about 2.2 μ. In the center of each A band is a region termed the H zone.

EXPERIMENTAL OBSERVATIONS

Muscle contractions can be studied experimentally under two special conditions: (1) *isometric*, in which muscle length stays constant and force is developed, and (2) *isotonic*, in which a muscle shortens against a constant force.

ISOMETRIC RESPONSES

Using the apparatus diagrammed in Figure 2–23, isometric contractions can be studied at a variety of initial lengths. A small, thin muscle is immersed in an oxygenated physiologic salt solution maintained at constant temperature. One end is fixed and the other is attached to an isometric transducer that measures the force (P) developed by the muscle. The muscle can be stimulated by external electrodes and an intracellular microelectrode records the muscle's membrane potential. The muscle is stimulated at its resting length L_0 (sarcomere spacing 2.2 μ) and at various lengths expressed as fractions of L_0. Consider first the response to a single brief stimulus with the fiber length at L_0. The electrical and tension records are shown in Figure 2–24. Note that the action potential lasts one msec or so and is completely inscribed before any mechanical response begins. The delay between the depolarization of the membrane and the onset of force development is 5–18 msec and is called the *latent period*. The mechanical response is an isometric twitch that lasts 50–200 msec in mammalian muscle at 37 C. The time from the beginning of force development to the peak response is the *contraction time* and the re-

Fig 2–23.—Diagram of apparatus to study isometric responses of skeletal muscle at different muscle lengths. L_0 = resting length.

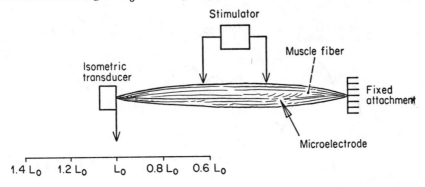

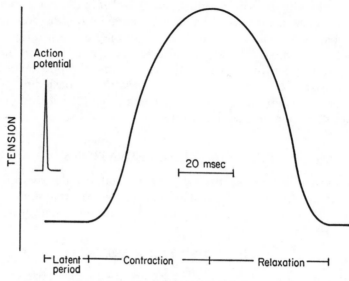

Fig 2–24.—Isometric twitch (see text).

maining part of the twitch cycle is the *relaxation time*. The force developed can be surprisingly large. Thus a 100-mg frog sartorius muscle can exert a force 1,000 times its weight, i.e., 100 gm. If the muscle is stimulated more than once, repetitive mechanical responses occur with the pattern shown (Fig 2–25) and, when the frequency of stimulation is very rapid, the responses fuse to form a *tetanus*. It should be noted that the mean force, resulting, for example, from asynchronous fluctuation, increases with frequency, and this suggests that one way the body can regulate the strength of muscle contraction is by altering the frequency of nerve action potentials transmitted to the muscle.

Let us now consider the effect of initial length on the mechanical responses to tetanic stimuli starting with the fiber at a much shorter length than is usual, e.g., 0.6 L_0. Before stimulation the isometric transducer records zero force, and stimulation produces only a small force development. As initial length is

Fig 2–25.—Responses of skeletal muscle or fiber to single and repeated electric stimuli.

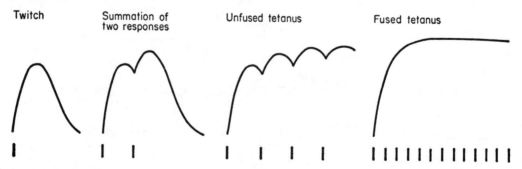

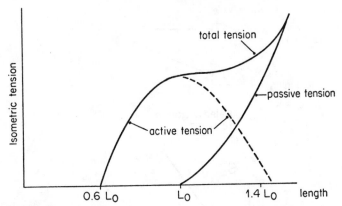

Fig 2–26.—Isometric tension developed by skeletal muscle as a function of its initial length (L_o).

increased to L_o, the force developed during stimulation increases to a maximum but still no force is recorded in the resting condition (Fig 2–26). When the fiber is stretched beyond L_o, the transducer registers a *passive* force probably transmitted by elements in the connective tissue around the fiber and by the sarcolemmal membrane. These elements are collectively termed the "parallel elastic elements." The passive force increases markedly with increasing stretch beyond L_o. When stimulated, the muscle generates additional force beyond that developed passively, but the increment becomes progressively smaller and, when the sarcomere length exceeds 3.6 μ, the *active* tension is practically zero. The bell-shaped curve of *active* tension vs fiber length is termed the *length-tension* curve and is explained later.

ISOTONIC CONTRACTIONS

The apparatus shown in Figure 2–27 can be used to study both isometric and isotonic contractions as well as the effects of "releasing" the muscle at

Fig 2–27.—Apparatus to study force development and shortening of skeletal muscle (see text).

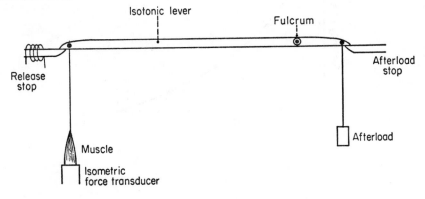

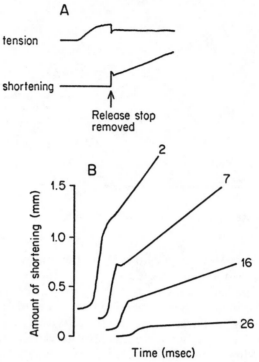

Fig 2–28. — **A,** changes in force and shortening when a muscle is first allowed to contract isometrically and then is released so that it shortens with a constant after-load. **B,** part of the family of curves showing the time course of shortening at different afterloads that have relative values indicated by the numeral on each curve.

Fig 2–29. — Force-velocity relation. P_o = isometric force.

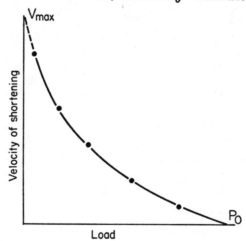

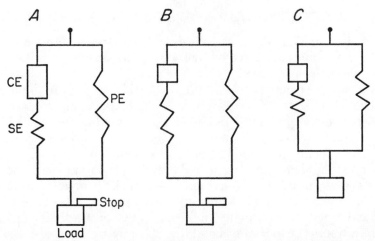

Fig 2–30.—Model showing behavior of contractile element *(CE)* and series elastic *(SE)* and parallel elastic *(PE)* elements. *SE* and *PE* are represented as springs. **A,** resting state. **B,** isometric contraction. *CE* shortens and extends *SE*. Overall length unchanged. **C,** stop is removed. *SE* and *PE* shorten and thereafter *CE* shortens, while *SE* remains constant.

different stages of an isometric response. One end of a muscle is attached to an isometric transducer and the other to an isotonic lever, the displacement of which can be measured (e.g., by a photocell). On the other side of the fulcrum, various loads can be placed. A stop is placed so that the load does not stretch the muscle. Since the muscle does not "see" the load until contraction begins, it is said to be afterloaded. Another stop, the release stop, is placed so that the muscle cannot shorten until the stop is released at predetermined times. The experiment is performed as follows. The muscle is tetanized and develops *isometric* tension. When this has reached a maximum, the release stop is removed. This causes a rapid reduction in length and tension, after which the muscle maintains a constant tension as it shortens with constant velocity (Fig 2–28, A). The experiment is repeated with different afterloads, thus obtaining the family of curves shown in Figure 2–28, B. By plotting the constant shortening velocities against afterload, a *force-velocity curve* (Fig 2–29) is obtained. The intercept (P_0) on the abscissa represents isometric force, whereas the extrapolated intercept on the ordinate is the maximum possible shortening velocity at zero load (V_{max}). The initial shortening when the release stop is removed can be explained by postulating a "series elastic element" in series with the contractile proteins. This elastic element partly resides in the tendons and partly in the contractile proteins themselves. When the contractile proteins shorten the series, elastic element is stretched. When the stop is released, the series elastic shortens to a new equilibrium length, and thereafter the contractile elements shorten at constant velocity. A simple model is shown in Figure 2–30, although it should be emphasized that, as measurements are refined, the model may have to be altered to account for new findings.

ULTRASTRUCTURE OF SKELETAL MUSCLE

As previously explained, the myofibrils are composed of alternating A and I bands and the I bands are bisected by Z lines. The sarcomere is the structure between two adjacent Z lines.

The A bands are composed of interdigitating *thick* and *thin filaments* (Fig 2–31). The thick filaments extend from one end of the A band to the other and are 1.6 μ long and 10–12 nm in diameter. Protruding from the thick filaments are cross bridges, which are 10 nm in diameter and 14 nm apart. The middle of the thick filaments (pseudo-H zone) is free of cross bridges over an interval of 0.2 μ. The thin filaments are attached to and project from the Z lines. There are twice as many thin filaments as thick. Each thick filament is surrounded by six thin filaments and each thin filament is surrounded by three thick filaments. The A band length is constant at 1.6 μ irrespective of the degree of stretch of the muscle fiber. On the other hand, the I bands are short at short muscle lengths and become longer as the muscle is extended. This is because the thin filaments move into the A band as the muscle shortens and out of the A bands as the muscle is stretched.

BIOCHEMISTRY OF MUSCLE PROTEINS

MYOSIN.—This is the protein that constitutes the think filaments. It is 150 nm long and has an α-helical tail and a double globular head. As shown in the diagram (Fig 2–32), it can be split by the protease, trypsin, into two portions, a helical region 90 nm long called light meromyosin (LMM, MW=150,000) and a fragment called heavy meromyosin (HMM, MW=340,000). HMM can be further split by the protease papain into two fragments, S_1 and S_2. S_1 contains the globular heads.

Myosin has two important properties: (1) it can hydrolyze ATP and (2) it can bind to the protein, f (fibrous)-actin, of the thin filament. Both properties are entirely localized to the S_1 fragment.

ACTIN.—This is the protein of the thin filaments. It can exist in two forms,

Fig 2–31.—Arrangement of thick and thin filaments within the sarcomere. *Arrows* indicate the arrangement of filaments as they would be seen if *transverse* sections were made through the sarcomere at the levels shown. (After Aidley, D. J.: *Physiology of Excitable Cells* [London: Cambridge University Press, 1971]. Used by permission.)

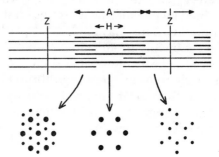

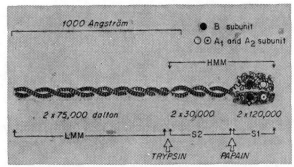

Fig 2–32.—Morphology of the myosin molecule and its constituents. The molecule is pictured as two main or heavy chains, which are partly helical, partly not. The helical sections are intertwined to form a supercoiled rod or tail, a major part of which is contained in the A-band rod of the sarcomere. It is shown how selective digestion with trypsin or papain gives rise to preferred fission products. The nonhelical parts of the heavy chain form the head-part of the molecule, or the S_1 frequent. In the head, where the ATP-ase and actin-binding sites are located, there are also associated small subunits or light chains. In myosin from fast muscle, as illustrated here, there are three subunits, two of which have large areas of identity in primary structure. There is no universal nomenclature, but they have been indicated here as A_1 and A_2, in distinction to the B units.

globular (g) actin (MW=46,000) and f actin, which is a polymer of g actin. F actin is a double-stranded superhelix made up of beadlike structures 5.5 nm in diameter and is the form in which it exists in the thin filaments. It does not hydrolyze ATP but binds to myosin to form a complex, *actomyosin*, which shrinks in the presence of Mg-ATP, at the same time splitting Mg-ATP and at a rate much faster than myosin alone.

TROPONIN AND TROPOMYOSIN.—These proteins are located on the thin filaments and regulate the interaction between actin and myosin.

MOLECULAR BASIS OF CONTRACTION: SLIDING FILAMENT HYPOTHESIS

The sliding filament hypothesis holds that the cross bridges interact with the thin filaments and pull them toward the center of the sarcomere. This interaction causes the thin filaments to slide past the thick filaments and causes shortening, force generation and splitting of ATP. A tentative scheme is as follows (Fig 2–33).

At rest, myosin exists as $M \cdot ADP \cdot P_i$ (formed by myosin reacting with ATP), and the cross bridge (the S_1 fragment) is at right angles to the long axis of the thick filament. When the muscle is activated, molecular rearrangement takes place in the thin filament allowing the S_1 fragment to bind to actin. When this binding occurs, the S_1 head rotates and induces a sliding movement of the attached thin filament. Thereafter ADP and P_i (inorganic phosphate) dissociate from the actin-myosin complex and, when additional ATP is present, are replaced by ATP. Upon ATP binding to the acto-S_1 complex, the actin and S_1 dissociate. The bound ATP is then hydrolyzed to ADP and P_i and the S_1

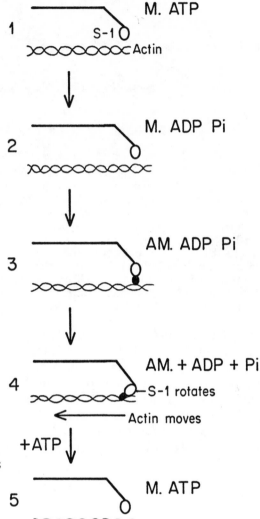

Fig 2–33.—Example of tentative molecular hypothesis to explain the sliding of actin on myosin during contraction (see text). M = myosin; A = actin; P_i = inorganic phosphate; $S-1$ = globular head of heavy meromyosin.

head returns to its right-angled position on the myosin molecule so that it is ready for another cycle. If ATP is absent, as occurs after prolonged local or general arrest of the circulation, the actin-myosin complex cannot dissociate and *rigor* develops. This accounts for the stiffening of corpses after death (rigor mortis).

The sliding filament hypothesis explains: (1) why the length of the A-band stays constant and the lengths of the thick and thin filaments remain unchanged when the sarcomere shortens (the filaments do not contract, they just slide past each other) and (2) the shape of the curve relating active tension to muscle length (see Fig 2–26). At L_o, sarcomere length 2.2 μ the maximum number of cross bridges is utilized so that force generation is maximal at this

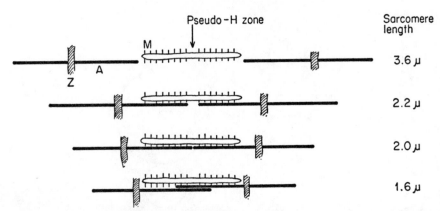

Fig 2–34.—Diagram to explain the shape of the length-active tension curve of skeletal muscle (see text).

length. At sarcomere lengths between 2 and 2.2 μ, the thin filaments are crossing the pseudo-H zone where there are no cross bridges; hence there is no change in force. This accounts for the plateau at the top of the curve. For sarcomere lengths below 2 μ, the thin filaments begin to overlap in the central region of the A band and progressively interfere with cross-bridge linkages. For sarcomere lengths greater than 2.2 μ the thin filaments interact with fewer and fewer cross bridges until at 3.6 μ there is no interaction (Fig 2–34). This explains why active tension declines when the sarcomere lengths are less than 2 μ or greater than 2.2 μ.

EXCITATION-CONTRACTION COUPLING

Over the years increasing evidence has accumulated that the intact muscle fiber can only contract when its intracellular free calcium ion concentration rises above a certain threshold value. Where does the calcium needed to increase the intracellular concentration come from? How is it regulated? And how do changes in calcium ion concentration lead to contraction or relaxation? To answer these questions, we now have to consider two important regulatory proteins, tropomyosin and troponin, and also the functions of the sarcoplasmic reticulum, which acts as a calcium store.

TROPOMYOSIN AND TROPONIN.—The thin filaments contain two proteins in addition to actin. *Tropomyosin* is a helical protein about 40 nm long and 1.7 nm wide with a molecular weight around 64,000. It exists as filaments, of which one is present in each of the two grooves of the actin molecule. Troponin is a globular molecule composed of three subunits designated TnT, TnI

Fig 2–35.—Arrangement of regulatory proteins, tropomyosin and troponin, on the thin filament.

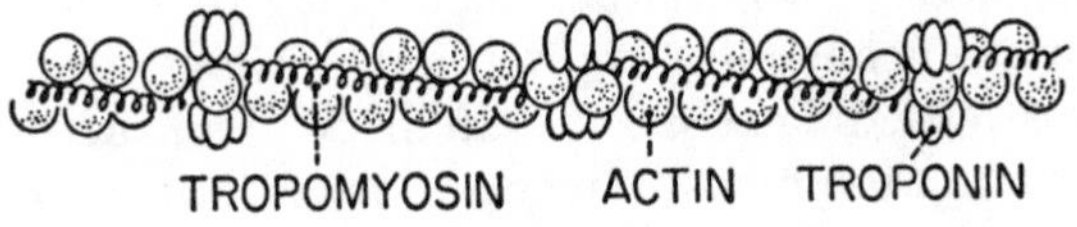

TROPOMYOSIN   ACTIN   TROPONIN

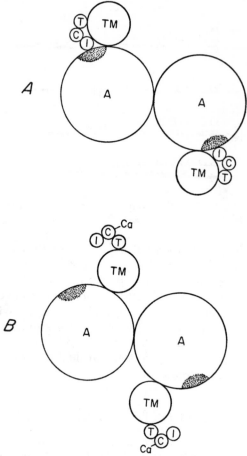

Fig 2–36.—Depiction of how an increase in free intracellular calcium might cause conformational change in the troponin-tropomyosin complex, enabling actin to interact with myosin. **A,** before activation, the site capable of binding myosin *(shaded area)* is covered by the troponin-tropomyosin complex, **B,** when intracellular calcium ion concentration rises above a critical level, calcium is bound by TnC. This causes a conformational change in the troponin-tropomyosin complex, which uncovers the binding site for myosin. Myosin is then free to interact with actin and thus initiate contraction. Troponin is composed of three subunits designated TnT, TnI and TnC. *Ca* = calcium ion; *A* = actin; *TM* = tropomyosin; *I* = TnI; *C* = TnC; *T* = TnT. (After Potter, J. D., and Gergely, J.: Biochemistry 13:2697, 1974.)

and TnC. The largest subunit, TnT, binds to tropomyosin, the smallest, TnC, has a high affinity for calcium and TnI binds to actin. The troponin complex is found at 40 nm intervals along the thin filament (Fig 2–35).

The troponin-tropomyosin complex can be regarded as an on-off switch that regulates contraction. A possible mechanism for its action is as follows. When free calcium ion concentration in the cell is low, the muscle is relaxed because tropomyosin lies in a position such that it blocks active sites necessary for interaction between the myosin heads and actin filaments. When internal calcium ion concentration rises, it is bound by TnC and this causes a conformational change in the troponin such that the binding of TnI to actin is weakened. This causes TnI to move away from actin and allows tropomyosin to move from its original position, uncovering the active site at which binding between actin and the S_1 head of myosin occurs (Fig 2–36).

We now have to consider how an action potential can bring about a rise in free calcium ion concentration within the cell. It was found many years ago that skeletal muscle can be made to contract for long periods in physiologic salt solutions containing zero calcium. This suggested that muscle cells have an appreciable intracellular calcium store. This has been located in the SR that surrounds the myofibrils. The SR is composed of longitudinal tubules that open into expansions called terminal cisterns near the Z lines. At this level the surface membrane is invaginated to form transverse tubules, each about 20 nm in diameter (Fig 2–37). The terminal cisterns lie on either side of the transverse tubules but are separated from them by a gap of 10–15 nm. It appears that the transverse tubules carry the surface membrane into the interior of the cell into close association with the calcium store. It is suggested that the passage of the action potential down the transverse tubule causes calcium to be released from the terminal cistern into the sarcoplasm. During relaxation it is pumped back by the SR membrane into the tubules and cisterns of the SR.

We are now in a position to give a reasonably complete description of excitation-contraction coupling. The action potential is propagated down the T tubule and by an unknown mechanism causing the terminal cistern to release Ca^{2+}. Because the SR has relatively few binding sites and troponin TnC has a high affinity for Ca^{2+}, troponin binds Ca^{2+}, which causes conformational changes that allow actin to bind with myosin cross bridges. This leads to ATP splitting and to a sliding of the thin filaments toward the center of the sarcomere, with consequent muscle shortening or force development. At the same time, the rise in Ca^{2+} concentration activates the SR pump, which splits ATP

Fig 2–37.—Relationships between surface membrane, transverse tubules, sarcoplasmic reticulum and thick and thin filaments.

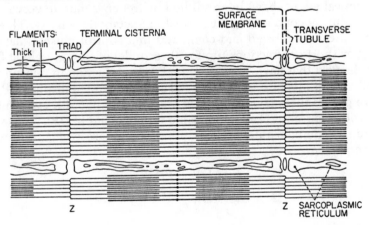

and pumps Ca^{2+} out of the sarcoplasm into the SR. This lowers Ca^{2+} ion concentration in the sarcoplasm, calcium dissociates from TnC and this causes tropomyosin to block the sites of interaction of actin and myosin so that the muscle relaxes.

CHEMICAL AND THERMAL CHANGES IN CONTRACTING MUSCLE

The metabolism of resting muscle produces heat as a byproduct. From the point of view of the muscle itself, this heat could be regarded as waste but from the point of view of the organism as a whole it is an important contribution to the maintenance of constant body temperature in homeothermic animals. In fact, muscle is responsible for more than 20% of the body's basal metabolic heat production.

When muscle is stimulated, heat production increases considerably. Several components have been distinguished.

1. *Activation heat.* This is the heat produced in association with the release and reuptake of calcium. It can be demonstrated even when contraction is prevented by stretching the sarcomeres to the point where the thin and thick filaments no longer overlap.

2. *Initial heat.* This is the heat produced during muscle contraction and relaxation and includes the activation heat. If the muscle is allowed to shorten, an additional amount of heat, the *shortening heat,* is generated. For the most part initial heat is produced by the enthalpy changes accompanying ATP hydrolysis.

3. *Recovery heat.* After a contraction, the rate of heat production of muscle remains higher than at rest for a period that is relatively long compared to the length of the preceding contraction.

Much effort has been expended in attempting to correlate heat production with mechanical and chemical events occurring during contraction. Mammalian muscle contractions are too fast to achieve these correlations and most of the experimental studies have been performed on cooled frog muscle, which contracts more slowly.

The chemical events have been studied by freezing muscle very rapidly at some point during contraction, thus arresting metabolic activity. The muscle is then chemically analyzed and compared to a control muscle that did not contract. For short contractions (up to 10 seconds) of frog muscle at 0 C, the only net changes that have been observed are a reduction in phosphoryl creatine (PC) and an increase in creatine (C) and inorganic phosphate (P_i). However, if we poison the enzyme, creatine phosphokinase (CPK), which catalyzes ATP synthesis from PC and adenosine diphosphate (ADP), a net splitting of ATP can be demonstrated. These observations are compatible with the reaction sequence:

$$(1)\ ATP \rightarrow ADP + P_i$$

$$(2)\ PC + ADP \xrightarrow{\quad CPK \quad} ATP + C$$

Adding (1) and (2), we see that the net effect is

$$PC \rightarrow C + P_i$$

Thus we believe that ATP is the primary source of energy during contraction and that the ATP level is prevented from falling by continuous resynthesis from PC (in the presence of CPK). Following contraction, PC is gradually restored by processes such as the breakdown of glycogen or fatty acids. These processes are responsible for production of the recovery heat.

NERVE-MUSCLE RELATIONSHIPS

Each nerve axon branches within the muscle to supply a number of muscle cells. The motor neuron, its axon and the muscle fibers that it supplies constitute a *motor unit*. Each unit reacts on the all-or-none principle, since an action potential passing down an axon will maximally activate all the fibers in the unit. Muscles that exert very fine control, e.g., extraocular eye muscles or laryngeal muscle, have small motor units in relatively large number so as to allow accurate gradation of developed force by recruiting additional small units. However, a hundred or so units will give all the gradation needed, so that, when large forces have to be developed, this is usually achieved by having very large units. Thus, a laryngeal muscle may have as few as two muscle cells per unit, whereas there may be several thousands per motor unit in the gastrocnemius. When muscles enlarge as a result of exercise, hypertrophy of individual fibers occurs but neither the total number of cells nor the size of the motor unit changes.

Another aspect of motor unit organization is that sustained contractions at fractional force development, such as those that occur in postural muscles, can be maintained by the activity of units in rotation. This prevents fatigue from developing in any one unit. When "fatigue" occurs during exercise, for example, it is not of the absolute kind, owing to exhaustion of energy-yielding metabolites but rather to circulatory factors or to psychologic causes.

The action potentials of the muscle cells are closely synchronized and temporal dispersion is no longer than 10 msec because all the cells are supplied by branches from the same axon and the conduction paths do not vary greatly. Other units, however, may be out of phase so that surface electrodes "see" the average activity of a number of units.

ELECTROMYOGRAPHY

The recording of electric activity of muscles by means of surface or needle electrodes is called electromyography and is often useful in investigating muscle diseases, as shown by the following examples.

Normally, relaxed muscle is electrically silent but, if axons supplying a muscle degenerate, action potentials may develop asynchronously in the muscle fibers of affected motor units. The cause is obscure. The observed potentials are random, 1-2 msec in duration and less than 200 μ in amplitude.

They are associated with fine visible muscle contractions, termed "fibrilla-tion," and are seen in poliomyelitis and in nerve injuries.

Fasciculation is the spontaneous involuntary discharge of motor units, and the accompanying contractions are coarser. The electrogram correspondingly shows larger (several millivolts) and longer (8–10 msec) potentials than in fibrillation. They are most characteristically seen in chronic destructive dis-eases of motor neurons.

SPECIALIZATION OF MOTOR UNITS

Histochemical, biochemical and functional criteria have shown the exis-tence of several types of muscle fiber. The muscle fibers in a single motor unit are all of the same type. Some units consist of relatively large pale or "white" muscle fibers low in mitochondria and oxidative capacity and rich in glycolyt-ic enzymes; their contractions are strong, fast and fatigable. Other units con-sist of smaller "red" fibers that are richer in myoglobin and oxidative enzymes and produce contractions that are weaker and slower than the white fibers and not fatigable. Certain muscles show marked predominance of a particular fiber type, e.g., white fibers predominate in the gastrocnemius, red in the so-leus. This has aided in differentiating the functional characteristics of these fibers. A third type of smaller fiber even richer in mitochondria has been de-scribed.

The different types of motor unit have been investigated in leg muscles of cats. Single motor unit stimulation has been achieved by dissecting a central root filament to isolate a single axon leading to the muscle that it supplied. Contractile force, speed and susceptibility to fatigue have been measured and correlated with histochemical studies of the muscle fibers. It was clearly es-tablished that a motor unit is not a separate strand within a muscle but occu-pies a diffuse territory interwoven with other units. This type of study also led to a more explicit identification of the fiber types described above. These are now referred to as FF (fast, fatigable), S (slow, nonfatigable) and FR (fast, re-sistant to fatigue; also called "fast red").

Nature of Motor Unit Differentiation

The finding that a motor unit contains only one muscle cell type could have one of two explanations.

1. Muscle cell differentiation occurs explicitly during development fol-lowing genetically coded information, and nerve fibers seek out the muscle cells whose future characteristics they recognize as appropriate to their place in the neurologic wiring diagram.

2. Neurons have a specific differentiating influence that causes the muscle cells to acquire the characteristics of a particular type, once innervation has occurred. In this case the genetic code in the muscle cell allows several possi-ble choices, the final pattern being determined by the innervation.

The second alternative appears to be correct. The cat soleus is a slow (S) muscle, the flexor digitorum longus fast (F). When the nerve supply to one was disconnected and repositioned so as to innervate the other, the characteristics of the muscle cells were reversed. Thus by changing innervation, S units could be converted to F units and vice versa. The change in characteristics was accompanied by changes in myosin ATPase, SR (sarcoplasmic reticulum) and lactic dehydrogenase isoenzymes. This appears to indicate that innervation profoundly alters gene expression.

The nerve not only determines the functional characteristics of the fibers it innervates but is required for general maintenance of muscle since severe atrophy, mainly due to disuse, occurs after denervation.

SMOOTH MUSCLE

Smooth muscle (also known as unstriated, plain, involuntary or visceral) differs in structure and function from both skeletal and cardiac muscle.

STRUCTURE.—Smooth muscle cells are tapering narrow cells (2–10 μ in diameter, up to 100 μ long), which do not show crossbanding when viewed under the light microscope. The surface membrane shows numerous small invaginations termed *caveolae*, which considerably increase the cell surface. The SR is not nearly so well developed as in striated muscle, and its amount varies considerably among smooth muscles. Electron microscopy shows many electron-dense areas, "dense bodies," in the surface membrane, and the myofilaments appear to be attached to these. Both thick and thin filaments have been described but their precise relationships have not been defined. Myosin, actin and tropomyosin have all been identified in smooth muscle, but troponin has not been isolated. It is possible that regulation of smooth muscle activity by calcium is mediated by a protein associated with myosin rather than with actin.

INNERVATION.—Smooth muscle cells are innervated by autonomic nerves, and there are no end-plate structures. The nerve terminals show numerous varicosities or "nodes" that contain vesicles in which the neurotransmitter, either acetylcholine or norepinephrine, is stored. The gap between nerve terminal and the nearest smooth muscle varies considerably among tissues ranging from 20 nm to several microns.

TYPES OF CONTRACTION.—On the whole, smooth muscle contractions tend to be slow and sustained and are not under voluntary control. Unlike skeletal muscle, many smooth muscles contract spontaneously and rhythmically even when denervated, and neural activity serves to enhance or inhibit contractions rather than to initiate them. Another difference from skeletal muscle is that excitation in some smooth muscles can be conducted from cell to cell through nexuses that appear to have low electric resistance.

EXCITATION-CONTRACTION COUPLING.—These processes have been less well worked out than in skeletal muscle. The problems are that smooth muscle varies from tissue to tissue and the cells are small so that microelectrodes are difficult to insert. In general, the resting potential is between −30 and −60 mv, i.e., lower than in skeletal muscle, and the cells are more permeable to sodium. Activation may be associated with graded depolarizations with or without action potentials. Moreover, many types of smooth muscle will contract even when their cells are completely depolarized. This indicates a non-electric or "pharmacomechanical" link between excitation and contraction. Unlike skeletal muscle, which in situ responds only to acetylcholine released from its motor nerve endings, smooth muscle will respond to a number of stimuli. For example, vascular smooth muscle may respond to epinephrine, norepinephrine, acetylcholine, serotonin, angiotensin, prostaglandins, variations in Po_2, Pco_2 and pH and many other stimuli. In all cases, however, the stimulus leads to an increase in intracellular free calcium ion concentration. This "activator calcium" often enters the cell from the extracellular fluid rather than from the SR, as is the case with skeletal muscle.

CARDIAC MUSCLE

Cardiac muscle fibers are similar to those of skeletal muscle in that they are striated and contain myofibrils, mitochondria and an SR. The myofibrils are also made up of sarcomeres like those of skeletal muscle. There are, however, many important structural and functional differences.

STRUCTURAL ASPECTS

Cardiac muscle cells are much shorter than skeletal muscle cells; they contain only a single nucleus and are *branched*. Their cytoplasm is slightly more abundant and the mitochondria are larger and more numerous. The cells come into very close relationship with each other at specialized junctions termed *intercalated disks*.

FUNCTIONAL ASPECTS

Skeletal muscle cells normally contract only when a nerve action potential releases acetylcholine at the motor end-plate. On the other hand, rhythmic contractions of the heart are initiated by spontaneous generation of action potentials in specialized pacemaker cells. These action potentials are conducted to all the heart cells by specialized conducting tissue and by direct spread of the action potential from cell to cell across the low-resistance intercalated disks. Accordingly, the heart behaves as a functional syncytium, i.e., when an impulse arises in one part of the heart, the entire heart is caused to contract. This is a marked difference from skeletal muscle where cell to cell conduction is not possible and only a small portion of the cells of a skeletal muscle may be activated during a weak contraction. Greater force is developed by recruitment of additional motor units, as described previously. This method of varying total developed force is not available to heart muscle.

A fundamental feature of skeletal muscle is that its action potential and refractory period are very short (1–2 msec) and the muscle will respond to additional action potentials during its contraction so that a fused or unfused tetanus results (see Fig 2–25). In contrast, cardiac muscle cells have very long action potentials and refractory periods and cannot respond to additional stimuli until relaxation has occurred. This means that the heart cannot be tetanized, a fortunate result since the function of the heart is just as much dependent on its ability to relax, and therefore to fill with blood, as on its ability to contract.

The excitation-contraction coupling mechanisms of skeletal and cardiac muscle also differ. Skeletal muscle cells are large, and diffusion of "activator calcium" from the exterior to the vicinity of the contractile elements would take too long for the functional requirements of the tissue. The calcium required for contraction is accordingly obtained from the calcium stores of the highly developed SR. Cardiac muscle cells are much smaller, and much of the activator calcium diffuses across the membrane from the cell surface. Cardiac muscle accordingly quickly loses its contractile function when placed in calcium-free medium.

LENGTH-TENSION RELATIONSHIP

Like skeletal muscle, the isometric force developed by cardiac muscle is length dependent due to the variable overlap between thick and thin filaments (see Fig 2–34). There are important differences between the two tissues, however. Skeletal muscle operates over a much wider range of sarcomere lengths. Attempts to stretch cardiac sarcomeres to lengths greater than 2.5 μ causes their disruption and they cannot be shortened to less than 1.7 μ. On the other hand, skeletal muscle fibers can be readily stretched over the range 1.4 μ to 3.4 μ. Another major difference is that skeletal muscle often exhibits negligible passive tension at the length that gives maximal active contractions, whereas cardiac muscle exhibits a passive tension that is equal to 15% of the total tension at this length and rises exponentially with slight additional stress. In other words, cardiac muscle is "stiffer" than skeletal muscle. An additional striking feature is that the length-tension curve for cardiac muscle shows no plateau, and active tension falls off much more rapidly when sarcomere length is varied from the optimal.

The reasons for these differences remain unclear. It is generally believed that the fundamental contractile mechanisms in cardiac and skeletal muscle are the same and that the observed differences in mechanical behavior are due to variations in elastic properties of the tissues, i.e., to differences in the series and parallel elastic elements. These presumably depend on differences in cellular and tissue structure, composition and geometry. Although we cannot explain the cardiac length-tension curve as satisfactorily as for skeletal muscle, we shall see in Chapter 4 (section on regulation of cardiac function) that it is of fundamental importance in regulation of the pumping action of the heart.

FORCE-VELOCITY RELATIONSHIP

Like skeletal muscle (see Fig 2–29), cardiac muscle shows an inverse relationship between force and velocity. Mathematical extrapolation of the curve obtained by plotting velocity of shortening against load allows determination of V_{max}, the maximum velocity of shortening of the contractile element of the unloaded muscle. The V_{max} was thought to be relatively independent of pre-load and therefore to have potential value as an index of the contractile state of the muscle. Because of the potential clinical usefulness of such an index, many studies were undertaken in an attempt to determine V_{max} for the intact heart under varying conditions of performance. However, measurements were difficult and their clinical usefulness doubtful, and the very validity of V_{max} as sometimes used has been strongly questioned.

CONTRACTILITY

The force developed by cardiac muscle is profoundly influenced by many factors that have no or little effect on skeletal muscle. For example, the hormone norepinephrine produces a dose-dependent increase in the isometric force developed by an isolated strip of cardiac muscle (Fig 2–38). Not only does peak force increase but time to peak is shortened, rate of force development is increased and duration of the twitch is reduced. These changes reflect a change in the contractile or *inotropic* state of the muscle, and norepinephrine is said to have a positive inotropic effect and to increase *"contractility."* Conversely, lack of oxygen, certain types of anesthesia and various drugs reduce the peak force and rate of force development of cardiac muscle and are said to have a negative inotropic action and to reduce contractility.

Fig 2–38.—Effect of norepinephrine on time course of tension development by cat papillary muscle. (From Koch-Weser, J., in Ross, G., Kattus, A. A., and Hall, V. E. [eds.]: *Cardiovascular Beta Adrenergic Responses* [Berkeley: University of California Press, 1970]. Used by permission.)

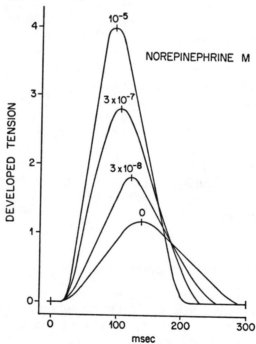

Efforts to assess the contractility of the intact heart are beset with difficulties. No index of a fundamental nature has emerged. Instead, reliance is placed on various empirical indices, which are discussed in Chapter 4 (regulation of cardiac function).

BIBLIOGRAPHY

Aidley, D. J.: *The Physiology of Excitable Cells* (London: Cambridge University Press, 1971).
Bulbring, E., and Shuba, M. F.: *Physiology of Smooth Muscle* (New York: Raven Press, 1976).
Carlson, F. D., and Wilkie, D. R.: *Muscle Physiology* (Englewood Cliffs, N. J.: Prentice-Hall, 1974).
Cohen, C.: The protein switch of muscle contraction, Sci. Am. 233:36, 1975.
Hagiwara, S.: Ca spike, Adv. Biophys. (Tokyo) 4:71, 1973.
Hille, B.: Ionic channels in nerve membranes, Prog. Biophys. Mol. Biol. 21:1, 1970.
Hubbard, J. I.: Microphysiology of vertebrate neuromuscular transmission, Physiol. Rev. 53:674, 1973.
Junge, D.: *Nerve and Muscle Excitation* (Sunderland, Mass.: Sinauer Associates, Inc., 1976).
Katz, B.: *Nerve, Muscle and Synapse* (New York: McGraw-Hill Book Co., 1966).
Kuffler, S. W., and Nicholls, J. G.: *From Neuron to Brain* (Sunderland, Mass.: Sinauer Associates Inc., 1976).
Lester, H. A.: The response to acetylcholine, Sci. Am. 236:106, 1977.
Murray, J. M., and Weber, A.: The co-operative action of muscle proteins, Sci. Am. 230:58, 1974.

3

Autonomic Nervous System

GORDON ROSS, M.D.

THE AUTONOMIC nervous system is responsible for regulating the contraction of smooth and cardiac muscle and the secretion of glands and thereby contributes notably to maintaining the constancy of the internal environment (homeostasis). It is involved in the regulation of body temperature, heart rate and output, arterial and venous pressures, regional blood flow, capillary filtration and exchange, airway resistance, gastrointestinal motility and secretion, ureteral and bladder motility, several aspects of sexual function and the secretion of most glands. Most of these activities operate outside the realm of consciousness and direct voluntary control, hence the term autonomic, although they can be strikingly influenced by psychologic influences, e.g., the blush of shame, pallor of fear, sweating of anxiety and diarrhea of nervousness.

Many of the functions listed above can proceed independently of neural control. Thus the heart continues to beat and the blood vessels regulate tissue blood flow appropriately even when deprived of autonomic nervous control either by surgery or by drugs. However, this is true only under favorable environmental conditions and, without the autonomic nervous system, the organism fails to react effectively to stresses such as severe exercise, excessive heat, and hemorrhage. Thus, its adaptability and power of survival are markedly restricted.

The distinction between somatic and autonomic nervous functions is not really very sharp. For example, many aspects of skeletal muscle control occur without reaching consciousness. Conversely, micturition, although under voluntary control, is regulated by autonomic nerves.

Because the autonomic system can so profoundly affect the activities of the major organs, it is appropriate to consider its general characteristics at this stage in the book in order to provide a background for the understanding of visceral regulatory mechanisms to be discussed in subsequent chapters.

The autonomic nervous system is divided anatomically and functionally into two major divisions, sympathetic and parasympathetic. As a generalization, the sympathetic system activates a variety of functions that reach their maximal intensity in emergencies and stresses, in reactions characterized as "fight, flight and fright," though on a more modest scale they play a role in ordinary regulations as well. Most sympathetic reactions, e.g., increased heart rate and cardiac output, increased thermogenic metabolism and mobilization of body fuel stores, are energy-consuming activities. On the other hand, parasympathetic reactions tend to be energy conserving, e.g., slowing of heart rate. Not all reactions fit this scheme but nevertheless it is a useful way of looking at autonomic function.

Each division of the autonomic nervous system consists of *afferent* fibers carrying impulses *toward* the central nervous system (CNS) and *efferent* fibers conveying impulses *away from* the CNS to the effector organs.

SYMPATHETIC NERVOUS SYSTEM

SYMPATHETIC EFFERENT PATHWAYS

The sympathetic outflow originates in neurons located in the lateral horns of spinal cord gray matter between the first thoracic and second lumbar segments. Each of these neurons gives off a myelinated axon, which passes in an anterior spinal root and *white ramus communicans* to synapse with cells in a sympathetic ganglion (Fig 3–1). This axon is termed the *preganglionic*

Fig 3–1.—Distribution of sympathetic outflow. Two sympathetic neurons are shown. The axon of one synapses in a paravertebral ganglion; the postganglionic fiber is distributed via a spinal nerve to blood vessels, sweat glands and erector pili. The other axon synapses in a prevertebral ganglion and the postganglionic fiber is distributed by autonomic nerves to visceral smooth muscle. (After Pick, J.: *The Autonomic Nervous System* [Philadelphia: J. B. Lippincott Co., 1970]. Used by permission.)

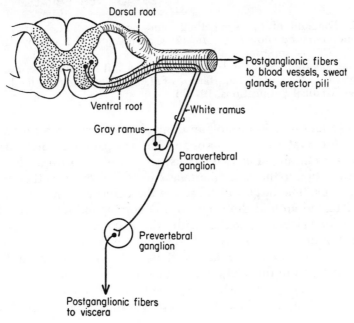

Dorsal root

Postganglionic fibers
to blood vessels, sweat
glands, erector pili

Ventral root

White ramus

Gray ramus

Paravertebral
ganglion

Prevertebral
ganglion

Postganglionic fibers
to viscera

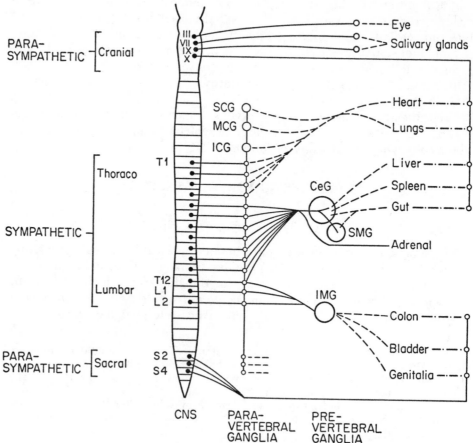

Fig 3–2.—Diagram of the sympathetic and parasympathetic nervous systems. Preganglionic fibers are shown by *solid lines*, postganglionic fibers by *dashed lines*. *SCG, MCG, ICG* = superior, middle and inferior cervical ganglia, respectively; *CeG* = celiac ganglion; *SMG* and *IMG* = superior and inferior mesenteric ganglia, respectively. (After Ganong, W. F.: *Review of Medical Physiology* [6th ed.; Los Altos, Calif.: Lange Medical Publications, 1973].)

fiber. Each ganglion cell gives off an unmyelinated axon, the *postganglionic* fiber, which innervates an effector organ. The sympathetic ganglia are collections of neurons arranged as two chains, one on either side of the vertebral column. In addition to these *paravertebral* ganglia, there are other collections of ganglion cells. The largest of these are the celiac and mesenteric ganglia (Fig 3–2) situated around the origin of the corresponding arteries from the aorta. These *prevertebral* ganglia, with others, contain the cell bodies that regulate gastrointestinal, urinary and reproductive functions.

The preganglionic fibers can take various courses after leaving the spinal cord. Some synapse in the sympathetic ganglion corresponding to the spinal segment, some pass up or down the sympathetic chain and synapse in more distant paravertebral ganglia and others pass through the sympathetic chain

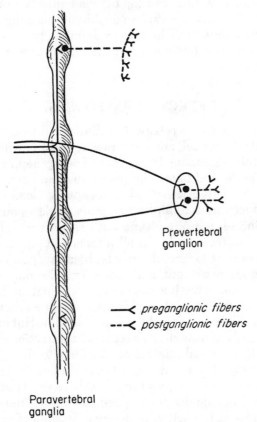

Prevertebral
ganglion

—< *preganglionic fibers*
---< *postganglionic fibers*

Paravertebral
ganglia

Fig 3–3.—Various pathways for sympathetic preganglionic axons. Three fibers are shown: one passes through the paravertebral ganglion at the same segmental level and synapses in the prevertebral ganglion; one descends in the sympathetic chain, which it leaves one segment lower; one sends branches up and down the sympathetic chain and synapses at various levels within it. (After Pick, J.: *The Autonomic Nervous System* [Philadelphia: J. B. Lippincott Co., 1970]. Used by permission.)

and synapse in a prevertebral ganglion (Fig 3–3). The presynaptic fibers may branch extensively, and a single axon may therefore make synaptic contacts with many ganglion cells, providing wide "divergence" of excitation. Some of the postganglionic fibers enter the spinal nerve via a *gray ramus communicans* and innervate effector cells within the area of the spinal nerve's distribution. Other fibers form sympathetic nerve plexuses that innervate their target organs. Again, a single postganglionic fiber may branch repeatedly and thereby control hundreds of effector cells.

SYMPATHETIC NEUROTRANSMISSION

Preganglionic nerve terminals are separated from ganglion cells and their dendrites by a small gap. This gap cannot be bridged electrically, and excitation is conveyed from one cell to another by release of a chemical neurotrans-

mitter, acetylcholine. There also is a gap between postganglionic nerve terminals and effector cells. This gap varies considerably among different tissues,
e.g., it is only a few nanometers in the vas deferens but may be hundreds of
microns in the aorta. The transmitter at the neuroeffector junction usually
is norepinephrine.

EFFECTOR RESPONSES

After the neurotransmitter is released, it diffuses through the gap between
nerve terminal and effector cell and combines reversibly with *receptors* located on the effector cell membrane. In the case of norepinephrine, there are two
types of receptor, *alpha* and *beta*. The response of any organ to norepinephrine depends on the relative activity of its receptors. Combination of norepinephrine with smooth muscle alpha receptors evokes constriction in most
tissues; combination of norepinephrine with smooth muscle beta receptors
causes relaxation. Cardiac muscle, for all intents and purposes, has only beta
receptors and, in contrast to smooth muscle, beta receptor activation by norepinephrine increases peak contractile force and the rate of force development. Epinephrine, one of the hormones of the adrenal medulla, can activate
both alpha and beta receptors. Many drugs influence adrenergic receptors.
For example, *phenylephrine* is an alpha receptor stimulant or *agonist*. It is included in many brands of nose drops because its activation of alpha receptors
of dilated vessels in the nasal mucosa causes them to shrink and thereby enlarge the nasal airway. *Isoproterenol*, a beta adrenergic agonist, is used in
treatment of asthma since it induces bronchodilation and thereby reduces airway resistance. Some antagonist drugs specifically block one or another receptor and thereby prevent the action of agonist. Studies of organ or tissue responses to these agonists and antagonists enable the types of receptor present
to be determined. Thus, a tissue response is characterized as alpha when the
order of potency is norepinephrine > isoproterenol, whereas the order of potency for beta receptor stimulation is isoproterenol > norepinephrine. Moreover, the beta receptor blocking agent *propranolol* prevents the action of all
adrenergic agents on beta receptors but does not influence their effects on alpha receptors. In contrast, *phenoxybenzamine* blocks the effects of adrenergic
agents on alpha receptors without influencing their effects on beta receptors.

Table 3–1 summarizes the adrenergic receptor types in the various tissues
and the responses produced by their activation.

NATURE OF BETA ADRENERGIC RECEPTOR

In the last few years it has become clear that the beta adrenergic receptor is
intimately related to a membrane-bound enzyme, adenylate cyclase. Adenylate cyclase catalyzes conversion of ATP to 3′,5′-cyclic adenosine monophosphate (cAMP). Administration of this agent or its more lipid-soluble derivative, dibutyryl cAMP, reproduces many of the effects of beta adrenergic
receptor stimulation. Moreover, noncatecholamines that increase cAMP produce effects resembling those of beta receptor activation. For example, gluca-

TABLE 3–1.—AUTONOMIC CONTROL OF VARIOUS TISSUES AND ORGANS

ORGAN	AUTONOMIC CONTROL°	NEUROEFFECTOR TRANSMITTER° S	PS	ADRENERGIC RECEPTOR	SYMPATHETIC EFFECTS†	PARASYMPATHETIC EFFECTS†
Pupil	S, PS	NE	ACh	α	Dilatation	Constriction
Ciliary muscle	S, PS	NE	ACh	β	Relaxation	Contraction
Lacrimal gland	PS	–		–	–	Secretion
Heart	S, PS	NE	ACh	β	Rate ↑, force ↑, conduction velocity ↑	Rate ↓, conduction velocity ↓
Blood vessels						
Cerebral	S, PS	NE	ACh	α	Constriction (slight)	Dilatation (slight)
Coronary	S, PS	NE	–	α β	Constriction (slight)	Dilatation (slight)
Gut	S	NE	–	α β	Constriction	–
Renal	S	NE	–	α	Constriction	–
Skin	S	NE	–	α	Constriction	–
Skeletal muscle	S	NE & ACh	–	α β	Constriction, dilatation	–
Gut	S, PS	NE	ACh	α β	Motility ↓, secretion ↓, sphincter tone ↑,	Motility ↑ secretion ↑, sphincter tone ↓,
Salivary glands	S, PS	NE	ACh	α	thick secretion	thin secretion
Bronchi	S, PS	NE	ACh	β	Dilatation, secretion ↓	Constriction, secretion ↑
Ureter	S, PS	NE	ACh	α	Increased tone & motility, detrusor relaxes, sphincter contracts	Increased tone & motility, detrusor contracts, sphincter relaxes
Bladder	S, PS	NE	ACh	α β		
Genitalia	S, PS	NE	ACh	α	Ejaculation	Erection
Sweat glands	S	ACh	–	–	Sweating ↑	–

°S = sympathetic; PS = parasympathetic; NE = norepinephrine; ACh = acetylcholine.
† ↑ = function is increased; ↓ = function is decreased.

gon, which increases myocardial cAMP, also has positive inotropic and chronotropic effects.

There appear to be at least two types of beta receptor. These have been designated beta-1 and beta-2, on the basis of their responses to synthetic agonists and antagonists. Beta-1 receptors are present in heart muscle, beta-2 receptors in vascular, bronchial and other smooth muscles. The agonist *salbutamol* is almost as potent as isoproterenol in stimulating beta-2 responses, e.g., bronchial dilatation, but far less potent than isoproterenol in activating beta-1 responses, e.g., increased heart rate. Again, small doses of the beta-blocking agent, *practolol*, block beta-1 (cardiac) responses but not beta-2 (smooth muscle) responses. Another interesting aspect is that norepinephrine is almost as potent as epinephrine in stimulating beta-1 receptors but is far less potent as a beta-2 stimulator. These differences between beta-1 and beta-2 receptors may have therapeutic significance. Thus salbutamol may prove to be a better drug than isoproterenol for the management of bronchial asthma, since effective bronchodilator (beta-2) doses produce relatively little cardiac stimulation (beta-1). Again, a beta-1 selective blocking agent might be more suitable than a nonselective agent such as propranolol when beta adrenergic blocking agents are administered in patients with a history of asthma.

Nature of Alpha Adrenotropic Receptor

Stimulation of alpha receptors causes contraction of some smooth muscles (e.g., vascular) and relaxation of others (e.g., intestinal). Glandular secretion is usually inhibited. Adrenergic contraction of smooth muscle involves both electric and nonelectric types of excitation. The former may be associated with graded depolarization, action potentials or both, depending on the type of tissue and the experimental preparation. Nonelectric activation can be demonstrated in tissues that have been depolarized by high external K^+ concentrations. Its physiologic importance is controversial. Presumably alpha receptor activation under these circumstances causes an increase in free intracellular calcium ion either by increasing the permeability of the cell membrane to Ca^{2+} or by releasing Ca^{2+} from an internal store. In some muscular tissues that are relaxed by alpha receptor stimulation, hyperpolarization occurs. Nonelectric types of inhibition also have been reported, but the physiologic role of this mechanism has not yet been established. Some authorities have suggested that the alpha receptor is related to adenylate cyclase and that receptor occupation *inhibits* the activity of this enzyme and *reduces* cAMP formation. The relationship between cAMP, membrane potential and calcium fluxes remains obscure.

Synthesis and Release of Catecholamines

Norepinephrine is synthesized in the sympathetic nerve endings from the essential amino acid tyrosine. The intermediate steps and the enzymes involved are as follows:

Tyrosine → Dihydroxyphenylalanine → Dopamine → Norepinephrine

Tyrosine   *Dopa*   *Dopamine β-*

hydroxylase   *decarboxylase*   *hydroxylase*

Norepinephrine is stored in vesicles in localized swellings, or *varicosities,* which occur in great numbers along the nerve endings. Several types of vesicle have been distinguished by electron microscopy. Some contain small granules, some contain large granules and some are agranular. Norepinephrine is mainly stored in the small granular vesicles along with the enzyme dopamine β-hydroxylase (DBH).

When an impulse passes along the axon, the vesicle fuses with the axonal membrane and then discharges its contents, including norepinephrine and DBH, by the process of *exocytosis.* The released norepinephrine accumulates in the space between nerve terminal and effector cell: some combines with receptor sites on the effector, some is taken up again by the nerve terminals and the remainder is destroyed by the enzymes monoamine oxidase (MAO) and catechol O-methyl transferase (COMT). MAO is located in mitrochondrial membranes, whereas COMT is found free in the cytoplasm. The principal metabolite of these enzymes is 3-methoxy, 4-hydroxy mandelic acid (vanillylmandelic acid, VMA), which is inactive and is excreted in the urine. When nerve stimulation ceases, the effector response rapidly declines because of re-uptake and enzymatic destruction of norepinephrine. The re-uptake mechanism is quantitatively the most important and conserves transmitter.

REGULATION OF NOREPINEPHRINE SYNTHESIS AND RELEASE.—Several

Fig 3–4.—Feedback control of norepinephrine (*NE*) release. **A,** *NE* feedback via α receptors on the nerve terminal membrane; **B,** prostaglandin (*PG*) feedback; **C,** *NE* feedback within the nerve terminal via modulation of tyrosine hydroxylase (*TyH*) activity. ⊖ = negative feedback.

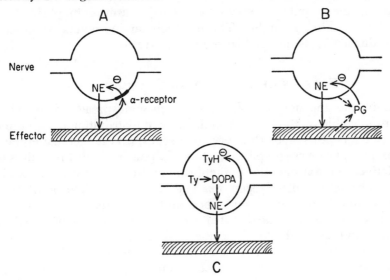

feedback mechanisms regulate norepinephrine synthesis and release (Fig 3–4). They include (1) inhibition of tyrosine hydroxylase activity and synthesis by norepinephrine, (2) inhibition of release by norepinephrine and (3) inhibition of release by prostaglandin.

The activity of tyrosine hydroxylase is inhibited by high concentrations of norepinephrine in the nerve terminal. During periods of increased nerve activity, norepinephrine concentration declines, tyrosine hydroxylase activity increases and more transmitter is formed. This type of regulation develops very rapidly and is reinforced, when sympathetic discharge is prolonged, by increased synthesis of tyrosine hydroxylase.

The nerve terminals possess alpha adrenergic receptors, which inhibit norepinephrine release when stimulated. Thus norepinephrine tends to inhibit its own release and so conserves transmitter.

In many tissues, sympathetic activity is associated with release of prostaglandin E, probably from the effector cells. This substance inhibits the amount of norepinephrine subsequently released by nerve stimulation.

ADRENAL GLANDS

The adrenal medullae may be looked upon as specialized sympathetic ganglia whose cells have no postganglionic fibers but which secrete their transmitter substances directly into the blood. The adrenal medullary secretion is a mixture of catecholamines, approximately 80% norepinephrine and 20% epinephrine. Synthesis of norepinephrine has already been outlined. Epinephrine is formed by N-methylation of norepinephrine by the enzyme phenylethanolamine N-methyltransferase. Both hormones are released by acetylcholine when the adrenal preganglionic fibers are activated. There is a continuous small secretion of catecholamines by the adrenal medulla. Plasma norepinephrine in the basal state is 1 ng/ml (but much of this is derived from the norepinephrine output of sympathetic nerves). Plasma epinephrine is considerably lower, about 0.2 ng/ml. These catecholamine levels are greatly augmented during stress-induced adrenergic activation.

EFFECTS OF CIRCULATING CATECHOLAMINES

Epinephrine and norepinephrine produce widespread effects. These are partly predictable on the basis of (1) the known distribution of alpha and beta adrenergic receptors among the various tissues, (2) the fact that epinephrine and norepinephrine are potent activators of alpha and beta-1 receptors and (3) the relatively potent effect of epinephrine and the relatively weak effect of norepinephrine on beta-2 receptors. The lack of complete predictability is due to modification of the direct effects on the adrenergic receptors by other influences. For example, the direct effect of norepinephrine on the sinus node is to cause an increased rate of pacemaker discharge and heart rate increases. This is demonstrable in isolated hearts. On the other hand, when norepinephrine is injected intravenously in a conscious animal, cardiac *slowing* often results. This is because norepinephrine causes a rise in blood pressure, which induces reflex slowing via the vagus nerves.

The effects of injected epinephrine include increased heart rate, stroke volume and cardiac output; elevated (large doses) or reduced (small doses) arterial pressure; constriction of veins; reduced blood flow to kidneys, gut and skin; muscle vasodilatation (small doses) or constriction (large doses); bronchial dilatation; intestinal relaxation; splenic contraction; variable effects on uterine contractions; restlessness and anxiety; elevated blood glucose and free fatty acids (FFA); and increased oxygen consumption.

Norepinephrine has somewhat similar effects but, because of its minimal ability to stimulate beta-2 receptors, it does not cause skeletal muscle vasodilatation and produces a greater rise in arterial pressure. Its bronchodilator and other smooth muscle actions are less prominent than those of epinephrine and its metabolic effects are smaller.

PARASYMPATHETIC NERVOUS SYSTEM

The efferent limb of this system is organized like its sympathetic counterpart and consists of pre- and postganglion neurons. However, the cell bodies of the preganglionic neurons are located in the brain stem and in the sacral part of the spinal cord. This arrangement is often referred to as the *craniosacral* outflow and contrasts with the *thoracolumbar* outflow of the sympathetic system. A further anatomical difference between the two is that the parasympathetic ganglia are not arranged in chains but are found in or near the tissues that they innervate. The parasympathetic efferent limb therefore consists of relatively long preganglionic and short postganglionic fibers, whereas the reverse is the case for the sympathetic system.

The cell bodies of the cranial preganglionic fibers are located in the brain stem and their axons are distributed with the cranial nerves. When activated, these fibers cause pupillary constriction; accommodation of the ocular lens; secretion of lacrimal and salivary glands; slowing of the heart, constriction of the bronchi; and increased motility of the stomach, small intestine, gallbladder and part of the colon. Gastrointestinal and pancreatic secretion is enhanced. The sacral outflow is distributed by the pelvic nerves to the colon and rectum and to the genitourinary system. Stimulation of appropriate fibers of the pelvic nerves causes contraction of the urinary bladder and relaxation of its sphincter, contraction of the lower colon and rectum and erection of the penis or clitoris (see Table 3–1).

CHOLINERGIC TRANSMISSION

The transmitter that conveys excitation from the preganglionic to the postganglionic fiber throughout the autonomic nervous system is acetylcholine. This agent is also the transmitter between parasympathetic postganglionic terminals and the effector cells that they supply. It is also the transmitter at certain *sympathetic* postganglionic terminals, e.g., sympathetic cholinergic fibers to skeletal muscle blood vessels and to sweat glands.

Acetylcholine is synthesized from acetate and choline in the nerve cells. Choline acetylase, coenzyme A and ATP are required in the synthetic process. Acetylcholine is stored in high concentration in *agranular vesicles* of the terminal varicosities of cholinergic nerves and is released by the passage of an action potential to diffuse into the tissue space between nerve terminal and effector cell. When nerve stimulation is stopped, effector responses rapidly terminate because of hydrolysis of acetylcholine by the enzyme acetylcholinesterase, which is found in both pre- and postjunctional cells.

CHOLINERGIC RECEPTOR

Receptors that bind acetylcholine are present on the surface of skeletal, cardiac and smooth muscle cells and of glandular cells. After receptor activation, a sequence of reactions causes contraction or relaxation of muscle or glandular secretion. When contraction occurs, it is usually preceded by depolarization with or without action potential generation, although even fully depolarized smooth muscle cells respond to acetylcholine. Thus acetylcholine *can* induce contraction by a nonelectric or pharmacomechanical coupling mechanism, but whether this is physiologically important is not known. Many synthetic substances can activate or block cholinergic receptors but there is considerable variation in the blocking action at different sites. As an example, atropine blocks smooth muscle cholinergic receptors in doses that have little effect on ganglionic or skeletal neuromuscular transmission. On the other hand, hexamethonium blocks ganglionic transmission but has little effect on smooth muscle responses to acetylcholine.

CONTROL OF AUTONOMIC FUNCTIONS

In normal animals the activity of the preganglionic neurons of the sympathetic and parasympathetic divisions of the autonomic nerve system is regulated by higher centers located in the hypothalamus and cerebral cortex. Stimulation of areas in the lateral and posterior hypothalamus increases heart rate, blood pressure and peripheral vasoconstriction. On the other hand, stimulation of areas in the midhypothalamus slows the heart and increases gastrointestinal motility. However, the parasympathetic and sympathetic representations are by no means discrete, and considerable variation and overlap occur.

Autonomic functions are also represented in the cerebral cortex, often in association with corresponding motor functions. Thus stimulation of the motor areas for the face and tongue produces salivation; activation of cortical areas 4 and 6, as would occur in exercise, increases heart rate, cardiac output and systemic arterial pressure. Gastrointestinal function is also influenced by stimulation of various cortical areas including the frontal lobes and sigmoid gyrus.

AUTONOMIC REFLEXES

So far only the efferent component of the autonomic system has been discussed, but it should be realized that the rate of efferent impulse discharge is largely under reflex control and is dependent upon incoming information transmitted along autonomic and somatic afferent nerves. A few examples will illustrate this point. Impulses from arterial baroreceptors conveyed by carotid sinus and vagal afferents signal information concerning the level of arterial pressure and modulate the discharge of vagal and sympathetic neurons that regulate heart rate, cardiac function and peripheral resistance. Afferents from the lung signal their degree of stretch and influence the respiratory center. Acute distention of the bowel may increase heart rate. Very intense stimuli, e.g., in renal or gallbladder colic, may trigger a vagal response, with slowing of the heart, a fall in blood pressure and sweating. Somatic afferents also influence autonomic functions. Thus changes in skin temperature, muscle and joint movements and painful stimuli can alter cardiovascular function. For example, cutaneous pain often increases heart rate and blood pressure. On the other hand, deep pain from muscle afferents can reduce heart rate and arterial pressure.

GENERAL ASPECTS OF AUTONOMIC FUNCTION

Parasympathetic activity tends to be relatively localized. In contrast, the sympathetic nervous system produces diffuse effects in many organ systems, partly because it tends to be activated as a whole and partly because increased secretion of adrenal medullary catecholamines is an inevitable accompaniment of sympathetic discharge. The increased blood levels of epinephrine and norepinephrine cause sympathomimetic effects on all tissues. Thus during stresses such as exercise, there may be increased heart rate and cardiac output, a rise in blood pressure, redistribution of blood flow, inhibition of gastrointestinal motility and increased carbohydrate and fat metabolism. Nevertheless, there may be considerable variations among different organs in the degree of response. This may depend on both central and peripheral factors. Thus at certain levels of stimulation those neurons activating a certain portion of the overall response, e.g., intestinal vasoconstriction in exercise, may not be activated but can be *recruited* by more intense stimulation. Peripheral factors include variations in density of sympathetic innervation and in intensity of alpha adrenergic activity of the effector cells. There is a reciprocal interaction between the sympathetic and parasympathetic systems. For example, during exercise vagal activity diminishes at the same time that cardiac sympathetic activity increases. Consequently the concentration of acetylcholine in the sinus node declines as the concentration of norepinephrine rises. Both these changes contribute to the increased heart rate that ensues. Many other examples occur and will be found in the subsequent sections on organ system physiology.

SPECIFIC AUTONOMIC INFLUENCES ON INDIVIDUAL ORGANS AND FUNCTIONS

The following brief list is presented in order to indicate the manifold effects autonomic nerves can produce throughout the body.

THE EYE AND ORBIT.—The diameter of the pupil is determined by the antagonism between the radial muscle fibers causing opening and the circular musculature causing closing. The radial fibers are innervated by sympathetic fibers from the superior cervical ganglion, whereas the circular fibers are innervated by parasympathetic fibers from the ciliary ganglion. Focusing of the lens is also regulated by parasympathetic fibers from this ganglion, which controls the ciliary muscle. When this muscle contracts, the lens becomes more spherical and enables near objects to be focused on the retina. Secretion of tears is controlled by parasympathetic fibers from the sphenopalatine ganglion.

SALIVARY GLANDS.—These receive a dual innervation. Both parasympathetic and sympathetic fibers increase salivary secretion, and this is one of the few instances when the sympathetic and parasympathetic divisions do not produce antagonistic effects.

THE HEART AND BLOOD VESSELS.—Most blood vessels have only sympathetic innervation. Sympathetic stimulation tends to reduce blood flow (arteriolar constriction) and blood volume (venous constriction). The magnitude of the blood flow change varies in different organs because of complex interaction of neural and metabolic factors. The neural factors include density of innervation, amount of transmitter release, width of the "synaptic cleft" and activity of alpha receptors. Intense vasoconstriction can be produced in skin, intestine, kidney and resting skeletal muscle. Little or no constriction occurs in the coronary vessels because of a paucity of alpha receptors and the accumulation of vasodilator metabolites associated with concomitant myocardial stimulation. Metabolites also account for the fact that sympathetic stimulation has a much smaller effect on blood flow in exercising than in resting skeletal muscle. Sympathetic control of cerebral blood flow, while controversial, appears to be slight.

The heart has a dual innervation from the vagus and sympathetic nerves. The vagus slows the heart and depresses conduction, whereas the sympathetic nerves increase heart rate, conduction and myocardial force.

BRONCHI.—A dual innervation is present. Vagal stimulation causes bronchoconstriction and increased bronchial secretion, whereas sympathetic activation causes bronchodilation and decreased secretion.

GASTROINTESTINAL TRACT.—There is a dual innervation and a complex interaction with gastrointestinal hormones. In general, vagal stimulation increases motility and secretion, whereas sympathetic stimulation causes the opposite effects.

SKIN.—Skin blood vessels are under sympathetic control, which plays an important role in thermoregulation (see section on cutaneous circulation, Chapter 4). The sweat glands are also controlled by sympathetic fibers that are exceptions to the usual rule in that the postganglionic transmitter is acetylcholine rather than norepinephrine.

URINARY SYSTEM.—The sympathetic system may cause considerable renal vasoconstriction in emergency situations and this may lead to diminished urine flow. Stimulation of pelvic parasympathetic nerves increases ureteral motility and participates in the act of micturition by causing contraction of the bladder and relaxation of its sphincter. Sympathetic stimulation has opposite effects.

SEXUAL FUNCTIONS.—Erection of the penis and clitoris is due to arteriolar dilatation and venous constriction in the vessels supplying the erectile tissue. These vascular changes are induced by stimulation of sacral parasympathetic nerves. The vessels are also innervated by sympathetic nerves, which produce opposing effects. Ejaculation is under the control of sympathetic nerves, which induce contractions of the smooth muscle of the male genital tract.

BIBLIOGRAPHY

Axelrod, J.: Neurotransmitters, Sci. Am. 230:58, 1974.
Bennett, M. R.: *Autonomic Neuromuscular Transmission* (London: Cambridge University Press, 1972).
Pick, J.: *The Autonomic Nervous System* (Philadelphia: J. B. Lippincott Co., 1970).

4

The Cardiovascular System

GORDON ROSS, M.D.

THE CIRCULATION: GENERAL CONSIDERATIONS

DISEASES OF THE HEART and blood vessels constitute the leading cause of disability and death in the Western world. Every year more than 600,000 people in the United States die as the result of coronary artery disease. Many more suffer serious symptoms that limit their capacity to work or to lead a normal life. Over 25 million Americans have abnormally high blood pressure. More than one million persons experience a "stroke" with paralysis or other severe neurologic disturbance because of inadequate blood flow to the brain. These facts dramatically emphasize the need for a thorough understanding of the physiology of the circulation. Beyond its clinical significance, however, the system has great fascination in the numerous examples that show how beautifully the structure of its components is related to their function and how effectively the intricate interactions and integration of its parts produce a smoothly functioning whole.

This section presents an overview of the circulation. Subsequent sections will discuss the individual functions of the heart, arteries, capillaries and veins. Finally, the integrated response of the entire system to various stresses will be described.

FUNCTIONS OF THE CIRCULATION

Life involves a constant exchange of materials between the organism and its environment. In single-celled creatures this exchange occurs through diffusion but, as we have previously seen (Chapter 1), this process is only adequate when the diffusion path is short. A spherical cell 7 μ in diameter exposed to 100% oxygen becomes 90% saturated with this gas in about 0.005 second. On the other hand, a spherical tissue mass 7 cm in diameter would require 6 days to reach equal saturation. This is because the diffusion path from surface to center is 10,000 times longer and the rate of change of concentration is inversely proportional to the square of the diffusion distance (Chapter 1). Multicellular organisms have overcome the problem of size by developing huge, specialized, infolded exchange surfaces (e.g., lung, gastrointestinal tract and kidneys) and a transport system, the circulation, to convey materials between these surfaces and the rest of the body. The materials are carried in the blood to a series of minute branching tubes, the capillaries, which permeate all metabolically active tissues. The capillary walls are less than 1 μ thick, and the capillaries are so numerous that the distance between the blood and tissue cells is small enough to permit adequate diffusional exchange. For example, a cubic millimeter of cardiac muscle contains about 3,800 capillaries and no muscle cell is farther than 20 μ from the blood. Because of their vital exchange function, discussed further in this chapter, the capillaries constitute the chief organ of the circulation. The heart, arteries and veins can be regarded merely as specialized plumbing designed to maintain optimal blood flow through the capillaries.

The circulation has only one function—transport. It carries (1) oxygen from the lungs and nutrients from the gut to the tissues, (2) excretory products from the tissues to the lungs and kidneys and to a lesser extent to the gut and skin, (3) regulatory substances from the endocrine organs to their target tissues and (4) heat between the body's core and its surfaces.

DESIGN OF THE CIRCULATION

The circulation is maintained by the heart, which functionally consists of two pumps that feed separate circuits. Each pump consists of a thin-walled priming chamber or *atrium* and a thicker-walled chamber, the *ventricle,* which provides the main propulsive force to the blood. The right ventricle pumps blood through the *pulmonary circulation* of the lung to the left atrium, whereas the left ventricle pumps blood to all the other tissues of the body through the *systemic circulation* to the right atrium (Fig 4 – 1). In each circulation, blood is conveyed from the ventricle to the capillaries through a system of branching tubes, the *arteries,* and is returned to the heart through another system of branched tubes, the *veins.*

The circulation of blood through the heart and blood vessels will function effectively only if the flow is unidirectional and if the outputs of the right and left sides of the heart are equal. Unidirectional flow is ensured by valves within the heart and in some of the veins. Equality of outputs of the right and left ventricles is necessary in order to maintain optimal pressures and volumes throughout the circulatory system. The blood vessels that link the right and left sides of the heart are distensible and, if the output of the left ventricle were to fall below the right, blood would progressively accumulate in the pulmonary circulation. If the right side of the heart pumped less than the left, blood would accumulate in the systemic circulation. Furthermore, relative failure of either pump would cause important changes in various circulatory pressures. Failure of the right side of the heart would increase the pressure within the systemic veins and capillaries, whereas failure of the left side of the heart would increase pulmonary venous and capillary pressures. These raised pressures would tend to cause fluid to leave the vascular system and to accumulate in systemic tissues or the lung (see section on cardiac failure).

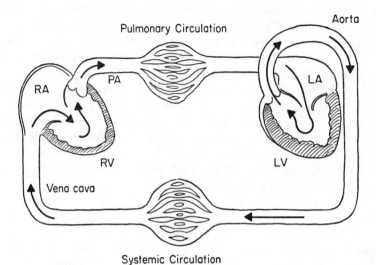

Pulmonary Circulation

Aorta

Systemic Circulation

Fig 4–1.—Diagram of the circulatory system. The left ventricle *(LV)* pumps blood through the systemic circulation to the right atrium *(RA)*. The right ventricle *(RV)* pumps blood through the pulmonary circulation to the left atrium *(LA)*. One-way flow is ensured by valves between the atria and ventricles and at the base of the aorta and main pulmonary artery *(PA)*.

The necessary equality of left and right outputs is achieved by the Frank-Starling mechanism, which is discussed in the section on regulation of stroke volume (p. 136).

Systemic circulation is a composite of many circuits supplying tissues and organs with very different exchange requirements. These are relatively constant for some tissues but enormously variable in others. The brain consumes about 50 ml oxygen per minute, skeletal muscle approximately 45 ml per minute and the splanchnic organs (gut, spleen, liver) together about 75 ml per minute. During severe muscular exercise, muscle oxygen usage may increase more than 20-fold, whereas the oxygen consumption of the brain remains practically unchanged and gut, spleen and liver oxygen consumption may actually decline. The design of the circulation has to take into account these differing and variable requirements and the consequent need to vary blood flow to each of the several areas independently of the others. To see how this is achieved, it is necessary to look more closely at the arrangement of the vessels in the organs and tissues.

Each vascular bed is ultimately supplied from the aorta by one or more arterial branches, which divide successively into smaller and smaller vessels. The smallest branches (radii $10-100$ μ) are called *arterioles* and have relatively thick smooth muscle coats, which regulate their diameter. They supply capillary networks through which blood flows into small veins termed *venules* and then into successively larger veins, which eventually open into one of two large vessels, the *venae cavae*. The *superior vena cava* receives blood from the head, neck, arms and upper trunk and the *inferior vena cava* collects blood from the lower trunk, abdominal organs and legs. Both venae cavae empty into the right atrium.

The various vessels constitute a resistance to blood flow and, consequently, pressure falls continuously from the arterial to the venous side of the vascular bed. The large arteries and veins offer little resistance to flow and hence only small pressure reductions occur between aorta and arterioles and between venules and vena cava. The small vessels, arterioles, capillaries and venules, offer a much larger resistance to flow (Fig 4–2). The greatest pressure drop occurs across the arterioles, which are therefore often called the "resistance" vessels because they play a dominant role in regulating resistance and blood flow in the tissues that they supply.

The average flow (Q) through an organ or tissue depends on the difference in blood pressure (ΔP) between its arteries and veins and the resistance (R) of its blood vessels, i.e.,

$$\text{flow } (Q) = \frac{\text{pressure difference } (\Delta P)}{\text{resistance } (R)}$$

Mean aortic pressure in a young, healthy adult is about 90 mm Hg and vena caval pressure about 6 mm Hg. The pressure difference or "gradient" across

Fig 4–2.—Pressures across a typical vascular bed. Note that pressure drops progressively from aorta to vena cava and that the greatest pressure drop is across the arterioles.

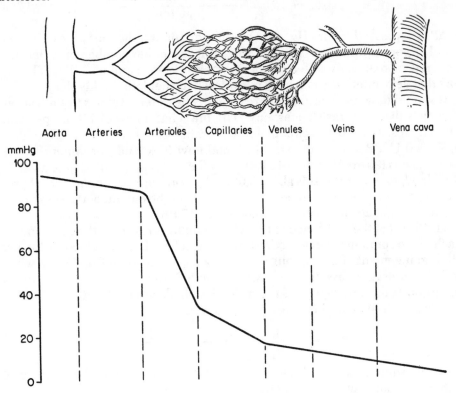

the systemic vascular beds is therefore around 84 mm Hg. Theoretically it would be possible to double the blood flow through an organ or tissue by doubling the pressure gradient across it or by halving the resistance. However, the arteries of man are so constructed that, although they will withstand a mean pressure of 90 mm Hg for many decades, even small increments of pressure above normal will produce degenerative changes in their walls. Higher arterial pressures also require greater expenditures of energy by the heart to generate them. To prevent these adverse effects, the body uses specific controlling mechanisms to keep arterial pressure within narrow limits (see section on the arteries). Pressure in the large veins near the heart *(central venous pressure)* is also maintained at a fairly constant value. Thus the pressure gradients across the vascular beds do not vary much in health, and changes in flow are brought about mainly by altering resistance, particularly in the arterioles.

A rough estimate of vascular resistance can be obtained by using the formula given above, which is analogous to Ohm's law in electric circuits.

Electric circuits (Ohm's law):

$$\text{resistance } (R) = \frac{\text{electromotive force } (E)}{\text{current } (I)}$$

Hydraulic circuits:

$$\text{resistance } (R) = \frac{\text{pressure gradient } (\Delta P)}{\text{flow } (Q)}$$

If ΔP is expressed in mm Hg and Q in ml per minute, the unit of resistance is mm Hg per milliliter per minute. Since this is a rather cumbersome designation, it is more usually called the peripheral resistance unit or (PRU). The factors that determine resistance are considered further later in this chapter.

It should now be clear that the reason for the different flows to the various tissues is that their resistances vary. The brain has a flow of 750 ml per minute. The resistance of the circuit, aorta $\rightarrow$ brain $\rightarrow$ vena cava is therefore (90 − 6)/75 = 0.11 PRU. In contrast, skin blood flow is 200 ml per minute and its resistance is (90−60)/200 = 0.42 PRU. For the systemic circulation as a whole, the blood flow rate at rest is about 5,000 ml per minute. Mean right atrial pressure is around 4 mm Hg. Hence the resistance of the circuit, aorta $\rightarrow$ systemic tissues $\rightarrow$ right atrium, is (90 − 4)/5,000 = 0.017 PRU.

The fact that the resistance of the whole circuit is less than the resistance of each of the components may seem surprising at first. It is explained by the parallel arrangement of the various vascular beds, as shown in Fig 4–3, in which the high pressure vessels are shown on the left, the low pressure vessels on the right. If the resistances of the individual beds are $R_1, R_2, R_3 \ldots R_n$, then the total resistance (R_T) is given by the formula:

$$\frac{1}{R_T} = \frac{1}{R_1} + \frac{1}{R_2} + \frac{1}{R_3} + \cdots \frac{1}{R_n}$$

Let us suppose a hypothetical circuit consists of 20 parallel circuits, each with a resistance of 2 PRU; then

$$\frac{1}{R_T} = \frac{10}{2}$$

hence

$$R_T = 0.2 \text{ PRU}$$

Thus total resistance is less than that of any of its parts. In contrast, when resistances are arranged in series, their total resistance is given by their sum,

$$R_T = R_1 + R_2 + R_3 + \dots R_n$$

If the ten resistances of our hypothetical circuit were linked in series, the total resistance would be 20 PRU or 100 times greater than the total resistance of the same elements arranged in parallel. Similarly, if the various vascular circuits of the body were linked in series rather than in parallel, the resistance to

Fig 4–3.—Parallel arrangement of the major vascular beds of the systemic circulation. *LV* = left ventricle; *RA* = right atrium. The pressure gradient across each bed in this particular case is 90 − 6 = 84 mm Hg.

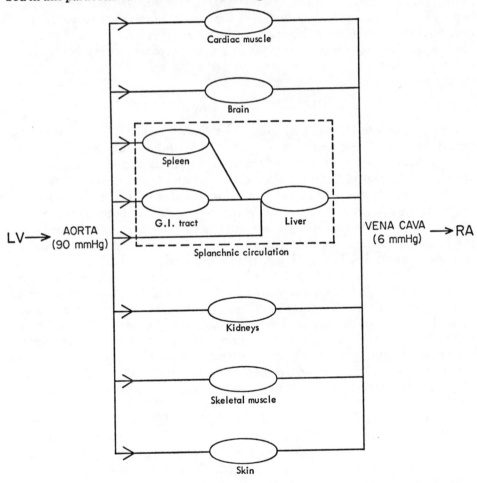

flow would be enormous, and a pressure of many thousands of mm Hg would be required to maintain a flow rate adequate to meet the requirements of all the tissues.

A major advantage of the parallel arrangement is that flow can be made to vary widely in individual vascular beds without causing a severe disturbance of the circulation as a whole. For example, if one of the 10 parallel resistances in our example increased to infinity, the total resistance would only increase by 11%. On the other hand, if any one of a number of resistances connected in series were to become infinite, the total resistance would be infinite and the circulation would stop. The resistances of the arteries, arterioles, capillaries, venules and veins in any individual organ or tissue *are* linked in series (see Fig 4–2), and block of either artery or vein, e.g., by a thrombus, stops the circulation through that tissue. Some arteries, e.g., those situated around joints, are liable to obstruction by positional changes. In such cases the vascular bed is usually supplied by several *anastomosing* vessels so that if one artery is obstructed, flow continues through the others. The major arteries supplying the brain anastomose freely to form an arterial circle, the circle of Willis (see section on cerebral circulation), and the brain is therefore protected against the effects of obstruction of any one or more of the constituents of the circle. Unfortunately, the major coronary arteries have few anastomoses, and serious consequences occur when a major coronary artery becomes occluded by disease.

REGULATION OF RESISTANCE

The principal factor regulating local vascular resistance is the caliber of the arterioles. This is influenced by (1) agents released into the tissue spaces by local metabolism; these usually cause arteriolar dilatation, reduced resistance and increased flow; (2) chemicals released from remote organs (hormones); and (3) transmitter substances released from nerves. The relative significance of each of these factors varies among different organs and will be discussed in relation to the regional circulations. The important general point to grasp now is that blood flow to an organ may be altered by local and by remotely controlled factors. Local regulation enables an organ to adapt its own blood supply to meet its metabolic demands. For example, light exercise increases blood flow to the exercising muscle and this is almost entirely local accumulation of vasodilator metabolites. The remote or neurohumoral factors can adjust the total and regional vascular resistances to ensure the well-being of the body as a whole even at the expense of some of the local circulations. For example, during hemorrhage, neurohumoral factors produce widespread vasoconstriction and diminished blood flow in muscle, skin, kidney and gut. This serves to maintain arterial blood pressure and adequate flow to the brain and heart.

The blood vessels of the various vascular beds differ in their degree of innervation and in their responsiveness to released neurotransmitter. There is also evidence that the neuron pools in the brain stem that control different vascular areas can be selectively activated. The effects of the transmitter may also be profoundly modified by the metabolic state of the tissues.

Because of these interactions the neural and hormonal regulatory mechanisms produce effects that vary from region to region. Thus excitation of the sympathetic nervous system produces coronary vasodilatation; variable vasoconstriction in kidney, skin, muscle and gut; and relatively slight effects in cerebral and pulmonary circulations. Often the neurohumoral effects conflict with local demands. For example, the increased sympathetic nerve activity that accompanies muscular exercise tends to constrict muscle vessels, whereas local metabolic changes tend to dilate them. The various interactions between the local and remote mechanism constitute some of the most interesting aspects of circulatory physiology, and several examples will be discussed in this chapter.

The ability of tissues to alter the resistance of their blood vessels is related to the variability of their exchange requirements. Thus skeletal muscle must exchange oxygen and carbon dioxide at greatly increased rates during exercise and its vascular resistance can fall to 5% of the resting value. Heat exchange across the skin is also very variable. For example, in hot climates heat must be lost rapidly from the body surface to prevent body temperature from rising. This calls for increased skin blood flow and a reduction in resistance of skin vessels. Under extreme conditions cutaneous vascular resistance may fall to less than 5% of its value at comfortable external temperatures. In contrast, the brain has relatively constant total exchange requirements and its vascular resistance changes little.

Although an increase in blood flow is the most important means of meeting increased metabolic demands, other mechanisms also exist. Extraction of substances from arterial blood may increase. For example, resting skeletal muscle extracts 20% of the oxygen from the blood flowing through it, whereas exercising muscle may extract as much as 90%. Some tissues can derive energy anaerobically for relatively long periods and can function with the circulation completely arrested. This can be demonstrated by pumping up a blood pressure cuff around the arm sufficiently to obliterate the pulse at the wrist. The forearm and hand muscles are able to contract for several minutes despite the local circulatory arrest. In contrast, the brain and the heart have little capacity for anaerobic metabolism. Function becomes impaired within seconds and is irreversibly lost within minutes of cessation of blood flow.

VARIABILITY OF THE CARDIAC PUMP

It has been stressed in preceding paragraphs that a major characteristic of mammalian circulation is the relative constancy of arterial and central venous pressures. It has also been emphasized that changes in blood flow through the various tissues and organs are determined by changes in vascular resistance. If overall systemic resistance falls, however, the goal of maintaining arterial pressure constant can only be achieved by increasing the output of the heart. Let us suppose that a resting subject has a cardiac output of 5,000 ml per minute with a skeletal muscle blood flow of 850 ml per minute. If the pressure gradient from aorta to right atrium is 85 mm Hg, total peripheral resistance (TPR) is 85/5,000 = 0.017 PRU and muscle resistance is 85/850 = 0.1 PRU. When this subject gently exercises, the resistance of the arterioles within the exercising muscles falls and flow increases, let us say to double its original value, i.e., to 1,700 ml. If we assume that cardiac output and resistances of the other vascular circuits do not change, arterial pressure will fall to approximately 68 mm Hg (the student might try to verify this by going through the calculations himself). In order to maintain pressure constant, cardiac output must rise or the vascular resistance of other organs must increase. In exercise, changes in both variables occur. Cardiac output and splanchnic and visceral vascular resistances increase, whereas muscle resistance falls. Similar considerations apply to other situations in which circulatory demand is increased; for example, during exposure to high external temperatures, skin vascular resistance falls, skin flow and cardiac output increase, visceral resistances rise and pressure does not change significantly.

The ability of the heart to vary its output is crucial to the functioning of the circulation. If this ability is impaired, the capacity of the body to cope with physiologic stresses becomes severely limited. This is clearly shown in the greatly reduced exercise tolerance of patients with heart failure (see section on abnormalities of cardiovascular function).

The heart of a healthy adult at rest beats 50–100 times per minute with a stroke volume, defined as the output of each ventricle per beat, of 50 to 100 ml. Cardiac output could obviously be increased either by increasing heart rate or stroke volume, or both. Moderate increases in cardiac output are achieved mainly by increased rate. However, rates above 200 per minute are rare even in strenuous exercise, and an increase in stroke volume is obligatory when outputs exceeding approximately 300% of the resting value are required. Heart rate is mainly controlled by the autonomic nervous system (Chapter 3). Increased sympathetic activity accelerates the heart, whereas increased parasympathetic (vagal) activity slows it. At rest, parasympathetic control is dominant. When circulatory demand is increased, vagal discharge is inhibited and sympathetic discharge is increased. This reciprocal interaction permits a smooth control of heart rate over a wide range. The force of cardiac contraction is mainly controlled by sympathetic nerves, and the vagi have only small direct effects on ventricular force.

In addition to its neural control, the heart is influenced by epinephrine and norepinephrine released from the adrenal glands. Both substances directly increase myocardial force. Their effects on rate are variable and depend on the interaction between their direct effects on the cardiac pacemaker and reflex effects induced by changes in arterial pressure.

FUNCTION OF THE VEINS

Blood leaves the capillary beds at a pressure of about 12–15 mm Hg when the body is supine. The mean pressure in the right atrium is about 4 mm Hg. The pressure gradient returning blood to the heart through the veins is therefore 10–15% of the total gradient from aorta to right atrium. Thus venous resistance is only 10–15% of total vascular resistance. The veins are therefore low pressure, low resistance conduits. They are frequently exposed to external pressures higher than their intraluminal pressures, e.g., when they are located within contracting muscle. To prevent backflow when this occurs, many veins contain valves that permit blood to flow toward the heart but not in the reverse direction. An important difference between arteries and veins is that veins are very distensible, i.e., small changes in transmural pressure cause substantial changes in volume. When a person stands, the pressure in the leg veins may increase considerably. This causes them to dilate to such a degree that they can accommodate several hundred milliliters of extra blood. The return of blood to the heart is therefore impaired, cardiac output is reduced, arterial pressure tends to fall and the subject may feel dizzy and may even faint. Because of their ability to contain large and variable amounts of blood at low pressures, the veins are often referred to as "blood reservoirs" or "capacitance vessels." The reservoirs can be emptied by active contraction of the venous smooth muscle under the influence of sympathetic nerve stimulation or circulating epinephrine and norepinephrine. They will also empty passively if atrial pressure falls or if the inflow of blood into the veins is reduced by arteriolar constriction.

We shall see in other sections of this chapter that venous pressure is of great importance for two reasons: (1) it is a major determinant of capillary pressure and consequently of blood and extracellular fluid volume, and (2) pressure in the great veins near the heart, central venous pressure, regulates cardiac filling and is one of the major factors that influence cardiac output.

CIRCULATORY RESERVE

As with most physiologic systems, the circulation has a number of functional and structural characteristics that protect it from the consequences of disease or from damage to its parts. As an example, we might consider the effects of a plaque of atheroma that is producing a progressive narrowing (stenosis) of the femoral artery. The stenosis increases the resistance of the artery so that pressure downstream from the plaque falls, thus reducing the effective pressure gradient across the femoral capillaries. Nevertheless, resting flow is not reduced because local compensating mechanisms produce arteriolar dilatation, which reduces the total resistance of the bed. This adaptation, however, reduces the *vasodilator reserve*, and flow cannot be increased sufficiently to meet metabolic demands during exercise. Metabolites accumulate within the tissues and produce a cramping muscular pain, which usually causes the subject to stop exercising. When stenosis is slight, the vasodilator reserve is considerable and pain may occur only during heavy exercise, but as the stenosis progresses, more and more of the reserve is used up to maintain resting flow and eventually light exercise induces pain. Ultimately the narrowing may become so severe that the vascular bed is maximally dilated at rest and the vasodilator reserve is zero. Blood flow then becomes totally pressure dependent, and progression of the stenosis causes muscle and skin tissues to degenerate as the effective perfusion pressure falls.

Some organs and tissues possess small vascular channels that connect the major arteries that supply them. These connecting vessels are called *collaterals* and do not normally carry significant flow. If one of the major arteries be-

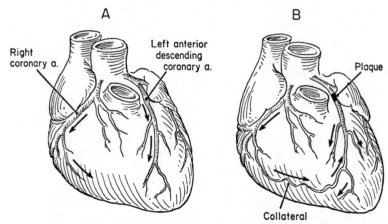

Fig 4–4. – **A,** diagram of normal coronary arteries. **B,** after the development of an atheromatous plaque, which obstructs the left anterior descending coronary artery. A small vessel, too small to be seen in **A,** has enlarged to form an effective collateral artery, which enables the anterior descending system to be filled retrogradely. In the absence of this collateral, the heart muscle in the area of distribution of the obstructed artery would have died.

comes blocked, the collateral circulation develops, i.e., the small collaterals dilate and enlarge to such an extent that they can carry the flow of the occluded vessel (Fig 4–4). If the initial narrowing develops in a coronary artery, the patient's life may depend upon the existence of potential collateral channels and the rapidity of their development to fully fledged collateral arteries.

The heart also possesses several structural and functional reserve mechanisms (see section on regulation of cardiac function). Consequently, minor functional impairment produced by cardiac disease may be manifested only when the heart is stressed, e.g., by exercise, heat or high altitude. If symptoms or signs appear at rest, severe disease is likely to be present since cardiac reserves must have been largely used up.

THE HEART: ANATOMY AND ELECTROPHYSIOLOGY

The heart is a muscular organ that weighs approximately 250–350 gm in the adult. It has four chambers: two atria and two ventricles. The atria are separated from the ventricles by a fibrous ring, bridged only by a thin band of specialized conducting tissue, the *atrioventricular* (AV) *bundle.* The right atrium receives blood from the superior and inferior venae cavae and empties into the right ventricle through a one-way valve called the *tricuspid valve* because of its three leaflets. The right ventricle has a crescent-shaped cavity and a relatively thin wall, 3–4 mm thick. Its internal surface is trabeculated and there are three *papillary* muscles, which project into the cavity and are connected to the tricuspid valve leaflets by tough fibrous cords, the *chordae tendineae.* The leaflets are attached circumferentially to the fibrous ring between the right atrium and ventricle and hang down into the right ventricular cavity.

The right ventricle ejects blood into the pulmonary artery through a valve consisting of three delicate semilunar membranous valve leaflets. This valve prevents backflow when the right ventricle relaxes.

The left atrium receives blood from the pulmonary veins and empties into the left ventricle through the *mitral valve*. This valve is attached circumferentially to the fibrous ring between the left atrium and ventricle and has two leaflets that hang down into the left ventricle.

The left ventricle has the shape of a prolate spheroid (egg shaped). Its free wall is 8–12 mm thick, three times thicker than the right. Two prominent papillary muscles, one anterior and one posterior, project into the left ventricular cavity and each is attached by chordae tendineae to both mitral leaflets. The left ventricle ejects blood into the aorta through the aortic valve, which is similar in structure and function to the pulmonary valve.

The atria are separated from each other by a thin muscular wall, the *interatrial septum*. The ventricular cavities are separated by the *interventricular septum*, which is thick and muscular except in its uppermost part where a small portion is membranous.

The thin-walled crescentic right ventricle is well suited for expulsion of blood into the low resistance pulmonary circuit, whereas the thick-walled spherical left ventricle is well adapted for generating the high pressures required to drive blood through the high-resistance systemic circuit. It is of interest that when resistance to right ventricular ejection is chronically increased, for example by narrowing of the pulmonic valve, the right ventricular wall becomes much thicker and the cavity becomes more spherical so that it resembles the left ventricle.

The heart cavities and valves are lined by endothelium, the *endocardium*, which is continuous with that lining the blood vessels. The outer surface of the heart is enclosed by the *pericardium*, which consists of two layers, visceral and parietal. The thin visceral layer, the *epicardium*, closely invests the outer layer of the myocardium and is separated from the thicker, tougher parietal pericardium by a narrow cavity containing 10–15 ml fluid, which lubricates movement of the heart within the pericardium. It has been suggested that the pericardium maintains the heart in an optimal functional position and also limits the degree of cardiac dilatation in some disease states. However, it is congenitally absent in some persons and can be removed surgically in others without apparent harm.

THE CONDUCTING SYSTEM

In order for the heart to function effectively as a pump, its chambers must be excited rhythmically and in the proper sequence. The ventricular muscle cells also must contract almost synchronously in order to develop the necessary expulsive force most effectively. These requirements are met by specialized tissues that initiate and conduct the rhythmic depolarizations that induce contractions of the myocardial cells. These tissues are the sinus node, AV node, internodal tracts, AV bundle, its major branches and terminal ramifications (Fig 4–5).

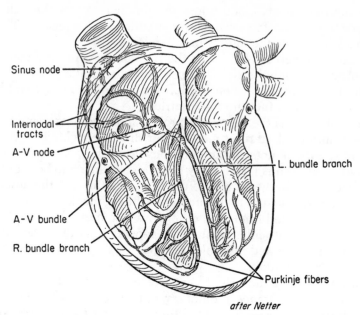

after Netter

Fig 4–5.—The pacemaker and conducting tissues of the heart. (Adapted from an original painting by Frank H. Netter, M.D., for *The Ciba Collection of Medical Illustrations,* Vol. 5. Copyright, Ciba-Geigy Corporation.)

The *sinus node* is a small piece of tissue 1–2 cm long, shaped like a flattened ellipse, located just beneath the epicardium at the junction of the superior vena cava and right atrium. It contains *pacemaker* cells and *transitional* cells in a dense collagen framework.

The *AV node* lies beneath the endocardium of the right atrium just above the insertion of the tricuspid valve. Like the sinus node, it contains a meshwork of pacemaker cells and numerous transitional cells embedded in collagen. The AV node receives the terminations of the *internodal tracts* superiorly. Inferiorly the specialized conducting cells of the node become oriented longitudinally to form a bundle of parallel fibers, the AV bundle.

The *AV bundle* or *bundle of His* passes medially and inferiorly into the interventricular septum. It is approximately 1–2 cm long and divides into a slender *right bundle branch* and a much wider, flattened sheet of fibers, the *left bundle branch*. The left bundle branch passes down the interventricular septum in the direction of the apex. It lies just beneath the endocardium of the left ventricle and soon separates into two major divisions. The anterosuperior division passes toward the anterior papillary muscle and provides branches that activate the anterior and anterolateral portions of the ventricle. The posteroinferior division passes toward the posterior papillary muscle, and its branches activate the posterior and posterolateral regions of the ventricle. The right bundle branch passes down the right side of the interventricular septum and branches diffusely to supply all parts of the right ventricle.

The *internodal tracts* are three tracts of conducting tissue that connect the sinus and AV nodes. Because of their location, they are named the anterior, middle and posterior internodal tracts.

BLOOD SUPPLY TO THE HEART

The heart is supplied by two coronary arteries, which are the first branches of the aorta and arise just above the aortic valve. The main vessels run on the surface of the heart and send numerous penetrating branches into the depths of muscle. The myocardial capillaries are very numerous, with a 1:1 ratio between muscle cells and capillaries. Coronary venous blood is collected by several veins, the largest being the coronary sinus, which runs in the posterior AV groove and opens into the right atrium. A more detailed description is given in the section on the coronary circulation.

NERVES OF THE HEART

The heart receives fibers from both the parasympathetic and sympathetic divisions of the autonomic nervous system (Chapter 3). The parasympathetic fibers are carried in the vagus nerves, and the sympathetic fibers in several nerves composed of axons of cells located in the upper-thoracic and inferior cervical ganglia. Various nerves from both divisions are comingled on the surface of the heart, where they form various plexuses. The parasympathetic fibers mainly supply the atria, the sinus node, the AV node and conducting tissues, whereas the sympathetic fibers supply all parts of the heart. Because of this distribution, the vagus mainly affects heart rate and conduction, whereas the sympathetic nerves influence force of contraction as well as rate and conduction. Many sensory fibers also originate in the heart. Some participate in cardiovascular reflexes. Others, when appropriately stimulated, give rise to "cardiac pain" (see section on the coronary circulation).

MICROSCOPIC ANATOMY OF THE HEART

All cardiac cells contain a nucleus, mitochondria, myofibrils and sarcoplasmic reticulum (SR), but nevertheless several distinct cell types can be discerned. These are the P cells, transitional cells, Purkinje cells and atrial and ventricular muscle cells.

P CELLS.—These are ovoid cells, $5-10$ μ in their longest diameter, found in the sinus and AV nodes. They contain few myofibrils or mitochondria and little glycogen. Their cytoplasm therefore appears relatively empty. The SR and T system are poorly developed. They make infrequent contacts with surrounding cells and there are no intercalated disks. They resemble embryonal cardiac cells that have not acquired the specialized structural attributes for strong contraction. They are *self-exciting* and those in the sinus node are normally responsible for initiation of each heart beat.

TRANSITIONAL CELLS.—These are intermediate in structure between P cells and working myocardial cells. The P cells do not appear to make direct contacts with regular myocardial or Purkinje cells except through the intermediary transitional cells, which therefore have the important function of conducting excitation from the P cells to all other tissues of the heart.

ATRIAL CELLS.—These are oval and unbranched. They are 20–30 μ long and 6–8 μ wide. Many of the cells lack a T system but contain numerous myofibrils and mitochondria. They make unspecialized side to side contacts with adjacent myofibrils and also make end to end contacts by means of intercalated disks.

VENTRICULAR CELLS.—These are long cells, up to 100 μ, with an average diameter of 10–15 μ. They rarely make the side to side contacts typical of atrial cells but have well-developed intercalated disks. They are packed with myofibrils and contain numerous mitochondria. The SR and T system are well developed.

PURKINJE CELLS.—These are broad cells, up to 30 μ in diameter and 60 μ in length. There is no T system, and myofibrils and sarcoplasmic reticulum are sparse. The intercalated disks are complex and extensive.

Transitional cells with characteristics intermediate between those of the main cell types also are found.

The structural differences between the various cells are clearly related to their functional role. The P cells are pacemakers. They initiate excitation and are not structurally adapted for tension development since they have few myofibrils, few mitochondria and no T system. Similarly, the Purkinje cells, which must conduct rapidly, have wide diameters and well-developed intercalated disks with low electric resistance, which enables conduction to be propagated rapidly from fiber to fiber. The ventricular cells, which provide the main pumping force, have prominent myofibrils and mitochondria; the T system and SR, which play an important role in excitation-contraction coupling, are highly developed.

EXCITATION OF THE HEART

The heart exhibits *intrinsic rhythmicity* and beats independently of external influences such as neural or hormonal activity. The beat is initiated by spontaneous generation of action potentials by a group of P cells in the sinus node. These action potentials are conducted across the atria via the internodal tracts with a velocity of 0.5–1 m/sec to the AV node. The atrial muscle is also excited by conduction from one regular "working" cell to another via the side to side contacts. AV nodal conduction is slow, and the impulse is delayed 0.1–0.2 second before being transmitted rapidly at 1–4 m/sec down the AV bundle and Purkinje cells to the regular myocardial cells. The septum is activated first, followed by the apical regions and finally the base. The subendocardial muscle is activated before the subepicardial. The excitation wave initiates contraction of the muscle over which it passes so that the right atrium contracts slightly before the left one. The delay at the AV node is important because it allows time for the atria to empty into the ventricles before they in turn contract. The ventricular contractions are not perfectly synchronous and onset of left ventricular contraction precedes that of the right by 10–30 msec.

An insight into the mechanisms of cardiac excitation can be gained by comparing the electric transmembrane potentials of regular myocardial cells and pacemaker cells.

NONPACEMAKER POTENTIALS

The potentials from a ventricular muscle cell are shown in Figure 4–6. They are due to an exceedingly complex series of changes in ionic conductances, which have not been fully worked out, and it should be recognized that the following description is simplified. Five phases are described.

PHASE 4.—The resting potential. This is stable at about −90 mv (inside negative). It is mainly determined by ionic concentration gradients across the cell membrane and by ionic permeabilities, as expressed by the Goldman equation (Chapter 2). Because potassium conductance (g_K) is high and about 100 times greater than sodium conductance (g_{Na}), the resting potential is not far removed from the equilibrium potential for potassium. As will be explained later, an electrogenic sodium pump also makes a small contribution to the resting potential.

PHASE 0.—The cell can be excited either by a stimulus conducted to it from other cells or by one applied experimentally. When the cell is activated, g_{Na} rapidly increases to several hundred times its resting value. This allows sodium ions to move into the cell down their electrochemical gradient, and the membrane potential changes very rapidly from −90 mv to about +15 mv. This rapid upstroke of the action potential is phase 0. The rate of change of potential is dependent on the level of the resting potential. The more negative the resting potential the greater the rate of rise of phase 0. Furthermore, the velocity of propagation of the action potential increases with the slope of phase 0. The resting membrane potential is therefore an important determinant of conduction. Reduced coronary blood flow, impaired cellular metabolism or elevation of interstitial potassium all reduce membrane potential, slow conduction and may lead to serious disturbances of cardiac rhythm.

PHASES 1 AND 2.—After phase 0, the positive potential declines slightly (phase 1) and then plateaus at around the zero potential level (phase 2). This

Fig 4–6.—Action potentials from a ventricular myocardial cell (A) and from a pacemaker cell (B).

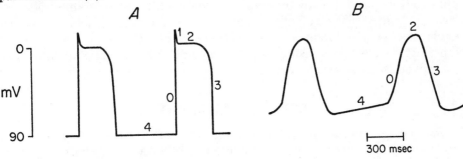

plateau is unique for heart muscle and may be maintained for several hundred milliseconds. During phase 1, g_{Na} decreases rapidly but nevertheless remains at several times its resting level. During phase 2, potassium conductance *falls* in marked contrast to nerve, and calcium conductance slowly increases. These changes together hold the membrane in a depolarized state. The slow calcium current is partly responsible for increasing the intracellular concentration of free calcium ion to levels that initiate and maintain contraction.

PHASE 3.—This is mainly due to a relatively rapid increase in g_K to resting levels. A further reduction in g_{Na} to its resting value also contributes.

During the action potential the cell gains a small amount of sodium and loses potassium. Between action potentials the ionic concentrations are restored by an ion pump associated with the enzyme, Na/K-activated ATPase, in the surface membrane. The activity of the pump is enhanced by increased intracellular sodium or increased extracellular potassium concentrations. The pump is therefore augmented during phase 4. It causes the extrusion of three sodium ions from the cell for every two potassium ions taken up from the extracellular fluid. This causes a net loss of positive charge, i.e., the pump is electrogenic and its activity is responsible for a small part of the resting membrane potential.

PACEMAKER POTENTIALS

Potentials from a pacemaker cell of the sinus node (see Fig 4–6) differ significantly from those just described. The most important difference is that the pacemaker potential is not stable during phase 4 but shows a gradual spontaneous depolarization. When this reaches a critical threshold value of about −55 mv, an action potential is produced. Similar spontaneous phase 4 depolarizations, but at slower rates, occur in the potential pacemaker cells of the AV node and Purkinje system. The spontaneous depolarization is caused mainly by a slowly declining g_K and to a smaller extent by increasing g_{Na}.

VARIATIONS IN ACTION POTENTIALS AMONG DIFFERENT CARDIAC CELLS

Atrial action potentials differ from those of the ventricle in having shorter and less well-maintained plateaus. This can be correlated with the shorter contracton period of the atria. The Purkinje cells have large diameters, and phase 0 is very rapid. They therefore conduct rapidly. The Purkinje cells also have the longest plateau of all cardiac cells. The sinus and AV nodal pacemakers have a low membrane potential, about −55 mv, at the end of phase 4 and therefore have a slowly rising phase 0 and slow cell to cell conduction. The plateau is also poorly sustained.

REFRACTORY PERIOD

While the membrane potential remains more positive than −45 to −50 mv, no stimulus, however great, will generate another action potential. During this period, termed the *absolute refractory period*, the myocardial cell is unre-

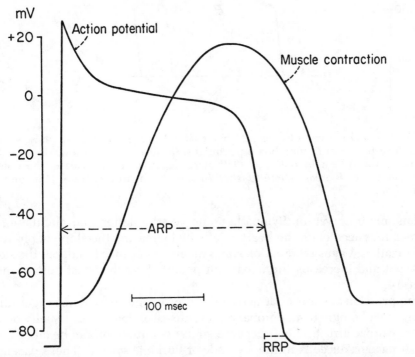

Fig 4–7.—Relationship between the ventricular cell action potential and the subsequent contraction. *ARP* = absolute refractory period; *RRP* = relative refractory period.

sponsive to stimuli no matter how strong. Following this, the cell is relatively refractory, i.e., it responds only to stimuli stronger than those normally required.

RELATION OF ACTION POTENTIAL TO MECHANICAL CONTRACTION.—Contraction begins soon after onset of the action potential and reaches its peak toward the end of the plateau phase (Fig 4–7). Thus the contraction phase of a cardiac muscle twitch occurs entirely within the absolute refractory period. Consequently, the heart, unlike skeletal muscle, cannot be tetanized.

REGULATION OF HEART RATE

The rate of firing of the sinus node can be increased in three ways (Fig 4–8): (1) if the rate of phase 4 depolarization is increased, heart rate will increase because the threshold potential will be reached earlier; (2) if the threshold potential becomes more negative, other factors remaining constant, phase 4 depolarization will trigger an action potential at an earlier time; and (3) if the membrane potential at the onset of phase 4 becomes less negative, it will be closer to the firing threshold and the heart will accelerate. The reverse of these changes causes cardiac slowing. Many factors can influence these basic mechanisms, thereby changing heart rate, but the most important effects occur through neural and humoral activity. The sinus node has an extremely rich

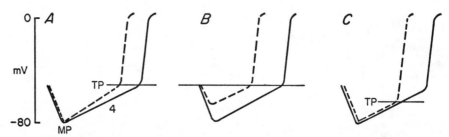

Fig 4–8.—Three ways to increase heart rate: **A,** make the slope of phase 4 steeper; **B,** reduce the level of membrane potential *(MP)* reached at the end of phase 3; and **C,** make the threshold potential *(TP)* more negative. In each case, the potentials associated with increased rate *(dashed line)* are compared to those at a control rate *(solid line).*

autonomic innervation. Sympathetic nerve stimulation and circulating catecholamines accelerate the heart chiefly by increasing the slope of phase 4 depolarization. Vagus stimulation slows the heart because it reduces the slope of phase 4 and increases the maximum potential reached just before onset of phase 4.

Suppression of sinus node activity by physiologic (e.g., reflex vagal stimulation) or pathologic (e.g., coronary artery disease) conditions usually does not cause cardiac arrest. This is because sinus node function can be taken over by other pacemaker cells in the AV node or Purkinje system. These pacemakers have an intrinsically slower rate and do not normally influence the heart rate because they are discharged by arrival of a conducted impulse from the sinus node. When the latter is suppressed or the conduction pathway is blocked, the intrinsic slower activity of these subsidiary pacemakers becomes manifest and the heart beats at a slower rate.

CONDUCTION OF ELECTRIC ACTIVITY FROM THE SINOATRIAL NODE

Conduction occurs from cell to cell through the atria and more rapidly through the internodal tracts that link the sinus and AV nodes. Conduction is slowed in the AV node because of the relatively small diameter of these cells, which also have lower resting potentials and a correspondingly slow rising phase 0. Conduction appears to be decremental in some parts of the node. The velocity of the depolarization wave increases after entering the AV bundle because of the special structural characteristics of the Purkinje cells and because these cells have high membrane potentials at the beginning of phase 0. Excitation is then conveyed by the AV bundle and by its branches in an orderly sequence within the ventricles from apex to base and from endocardium to epicardium.

IMPORTANCE OF PACEMAKER AND CONDUCTING SYSTEM

Numerous clinical disorders demonstrate the importance of the normal conducting sequence. For example, in complete heart block the conducting system is blocked in the vicinity of the AV node and the sinus impulse is not conducted to the ventricles. The atria continue to beat at the rate set by the sinus. On the other hand, the ventricular rate is set by a slower Purkinje pacemaker, which may discharge at a rate as low as 40 per minute. This rate tends to be fixed because Purkinje pacemakers are relatively unaffected by neural or hormonal activity. The slow rate may be inadequate to maintain cardiac output and the patient is liable to have attacks of fainting and may die suddenly. Impaired areas of conduction anywhere in the heart may also set the scene for the phenomenon of re-entry (see section on the ECG), which may induce various disorders of rhythm including fatal ventricular fibrillation.

THE ELECTROCARDIOGRAM

The waves of depolarization and repolarization that sweep over the atria and ventricles with each heart beat produce a continuously changing electric field that extends to the body surface. An electrode placed on the skin therefore undergoes variations of potential related to cardiac activity. These variations constitute the ECG, which is amplified and recorded on an oscilloscope or on chart paper ruled in millimeter squares moving at a speed of 25 mm per

Fig 4–9.—A, electrode connections for standard bipolar limb leads. The *solid circles* show the actual position of the electrodes. The *dashed lines* indicate that the limbs may be considered as cables connecting the electrodes to areas of the trunk (*crosses*). **B**, lines joining any two crosses constitute a *lead axis*. Einthoven suggested that the three standard lead axes could be considered an equilateral triangle. Note the polarity of the electrodes. The left arm is positive when connected to record lead I but negative when connected to record lead III.

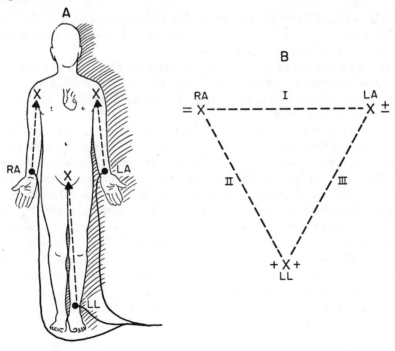

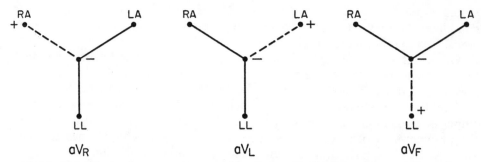

Fig 4–10.—Electrode.connections and lead axes *(dashed lines)* for augmented unipolar limb leads aV_R, aV_L and aV_F (see text).

second. The appearance of the ECG will depend on the position of the surface electrodes and their connections. Conventionally, 12 *leads* are recorded: three bipolar limb leads, three unipolar limb leads and six unipolar chest leads. Their names and electrode connections are as follows:

BIPOLAR OR STANDARD LIMB LEADS (Fig 4–9).—These record differences of potential between two electrodes, each of which undergoes variation in potential.

Lead I.—This records the difference in potential between electrodes on left and right arms. The connections to the recorder are such that when the left arm potential is positive with respect to the right, an upward deflection is obtained.

Lead II.—The electrodes record differences in potential between right arm and left leg. When the arm electrode is positive with respect to the leg, an upward deflection occurs.

Lead III.—The differences in potential between electrodes on the left arm and left leg are recorded. When the arm electrode is positive with respect to the leg, an upward deflection is obtained.

UNIPOLAR LEADS.—These record differences in potential between an electrode that is undergoing variations of potential and an indifferent electrode that has a relatively small fixed potential. There are three unipolar limb leads and six unipolar chest leads.

UNIPOLAR LIMB LEADS.—Originally the point of fixed potential was obtained by connecting all three limb electrodes (left arm, right arm, left leg). This common connection formed a reference point with essentially zero potential. The unipolar left arm lead (designated V_L) was then obtained by connecting the left arm electrode to one pole of the recorder galvanometer and the other pole to the common reference point. Unipolar right arm (V_R) and left leg (V_F) leads were similarly obtained. It was subsequently found that the amplitude of the recorded potential was augmented if the connection to the limb whose potential was being recorded was omitted from the combined electrode. Limb leads of this type were therefore termed *augmented limb leads* and designated by the prefix *a*. Thus, leads aV_R, aV_L, aV_F, respectively, record the potential between an electrode on the right arm, left arm or left leg

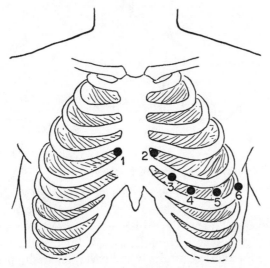

Fig 4–11. – Electrode positions for unipolar chest leads V_1–V_6.

and a point obtained by connecting the other limbs, as described (Fig 4–10).

UNIPOLAR PRECORDIAL (CHEST) LEADS. – Six leads, designated V_{1-6}, are usually obtained. They record the potential difference between an exploring chest electrode and an indifferent electrode obtained by joining the electrodes from left arm, right arm and left leg. Chest electrode positions (Fig 4–11) are as follows:

V_1 – 4th intercostal space immediately to right of sternum
V_2 – 4th intercostal space immediately to left of sternum
V_3 – midway between V_2 and V_4
V_4 – 5th intercostal space in left midclavicular line
V_5 – 5th intercostal space in left anterior axillary line in same horizontal plane as V_4
V_6 – 5th intercostal space in left midaxillary line in same horizontal plane as V_4

NORMAL ECG

Although the appearance of the ECG varies among the different leads, three components usually are seen. These correspond to atrial depolarization (P wave), ventricular depolarization (QRS complex) and ventricular repolarization (T wave). By definition, any positive (upward) deflection of the QRS complex is an R wave. A negative deflection that precedes the R wave is a Q wave, and a negative deflection that follows R is an S wave. A normal ECG pattern is shown in Figure 4–12.

It is important in clinical practice to memorize the normal duration of certain intervals. The *P–R interval* is the time taken for the impulse to travel from the sinus node to ventricular muscle. Prolongation of this interval indicates an abnormality of the conducting system. The *Q–T interval* indicates

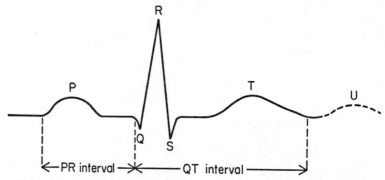

Fig 4–12.—The ECG, its waves, complexes and intervals (see text).

the duration of ventricular systole; it becomes shorter with increasing rate and also is affected by age and sex. The following values for the various waves and intervals are those for average adults with heart rates in the physiologic range.

P wave	0.07–0.14 second
P–R interval	0.10–0.21 second
QRS complex	0.06–0.10 second
Q–T interval	0.28–0.44 second

It should be noted that depolarization of the sinus and AV nodes is not recorded on a conventional ECG because the small mass of the tissues is inadequate to produce surface field changes large enough to generate recordable potentials. Atrial repolarization is buried in the QRS complex. A small U wave of obscure origin may follow the T wave in some leads.

Fig 4–13.—*ECG* and idealized His bundle electrogram *(HBE).*

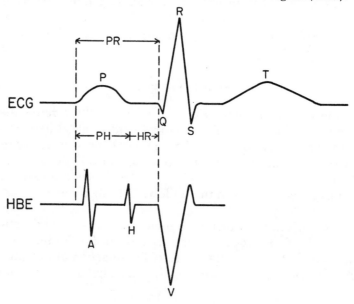

Conduction through the AV node and AV bundle can now be studied relatively easily by the recently introduced technique of *His bundle recording*. A special electrode catheter is introduced into the right side of the heart and, when this is suitably positioned, a His bundle electrogram (HBE) is obtained. Three main groups of deflections, A, H and V, are obtained corresponding to atrial, His bundle and ventricular activation. A comparison of the ECG and HBE often helps clinically in assessment of impaired conduction. Figure 4–13 shows that the P–R interval can be subdivided into three parts, the PA interval, which measures intra-atrial conduction time, the AH interval, which approximates conduction time through the AV node (normally 0.119 ± 0.038 second), and the HV interval, which measures His-Purkinje conduction time (normally 0.043 ± 0.012 second). Prolongation of AH with normal HV indicates conduction block localized to the AV node or above whereas prolongation of HV with normal AH indicates impaired conduction between the AV bundle and the ventricular muscle.

GENESIS OF THE ECG

The ECG is a record of the variation with time of surface potentials occurring in the continuously changing fields generated by the electric activity of the heart. Rigorous analysis of these fields is exceedingly complex. However, by greatly simplifying the situation, we can gain some insight into the factors that determine the pattern of the ECG in the various leads.

Let us consider two equal charges, one positive and one negative, located in a simple homogeneous volume conductor. These charges constitute a dipole. This dipole has magnitude, direction and sense, i.e., it is a vector quantity, which can be represented by an arrow. The length and direction of the arrow indicate the magnitude and orientation of the dipole. Its "sense" will also be indicated if we define the head of the arrow to indicate positivity and its base, negativity. The potentials recorded by any lead will depend on the magnitude of the "projection" of the dipole on the *lead axis*, i.e., on the line joining the two electrodes of the lead. The projection is determined by drawing perpendiculars from the vector arrow representing the dipole to the axis (Fig 4–14). When the vector is parallel to the lead axis, a maximum potential is obtained and when it is perpendicular to the lead axis, the surface potential is zero.

The electric activity of the heart can be represented as a continuously varying dipole. This is best illustrated by considering the activation of the ventri-

Fig 4–14.—Illustrating the "projection" of a cardiac vector onto a lead axis (*horizontal dashed line*). Perpendiculars are drawn from the vector *arrow* to the axis. The projection is greatest when the vector is parallel to the axis, and zero when the vector is at right angles to the axis.

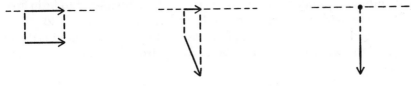

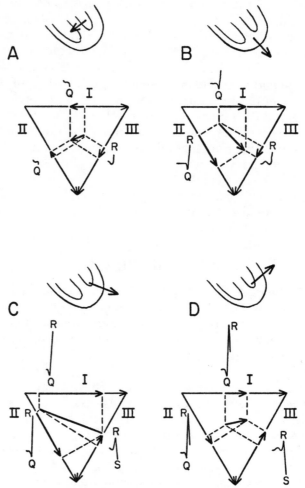

Fig 4–15.—Genesis of the QRS complex. **A,** the septum depolarizes first. Projections of the vector representing this depolarization cause Q waves in leads *I* and *II* and an *R* in lead *III* (*arrowheads* on vectors and lead axes indicate positivity). **B,** the apex is now being depolarized, causing *R* waves to appear in *I* and *II*, while the *R* in *III* enlarges. **C,** the lateral wall of the left ventricle is depolarized. Projections of this vector now produce an *S* in *III*, whereas *R* increases in *I* and *II*. **D,** the basal lateral wall is now depolarized. The *R* wave declines in *I* and *II* but the vector orientation is such that *S* deepens in *III*. When the entire ventricle is depolarized, the vector disappears and all traces return to the isoelectric line (not shown). (Adapted from an original painting by Frank H. Netter, M.D., for *The Ciba Collection of Medical Illustrations*, Vol. 5. Copyright, Ciba-Geigy Corporation.)

cles and potentials, which would be picked up by standard bipolar leads I, II, III. As Einthoven pointed out, these leads can be considered to form an equilateral triangle with vertices at the left and right shoulders and the lower trunk (see Fig 4–9). The first part of the ventricles to be activated is the upper septum, and the wave of depolarization sweeps from left to right. At this time, the electric state of the heart can be represented by the dipole shown in Figure 4–15, A. The projections of this dipole on the various lead axes indicate that a small Q wave will be inscribed in leads I and II and a small R wave in lead III. A few moments later, activation will involve the right ventricle and apical parts of the left ventricle. At this stage the activity can be represented as a larger dipole, pointing downward and leftward. The projections of the vector now indicate that leads II and III will show tall R waves and lead III a small R wave (Fig 4–15, B). A little later the depolarization wave spreads over the free wall of the left ventricle. The R in II and III becomes larger, but the orientation of the vector is such that an S appears in III (Fig 4–15, C). Finally, the base is activated and this is represented by a smaller vector pointing leftward and slightly up. Projections to the lead axes show smaller positive potentials in leads I and II and a deeper S in lead III (Fig 4–15, D). Finally, when the ventricles are completely depolarized, all the ventricular muscle will be at the same potential, i.e., the dipole will vanish and the ECG will be isoelectric in all leads, i.e., the S–T segment will be horizontal.

Activation of the ventricles can thus be represented as a sequence of dipoles of continuously changing amplitude and direction. Some of these dipoles are indicated in Figure 4–16, A, and each is shown originating from a single point. The numerals indicate the order in which they arise. Each arrow is an *instantaneous* vector. When the tips of the arrows are joined, a "*vector loop*" is formed. If the instantaneous vectors are averaged, a *mean QRS vector* for the whole ventricular activation cycle is derived. The mean QRS vector shown in Figure 4–16, B, tells us that overall activation of the ventricles can be represented by a vector pointing downward and to the left. Its direction in the frontal plane can be more precisely determined using a triaxial reference system derived from the Einthoven limb leads. If the axes of these leads are moved appropriately so that they intersect, the system shown in Figure 4–16, C, is formed. The normal mean frontal plane vector QRS usually lies between −30 and +110 degrees. Left axis deviation is said to be present when the axis

Fig 4–16.—**A,** a few of the *instantaneous* frontal plane cardiac vectors (*solid arrows*) that occur in sequence during ventricular activation. The mean cardiac vector (*dashed arrow*) and the QRS vector loop also are shown. **B,** the mean QRS vector in relation to Einthoven's triangle. **C,** the mean QRS vector in relation to a triaxial reference system formed when the sides of Einthoven's triangle are moved so as to intersect. The axis of the QRS vector shown is approximately +50 degrees.

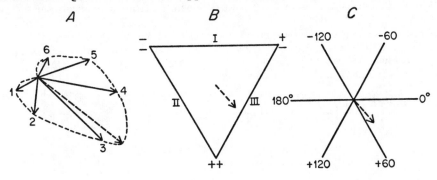

is more negative than −30 degrees and right axis deviation exists when the axis is more positive than +110 degrees.

So far we have considered the vectors only in relation to leads I, II, III, which are all in the frontal plane, i.e., a plane parallel to that containing the anterior surface of the body. In order to determine whether the mean vector is oriented anteriorly or posteriorly, we can utilize the unipolar chest leads, which can be considered as lying in a horizontal plane at right angles to the frontal plane.

Normally the vector points somewhat posteriorly so that predominantly negative (S) waves are seen in V_1 and V_2, and predominantly positive waves (R) are seen in V_5 and V_6.

The directions of the mean P wave and T wave vectors can be determined in the same way as the QRS vector. Repolarization of the ventricles, for reasons not yet clearly defined, first begins in the epicardium and later spreads to the endocardium. Because this is the reverse of the direction of spread of the preceding depolarization, the mean vector of the T wave has the same general direction as the QRS complex.

IMPORTANCE OF THE ECG

The ECG is of inestimable value as a diagnostic aid. It is used to reveal disorders of rhythm, hypertrophy of one or more chambers, and injury or death of cardiac muscle. Because the electric activity of the heart is susceptible to changes in the extracellular ionic environment, the ECG may also reveal abnormalities of electrolyte regulation. Examples of these conditions will now be briefly discussed.

DISORDERS OF RHYTHM

These arise by two principal mechanisms—abnormal pacemaker activity and abnormal conduction.

ABNORMAL PACEMAKER ACTIVITY

This may arise if the sinus node is depressed or if the automaticity of other pacemaker cells is enhanced. When the sinus node is depressed, e.g., by excessive vagal activity, lower pacemakers "escape" from sinus domination, and

Fig 4–17.—Escape beat. The sinus node failed to discharge after the third beat. The resulting pause was terminated by discharge of nodal pacemaker (4th complex). (From Conway, N.: A Pocket Atlas of Arrhythmias [Chicago: Year Book Medical Publishers, Inc., 1974]. Used by permission.)

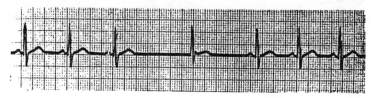

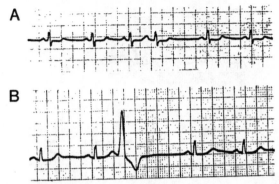

Fig 4–18.—A, the fourth beat is a premature atrial beat caused by the increased automaticity of an atrial pacemaker. Note the relatively normal QRS. **B,** the third beat is a premature ventricular beat. The QRS is wide and the P wave cannot be seen. (From Conway, N.: *A Pocket Atlas of Arrhythmias* [Chicago: Year Book Medical Publishers, Inc., 1974]. Used by permission.)

one or more *escape beats* occur (Fig 4–17). Increased automaticity of nonsinus pacemakers may occur as the result of a variety of stimuli including increased sympathetic nerve discharge, increased circulating catecholamines, use of various drugs and myocardial ischemia. Beats that originate from a pacemaker other than the sinus node are termed *ectopic beats* and can be further described, according to their origin, as atrial, junctional (arising from the AV node or bundle) or ventricular. Atrial ectopic beats are conducted abnormally through the atria, but conduction through the AV bundle and ventricles usually is normal. The P wave is therefore abnormal, whereas the QRS is unaffected (Fig 4–18, A). Ventricular ectopic beats are conducted abnormally through the ventricles, producing a wide, bizarre QRS (Fig 4–18, B). Single premature ventricular ectopic beats are followed by an abnormally long pause because the next sinus beat falls within the refractory period of the ectopic beat; hence the ventricles cannot be activated until the following sinus impulse is conducted. The pause is often termed the "compensatory" pause.

Although abnormal pacemaker activity is common in a variety of diseases, it is not infrequent in perfectly healthy hearts and there can be few of us whose hearts have not skipped a beat at one time or another.

ABNORMAL CONDUCTION

Partial block of the conducting system may occur in several types of heart disease or as the result of use of drugs such as digitalis. If the slowing of conduction is in the vicinity of the AV node or His bundle and its branches, the ECG shows a prolonged P–R interval (Fig 4–19, A) or a "dropped beat," i.e., a failure of an impulse to be conducted from the atrium to the ventricle (Fig 4–19, B). In complete heart block no impulses are conducted to the ventricles, which are then controlled by pacemakers in the His-Purkinje system that have low intrinsic rates. Under these circumstances, the atria and ventricles beat at different rates (Fig 4–19, C).

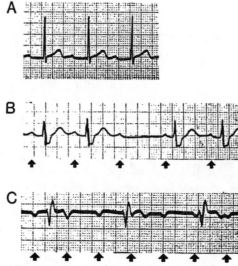

Fig 4–19. – **A,** first degree heart block. The P–R interval is prolonged and is approximately 0.3 second (upper limit of normal 0.2 second). **B,** second degree heart block. The third P wave is not conducted to the ventricles and so a QRS is "dropped." **C,** complete heart block. None of the P waves is conducted to the ventricles, which are activated by an ectopic ventricular focus discharging at 45 per minute. The atrial rate is 110 per minute, and the contractions of the atria and ventricles are completely dissociated. (From Conway, N.: *A Pocket Atlas of Arrhythmias* [Chicago: Year Book Medical Publishers, Inc., 1974]. Used by permission.)

RE-ENTRY

This is a common cause of rhythm disorders and occurs when the depolarization wave traverses a "circular" path with a conduction velocity sufficiently low that cells that were activated by the initial passage of the impulse have passed through their refractory period and can be reactivated by the returning or "re-entering" impulse. For re-entry to occur, *unidirectional conduction* block must be present somewhere in the pathway. This can be readily understood by reference to Figure 4–20.

Fig 4–20. – Re-entry (see text).

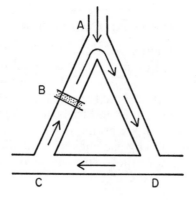

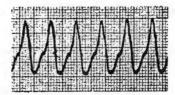

Fig 4–21.—Ventricular tachycardia. This serious rhythm disturbance may be caused by the rapid discharge of a ventricular ectopic focus or by re-entry. Note the broad QRS complexes. The rate is about 170 per minute. (From Conway, N.: *A Pocket Atlas of Arrhythmias* [Chicago: Year Book Medical Publishers, Inc., 1974]. Used by permission.)

Let us suppose we have a bundle of Purkinje fibers that divides at A into two branches connected by ventricular muscle CD. Let us assume that an abnormality exists at B that blocks conduction in the direction A→C but permits conduction from C→A (such abnormalities have been demonstrated experimentally). The excitation wave will then be conducted from A→D→C and back to A. If the conduction through the abnormal area B is sufficiently slow, the impulse will emerge from B after the refractory period of A has ended. A will then be re-excited and a *premature ventricular contraction* will result because of the widespread conduction of the re-entrant impulse. When conditions (length, conduction velocity) in the loop ADCA are appropriate, an impulse may repeatedly traverse the loop, continuously exciting nearby fibers and so cause an abnormally rapid regular ventricular response, termed *ventricular tachycardia* (Fig 4–21). Similar re-entry mechanisms in the atria may produce atrial premature beats or atrial tachycardia.

Ventricular Fibrillation

In this disorder the ventricular muscle is irregularly and frequently excited in such a way that coordinated contractions are lost. The ventricles become a writhing mass of muscle, which has aptly been compared to a bag of worms. Expulsion of blood fails to occur and circulation comes to a halt. This disturbance of rhythm is most commonly induced by coronary artery disease. As a result of inadequate oxygen delivery, some myocardial cells are inexcitable and others are partly depolarized and conduct slowly. Still others are normal and conduct normally. Other cells may discharge at very rapid rates due either to re-entry or to increased automaticity. The combination of these abnormalities sets the scene for development of fibrillation, since impulses may be conducted along constantly changing pathways with variable conduction

Fig 4–22.—Ventricular fibrillation (see text). (From Conway, N.: *A Pocket Atlas of Arrhythmias* [Chicago: Year Book Medical Publishers, Inc., 1974]. Used by permission.)

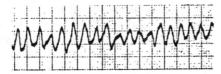

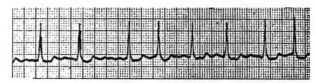

Fig 4–23.–Atrial fibrillation. (From Conway, N.: *A Pocket Atlas of Arrhythmias* [Chicago: Year Book Medical Publishers, Inc., 1974]. Used by permission.)

rates. The ECG reflects this chaotic state (Fig 4–22). The only effective treatment is to use a defibrillator to administer a D–C shock to the myocardium through the chest wall. This depolarizes all the ventricular fibers and provides an opportunity for the S–A node to re-establish itself as the pacemaker.

ATRIAL FIBRILLATION

This occurs from electrophysiologic causes similar to those that produce ventricular fibrillation. The AV node is then irregularly bombarded with impulses that are irregularly conducted to the ventricle, depending on whether or not they arrive when the AV nodal cells are refractory. Commonly the ventricular response is rapid and irregular and the efficiency of the pump is impaired. Indeed, the onset of atrial fibrillation may precipitate failure in abnormal hearts, e.g., in mitral stenosis. The ECG shows replacement of the P waves by irregular fibrillation waves (Fig 4–23). The ventricular complexes, though irregularly spaced, need not be grossly abnormal, although they usually reflect the underlying cardiac abnormality responsible for the fibrillation, e.g., mitral valve disease, ischemic heart disease, hypertension.

DISORDERS OF MUSCLE

VENTRICULAR HYPERTROPHY

When the mass of muscle in one of the ventricles is increased, the orientation of the vector loop is altered. In left ventricular hypertrophy the mean cardiac vector is rotated upward, to the left and backward. Left axis deviation is present and tall R waves are seen in leads V_5 and V_6 and deep S waves in V_1 and V_2. When the right ventricle hypertrophies, the mean cardiac vector is rotated to the right and anteriorly. This produces tall R waves in V_1 and V_2 and deep S waves in V_5, V_6 and lead I.

MYOCARDIAL ISCHEMIA

Ischemia means inadequate blood flow. When this occurs in heart muscle, a crushing pain, *angina pectoris,* is often felt in the chest. At this time one or more ECG leads may show abnormal S–T segments, probably caused by differences in the diastolic membrane potentials of injured and healthy myo-

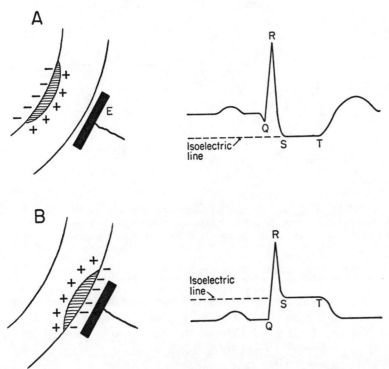

Fig 4–24.–**A,** subendocardial injury. During diastole the electrode *E* is in a zone of positivity and the ECG trace is therefore above the isoelectric line. During systole the whole ventricle is depolarized and the ECG returns to the isoelectric line. This is read as S–T segment depression. **B,** when the injury is subepicardial, *E* is in a zone of negativity during diastole and the ECG is below the isoelectric line but returns to it when the entire ventricle is depolarized. This is read as S–T segment elevation.

cardium (Fig 4–24). Mild ischemia may not produce changes in resting membrane potential and in these cases the S–T segment is normal but, since repolarization is almost always impaired, T wave abnormalities are often seen.

MYOCARDIAL INFARCTION

This means death of a portion of myocardium and it changes the orientation of the vector loop. The direction and magnitude of the shift depend on the size and location of the infarct. As an example we can consider the effects of a large inferior infarct on the standard limb leads. Because of the loss of inferior forces, the mean cardiac vector now points upward instead of downward, causing negative deflections (Q waves) to appear in leads II and III, which previously showed R waves (Fig 4–25). In addition, S–T segment shifts and T wave changes are usually seen because areas of cell injury and ischemia occur around the infarcted region.

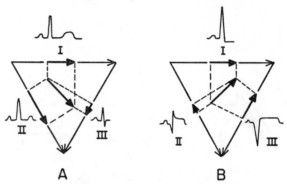

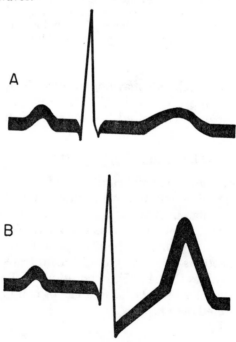

Fig 4–25.—Inferior infarction. **A,** normal heart. Vector forces are normal. **B,** after inferior infarction. The loss of inferior forces causes the initial vector to be more superiorly directed. Hence, Q waves appear, as shown in leads *II* and *III.* Because of areas of injury surrounding the dead zone, S–T segment elevation also is seen in *II* and *III.*

ELECTROLYTE ABNORMALITIES

CALCIUM.—An increase in external calcium ion concentration shortens the plateau phase of the myocardial action potential and consequently reduces the Q–T interval of the ECG. Conversely, reduction in external calcium ion concentration prolongs the Q–T interval.

Fig 4–26.—Hyperkalemia. **A,** normal; serum K^+ = 4.5 mEq/L. **B,** hyperkalemia; serum K^+ = 7.4 mEq/L. Note prolonged P–R interval, S–T segment depression, and tall, "tented" T waves.

POTASSIUM.—An increased external potassium ion concentration depolarizes cardiac muscle cells, reduces the slope of phase 0 and slows conduction. The sinus node is more resistant to these effects than the other pacemaker and conducting cells. Therefore, hyperkalemia will prolong the P–R interval (prolonged conduction between sinus and ventricles) and widen the QRS complex (slow intraventricular conduction) without necessarily altering the heart rate. Tall, narrow, "tented" T waves also occur because hyperkalemia accelerates repolarization (Fig 4–26).

A reduction in extracellular potassium (hypokalemia) reduces T wave amplitude and, for unknown reasons, enhances the U wave.

THE CARDIAC CYCLE

The recurring depolarization and repolarization of the myocardium induce rhythmic contractions and relaxations that cause cyclic changes in the pressures and volumes of the heart chambers. These are associated with movements of the AV and semilunar valves, which prevent backflow. A thorough knowledge of these events is indispensable for understanding the pumping action of the heart and for diagnosis and management of many types of heart disease. Figures 4–27 and 4–28 should be carefully studied in conjunction with the following description.

Fig 4–27.—Timing of the contraction and relaxation of the four cardiac chambers in a heart beating at 60 per minute. *Shaded areas* indicate periods of contraction. The lines *M, T, A* and *P* mark the closure of the mitral, tricuspid, aortic and pulmonary valves.

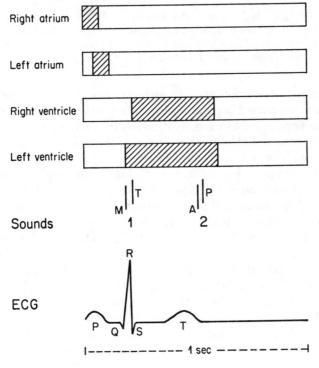

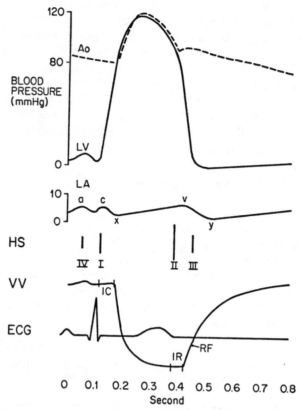

Fig 4-28.—Pressure and volume changes in the left side of the heart. *Ao, LV* and *LA* are pressures in the aorta, left ventricle and left atrium. *I, II, III,* and *IV* are the respective heart sounds *(HS). VV* = left ventricular volume; *IC* and *IR* = isovolumic contraction and relaxation periods, respectively; *RF* = rapid filling phase.

Atrial depolarization is followed by atrial contraction, which causes a rise in atrial pressure, increased separation of the AV valve leaflets and ejection of blood into the ventricles. Since activation begins on the right, the right atrium contracts before the left. The further spread of depolarization induces ventricular contraction. The left ventricle begins to contract before the right and so left ventricular pressure begins to rise slightly before right ventricular pressure rises and the mitral valve closes before the tricuspid. The rising pressures soon lead to opening of the semilunar valves and, because pulmonary artery pressure is lower than aortic, the pulmonic valve opens first and right ventricular ejection begins before the left. During the period between closing of the AV valves and opening of the semilunar valves, ventricular volume is constant and this period is therefore termed *isovolumic contraction* (see Fig 4–28). When repolarization begins, the ventricles relax, causing ventricular cavity pressures to fall below those in the aorta and the pulmonary artery, and the semilunar valves close, aortic before pulmonic. The ventricles then relax isovolumically until their pressures fall below those in the atria.

The AV valves then open, the tricuspid before the mitral, and the ventricles fill, rapidly at first and then more slowly until the next atrial contraction.

PRESSURES WITHIN THE HEART

The atrial pressure curve shows several fluctuations during each cardiac cycle. These are conventionally described as the a, c and v waves and the x and y descents (see Fig 4–28). The a wave is produced by atrial contraction. The c wave is the slight increase in atrial pressure caused by ventricular contraction, which makes the AV valve leaflets bulge into the atria. The x descent is caused by the pulling down of the AV valves by the contracting ventricle. The v wave is due to atrial filling and reaches a peak in late ventricular systole while the AV valves are closed. The y descent is the fall in atrial pressure following the opening of the AV valves.

The pressure changes within the right atrium are communicated to the great veins and, when a person is supine with his back at a slight angle to the horizontal, the a and v waves and the x and y descents can be seen in the jugular veins when the neck is carefully examined. When venous pressure is raised, as in heart failure, the jugular veins become more prominent and the pulsations can be observed with the trunk at a 45 degree angle or even when the subject is sitting with the trunk at a 90 degree angle.

TABLE 4–1.—PRESSURES IN
HEART AND GREAT VESSELS OF
NORMAL RECUMBENT ADULTS
(mm Hg)

SITE	MEAN	RANGE
Left atrium		
Maximum	13	6–20
Minimum	3	−2 to +9
Mean	7	4–12
Left ventricle		
Systolic	130	90–140
End-diastolic	7	4–12
Right atrium		
Maximum	7	2–14
Minimum	2	−2 to +6
Mean	4	−1 to +8
Right ventricle		
Systolic	24	15–28
End-diastolic	4	0–8
Pulmonary artery		
Systolic	24	15–28
Diastolic	10	5–16
Mean	16	10–22
Aorta		
Systolic	130	90–140
Diastolic	70	60–90
Mean	85	70–105
Venae cavae		
Maximum	7	2–14
Minimum	5	0–8
Mean	6	1–10

Atrial contraction causes the AV valve leaflets to move apart and blood is ejected into the ventricles. Since the atria and ventricles now communicate freely, the a wave can be seen in the ventricular as well as atrial pressure record. Ventricular contraction then causes ventricular pressure to rise steeply at first and then more slowly after the semilunar valves open. During ventricular relaxation, pressure falls steeply until the AV valves open. Thereafter, the atrial and ventricular cavities again form part of a common chamber with identical pressures until onset of the next ventricular contraction.

The normal pressures within the heart and great vessels are summarized in Table 4–1.

VENTRICULAR FILLING

Diastolic filling of the ventricles is mainly the passive consequence of the pressure gradient between the veins and ventricular cavities. Two phases are described. The rapid filling phase follows the opening of the AV valves and lasts for one quarter to one third of the total diastolic period. The slow filling phase occupies the remainder of diastole. Because the rapid filling phase is short, the heart rate can increase markedly without significantly reducing filling. Atrial contraction contributes relatively little to ventricular filling at normal heart rates, and the heart can function quite satisfactorily for many years even when effective atrial contraction is lost (as in atrial fibrillation) provided the ventricular rate is not unduly rapid. The contribution of atrial contraction to ventricular filling increases at faster heart rates.

VENTRICULAR EJECTION

When the semilunar valves open, blood is ejected into the aorta and pulmonary arteries. The maximum ejection rate is achieved very early in the ejection phase. Thereafter the ejection rate declines, slowly at first and then more rapidly until it is halted by semilunar valve closure. The ventricles eject only a portion of their contained blood with each beat. The *ejection fraction (EF)* is that fraction of the end-diastolic volume *(EDV)* that is ejected during systole, i.e.,

$$EF = \frac{SV \text{ (stroke volume)}}{EDF}$$

The *EF* is normally greater than 0.5. The volume remaining within the ventricle may be looked upon as a reserve that can be utilized when increased cardiac output is required.

FUNCTION OF THE VALVES

The AV valves are composed of thin, pliable membranous leaflets, which hang down into the ventricular cavities during diastole and cause no impediment to flow through the valve orifices. After atrial systole, the mitral valve leaflets begin to close and the leaflets come together just before onset of ventricular systole. The mechanism producing this early closure is obscure but may be associated with ventricular eddy currents. When the left ventricle contracts, the sealing process is completed and the closed valve is pushed into the atrium (causing the c wave in the atrial pressure pulse) until it is brought to a sudden halt by the chordae tendineae. Tricuspid valve motion is similar to that of the mitral.

The aortic and pulmonic valves each consist of three semilunar cusps that resemble pockets projecting into the arterial lumen. During diastole the pockets fill with blood and their free edges become closely approximated, completely preventing regurgitation of blood into the ventricles. During systole the cusps are flung toward the arterial wall, leaving a wide opening for ejection of blood from the ventricles.

ABNORMALITIES OF VALVE FUNCTION

The heart valves may function inadequately as the result of congenital abnormality or acquired disease. Valve leaflets may become fused, thickened and even calcified or ossified. The valve may then be unable to open fully and its orifice may be greatly narrowed (stenosed). In some cases the major disturbance is inability of the valve to close properly, and backflow (regurgitation) occurs. Both stenosis and regurgitation may be present together. When a valve is stenosed, the proximal chamber must generate a higher pressure to force the usual volume of blood through the valve at each beat, i.e., the proximal chamber experiences a *pressure overload.* On the other hand, regurgitant valves produce a *volume* overload of the chamber receiving the backflow. For example, in aortic regurgitation, the left ventricle receives its normal inflow through the mitral valve during diastole plus an additional volume of blood through the leaking aortic valve. Pressure and volume overloads both eventually lead to heart failure.

HEART SOUNDS

When the stethoscope is placed on the chest wall over the heart, two sounds are normally heard during each cardiac cycle. The *first heart sound* has a low pitch, is more prolonged than the second and begins immediately after the R wave of the ECG. It is probably mainly due to the sudden tensing of the AV valve leaflets when their movement toward the atria, caused by rising ventricular pressure, is halted by the chordae tendineae. Because the AV valves close asynchronously, the first sound may be audibly split, the first component representing mitral valve closure and the second tricuspid closure. The *second heart sound* is shorter and of higher pitch than the first and occurs at the end of the T wave of the ECG. It is due to closure of the semilunar valves and usually is split, the first component being due to aortic, the second to pulmonic valve closure. The split widens during inspiration because the increased inflow of blood to the right ventricle (see section on pulmonary circulation) prolongs right ventricular systole.

In young persons a low-pitched *third heart sound* often can be heard shortly after the second sound. It is probably caused by vibrations of the ventricular walls during the period of rapid ventricular filling that follows the opening of the AV valves. A fourth heart sound due to ventricular vibrations set up when the atria eject their blood into the ventricles is not heard in normal hearts but occurs during ventricular overload or when ventricular compliance is reduced (see section on clinical investigation of the heart).

HEART MURMURS

These are caused by vibrations set up within the heart and great vessels by turbulent blood flow. Turbulence develops when the velocity of flow exceeds a certain critical value. In vitro studies by Reynolds almost a century ago showed that the critical velocity (V_c) was dependent upon the viscosity (η) and density (ρ) of the fluid and on the tube radius (r) and that these variables were related by the formula

$$V_c = \frac{R\eta}{\rho r}$$

where R is a dimensionless number, the Reynolds number. This number var-

Fig 4–29.—Phonocardiograms. The *normal phono* shows the vibrations associated with the first, second, third and fourth heart sounds. *AS* shows the diamond-shaped ejection systolic murmur *(SM)* of aortic stenosis. *AI* shows decrescendo diastolic murmur *(DM)* of aortic insufficiency. (Adapted from an original painting by Frank H. Netter, M.D., for *The Ciba Collection of Medical Illustrations*, Vol. 5. Copyright, Ciba-Geigy Corporation.)

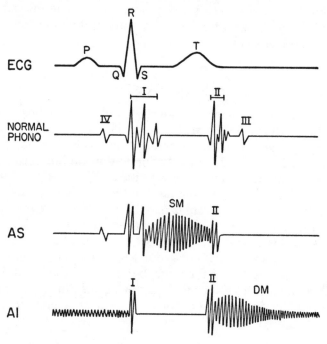

ies with the characteristics of the tube. Turbulence develops in a straight tube when R exceeds 1,000. In curved tubes the number is considerably smaller. Turbulent flow does not normally occur in the circulation except at the origin of the aortic arch and main pulmonary artery where high blood velocity, wide vessel diameter and curvature of the vessels enable the critical Reynolds number to be exceeded. Because of the turbulence, a soft murmur can often be heard over the aortic and pulmonary auscultation areas during ejection. This innocent murmur may be accentuated when cardiac output is high, e.g., in anxiety, fever or anemia. Very high blood velocities may be reached when blood flows through stenosed cardiac valves or arteries. The resulting turbulence produces a murmur, the location, quality and timing of which are often helpful in clinical diagnosis. For example, a harsh crescendo-decrescendo murmur confined to the period of left ventricular ejection and loudest over the aortic area suggests the possibility of aortic valve stenosis. A blowing diastolic murmur in the same area suggests aortic valve regurgitation. A continuous murmur suggests an abnormal communication between aorta and pulmonary artery, e.g., patent ductus arteriosus, with turbulent flow through the abnormal channel during both systole and diastole.

The heart sounds and murmurs, when present, can be characterized most effectively by *phonocardiography*. High fidelity microphones are positioned over various precordial areas and records are made on chart paper moving at high speed. The ECG is usually recorded simultaneously on the same chart for timing purposes. Examples of normal and abnormal phonocardiograms are shown diagramatically in Figure 4–29.

REGULATION OF CARDIAC FUNCTION

The only function of the heart is to pump blood. The volume ejected by each ventricle per minute is termed the *cardiac output* and varies with body size. The cardiac output per square meter of body surface is called the *cardiac index* and is approximately 3 L/sq m in males and 7–10% lower than this in females. The average basal cardiac output is about 5 L per minute in adults. It can be increased either by increasing heart rate or by increasing the volume ejected per beat (stroke volume) or by both mechanisms.

CONTROL OF RATE

Increasing the heart rate is the most important means of increasing cardiac output. Although it can be demonstrated experimentally that the heart has an intrinsic rhythm independent of external influences, the heart rate under physiologic conditions is regulated by the autonomic nervous system. Vagal control is dominant at rest and the sinus rate is kept relatively slow by the continuous release of acetylcholine from the abundant vagus nerve endings in the sinus node. During exercise, vagal activity is suppressed and sympathetic discharge is increased. This reduces acetylcholine concentration and increases norepinephrine concentration around the pacemaker cells. The heart rate is also influenced by changes in the blood levels of the circulating hormones, epinephrine and norepinephrine. The mechanisms by which these agents produce their effects on the pacemaker cells were described earlier in the section on anatomy and electrophysiology.

Other factors that increase pacemaker discharge rate include a rise in body temperature and stretch of sinus node tissue, but these factors are normally of subsidiary importance.

INTRINSIC HEART RATE

The influence of the autonomic nerves on the heart can be removed by blocking the receptors of the P cells. Atropine blocks cholinergic receptors and produces the equivalent of parasympathetic denervation. Propranolol blocks beta adrenergic receptors and produces a pharmacologic cardiac sympathectomy while at the same time blocking the cardiac effects of circulating epinephrine and norepinephrine. The heart rate after the administration of both blocking agents is termed the *intrinsic heart rate.* It is about 100 beats per minute in normal young adults and declines to about 80 beats per minute in the seventh and eighth decades. The intrinsic heart rate is of interest in patients with heart disease since it seems to be correlated with the intrinsic "contractility" of the myocardium. Slower intrinsic rates are often associated with greater impairment of myocardial function. The reason for this interesting correlation has not been established.

CONTROL OF STROKE VOLUME

The volume of blood ejected from the ventricles at each beat is mainly dependent on the force of myocardial contraction. In Chapter 2 we saw that isolated heart muscle contracts more forcibly when it is stretched over a certain range. It was also shown that when length is kept constant, force can be increased by various inotropic interventions. Similar considerations apply to the intact heart and, accordingly, there are two principal ways of regulating its pumping action at constant rate.

HETEROMETRIC REGULATION.—The first type involves a change in muscle fiber length secondary to alterations in filling pressure, or outflow resistance. Heterometric regulation is often referred to as the Frank-Starling mechanism because of the classic experiments of O. Frank and E. H. Starling. It was shown that increasing the venous inflow over a wide range in a heart free of neural or hormonal influences caused the heart to expel more blood so that outflow continued to match inflow. This adaptation was associated with an increased end-diastolic volume of the heart. Similarly when the resistance to outflow was raised, end-diastolic volume increased and the heart was able to expel the same amount of blood as before, despite the increased resistance. The explanation is that the increased fiber length that occurs with increased cardiac volume produces a more favorable overlap of actin and myosin filaments in the sarcomere. This enables more cross-bridge linkages to be formed during contraction and hence greater force development (Chapter 2).

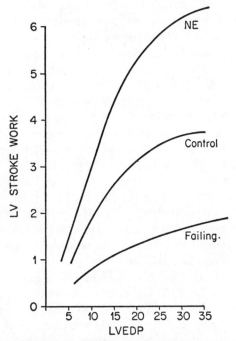

Fig 4–30.—Left ventricular *(LV)* function curves. Left ventricular end-diastolic pressure *(LVDEP,* mm Hg) is plotted against *LV* stroke work (arbitrary units). Norepinephrine *(NE)* shifts the curve upward and to the left, indicating improved function, whereas failure moves the curve downward and to the right.

The relationship between end-diastolic volume and cardiac function has been studied extensively by many investigators using modifications of Starling's heart-lung preparation. Sarnoff in particular has pointed out the usefulness of *ventricular function curves,* in which end-diastolic volume or a related variable is plotted against stroke volume or some other index of cardiac function. In clinical practice it is rather difficult to determine end-diastolic ventricular volume but cardiac catheterization makes it relatively easy to measure left ventricular end-diastolic pressure (LVDEP). Since pressure and volume are related, a clinical left ventricular function curve can be constructed by plotting LVDEP against stroke volume or stroke·work. These curves can be used to assess changes in cardiac status since a shift of the curve to the left indicates improved function, whereas a shift to the right indicates deterioration (Fig 4–30). Although clinically useful, the limitations of end-diastolic pressure as an index of end-diastolic volume or fiber length should be realized. Thus a given increment in LVDEP will produce a larger end-diastolic increase in volume and fiber length in a normally compliant ventricle than in a poorly compliant ("stiff") ventricle. This accounts for the fact that the thicker, less compliant left ventricle has a flatter function curve than the right-ventricle when end-diastolic pressure is plotted against stroke volume.

It will be noted that the function curves do not show a descending limb, as is the case for isolated cardiac muscle strips. This is because the pressures necessary to stretch the sarcomeres sufficiently to reduce cross-bridge linkages cannot be achieved in the intact animal.

The Starling mechanism is responsible for the equality of outputs of the two ventricles. If the output of one ventricle increases, the increased flow into the other ventricle stretches its fibers and its output rises until it matches the increased inflow. The Starling effect is also responsible for variations in cardiac output that occur with respiration. During inspiration the "negative" pressure in the chest becomes more negative and the transmural pressures of the cardiac chambers and intrathoracic blood vessels increase. The pressure gradient between the veins outside the chest and the right side of the heart also increases. Right ventricular filling and hence end-diastolic volume and output are therefore augmented. Part of the increased output is absorbed by the increased capacity of the pulmonary vessels resulting from their raised transmural pressure. However, after a beat or two, all the increased right ventricular output is delivered to the left ventricle, which is thereby slightly stretched, causing its output to increase.

HOMEOMETRIC REGULATION.—The second major type of regulation is termed homeometric because it involves factors that affect the contractile state of the heart muscle other than fiber length. This type of regulation is of major importance because there are many examples of activities, including muscular exercise, in which cardiac output is substantially augmented with no increase in diastolic heart size and presumably no change in initial fiber length. It is important to remember that ventricular muscle is a functional syncytium and contracts as a whole to the full extent to which it is capable under existing conditions. It cannot produce greater force by recruiting more motor units as skeletal muscle does, since it is in essence a single unit only. Accordingly, the only mechanism whereby the heart can increase its force of contraction at a given fiber length is an increase in its contractile or "inotropic" state. Several mechanisms have been described whereby such an increase can be brought about. These are the staircase effect, the Anrep effect and neural, humoral and chemical effects.

STAIRCASE EFFECT

This is most clearly shown in isolated papillary muscles. When contraction frequency is suddenly increased, the force of contraction and the rate of tension development increase in a stepwise fashion to reach a plateau after the next few contractions. A similar inotropic effect is observed after premature contractions (extra systoles) and has been termed "postextrasystolic potentiation." It has been suggested that these responses are due to a "lag in the sodium pump." When the frequency of excitation is increased, more sodium passes into the cell and, since sodium pump activity does not immediately increase, intracellular sodium concentration rises. This is believed to activate a sodium-calcium exchange mechanism, which brings an increased amount of calcium into the cell, thereby increasing force development.

ANREP EFFECT

When the outflow resistance in an isolated heart preparation is increased, stroke volume is maintained because end-diastolic volume rises and the Starling mechanism increases contractile force. However, it was noted by Anrep that end-diastolic volume soon returns to control levels but that, despite this, stroke volume is maintained. The initial heterometric regulation must therefore be followed by a length-independent inotropic effect. Its nature remains obscure, and its physiologic significance in man is not established.

NEUROHUMORAL FACTORS AND DRUGS

The effects of sympathetic nerve stimulation, epinephrine and norepinephrine on the isolated papillary muscle at constant length are similar. Peak tension, rate of tension development and rate of relaxation are increased and the time to peak tension and the twitch duration are reduced. Similar effects are seen in intact animals even when their hearts are paced to prevent rate changes. The rate of ventricular pressure development (dP/dt) and the rate of ventricular relaxation are both increased. Furthermore, the amount of work that can be performed at a given end-diastolic ventricular pressure is augmented, i.e., there is an elevation and shift to the left of the ventricular function curve (see Fig 4-30).

The mechanism of these effects has not been fully worked out. It has been established that a membrane-bound enzyme adenylate cyclase is activated and increases the conversion of ATP to cAMP. It has been suggested that in association with a protein kinase, cAMP facilitates the uptake of calcium into sequestering sites within the cell. This may account for the more rapid relaxation. Furthermore, the increased stores of calcium may make more calcium available for release in subsequent contractions. Acetylcholine, the neurotransmitter of the parasympathetic nervous system, has opposite effects, i.e., it reduces contractile force and the rate of tension development. The physiologic significance of this is doubtful since the ventricles, in contrast to the atria and pacemaker cells, receive a relatively sparse innervation from the vagus. There is little doubt that the principal effect of vagus stimulation is on rate rather than force.

Many drugs influence the inotropic state of cardiac muscle, presumably by producing alterations in excitation-contraction coupling mechanisms. The digitalis glycosides have been widely used for two centuries to increase myocardial performance in the failing heart. They are believed to act by inhibiting the sodium pump, thereby causing an increase in intracellular sodium. It has been suggested that this stimulates a "sodium-calcium" pump, which couples sodium efflux to calcium influx. As a result, the concentration of intracellular free calcium ion rises and contractility is enhanced. The staircase effect and the inotropic mechanism of digitalis therefore appear to have a similar molecular basis.

Figure 4-31 summarizes the factors that influence heart rate, stroke volume

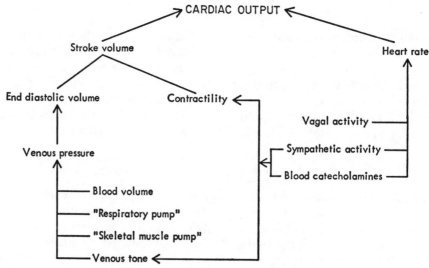

Fig 4–31.—Factors that influence cardiac output.

and cardiac output. For completeness this figure includes several variables that influence venous pressure (discussed further in the section on the veins).

EFFECT OF GEOMETRY OF THE VENTRICLE

The function of the ventricular muscle is to generate enough pressure in the ventricular cavity to expel an adequate amount of blood with each beat. The ventricles have a complicated shape, and the amount of tension (T) that must be generated at any point to produce a given pressure (P) is governed by the *law of Laplace*, which is expressed by the equation

$$T \left(\frac{1}{R_1} + \frac{1}{R_2} \right) = P$$

where R_1 and R_2 are the principal radii of curvature at that point. Thus the muscle fibers of a dilated heart must generate more tension to produce a given pressure than a smaller heart. This has several interesting consequences. During the ejection phase, the cardiac muscle fibers shorten and, because of the length-tension relationship, the tension that they produce falls off. At the same time, however, the tension required to maintain pressure is lessened because of the Laplace effect. In other words, pressure on the blood can be adequately maintained despite falling tension during the ejection phase because of the increased mechanical advantage conferred by Laplace's law. On the other hand, the Laplace relationship puts the cardiac muscle at a mechanical disadvantage when the ventricle dilates, as in heart failure. The dilatation is beneficial if the heart is operating on the ascending limb of the length-tension curve because the increased initial length enables a greater contractile force to be developed. However, some of this benefit is lost because more force is required to maintain a given pressure. Furthermore, the need to develop increased tension at the greater radius results in increased energy cost. If the ventricle is operating on the flat portion of the Starling curve, further dilatation has only a deleterious effect.

WORK OF THE HEART

The work *(W)* done by a force *(F)* in moving a point through a distance *(d)* is given by the expression

$$W = F \cdot d$$

In the circulatory system, however, we have to consider the work done by pressures in moving volumes. Since pressure *(P)* is force per unit area *(a)*, the equation above becomes $W = P \cdot a \cdot d$, and since $a \cdot d = $ volume *(V)*, it is clear that the work done by a ventricle in ejecting a volume *(V)* is given by $P \cdot V$ where P is the mean systolic ventricular pressure. During the ejection period, ventricular pressure varies continuously and the work done is therefore given more accurately by

$$W = \int P dV$$

These measurements cannot easily be obtained clinically but, as a close approximation, we can take mean systolic aortic pressure instead of integrating the ventricular systolic pressure curve. The work done by the ventricle at each beat (stroke work) is then given by the product of mean systolic aortic pressure and stroke volume.

ENERGY REQUIREMENTS AND EFFICIENCY OF THE HEART.—If a person attempts to lift a 400 lb weight from the floor, the muscles contract strongly, breathing becomes labored, the face flushes and sweat may appear on the brow. Obviously, that person is consuming energy. With great effort he or she may eventually be able to raise the weight a small distance. However, the external work performed will represent only a small fraction of the total energy turnover; most of the energy will be used to develop and maintain tension in the contracting muscle. In the same way, the mechanical work (pressure × volume) performed by the heart in expelling blood is a relatively small percentage of the energy consumption, most of which is used to generate tension in the ventricular muscle.

The total cardiac energy consumption of the heart can be estimated from its oxygen consumption, since each milliliter of oxygen used is energetically equivalent to 2.5 kg-m. An average oxygen consumption for the heart of a resting subject is 24 ml per minute, equivalent to 60 kg-m. The mechanical work of the left ventricle, assuming a cardiac output of 5 L, and a mean systolic arterial pressure of 100 mm Hg, is equivalent to 6.8 kg-m per minute. The mechanical work of the right ventricle, assuming a mean systolic arterial pressure of 15 mm Hg is 1.02 kg-m. Cardiac *efficiency*, i.e., the fraction of the total energy that appears as mechanical work is therefore 7.82/60 = 13%.

Mechanical work can be increased by increasing the cardiac output ("volume work") or arterial pressure ("pressure work"). Considerable increases in volume work can be achieved with relatively minor increments in

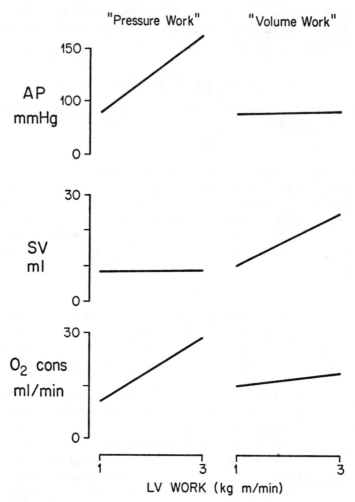

Fig 4–32. – Comparison of oxygen costs of "pressure" and "volume" work. *Pressure work* (stroke volume [SV] constant, aortic pressure [AP] increased) causes a proportional increase in oxygen consumption. *Volume work* (AP constant, SV increased) causes a much smaller increase in oxygen consumption.

oxygen consumption. For example, in one of Sarnoff's dog experiments, raising the cardiac output 700% with arterial pressure constant caused oxygen consumption to rise only 50%, i.e., efficiency increased. This is understandable if it is remembered that the major factor in myocardial oxygen consumption is tension development and this does not increase much in volume work. On the other hand, raising arterial pressure 100% with cardiac output constant (pressure work) produces a proportional increase in oxygen consumption (Fig 4–32).

Because of the close relationship between energy consumption and tension development, the "tension-time index" (TTI) has been used clinically to obtain an estimate of the heart's energy usage and oxygen requirements. The TTI per beat is defined as the product of the mean systolic arterial pressure and the duration of ejection. The TTI per minute is obtained by multiplying this product by the heart rate. This index, which preferably should be called the "pressure-time index," provides only an approximate index of tension development because it does not take into account the size and shape of the ventricle or the velocity of contractile element shortening, both of which play a major role in determining oxygen consumption. Experiments in dog hearts have shown that a 50% increase in myocardial oxygen consumption could be produced either by increasing *peak* tension development by 80% or by increasing the maximum velocity of contraction (V_{max}) by 40% (peak tension being maintained constant experimentally). These figures cannot be extrapolated to man but they do clearly indicate the significance of the velocity of contraction. Sympathetic nerve stimulation or norepinephrine would therefore increase oxygen consumption even if there were no change in heart rate or aortic pressure.

The major determinants of myocardial oxygen consumption *per beat* are therefore (1) tension development, which is determined by ventricular systolic pressure and by size and shape of the ventricle, and (2) the contractile state of the myocardium, which is mainly under neurohumoral control. Heart rate is a major determinant of oxygen consumption *per minute* and also slightly increases oxygen consumption per beat because of its positive inotropic effect. External work is a relatively minor factor. Oxygen also is required to support the cellular processes necessary to maintain the structural and functional integrity of the heart cells and to supply the energy requirements of depolarization, excitation-contraction coupling and the sodium pump. These requirements together account for about 20% of the cardiac energy expenditure in a resting subject.

ADAPTATION TO INCREASED VOLUME AND PRESSURE LOADS

The response to acute increases in volume or pressure loads has been previously discussed. Acute volume overloads are met by more forceful contractions owing to increased fiber length. The rise in oxygen consumption is disproportionately small relative to the increase in volume load, and cardiac efficiency increases. Acute pressure overloads are also met initially by cardiac dilatation but the muscle cells then return to their initial length with an enhanced contractile state, which enables them to meet the increased load (Anrep effect). This adaptation requires a proportional increase in oxygen consumption, and efficiency is unchanged. Clinical examples of volume overloads include large transfusions; high-output states, e.g., hyperthyroidism; beriberi and abnormal shunts such as ventricular or atrial septal defects. Examples of pressure overloads include pulmonic or aortic valve stenosis and systemic or pulmonary hypertension. The various overloads usually do not affect all the cardiac chambers uniformly. For example, in a left-to-right shunt through an atrial septal defect, the right side of the heart bears the brunt of the volume overload. In aortic stenosis the left ventricle shows the effects of pressure overload, whereas the right ventricular load is normal until left atrial pressure rises significantly.

When volume or pressure overloads persist for long periods, the cardiac muscle cells hypertrophy. They increase in size, and additional sarcomeres and mitochondria are formed. The increased load is therefore spread over a greater mass of contractile tissue. At the same time the connective tissue cells in the heart increase in number and more collagen is produced. The heart therefore becomes less compliant and, in long-standing or severe overloads, the end-diastolic ventricular pressure may rise considerably. This in turn causes impaired perfusion of the subendocardial muscle (see section on coronary circulation), with degeneration of muscle cells and more fibrosis so that a vicious cycle becomes established and the heart fails.

The amount of hypertrophy seems to be limited by the fixed DNA content of the cell, which restricts the capacity for RNA production and new protein synthesis. When the limit has been reached, further overload results in cell degeneration and fibrosis. This is aggravated by several factors including (1) failure of the vascularity of the heart to increase with hypertrophy and (2) increased diffusion distances and impaired myocardial perfusion, particularly in the subendocardium.

CLINICAL INVESTIGATION OF THE HEART

Much can be learned about the cardiovascular function of patients by taking a careful history and by traditional methods of examination, inspection, palpation and auscultation, augmented by electrocardiography and phonocardiography. Additional information can be obtained by specialized cardiologic procedures. These additional data are particularly important nowadays when so many formerly lethal cardiac abnormalities can be corrected by surgery. A thorough knowledge of the abnormal anatomy and physiology of the heart of a patient assists the surgeon in deciding whether to advise operation and in assessing the type and risk of the procedure needed. The account that follows is necessarily extremely sketchy. More advanced discussions of the various topics can be found in the references cited at the end of the chapter.

CARDIAC CATHETERIZATION

Insertion of a catheter into the heart chambers and great vessels enables pressures to be determined at these sites. Blood samples for analysis can also be withdrawn from different positions within the heart. Cardiac output can be measured and x-ray opaque substances (contrast material) can be injected to detect abnormalities in size or shape of the cardiac chambers or in the flow of blood through the heart. Contrast material can also be injected selectively into the coronary arteries to determine whether obstructive lesions are present. The information obtained at catheterization is often of vital importance in establishing the presence of heart disease, its nature and its severity.

CATHETERIZATION OF RIGHT SIDE OF THE HEART

A catheter is inserted into a vein in the right elbow or, alternatively, into a vein in the groin. It is then advanced under fluoroscopic guidance into the right atrium, right ventricle, pulmonary artery and distal part of the pulmonary arterial system until it becomes wedged in one of the small pulmonary arteries. The catheter is attached to a pressure transducer, and pressures are measured at each of the sites mentioned. Particular care is taken to determine whether abnormal pressure gradients are present across the AV or pulmonic valves. Pulmonary wedge pressure is equal to pulmonary capillary pressure and is therefore very close to left atrial pressure. This information is useful when the more risky direct determination of left atrial pressure is considered inappropriate. Blood samples are withdrawn from various sites within the heart and great vessels and their oxygen contents are determined.

As an example of the value of pressure and oxygen determinations from the right side of the heart, let us consider the findings from a patient with pulmonic valvular stenosis and from a patient with ventricular septal defect. Pulmonic valvular stenosis, i.e., narrowing of the pulmonary valve orifice, is often congenital. The main abnormalities are elevated right ventricular systolic pressure owing to right ventricular outflow obstruction and a systolic pressure gradient across the pulmonic valve. This condition may often be diagnosed without catheterization because the narrowed valve causes turbulent flow and therefore a prominent systolic murmur during right ventricular ejection. Long-standing obstruction produces right ventricular hypertrophy, which sometimes can be detected by the abnormally strong thrust of the right ventricle against the chest wall. The ECG and chest x-ray show the changes of right ventricular hypertrophy. Catheterization enables the degree of stenosis to be assessed, and the valve area can be calculated from a formula derived by Gorlin:

$$\text{valve area (sq cm)} = \frac{F}{K\sqrt{P_1 - P_2}}$$

where F is the volume flow rate (ml/sec) during the time the valve is open, $P_1 - P_2$ the pressure gradient across the valve and K a constant. The formula indicates that the pressure gradient is proportional to the square of blood flow rate across the valve orifice. This is important because when cardiac failure is present, the flow may be so low that the gradient across even a considerably stenosed valve may be insignificant. Thus it is mandatory to determine cardiac output when attempting to assess the severity of valve lesions.

In ventricular septal defect there is a hole in the ventricular septum through which blood flows during systole. The direction of this abnormal flow or "shunt" is usually from left to right because left ventricular pressure is higher. This abnormal blood flow is turbulent and frequently associated with a loud murmur that can be heard throughout systole. The diagnosis can be confirmed in several ways. A common method is to detect the sudden increase in oxygen content of the blood in the right ventricle owing to admixture of its deoxygenated blood with highly oxygenated blood from the left ventricle. Usually the oxygen saturation of the blood in the right atrium, right ventricle and pulmonary artery is very similar, but when a left to right shunt occurs, the oxygen saturation of blood samples drawn from sites downstream from the shunt is increased.

CATHETERIZATION OF LEFT SIDE OF THE HEART

A catheter is inserted into the brachial or femoral artery and advanced under fluoroscopic control to the origin of the aorta. It can then be manipulated through the aortic valve into the left ventricle and through the mitral valve into the left atrium. An alternative method of catheterization of the left side of the heart is to perform catheterization of the right side first. When the catheter is in the right atrium, a special needle is inserted through the catheter and used to perforate the atrial septum. The catheter is advanced over the needle and through the perforation into the left atrium. The needle is then withdrawn. Pressures and oxygen saturations are measured in the various chambers and, as on the right side, can be used to diagnose valvular disease and abnormal shunts.

ANGIOGRAPHY

Solutions of organic compounds containing iodine (which has a high atomic number) are opaque to x-rays. If they are injected into a heart chamber during cardiac catheterization and a cine x-ray taken, they will indicate the changes in size and shape of that chamber during the cardiac cycle. Abnormalities of the valves that cause obstruction to flow or permit backflow are readily revealed. The volume of the ventricles, stroke volume, ejection fraction and amount and uniformity of ventricular myocardial shortening can be estimated from films taken in one plane, provided certain geometric assumptions are made. Figure 4–33, A shows the outlines of the normal left ventricle in systole and diastole, as seen in the right anterior oblique projection (patient's right shoulder turned toward film). If the long axis at end-diastole is el cm and the short axis (drawn perpendicular to el at its midpoint) is d cm, then the end-

Fig 4–33.—Tracings of silhouettes of the left ventricular cavity obtained by angiography. **A**, normal ventricle. **B**, after myocardial infarction. Note the impaired systolic movement of the apical portion of the ventricle.

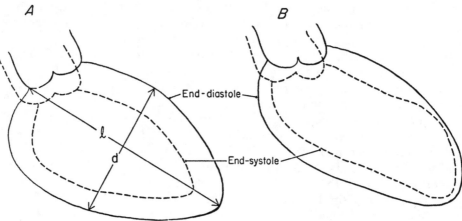

diastolic volume *(EDV)* of the ventricle, considered as a prolate spheroid, is given by the formula:

$$EDV = \frac{4}{3} \cdot \pi \cdot \frac{l}{2} \cdot \frac{d^2}{4}$$
$$= 0.524 \, l \cdot d^2$$

End-systolic volume *(ESV)* can be similarly calculated using end-systolic dimensions. Stroke volume *(SV)* is then *EDV − ESV*, and the ejection fraction is *SV/EDV*.

The normal ventricle contracts uniformly in all dimensions during systole but when myocardial function is locally impaired, e.g., following myocardial infarction, shortening of the affected area is diminished. This is readily seen when end-diastolic and end-systolic ventricular silhouettes are appropriately superimposed (see Fig 4–33, B). When contraction of a portion of ventricular muscle is lost completely, this area paradoxically may bulge out during systole. This abnormal motion is termed *dyskinesia* and is an important sign of myocardial damage.

DETERMINATION OF CARDIAC OUTPUT

This is a significant and valuable determination because it provides a direct measure of the transport function of the circulation. Two closely related methods are commonly used.

1. THE FICK PRINCIPLE.—This states that the total uptake or release of a substance by an organ is equal to the blood flow through the organ multiplied by the difference between arterial and venous concentrations of the substance. This principle can be applied to the uptake of oxygen by blood flowing through the lung. Let us suppose that a subject takes up 250 ml of oxygen from inspired air and that the pulmonary arterial oxygen content is 15 volumes/100 ml and pulmonary venous oxygen content is 20 volumes/100 ml. This means that each 100 ml of blood must have taken up 5 volumes of oxygen in its passage through the lung. In order for 250 ml of oxygen to have been taken up per minute, $50 \times 100 = 5{,}000$ ml blood must have traversed the lung per minute. Thus the cardiac output is 5 L. The Fick equation for oxygen uptake by the lung is as follows:

$$Vo_2 = Q \cdot (PV_{O_2} - PA_{O_2})$$

where Vo_2 is oxygen uptake per minute, Q is cardiac output and PV_{O_2} and PA_{O_2} are the oxygen contents of pulmonary venous and pulmonary arterial blood, respectively. In practice, $\dot{V}o_2$ is obtained by having the subject breathe air from a bag and determining the amount of oxygen used in a given time. PV_{O_2} is the same as the oxygen concentration of the blood in any systemic artery, since no significant change in oxygen content occurs as blood passes from the pulmonary veins through the left side of the heart into the systemic arteries. It is therefore readily obtained by analysis of blood from needle puncture of any accessible artery, e.g., radial, brachial or femoral. PA_{O_2} is obtained by withdrawing blood from the pulmonary artery during catheterization of the right side of the heart.

2. INDICATOR-DILUTION METHOD.—If a known quantity (I) of an indicator is added to a stationary volume of liquid and allowed to mix completely, the volume (V) of the liquid is given by the formula

$$V = \frac{I}{C}$$

where C is the concentration of the indicator after thorough mixing. This principle has been applied to measurement of cardiac output. An indicator substance is rapidly injected into a vein, its concentration is measured continuously at a more distant sampling site and its average concentration is determined. If the time (t) between the moment of first appearance and the moment of final disappearance of the indicator at the sampling site (passage time) is known, then cardiac output (C.O.) is given by

$$\text{C.O.} = \frac{I \cdot 60}{C \cdot t}$$

If I is measured in milligrams, the mean concentration (C) in mg/L, and the passage time (t) in seconds, then C.O. will be given as liters per minute.

Several types of indicators have been used. The ideal substance should be nontoxic, should mix instantaneously and completely, should remain within the circulation during the period of determination and should not alter cardiac output. Indocyanine green is a commonly used agent. Hot or cold solutions also have been employed as indicators, and their dilution can be followed by recording temperature changes at the distal site.

Whatever the nature of the indicator injected, the plot of its concentration

Fig 4–34.—Indicator-dilution curve showing the variation in arterial blood concentration of the indicator with time after a bolus intravenous injection. AT = appearance time; PT = passage time.

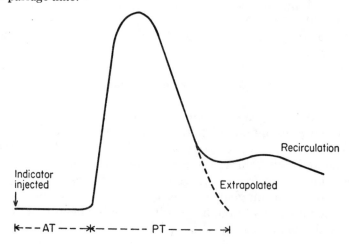

against time at the sampling site has characteristic features. Figure 4–34 shows the typical curve obtained when a bolus of indocyanine green is rapidly injected into a vein while blood is continuously withdrawn from a peripheral artery through a densitometer that continuously measures the dye concentration. After injection the indicator instantly mixes with the blood and is then distributed among the innumerable vascular pathways between the injection and sampling sites. Different portions of the dye therefore arrive at the sampling site at different times. The *appearance time* is the interval between the moment of injection and the first appearance of the indicator at the sampling site. Thereafter the concentration rises rapidly to a peak, as more and more dye arrives from the longer pathways. The concentration then declines but rises again as the fastest moving dye, having completed a first circulation, comes around again, producing a "recirculation" hump. This recirculation is the major difficulty of the method because it prevents determination of passage time and average dye concentration, which would have been achieved had only one circulation occurred. Fortunately, by taking advantage of the fact that the falling concentration of dye follows an exponential curve, it is possible to determine when the concentration of dye would have fallen to zero if no recirculation had taken place. To do this, concentrations on the downslope are replotted on semilog paper against time and the straight line so obtained is extrapolated to zero concentration. The points obtained by extrapolation can then be replotted on the original curve. The mean concentration of the dye during its first circulation and the passage time can then be determined and the cardiac output calculated using the formula given above.

The thermodilution method is an indicator dilution method that has been increasingly used recently. Cold is employed as the indicator. It has the great advantage that no significant recirculation occurs and so calculation of output is greatly simplified. Furthermore, frequently repeated determinations can be made. Special flow-directed pulmonary arterial balloon catheters are now available that carry a thermistor at the tip. These catheters can be introduced at the bedside even in severely ill patients. The catheter has three separate lumina and can be so positioned that 10 ml of cold (0 C) saline can be introduced into the central venous blood flow through one of the ports. The cold is "mixed" in the right ventricle and its "concentration" (temperature change) is continuously monitored by the thermistor, which lies in the pulmonary artery. The cold "dilution" curve can be integrated electronically and the computed cardiac output can be displayed by commercially available equipment within one minute. The formula for deriving C.O. is as follows:

$$C.O. = \frac{k(T_B - T_i)V \cdot 60}{\int_0^\infty T_B(t)dt}$$

where T_B = pulmonary arterial blood temperature
T_i = temperature of injected cold fluid
V = volume of injected fluid
t = time (seconds)

$k \doteq$ constant, which takes account of the different specific gravities and specific heats of blood and the injectate and the warming of the injectate as it passes through the catheter before emerging from the tip.

NONINVASIVE METHODS

In experienced hands cardiac catheterization is a very safe procedure but, nevertheless, complications occasionally develop. These include local hematomas, impairment of circulation of the limb in which the catheter is inserted and infection. More serious hazards are perforation of the heart, arrhythmias, myocardial infarction (especially during selective coronary angiography) and cardiac failure. Because of these risks, much effort has been expended in developing noninvasive methods of assessing the heart's performance. These methods have the additional advantage that they can be repeated at frequent intervals, they do not alter circulatory variables and they are comparatively cheap.

SYSTOLIC TIME INTERVALS

Considerable information regarding the timing of events in the cardiac cycle can be obtained by simultaneous recording of the phonocardiogram, ECG and carotid pulse wave. The latter is detected by a suitable transducer placed on the skin over the carotid artery. Three intervals can be defined and are shown in Figure 4–35. Electromechanical systole $(Q-S_2)$ is the interval between the beginning of Q and onset of the second heart sound. Left ventricular ejection time (LVET) is the time between the beginning of the carotid pressure upstroke and the incisura. The pre-ejection period (PEP) is determined by subtracting LVET from $Q-S_2$. These intervals are affected by the heart rate, as shown in Table 4–2, which gives average values for normal individuals at rates of 70 and 100 per minute. It is to be noted that the ratio PEP/LVET is not significantly affected by rate.

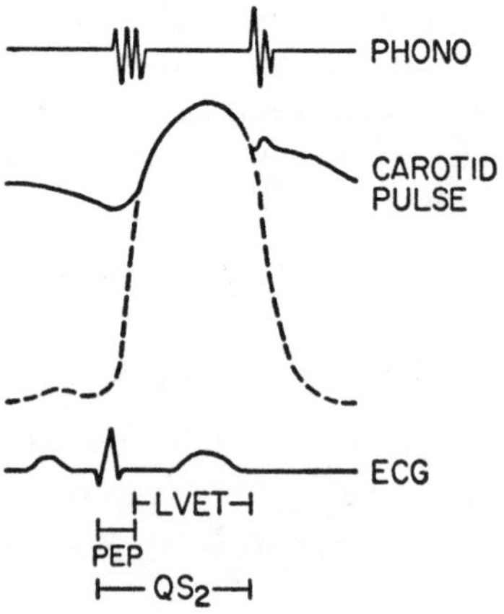

Fig 4–35. – Systolic time intervals (see text). *LVET* = left ventricular ejection time; *PEP* = preejection period; *QS₂* = time between Q and second heart sound.

TABLE 4–2.—AVERAGE VALUES
FOR SYSTOLIC TIME
INTERVALS

HEART RATE	70/MIN	100/MIN
Q–S$_2$ (msec)	399	336
LVET (msec)	294	243
PEP (msec)	105	93
PEP/LVET (%)	35	37

Q–S$_2$ = interval between Q and second heart sound; LVET = left ventricular ejection time; PEP = pre-ejection period.

In patients with impaired myocardial performance, the interval Q–S$_2$ changes little, but LVET shortens and PEP lengthens and so the PEP/LVET ratio increases. Apparently good correlations have been found between this ratio and other indices of myocardial function such as ejection fraction. The PEP/LVET ratio is less reliable in some conditions than in others and, because of numerous factors that influence both PEP and LVET, the ratio should be interpreted with caution. Nevertheless, it appears to be useful in some circumstances, particularly in following the progress of an individual patient.

ECHOCARDIOGRAPHY

Pulses of high frequency sound (2,000,000 Hz) are generated by a transducer placed on the chest wall over the heart. Commercial instruments vary, but the duration of each pulse is about 1 μsec and the pulses are emitted at rates of 200–2,000 per second. During the time when the transducer is not emitting pulses, it acts as a receiver. The ultrasonic beam passes through the chest and is reflected from interfaces between tissues of different acoustic density. The echoes are received by the transducer and converted into an electric signal, which is displayed on an oscilloscope. This is calibrated so that the interval between the transmitted signal and its echo indicates the distance in centimeters between the transducer and the point of reflection. If the reflecting surface is moving, its echo will move also. Suitable electronic processing enables the change in position of the reflecting surface with time to be recorded as a dot moving across the oscilloscope face. The trace can either be filmed or recorded on a moving chart to provide a permanent record. When the transducer is placed on the chest of a supine individual in the third or fourth intercostal space and directed, as shown in Figure 4–36, echoes can be obtained from the anterior and posterior walls of the right and left ventricles and the anterior and posterior leaflets of the mitral valve. The record can be used to estimate end-diastolic and end-systolic ventricular volumes and hence stroke volume, ejection fraction and cardiac output. It was shown previously that the volume (V) of the ventricle is given by the formula:

$$V = \frac{4}{3} \cdot \pi \cdot \frac{l}{2} \cdot \frac{d^2}{4}$$

where l is the length of the long axis and d is the radius at the midpoint of l Since l is approximately $2d$, the formula can be expressed

$$V = \frac{4}{3} \cdot \pi \cdot \frac{2d}{2} \cdot \frac{d^2}{4}$$
$$= 1.05d^3$$

Thus the volume of the ventricle is approximately equal to the cube of the diameter. When a standardized echocardiographic procedure is used, it is possible to determine the left ventricular diameter in diastole and systole reasonably accurately, and hence end-diastolic and end-systolic stroke volumes can readily be obtained.

As an example of the value of echocardiography in diagnosis of valvular disease, echoes from a normal and a stenosed mitral valve are compared in Figure 4–36, B. The normal anterior leaflet (upper trace) closes (posterior movement A–B–C) immediately following atrial systole. Ventricular contraction tightens the seal and shortening of ventricular dimensions moves the closed leaflets somewhat anteriorly (C–D) during systole. In diastole the anterior leaflet opens with an abrupt anterior movement (D–E) but then tends

Fig 4–36.—A, echocardiography. A transducer *(T)* is placed on the chest wall in the fourth interspace just to the left of the sternum. When angled as shown, its ultrasonic beam traverses and generates echoes from the free wall of the right ventricle, the interventricular septum and the mitral valve leaflets. *RV* = right ventricle; *LV* = left ventricle, *AMV* = anterior leaflet of mitral valve; *PMV* = posterior leaflet of mitral valve; *LA* = left atrium; *Ao* = aorta. (From Feigenbaum, H.: Prog. Vasc. Dis. 14:531, 1972. Used by permission.)
B, *upper trace* = echoes from normal mitral valve; *lower trace* = echoes from stenosed mitral valve.

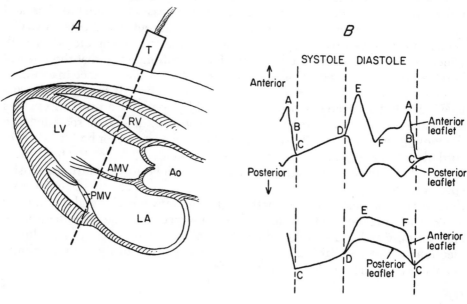

to move toward the closed position (E – F) again. Atrial systole again causes an abrupt anterior movement of the leaflet to point A and the cycle then begins again. The movements of the posterior leaflet during diastole are, as might be expected, almost mirror images of those of the anterior leaflet. The movements of the valve leaflets in mitral stenosis (see Fig 4–36, B, lower trace) are grossly abnormal. The closing movement (E – F) of the anterior leaflet during mid-diastole is greatly diminished, and no movement due to atrial systole can be discerned. Moreover, in diastole the posterior leaflet moves in the same direction as the anterior leaflet and the opening between the two is diminished. Clinically the E – F slope has been found to be a useful index of the severity of mitral stenosis. It is normally greater than 70 mm per second but may fall to less than 15 mm per second with severe stenosis.

Echocardiography is the most sensitive method available for detecting abnormal amounts of fluid in the pericardial sac since it readily demonstrates the fluid as an echo-free space behind the heart.

New instruments and applications are being rapidly developed and refined and the method will undoubtedly enhance the already proved value of echocardiography in assessing ventricular and valvular function.

RADIONUCLIDE IMAGING

Radioisotopes are being increasingly used in cardiology in a variety of applications, only two of which are now cited as examples.

1. ASSESSMENT OF CHAMBER SIZE AND LEFT VENTRICULAR FUNCTION. – A suitable isotope, e.g., ^{99m}Tc-pertechnetate, is injected intravenously. During its first pass through the heart, its presence is detected by an external high-speed scintillation camera. With computer assistance, images of the various chambers can be obtained showing systolic and diastolic shape and dimension, and the left ventricular ejection fraction can be calculated.

2. LOCALIZATION OF INFARCTS. – A myocardial infarct is an area of myocardial necrosis that develops as the result of impaired blood supply (see section on coronary circulation). The ^{99m}Tc-labeled pyrophosphates are preferentially concentrated in severely damaged myocardium and this can be recognized as a "hot spot" on the image produced by the scintillation camera.

HEMODYNAMICS

We have seen in previous sections that the function of the heart is to transfer energy to the blood, thus causing it to flow through the various capillary beds where the vital exchanges of oxygen, nutrients and metabolites occur. This energy transfer causes pressure changes in all parts of the circulation. The study of these pressures and flows is termed *hemodynamics*. It is of enormous clinical importance, and cardiac catheterization and intensive care monitoring are devoted to a considerable extent to the acquisition of hemodynamic data. Comparison of these with known normal values is useful in making a diagnosis and in following the course of a circulatory abnormality as well as in deciding and evaluating therapy.

The factors that govern pressure and flow in the circulatory system are exceedingly complex. The output of the heart is phasic and the rate of ejection during systole is not uniform. Furthermore the blood is ejected into a system of branching tubes that vary in resistance and distensibility from their origin to the periphery. Blood is a suspension of particles and does not follow Newtonian laws. Its flow characteristics in narrow tubes such as capillaries is quite different from that in large tubes such as the aorta. Much effort has been expended in an attempt to reduce these complexities to precise mathematical terms but many questions are still unresolved and controversial. Fortunately we can gain considerable insight into circulatory behavior from simple models despite the fact that they are rather far removed from the actual situation.

POISEUILLE'S LAW

Poiseuille, a French physician, studied the flow of liquids in glass tubes and observed that the volume flow rate through a tube was affected by the pressure difference between the ends of the tube, the viscosity of the fluid and the tube radius and length. The relationship between these factors is given by the following equation, known as Poiseuille's law:

$$Q = \frac{P\pi r^4}{8\eta l}$$

where Q = volume flow rate, P = pressure gradient across the tube, r = tube radius, l = tube length, and η = viscosity of the fluid. If this relationship is applied to the systemic circulation as a whole, Q will be equivalent to cardiac output and P to the difference between mean aortic pressure and mean right atrial pressure. If we pursue our previous analogy with electric circuits, P/Q will be equivalent to the resistance (R) of the circuit. The Poiseuille equation then becomes

$$\frac{P}{Q} = R = \frac{8\eta l}{\pi r^4}$$

This indicates that the resistance of the circulation is determined by the viscosity of the blood and by the geometry of the blood vessels, i.e., their lengths and radii.

VISCOSITY OF BLOOD

If syrup is made to flow through a tube under the influence of a pressure gradient, it will flow more slowly than water exposed to the same pressure gradient. This is because the "internal friction" between adjacent laminae of syrup is greater than the internal friction or viscosity of water. The viscosities of plasma and whole blood are, respectively, about 1.7 times and four times greater than water. The relatively high viscosity of whole blood is mainly due to its red cells. If a sample of blood is centrifuged, the cells will sediment out. The percentage volume occupied by the red cells after centrifugation is termed the *hematocrit:*

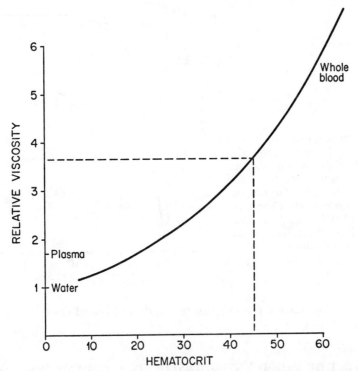

Fig 4–37. – Influence of hematocrit on the viscosity of whole blood relative to water. The coordinates for normal blood (hematocrit 45%) are indicated by *dashed lines*.

$$\text{hematocrit} = \frac{\text{volume of red cells in sample}}{\text{volume of sample}} \times 100$$

The normal value is about 45%. In some diseases there is an abnormally large number of red cells, *polycythemia*, and the hematocrit may rise to values as high as 60%. The relationship between hematocrit and relative viscosity is such that viscosity rises steeply for hematocrits above 50% (Fig 4–37) and at 60% may be about twice normal. The increased resistance to flow tends to cause systolic hypertension and to increase the work of the heart. It also may impair tissue perfusion and increase a tendency to thrombosis because of low flow rates in certain vessels.

FÅHRAEUS-LINDQUIST EFFECT

The viscosity of water is independent of the size of the tube through which it flows. On the other hand, the viscosity of whole blood declines markedly when the tube diameter is less than 1 mm (Fig 4–38). This is the Fåhraeus-Lindquist effect and its mechanism is not fully understood. It is at least partly due to the fact that a "sleeve" of cell-free and therefore low-viscosity plasma lies adjacent to the wall of all blood vessels and becomes *relatively* larger as diameter decreases. The Fåhraeus-Lindquist effect is obviously important

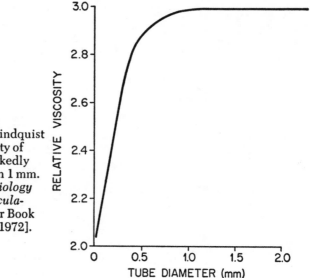

Fig 4–38.—Fåhraeus-Lindquist effect. The relative viscosity of whole blood declines markedly in tube diameters less than 1 mm. (From Burton, A. C.: *Physiology and Biophysics of the Circulation* [2d ed.; Chicago: Year Book Medical Publishers, Inc., 1972]. Used by permission.)

and beneficial because of the very large number of blood vessels with diameters less than 1 mm.

EFFECT OF BLOOD VESSEL DISTENSIBILITY; IMPORTANCE OF TRANSMURAL PRESSURE

When blood flows through a rigid tube, Poiseuille's law indicates that its flow rate will be linearly dependent on the pressure gradient between the ends of the tubes. Thus if a pressure gradient of 100 mm Hg causes a flow of 5 L per minute, this flow will occur regardless of the absolute pressure values at the ends of the tube. Thus 5 L per minute will flow if the pressures at the two ends of the tube are 200 and 100 mm Hg, or 700 and 600 mm Hg. Moreover, a reduction in gradient produces a proportionate reduction in flow. On the other hand, distensible tubes are significantly influenced by the absolute pressures because tube diameter increases when the pressures are raised and decreases when they are lowered. Let us consider a distensible tube that has a diameter of 0.8 cm when its transmural pressure (P_t) (i.e., pressure inside–pressure outside) is 50 mm Hg, 1 cm when P_t is 100 mm Hg and 1.1 cm when P_t is 200 mm Hg. Suppose a flow of 5 L per minute occurs when the pressure at one end of the tube is 150 mm Hg and the pressure at the other end is 50 mm Hg. The mean P_t in the tube will be 100 mm Hg and the mean diameter 1 cm. If the pressure at each end of the tube is reduced by 50 mm, the gradient will be unchanged, but the mean transmural pressure will fall to 50 mm Hg and the diameter will be reduced to 0.8 cm. Since flow is directly proportional to the fourth power of the tube radius, flow will fall to $0.8^4 \times 5 = 2.05$ L per minute. On the other hand, if pressures at the ends of the tubes are increased to 250 and 150 mm Hg, the pressure gradient will still be 100 mm Hg but the mean transmural pressure will be 200 mm Hg, the tube diameter

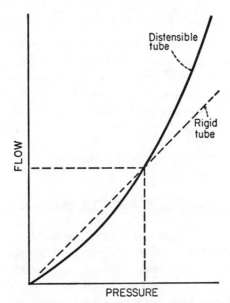

Fig 4–39.—Relationship between transmural pressure and flow in rigid and distensible tubes. If we start out from a pressure-flow status described by the *dashed coordinates,* flow in a rigid tube will change linearly in response to transmural pressure changes. On the other hand, the pressure-flow curve for a distensible tube will be curvilinear (see text).

will be 1.1 mm and flow will be $1.1^4 \times 5 = 7.26$ L per minute. Thus, increasing the pressure gradient across a distensible tube will cause flow to increase disproportionately, and reducing the gradient will cause flow to decrease disproportionately. Accordingly, the relationship between transmural pressure and flow is linear for rigid tubes but curvilinear and convex toward the pressure axis for distensible tubes (Fig 4–39).

Effect of Active Changes in Arteriolar Diameter

Under most physiologic conditions, blood viscosity and length of vessels remain fairly constant; the main factor that regulates vascular resistance is the radius of the various vessels. It has previously been emphasized (see Fig 4–2) that the arterioles constitute the main fraction of the total resistance. Thus the principal determinant of circulatory resistance (R) is arteriolar radius (r) and it is an extremely potent factor since $R \propto 1/r^4$. If the average arteriolar radius increases by 20%, resistance is halved; if it increases by 50%, resistance falls to one fifth; and if it doubles, resistance falls to one 16th of its initial value. It is therefore of crucial importance to understand the factors that control arteriolar diameter. We have seen that alterations in blood vessel diameter may occur passively in response to changes in transmural pressure, but a far more important cause is the active contraction or relaxation of the smooth muscle cells that surround the vessel lumen.

INSTANTANEOUS PRESSURES IN CARDIOVASCULAR SYSTEM

In this section we have mainly considered the factors that regulate average pressures and flows. However, the phasic contractions and relaxations of the heart produce pressures and flows that vary continuously from instant to instant. The instantaneous intracardiac pressure and volume changes and instantaneous central venous pressure curve have already been described (see section on the cardiac cycle). The instantaneous arterial pressure curve depends on volume and pattern of left ventricular ejection and also to a considerable extent on mechanical properties of the arterial wall (p. 165).

VASCULAR SMOOTH MUSCLE

In general, vascular smooth muscle cells are fusiform, 25–100 μ long and 1–5 μ wide. The surface of each cell is composed of a plasma membrane with a well-defined basement membrane. The plasma membrane shows many invaginations or *caveoli*, which increase its surface area considerably. The cell contains a central elongated nucleus, Golgi apparatus, smooth and rough endoplasmic reticulum, mitochondria and thick and thin myofilaments. The thick filaments are bundles of myosin molecules and the thin filaments appear to be fibrous actin. Contraction probably occurs by a sliding filament mechanism activated by an increase in intracellular calcium ion concentration (see section on muscle, Chapter 2).

The outer layers of the plasma membranes of adjoining muscle cells in the smallest arterioles and venules often show areas of fusion, termed *nexuses*. These are low-resistance junctions that enable excitation to be transmitted rapidly from cell to cell. Nexus formation is far less frequent in the smooth muscle of large conduit vessels such as the aorta.

Recently, unequivocal evidence of a sarcoplasmic reticulum of sufficient volume and organization to act as an intracellular store or sequestering site for calcium has been obtained in vascular smooth muscle, but its physiologic significance remains unclear.

INNERVATION

Most blood vessels receive adrenergic vasoconstrictor innervation from the sympathetic nervous system. Histochemical fluorescence methods have demonstrated that the adrenergic nerves are arranged as a network at the adventitiomedial junction. Only rarely do the nerve fibers pass into the media. Adrenergic innervation of the smaller arteries is denser than that of the large arteries, and innervation of veins is usually less dense than that of the corresponding arteries. The gap between the adrenergic nerve terminals and the nearest muscle cells is quite large, usually greater than 0.08 μ, and muscle cells in the deepest layer of the media of large vessels may be several microns from the nearest nerve. Activation of these distant cells may be achieved by slow diffusion of the transmitter in some vessels, whereas in others, particularly those

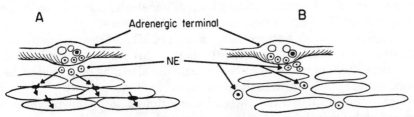

Fig 4–40.—Two types of adrenergic activation of smooth muscle. **A,** norepinephrine *(NE)* released from adrenergic terminals activates the superficial muscle cells. Excitation then spreads relatively rapidly from cell to cell via nexuses. **B,** *NE* diffuses through the vessel wall, activating each layer in turn.

with many nexuses, electric activity is transmitted cell to cell from the superficial layer adjacent to the nerve net (Fig 4–40).

The small blood vessels of skeletal muscle receive a dual sympathetic innervation, adrenergic—as already described—and cholinergic. The sympathetic cholinergic fibers are vasodilator and, in some species, are responsible for the anticipatory muscle vasodilatation that occurs before the onset of exercise.

Parasympathetic cholinergic fibers are distributed to arterial vessels only in relatively localized vascular areas such as the salivary glands and genitalia.

EXCITATION-CONTRACTION COUPLING

Resting smooth muscle cells contain higher concentrations of Na^+ and Cl^- than does cardiac muscle and the permeability to these ions is also greater. This causes the resting membrane potential to be less negative than cardiac muscle. It varies among different blood vessels but is usually between -40 and -60 mv. Steps in the sequence of events leading to contraction vary widely among different vessels, although in each case the end result is increased intracellular calcium ion concentration and actin-myosin interaction. In some vessels the excitatory stimulus leads to depolarization with or without action potentials. The depolarization is usually caused by increased permeability to Ca^{2+} rather than to Na^+, as in nerve and other types of muscle. Some vessels fail to show electric activity when excited but, nevertheless, intracellular Ca^{2+} rises and contraction occurs. This process has been variously termed "pharmacomechanical" or "nonelectric" coupling.

Most vessels that can be activated electrically also respond to vasoconstrictor stimuli after being depolarized by high external potassium ion concentrations. Both electric and nonelectric mechanisms must therefore be present in the same cell, and the relative importance of each when the cell is normally polarized remains to be determined.

The source of the activator calcium may be extra- or intracellular. The response of some vessels, e.g., cat mesenteric artery, to vasoconstrictor stimuli is greatly reduced by a few minutes' exposure to calcium-free solutions, and the activator calcium is therefore mainly extracellular. Other vessels, e.g., aorta and pulmonary artery, contract in calcium-free solution for several hours and therefore presumably have relatively large intracellular calcium stores that do not exchange freely with the extracellular fluid. Interestingly, there appears to be a rough correlation between the amount of SR and the ability to contract in calcium-free solution. This suggests that the SR may be an important calcium storage site. There is evidence that calcium also may be stored at the cell membrane and possibly also in the mitochondria.

Relaxation occurs when the intracellular calcium ion concentration is reduced either by expulsion of calcium across the cell membrane against the large concentration gradient or by sequestration in the SR, surface membrane or mitochondria.

VARIATIONS IN BEHAVIOR AMONG SMOOTH MUSCLES FROM DIFFERENT VESSELS

Vascular smooth muscle shows striking differences in behavior and reactivity depending on its source. At one end of the spectrum, the portal vein exhibits spontaneous rhythmic contractions associated with depolarization and bursts of action potentials that are conducted from cell to cell via nexuses. The contractions rapidly diminish in low-Ca^{2+} solutions. The vein contracts briskly when depolarized by high-K^+ solutions but, after a minute or two, it partially relaxes. At the other extreme, the aorta does not show spontaneous activity and the activation-contraction coupling mechanism appears to be mainly nonelectric. It contracts for prolonged periods in low-Ca^{2+} solutions and produces a sustained contracture in high-K^+ solution. Most vessels show properties intermediate between these extremes, the variations being so great as to defy classification. In general, the arterioles and venules have characteristics resembling those of the portal vein, whereas the larger vessels behave more like the aorta.

VASCULAR TONE

Some vessels show constriction even in the absence of external neural and chemical constrictor influences. This intrinsic tone fluctuates particularly in the smaller vessels and often produces oscillations of flow, with a period of 20–40 seconds. This behavior is termed *vasomotion* and its periodicity suggests the presence of pacemaker cells and propagation of electric changes throughout the tissue. Studies of the spontaneous rhythmic activity of the portal vein have shown that the pacemakers are not fixed and that the dominant pacemaker can move from one part of the vessel to another. Pacemaker frequency is increased when the muscle is stretched and this may in part explain the phenomenon of autoregulation, i.e., the tendency of flow through some tissues to remain constant despite alterations in perfusion pressure (see section on regional circulations). When pressure increases, the muscle is stretched, the pacemakers discharge more frequently and tone is augmented. The resulting increase in vessel resistance balances the raised pressure head and so flow remains constant.

The intrinsic level of tone is modified under physiologic conditions by numerous metabolic and neural influences. In some tissues the vessels are maintained in a state of partial constriction by the continuous activity of the sympathetic nervous system. These vessels are said to possess *neurogenic tone.* It can be readily demonstrated in skeletal muscle blood vessels by cutting the sympathetic nerves; blood flow immediately increases several-fold. In contrast, the gastrointestinal vessels have little neurogenic tone, and cutting the sympathetic nerves causes little change in flow. The variations in neurogenic and metabolic control of the vasculature of the various tissues will be further discussed when we consider the regional circulations.

THE ARTERIES AND ARTERIAL PRESSURE

The adult human aorta is about 2.5 cm in diameter and its wall is 2 mm thick. It has numerous branches, which successively divide and become smaller.

Structure of aorta and major branches.—The arterial wall is composed of three layers arranged in concentric sleeves. The innermost layer, or tunica intima, consists of endothelium resting on a basement membrane beneath which are collagen and elastic fibers, fibrocytes and modified smooth muscle cells. The middle sleeve, or tunica media, contains elastin and collagen fibers and smooth muscle cells. It is separated from the inner coat by the *internal elastic lamina.* The outer layer, or tunica adventitia, is loose connective tissue containing elastin and collagen. The walls of the large arteries are nourished by small blood vessels, the vasa vasorum, which form capillary networks within the adventitia and outer media.

Structure of smaller arteries and arterioles (Fig 4–41).—The tunica intima of the smaller arteries consists of an endothelium, which is separated from the tunica media only by the internal elastic lamina. The media contains relatively less elastin and collagen and more smooth muscle than do the large arteries. The smooth muscle cells of the arterioles are spirally arranged around the long axis of the vessel, with relatively few connective tissue elements between them. There are no elastic laminae in the smallest ar-

Fig 4–41.—Diagrammatic cross section of an artery (not drawn to scale).

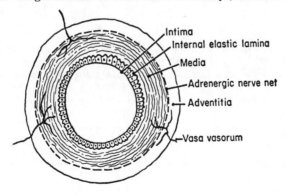

Intima
Internal elastic lamina
Media
Adrenergic nerve net
Adventitia
Vasa vasorum

terioles, and the endothelium with its basement membrane rests directly on the media.

PHYSICAL PROPERTIES OF ELASTIN AND COLLAGEN

Elastin is a fibrous protein that is easily stretched. Microscopic sections of constricted arteries show that the elastin fibers are coiled, whereas in dilated vessels the coils are straightened out. Collagen resists stretch much more than elastin but the arrangement of the fibers is slack and they do not begin to stretch until the vessel is significantly dilated.

ELASTIC DIAGRAM OF ARTERIES. — If fluid is injected into an isolated arterial segment and the resulting pressure changes are measured, the changes in arterial wall tension in response to stretch can be calculated. The Laplace relationship for a cylinder indicates that $T = Pr$ where T (dynes/cm) is the wall tension defined as the force that would act upon each centimeter of a hypothetical longitudinal split made in the wall of the vessel (Fig 4–42), P (dynes/sq cm) is the pressure difference across the wall and r is the radius calculated from the known volume of the segment and its length. When T is plotted against percentage increase in initial circumference, an "elastic diagram" of the vessel is obtained. The elastic diagrams of the aorta of a young adult and an elderly subject are shown in Figure 4–43. The curve from the younger individual shows that the vessel can initially be stretched considerably with little increase in wall tension. During this period the stretch occurs in the endothelium, smooth muscle and elastic fibers, which elongate readily with little tension development. Further stretch takes up the slack in the collagen fibers and it then becomes more and more difficult to dilate the vessel, so that small volume increments induce large pressure changes and correspondingly high wall tensions. In older persons much of the elastic tissue has degenerated. Collagen content is increased and tethering develops between collagen fibers, so that the artery wall becomes stiffer and develops high wall tensions with much smaller volume increments.

RELATION OF ARTERIAL WALL STRUCTURE TO FUNCTION

The thickness of an arterial wall is directly related to the tension that it must sustain. The tension depends on the transmural pressure and radius (Table

Fig 4–42. — The law of Laplace applied to blood vessels. T = wall tension, defined as the force that would act upon each centimeter of a hypothetical longitudinal split in the vessel wall; P = pressure difference across the wall; r = radius. (See text.) (From Burton, A. C.: *Physiology and Biophysics of the Circulation* [2d ed.; Chicago: Year Book Medical Publishers, Inc., 1972]. Used by permission.)

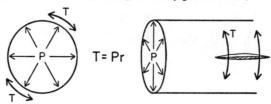

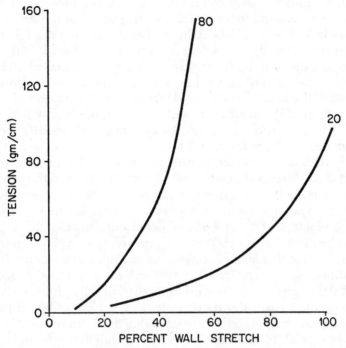

Fig 4–43.—Elastic diagrams of aortas from subjects aged 20 and 80 years.

4–3). It should be noted that the pressure in the arteriole is just under half that in the aorta but, because the arteriole has a much smaller radius, its wall tension is less than 1/1,000 that of the aorta. Consequently the wall of the arteriole is 1,000 times thinner than the aortic wall. It will also be noted that the aorta and its major branches contain much elastic tissue, whereas the small arteries contain relatively more muscle and less elastin and collagen. The elastic tissue in the aorta and its large branches has an important hemodynamic function. The left ventricle ejects 60–80 ml of blood into the arterial system during each systole. Some of the energy associated with this expulsion causes

TABLE 4–3.—RELATION OF ARTERIAL WALL TENSION
TO TRANSMURAL PRESSURE AND RADIUS

ARTERY	MEAN PRESSURE (MM HG)	RADIUS (CM)	WALL TENSION (DYNES/CM)	WALL THICKNESS (CM)	COMPOSITION
Aorta	85	1.2	136,000	0.2	Much elastic tissue
Small artery	75	0.3	30,000	0.06	Less elastic tissue than aorta; more smooth muscle
Arteriole	40	0.002	106	0.002	Very little elastic tissue; 1–3 layers of smooth muscle

blood to move from the arteries into the capillaries and veins. Much of the energy, however, is used to stretch the elastic tissue of the arterial system, which expands and "stores" blood during systole. During diastole the elastic recoil of the arterial walls exerts pressure on the contained blood and causes flow to continue through the systemic capillaries during diastole. Thus part of the energy released during cardiac muscle contraction is stored in the elastic tissue of the arterial system and is used to maintain diastolic flow, i.e., the phasic output of the heart is converted to continuous flow in the capillaries. The elastic properties of the arterial system greatly modify the pressure pulses produced by the ventricles. The arterial pressure pulse rises and falls more slowly than does ventricular pressure and arterial diastolic pressure is maintained at 60–90 mm Hg instead of falling to 0–10, as in the ventricles.

Storage of energy by the arterial walls reduces the total energy that the heart needs to expend to maintain a given flow rate. Let us suppose that the tissues demand a blood flow of 4,800 ml per minute and that this is being met by the heart beating at a rate of 60 per minute and ejecting 80 ml per beat. Let us suppose also that this 80 ml is ejected in one third of a second. If capillary flow is continuous, the capillary flow rate will be 80 ml per second. On the other hand, if the arterial system were composed of rigid pipes, no flow would occur through the tissues during diastole, and the entire 80 ml per second requirement would have to be delivered in one third of a second, i.e., at a rate of 240 ml per second. In order to deliver such a high flow rate, the mean pressure generated by the heart would have to be three times as great as in the continuous flow model and would require a larger energy expenditure. Another important consequence of the elasticity of the arterial system is that significant coronary flow can be maintained during myocardial relaxation. When the heart contracts, the deeper coronary vessels are exposed to pressures equal to or greater than aortic systolic pressure. The deeper myocardial cells can therefore be perfused only during diastole and an adequate diastolic pressure is therefore essential (see section on coronary circulation).

The arterial wall becomes stiffer with age and a given stroke volume then produces a higher systolic pressure and a lower diastolic pressure than in younger persons.

The diameter of all blood vessels is maintained in a state of equilibrium by two opposing forces, transmural pressure and wall tension. If transmural pressure increases, the vessel diameter tends to increase slightly, but at the same time wall tension increases and the vessel will remain stable because, at the new diameter, the raised transmural pressure is equally opposed by the increased wall tension. In certain diseases, including tertiary syphilis, the elastic tissues of the wall degenerate and the arterial diameter increases substantially even at normal transmural pressures. This is a perilous situation because once the artery begins to expand the wall tension required to maintain equilibrium rises even higher. Thus the weakened wall is progressively less able to maintain equilibrium and so dilatation progresses until rupture occurs.

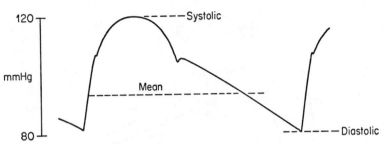

Fig 4–44.—Aortic pressure pulse.

ARTERIAL BLOOD PRESSURE

In most mammals mean arterial blood pressure is maintained at a relatively constant level between 75 and 100 mm Hg. Some fluctuation does occur. For example, blood pressure may fall 10–20 mm Hg during sleep and rise 20 mm Hg or more during severe exercise. Nevertheless it rarely changes to values outside the limits described. This constancy is essential for several reasons. First, arterial pressure is an important determinant of regional blood flow. This is particularly true of the brain, as shown by the fact that sudden falls in arterial pressure may produce fainting owing to cerebral ischemia. Second, arterial pressure is an important determinant of capillary filtration (see section on microcirculation) and influences glomerular filtration as well as general fluid exchange between plasma and the interstitial space. Third, arterial pressure is a major factor determining myocardial oxygen requirements. Fourth, abnormally high transmural pressures lead to degenerative changes in arteries resulting in eventual stenosis, thrombosis or rupture of major vessels with serious and often fatal results.

Arterial pressure changes continuously throughout each cardiac cycle (Fig 4–44). The highest pressure reached during systole is termed the *systolic arterial pressure* and the lowest pressure reached during diastole is called the *diastolic arterial pressure*. The *pulse pressure* is the difference between these two values. Mean arterial pressure is the average pressure during the cardiac cycle. It is *not* the arithmetic mean of the systolic and diastolic pressure because of the complex shape of the pressure curve. A good approximation of the mean pressure is usually given by the formula

$$\text{mean pressure} = \text{diastolic pressure} + \frac{\text{pulse pressure}}{3}$$

The shape of the instantaneous pressure curve and the values of the systolic, diastolic and pulse pressures are determined by stroke volume, rapidity of ejection and compliance of the arterial tree. As previously described, blood is stored in the aorta and its branches during systole. If the vessels are very compliant, a large volume can be stored with little rise in pressure. A reduction in compliance causes a higher systolic arterial pressure to be reached for a given stroke volume. It should also be obvious that the larger the stroke volume the higher the systolic arterial pressure.

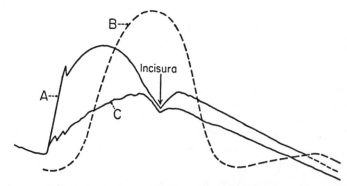

Fig 4–45.—Simultaneously recorded pressure pulses from the ascending aorta (*A*) and femoral artery (*B*) of a normal person. Note that the femoral pulse is delayed, the pulse pressure is greater and the contour is smoother, with a loss of the high frequency components seen in the aortic pulse. *C* shows the pressure curve in the ascending aorta of a patient with aortic stenosis. Note the slow rise and the small pulse pressure.

The arterial pressure curve usually shows a sharp upstroke owing to the rapid ejection of blood in the early ejection phase of the cardiac cycle. This upstroke becomes less steep when the ejection rate is reduced. In aortic stenosis, in which left ventricular outflow may be obstructed by a distorted, thickened and immobile aortic valve, the upstroke may rise very slowly and the pulse pressure is reduced (Fig 4–45). These are important diagnostic physical signs in this condition. The completion of left ventricular ejection is marked by aortic valve closure, which produces a small pressure wave, the *dicrotic* wave, which is responsible for the small notch, the *dicrotic notch* or *incisura*, commonly seen in aortic pressure records.

The pressure wave generated in the aorta is transmitted very rapidly throughout the arterial system and becomes considerably modified as it does so. The higher frequency components, e.g., the dicrotic notch, become damped and reflections of the pressure wave from the periphery summate with the primary wave. Some components of the wave travel at different rates than others. These factors together cause the peripheral arterial pressure pulse to have a higher systolic pressure, a lower diastolic pressure and a smoother contour (see Fig 4–45).

DETERMINATION OF ARTERIAL PRESSURE

Measurement of arterial pressure should form part of the routine examination of every individual. Hypertension is an insidious disease that may be symptomless for years while producing progressive degenerative changes in important arteries. Since effective treatments are now available, early detection may prevent the development of these changes. Arterial pressure is routinely measured by the cuff method. A cuff is placed around the upper arm and inflated until the radial pulse disappears. A stethoscope is placed over the brachial artery at the elbow and the cuff pressure is slowly reduced. When cuff pressure exceeds brachial artery pressure, the brachial artery is collapsed and no sound is heard. When cuff pressure falls just below systolic pressure, the collapsed artery is briefly snapped open during systole and the stethoscope at the elbow detects the sound produced by the jet of blood escaping under the cuff. As the cuff pressure is progressively lowered, more blood escapes under the cuff with each beat and the sounds become louder. When the cuff pressure falls just below diastolic pressure, the artery remains open throughout the cardiac cycle and the sounds become muffled and quickly disappear because continuous flow is silent. The sounds heard during blood pressure determination are termed the *Korotkoff* sounds after the Russian physician who first described them. The onset of the sounds indicates systolic pressure and the muffling indicates diastolic pressure.

The method is not accurate enough to monitor blood pressure in shock because the vessels may be so constricted that the sounds are difficult to hear and to interpret. Intensive care monitoring of blood pressure therefore requires arterial catheterization and electronic transducers.

REGULATION OF ARTERIAL PRESSURE

Mean arterial pressure is determined by the values of cardiac output and systemic peripheral resistance. These in turn are determined by the interaction of numerous variables, a few of which are indicated in Figure 4–46. The relative constancy of mean arterial pressure is achieved by continuous adjustments of these variables. One of the principal mechanisms involves reflex modulation of sympathetic and parasympathetic efferent nerve activity by nerve impulses arising in specialized receptors in the aorta and carotid arteries and their related vascular structures, the aortic and carotid bodies. There are two main receptor types, *baroreceptors* and *chemoreceptors*. The baroreceptors are sensitive to stretch, a function of pressure. The chemoreceptors respond to changes in the chemical composition of the blood.

The *baroreceptors* may be conveniently divided into high pressure and low

Fig 4–46. — Major determinants of systemic arterial pressure.

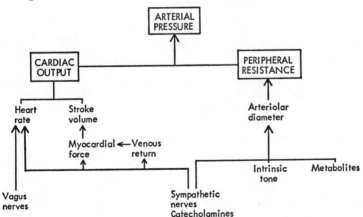

pressure groups. The high pressure receptors are found in the aortic arch and in the carotid sinuses, which are dilatations of the origins of the internal carotid arteries. They regulate heart rate, cardiac output and peripheral resistance and thereby maintain arterial pressure relatively constant. The low pressure receptors are located in the atria, particularly at their junctions with the venae cavae and the pulmonary veins. They also participate in regulation of arterial pressure but to a lesser extent than do the high pressure receptors. Their main function is to regulate blood volume.

HIGH PRESSURE RECEPTORS

These are undifferentiated nerve endings situated in the adventitia of the large arteries mentioned above. The carotid sinus terminals are branches of the carotid sinus nerves, which join the ninth cranial (glossopharyngeal) nerves. The other receptors are innervated by branches of the tenth cranial (vagus) nerves.

When pressure in the carotid sinus or aorta rises, the arterial walls and their receptors are stretched and the frequency of action potential discharge is increased. The receptor fibers have varying thresholds, i.e., some require a greater degree of stretch to initiate action potentials than others. Accordingly, as arterial pressure rises, a progressive recruitment of receptors occurs. The normal mean arterial pressure produces sufficient stretch to exceed the threshold of many of the receptors, which are therefore *tonically active* at physiologic pressures. At low pressures (<50 mm Hg) none of the receptors are engaged, whereas at high pressures (>200 mm Hg) all are active.

The receptors are sensitive not only to the level of mean pressure but to the

Fig 4–47.—Discharge *(vertical lines)* of a single baroreceptor fiber during a single cardiac cycle at different levels of aortic pressure. Note the absence of discharge at the lowest pressure, the greater frequency of discharge during the rising phase of the pressure pulse at immediate pressures and the continuous discharge at high pressures.

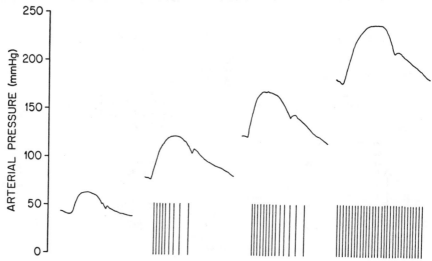

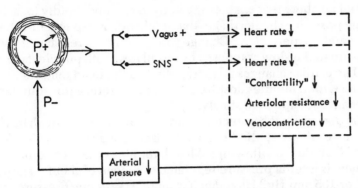

Fig 4–48.—Baroreceptor negative feedback regulation of arterial pressure. A rise in arterial pressure *(P+)* in a baroreceptor-region causes reflex changes in the circulation that tend to return *(P−)* baroreceptor pressure to its normal value. *SNS⁻* = reduced sympathetic nerve discharge; *vagus +* = increased vagus nerve discharge.

rate of change of pressure. Moreover, receptor discharge is increased more by rising pressures than by falling pressures, with an equal rate of change. Because of these properties the frequency of firing increases during systole and declines during diastole when arterial pressure is in the physiologic range. At very high pressures the receptors discharge continuously, and cyclic variations are not seen (Fig 4–47).

The baroreceptor fibers synapse with cells in the medulla oblongata and perhaps in higher levels of the brain stem. The precise anatomical connections are unknown and the nature of the central processing of the afferent impulses is poorly understood, although the end results of baroreceptor stimulation are relatively clear-cut. Increased impulse frequency in the carotid sinus nerve causes decreased sympathetic discharge and a reciprocal increase in vagal activity. Conversely, decreased impulse frequency in the carotid sinus nerve reduces vagal discharge and increases sympathetic activity. Consequently, elevated arterial pressure, which stretches the baroreceptor fibers, induces reflex changes that tend to restore arterial pressure to its previous level (Fig 4–48). These changes include cardiac slowing, owing to vagal activation and sympathetic inhibition, and reduced myocardial force and peripheral vasodilatation, owing to sympathetic inhibition only (it should be remembered that parasympathetic control of ventricular myocardium and most peripheral vessels is weak). On the other hand, a fall in arterial pressure reduces the degree of stretch of the baroreceptors and causes reflex sympathetic activation and vagal inhibition. Heart rate, cardiac output and peripheral resistance increase and arterial pressure tends to be restored. The entire system operates as a *negative feedback regulator* to maintain arterial pressure constant, i.e., a pressure change in one direction induces cardiac and vascular adjustments that produce an opposing pressure change. The maximum sensitivity of the system occurs at normal mean arterial pressures, an indication of its importance in the continuous regulation of arterial pressure.

The cardiac responses to alterations of baroreceptor activity are very rapid. Changes in heart rate and inotropic state occur within a single second. This is an important attribute that enables the body to respond rapidly to factors, such as changing posture or hemorrhage, that might otherwise reduce arterial pressure and cerebral perfusion. Changes in peripheral resistance, which are dependent upon the slow responses of vascular smooth muscle, take longer to develop. The cardiac changes adapt rapidly and the heart tends to return to its control state within minutes despite continuing baroreceptor stimulation. The vascular changes adapt more slowly.

Since the discovery of baroreceptors by Koster and Tschermak in 1902, they have been extensively studied in anesthetized and awake animals. The questions posed include the following: What determines the "set point" of the system, i.e., how is arterial pressure regulated at a mean of about 90 mm Hg rather than 65 or 105 mm Hg? How much change in systemic pressure is induced by a unit change in baroreceptor pressure under different conditions, i.e., what is the *gain* of the system and what controls it? How does the system react to phasic changes in baroreceptor pressure? How does the pattern of output (systemic arterial pressure change) relate to the pattern of input (baroreceptor pressure change)? Is the alteration of heart rate during baroreflex activity mainly a vagal or a sympathetic function? Do the pressure responses depend chiefly on alterations of cardiac output or of peripheral resistance, and does the relative importance of these factors vary under different conditions? Which regional blood flows are affected during baroreflexes and to what extent do the changes depend on the magnitude and time course of the baroreceptor stimulus? Is there any difference in function between aortic and carotid baroreceptors?

The answers to most of these questions are not available in man, and even in animals many aspects of baroreceptor function remain controversial. In general it appears that the carotid baroreceptors are more sensitive than the aortic and that changes in heart rate and blood pressure are mainly achieved by changes in vagal activity when mean arterial pressure is within the physiologic range. At low arterial pressures, changes in sympathetic activity assume greater importance and contribute increasingly to the production of cardiac acceleration, increased cardiac output and increased peripheral resistance. Vasoconstriction occurs mainly in the arterioles of skeletal muscle, but also to a smaller though significant extent in the renal and mesenteric arterioles. In dogs the aortic baroreceptors have a considerably higher threshold than do the carotid baroreceptors. They appear to play no significant role in counteracting hypotension but act with the carotid baroreceptors in opposing hypertension. The relative sensitivity of the aortic and carotid receptors in man is unknown.

LOW PRESSURE RECEPTORS

The atria contain myelinated nerve endings, especially at the junctions of the venae cavae with the right atrium and of the pulmonary veins with the left atrium. They are tonically active, and bursts of action potentials can be demonstrated with each cardiac cycle. The afferent fibers are vagal but the central connections have not been worked out in detail. Their activity affects heart

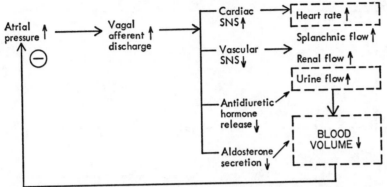

Fig 4–49.—Effects of increased atrial pressure on the circulation and on blood volume. A rise in atrial pressure causes reflex changes, which oppose (⊖) this increase. *SNS* = sympathetic nerve discharge.

rate, arterial pressure and regional blood flow, especially in the renal and splanchnic areas. However, the principal function of the low pressure receptors is concerned with the control of *blood volume.*

The effects on heart rate are variable. Distention of the pulmonary vein – left atrial junctions in dogs activates the sympathetic nerves to the sinus node and produces tachycardia. However, if the initial rate is high, similar distention produces cardiac slowing. The role of the atrial receptors in regulating heart rate in man is uncertain.

Distention of the atria, particularly the left atrium, inhibits sympathetic discharge and increases blood flow in renal and splanchnic vascular beds. Cardiac output changes little but arterial pressure falls. Urine flow increases, partly because of the increased renal flow but mainly because of a reflex reduction in secretion of antidiuretic hormone (ADH). Aldosterone secretion is also reduced and sodium excretion is accordingly increased (see Chapter 10). These changes together tend to reduce blood volume and atrial distention and therefore close a negative feedback loop (Fig 4–49).

OTHER CARDIAC RECEPTORS

In addition to the receptors discussed above, the heart and the pulmonary arteries of some animals contain many nonmyelinated fibers. These appear to be stimulated only by *excessive* volume changes or by certain synthetic chemicals such as veratridine or capsicain. Stimulation causes hypotension and vagal bradycardia. Their physiologic significance is not clear but it has been suggested that they may play a role in "vasovagal" fainting responses in man.

ARTERIAL CHEMORECEPTORS

Receptors sensitive to the chemical composition of the blood are located in the *carotid* and *aortic* bodies. The carotid bodies are small, very vascular structures approximately 6 × 2.5 × 1.5 mm located near the bifurcation of

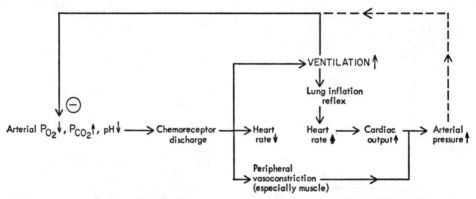

Fig 4–50.—Chemoreflex changes induced by a fall in arterial P_{O_2} or pH or a rise in P_{CO_2}. Ventilation is increased and this opposes ($\ominus$) these changes and tends to maintain the variables at constant levels. Cardiovascular changes also are produced that are of lesser importance but assist also in blood gas regulation. Heart rate rises in intact animals because of reflexes induced by increased lung movements, but in experimental situations with controlled ventilation the heart rate slows.

each common carotid artery. Each is composed of irregular groups of epithelioid cells resting against the endothelium of the numerous sinusoids that pervade the tissue. There is a rich innervation, with numerous afferent fibers that join the carotid sinus nerve and also many efferent sympathetic terminals of axons derived from cells in the superior cervical ganglion. Some parasympathetic ganglion cells are also present. The aortic bodies are more diffuse and consist of irregular small masses of epithelioid cells lying between the aorta and pulmonary artery. They also are very vascular and densely innervated. Their efferent nerve fibers are carried in the vagus nerve. The blood flow to the chemoreceptor tissue is enormous, about 2,000 ml/100 gm per minute, and is about 40 times the average flow rate per 100 gm brain tissue.

The chemoreceptors are very sensitive to reductions in arterial P_{O_2} and to a lesser extent to elevations of arterial P_{CO_2}. When P_{CO_2} is constant, the chemoreceptor discharge frequency begins to rise when arterial P_{O_2} falls to 60–65 mm Hg. With continuing reductions in P_{O_2}, the discharge rate increases rapidly and becomes maximal when P_{O_2} has reached 30 to 40 mm Hg. Increases in arterial P_{CO_2} and hydrogen ion concentration also stimulate the chemoreceptors and enhance the response to low P_{O_2}. The manner in which changes in blood chemistry alter chemoreceptor discharge is disputed. Some authorities believe that they interfere with the sodium pump and thereby induce depolarization of afferent nerve terminals. Others believe that they release acetylcholine, which induces action potential generation.

The main function of the chemoreceptors is to regulate ventilation (see section on control of ventilation, Chapter 5) but they also induce definite cardiovascular changes (Fig 4–50): Chemoreceptor stimulation causes widespread vasoconstriction, increased arterial pressure and tachycardia owing to increased sympathetic efferent discharge. The vasoconstriction is most pronounced in skeletal muscle vessels but is also significant in the mesenteric and renal vascular beds. Cutaneous vessels are not constricted. The tachycardia is attributable to a reflex induced by increased lung inflation secondary to stimulation of ventilation. When ventilation is controlled (e.g., by a pump), hypoxia and hypercapnia may induce cardiac slowing.

OTHER REFLEXES AFFECTING THE CIRCULATION.—Many sensory inputs affect the cardiovascular system apart from those discussed above. Pain of moderate severity frequently causes tachycardia, peripheral constriction and elevated arterial pressure. On the other hand very severe somatic pain (e.g., crush injuries) and some types of visceral pain (e.g., perforated peptic ulcer) may produce cardiac slowing and hypotension. Cooling of the skin constricts, whereas warming dilates the skin vessels (see section on cutaneous circulation).

Afferent impulses from moving joints and muscles may be partly responsible for the circulatory changes of exercise (see section on circulation in skeletal muscle), but their precise role remains unclear.

CENTRAL ORGANIZATION OF CARDIOVASCULAR REGULATION

If the spinal cord is cut across at the first cervical segment, blood pressure falls precipitously and stimulation of the baroreceptors no longer produces vascular changes. On the other hand, section at the level of the upper pons has little effect on arterial pressure, and cardiovascular reflexes are maintained. This suggests that the medulla is the most important area for integrating afferent impulses of relevance to the circulation and for producing appropriate outputs to regulate cardiovascular functions. This is supported by the fact that stimulation of certain loci ("pressor areas") within the medulla raises blood pressure, whereas stimulation of other medullary sites ("depressor areas") produces hypotension. The various pressor and depressor sites are scattered diffusely through the medulla and considerable overlap occurs. Some progress is being made in animals in unraveling the interconnections among these areas as well as the manner in which they are linked to the sympathetic and parasympathetic outflows. Little is known of the internal organization of these neurons in man.

The hypothalamus is responsible for integrating some of the more complex cardiovascular responses and may also modulate the activity of some of the medullary "centers." Electric stimulation of hypothalamic sites ("alarm centers") in certain animals increases heart rate and arterial pressure and redistributes blood flow. Renal and gastrointestinal vessels constrict because of increased sympathetic adrenergic activity, whereas skeletal muscle vessels undergo sympathetic *cholinergic* dilatation. Other hypothalamic centers are concerned with body temperature regulation, and their stimulation produces striking changes in skin blood flow. The hypothalamic-pituitary axis is also responsible for regulating the volume and osmotic pressure of the body's fluid compartments, and this function is influenced by impulses received from the baroreceptors, particularly the low pressure atrial receptors.

The cerebral cortex clearly can produce profound circulatory changes, as shown by emotional fainting and tachycardia. It is also involved in regulating cardiovascular adjustments to exercise.

The CNS, at various levels, constantly receives information from the baro-receptors and chemoreceptors as well as less specific information from the skin, musculoskeletal system and viscera. The individual inputs may call for conflicting responses, as for example during prolonged severe exercise at high temperatures. The optimal exercise response as far as the exercising muscle is concerned is vasodilatation in the muscle and vasoconstriction elsewhere. However, the optimal thermoregulatory response demands skin vasodilatation and muscle constriction. At the same time, painful stimuli from the aching muscles and joints demand attention and, if sweating is profuse, the atrial receptors will signal the need for appropriate vascular adjustments to volume depletion. The conflicting needs require complex interactions between the medullary, hypothalamic and cortical cardiovascular centers, with the need to facilitate some circuits and to inhibit others so that the optimal response to complex situations can be made. How this is done is unclear and will probably remain so for many years because of the difficulty of studying the system in man. The cardiovascular control mechanism remains a "black box." We know a little about what goes into it and quite a lot about what comes out but we know practically nothing of the circuitry within the box, let alone how it works. We do know, however, that it functions amazingly well, as shown by its ability to regulate the circulation to cope with a wide range of stresses including weightlessness, deep sea diving and exercise in arctic and tropical climates.

THE VEINS

Veins are the vessels that carry blood from the capillaries back to the heart. Systemic veins carry relatively deoxygenated blood from systemic organs and tissues to the right atrium; the pulmonary veins return oxygenated blood from the lung to the left atrium. Veins are thin walled because of their low wall tensions and can undergo much larger changes in volume than can arteries or capillaries. The venous system is about 20 times as compliant as the arterial

Fig 4–51.—Comparison of pressure-volume relationships of arteries and veins. (From Burton, A. C.: *Physiology and Biophysics of the Circulation* [2d ed.; Chicago: Year Book Medical Publishers, Inc., 1972]. Used by permission.)

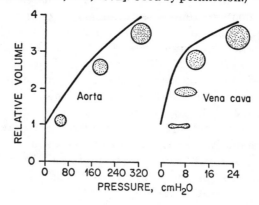

system over their respective ranges of physiologic pressures, i.e., for every mm Hg rise in pressure, the volume increment of veins is 20 times greater than that of arteries. This difference is partly attributable to the fact that veins are usually partially collapsed at low pressures and have flat elliptic cross sections. Increases in volume cause the ellipse to become rounder and then circular. During these changes pressure increments are small (Fig 4-51). Once the cross section becomes circular, the collagen fibers in the wall become taut and impose a much greater resistance to further distention, so that further changes in volume can be achieved only by relatively large increases in pressure. Arteries on the other hand have stiffer walls, the cross section is circular at low pressures and relatively large pressure changes are required from the outset to stretch the collagen and elastic tissues in their walls.

VEINS AS BLOOD RESERVOIRS

The cardiovascular system contains about 5 L of blood. Approximately 65% of this is in systemic veins and venules, 12% in arteries and arterioles, 8% in the heart, 10% in lung vessels and 5% in systemic capillaries. Thus the veins may contain about 3,500 ml of blood, which can be considered as a reservoir from which the heart is filled. The actual amount of blood contained in the reservoir is variable since it depends on the pumping action of the heart, degree of contraction of the smooth muscle in the vein walls and pressures inside and outside the veins. Two clinical examples of changes in reservoir level may be quoted. If the heart fails, blood "backs up" in the reservoir, and venous pressure and volume increase. On the other hand, severe bleeding may deplete the reservoir, the venous volume and pressure diminish, cardiac filling is impaired and this in turn may lead to reduction in arterial pressure.

VENOUS PRESSURE

Blood enters the systemic venous system via the venules at a pressure of 12-15 mm Hg when the body is horizontal. The systemic veins open into the right atrium where the average mean pressure is 0-5 mm Hg. The reference level for these pressures is a horizontal plane passing through the tricuspid valve at right angles to the frontal plane. It was chosen because the action of the heart keeps venous pressure at this level constant and largely independent of hydrostatic factors. Thus if pressure tends to rise, the right ventricle contracts more vigorously (Starling mechanism) and this causes venous pressure to fall back to its previous level. Conversely, if venous pressure tends to fall, the ventricle contracts less strongly, causing pressure to rise to its initial level.

Systemic venous pressure can be altered by factors that influence blood flow into and out of the venous system, by external pressures, by changes in blood volume and by the activity of venous smooth muscle. Pressure in the veins is partly due to the residual energy of left ventricular contraction after frictional losses in arteries, arterioles and capillaries. Consequently any factor that causes arteriolar constriction increases the energy loss and will decrease venous pressure and volume. Conversely, arteriolar dilatation tends to increase venous pressure and volume.

Systemic venous pressure is highly dependent on right atrial pressure. If this rises as the result of right ventricular failure or tricuspid valve disease, pressure in the venae cavae and their tributaries rises and systemic veins become distended. The elevated venous pressure may also cause increased capillary transudation, and fluid may accumulate in the abdominal cavity (ascites) or in subcutaneous tissues (edema).

External pressure changes produce marked alterations in venous pressure, volume and flow. When a subject breathes in, intrathoracic pressure falls and abdominal pressure rises. The intrathoracic veins therefore passively dilate, whereas the abdominal veins become smaller. The pressure gradient between veins outside and inside the chest increases and, accordingly, venous flow toward the heart is augmented. During expiration intrathoracic pressure rises and this impairs the flow of venous blood into the great veins within the thorax. Pressures in the veins outside the chest therefore passively rise until the requisite pressure gradient for flow is re-established. This is clearly demonstrated by the distention of neck veins that occurs during prolonged expiratory efforts, e.g., in singers sustaining a long note.

Veins can also be compressed and emptied by contractions of surrounding muscles and this is a major factor that helps to maintain venous return in exercise (see section on integrated responses of the circulation). The presence of valves in the limb veins ensures that external compression pushes blood only toward the heart.

EFFECTS OF POSTURE. — Gravity is a most important factor affecting venous function. When a person stands, the pressure in the veins of the feet is increased by the height of the column of venous blood between the feet and the heart. This depends on the height of the individual and may be equivalent to 80 mm Hg or more. Pressure in an ankle vein may therefore rise from 5 to 85 mm (Fig 4–52), substantially increasing venous volume. As much as 500 ml of blood may be sequestered in the veins below heart level. Of course, the arterial pressure in the feet also rises by a similar amount but, arteries being much less distensible than veins, their volume increment is relatively small. Since the pressure gradient from arteries to veins is unaltered, no direct changes in flow would be expected from the purely physical effect of postural change. However, pooling of blood in dilated veins reduces the return of blood to the heart. This may lead to such a reduction in cardiac output that arterial pressure falls and susceptible persons may faint. More usually the venous return is maintained by increased skeletal muscle tone, which reduces venous volume by external compression and acts as a secondary pump moving blood toward the heart. Also any tendency for arterial pressure to fall is usually counteracted by a reflex increase in sympathetic nerve activity engendered by the baroreceptors.

Although standing does not alter the pressure *gradient* across the vascular beds of the legs, it does increase the *transmural* pressure of all vessels including the capillaries. This increases the transudation of fluid across the capillary walls and explains the commonplace experience of a slight swelling of the

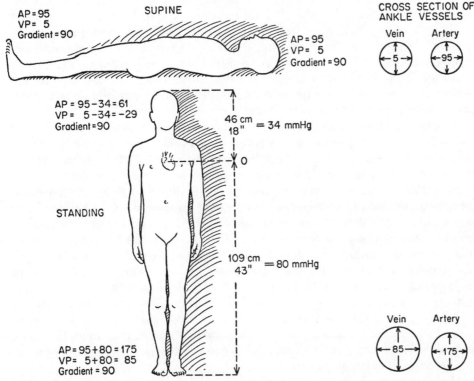

Fig 4–52.—Effects of posture on vascular pressures and volumes. In the supine position, vascular pressures in the feet and scalp are about equal. Cross sections of ankle vessels are shown. The numbers inside the vessels refer to *transmural* pressures. On standing, pressures in the ankle vessels increase by the height of the blood column between them and the heart (80 mm Hg). No change in arteriovenous pressure gradient (90 mm Hg) occurs, but the veins become much larger because of greater distensibility. Pressures in the scalp vessels are reduced as shown. *AP* = arterial pressure; *VP* = venous pressure.

feet after standing all day. This tendency to dependent edema is counteracted by the muscular contractions mentioned above, which tend to break up the long columns of venous blood. Thus ankle vein pressure that may be 80 mm Hg during quiet standing falls to 30–40 mm Hg during walking. Reflex sympathetic nerve activation also reduces capillary pressure by producing arteriolar constriction (see section on the microcirculation).

Veins above heart level also are affected by gravity. The pressure within a cerebral vein may be 5 mm Hg when the body is supine. In the upright position this is *reduced* by the height of the blood column between the vein and the heart, and cerebral venous pressure may fall to −30 mm Hg (30 mm Hg below atmospheric pressure) in adults. Arterial pressure will fall by a similar amount, e.g., from 90 to 60 mm Hg. The vessels in the brain are surrounded by cerebrospinal fluid, and the pressure within this fluid also falls by a corresponding amount so that *transmural* pressures of the brain vessels do not alter. However, the veins *outside* the skull will have subatmospheric pressures inside and atmospheric pressures outside and will therefore collapse. Consequently the superficial neck veins are not usually visible when a person sits with his back at a 45 degree angle with the horizontal. If venous pressure rises, e.g., in right ventricular failure, the level at which venous pressure becomes subatmospheric rises and the veins become easily visible at the base of the neck. The difference in height between the sternal angle and the point at which the neck veins disappear is a useful clinical measure of venous pressure and can be used to assess the severity and progress of cardiac failure.

NEURAL CONTROL OF VEINS. — Most veins possess a network of sympathetic nerve fibers. Stimulation of these nerves causes venous constriction that helps to maintain an adequate cardiac filling pressure during physiologic stresses such as exercise and in pathologic states such as hemorrhage.

VARICOSE VEINS. — When veins are exposed to high transmural pressures for prolonged periods, they become dilated and tortuous and are then termed varicose veins or varices. They are common in the legs and are the result of incompetence of venous valves. Normally the valves in the leg veins break up the hydrostatic columns of blood and, when the leg muscles contract, the deep veins are emptied toward the heart and so their internal pressure is lowered. When the valves are incompetent, blood may be driven in other directions, i.e., retrogradely or into the superficial veins. Pressure in these veins is thus maintained at an abnormally high level. Pregnancy aggravates this situation in a number of ways: blood volume is increased, the veins are dilated and raised intra-abdominal pressure caused by the enlarging uterus increases distal venous pressure.

Varicosities may also occur elsewhere, particularly in poorly supported veins such as the hemorrhoidal and spermatic veins. Abnormally raised abdominal pressure, e.g., straining due to constipation, prostatic obstruction or frequent coughing, favors the development of varicosities in these areas.

When portal venous pressure is raised, as in cirrhosis of the liver, vessels that drain into the portal system may become varicose. This is particularly common in the poorly supported lower esophageal veins, which may become very large, tortuous, thin walled and prone to rupture. Serious hemorrhage may result.

THE MICROCIRCULATION

The microcirculation is the circulation through vessels less than 100 μ in diameter. These include arterioles, capillaries and venules. The capillaries are the most important vessels of the circulation since the vital exchange of oxygen, carbon dioxide, nutrients and waste products between blood and tissues takes place across their walls.

Investigation of the microcirculation is beset with many difficulties. In vivo studies have been possible only in a restricted number of accessible tissues that can be illuminated sufficiently for microscopy. These tissues include the mesentery of various species: the bat's wing, rat cremaster muscle, hamster cheek pouch, rabbit's ear and human nail bed. The microcirculation in some of the viscera, e.g., stomach, spleen, liver and intestine also has been examined to a limited extent in animals, using transillumination methods.

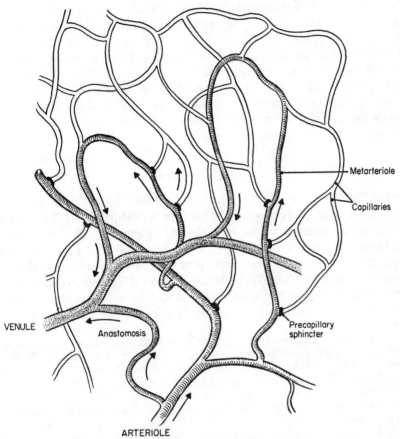

Metarteriole

Capillaries

VENULE

Anastomosis

Precapillary
sphincter

ARTERIOLE

Fig 4–53. — Microcirculation of the mesentery.

The mesentery was one of the earliest tissues to be studied. Its microcirculation is organized as shown in Figure 4–53. The arterioles are endothelial tubes surrounded by a continuous layer of smooth muscle. They give off branches, the metarterioles, which are narrower and have a discontinuous muscle coat. The metarterioles in turn give rise to the true capillaries, which form an anastomosing network with a very large surface area. The capillaries are 4–8 μ in diameter and have very thin walls composed only of endothelial cells and their basement membrane. Junctions of the true capillaries with the metarteriole are often surrounded by a single muscle cell, the precapillary sphincter. These sphincters regulate the number of open capillaries and therefore the exchange area but have less effect on blood flow because blood can still pass through the open metarterioles when the precapillary sphincters are closed. Blood from the capillary network passes into venules and then into veins. The smooth muscle of the microvessels in many beds is innervated by sympathetic nerves, which can therefore exercise a controlling influence on flow and exchange.

In some other tissues, e.g., skeletal muscle, anatomically distinct preferential channels and precapillary sphincters have not been identified. Instead the capillaries arise as multiple branches from lateral or terminal branches of arterioles. With this arrangement, the arterioles control both flow and exchange in the microcirculation.

The capillary bed in the skin, and certain other tissues, can be short-circuited by muscular vessels, *arteriovenous anastomoses,* which link arterioles and venules. These anastomoses are controlled by sympathetic nerves. When open, they divert blood from the exchange vessels, into the venous system. This is an important aspect of the thermoregulatory function of the skin (see section on the cutaneous circulation).

THE CAPILLARIES

The capillaries are endothelial tubes enclosed by a sleeve of basement membrane. In general the endothelial cell is very thin, about 0.1 μ, except in the region of the nucleus where it thickens to about 3 μ. The cytoplasm contains numerous vesicles, particularly near the cell surfaces. It has been conjectured that these vesicles may transport large molecules across the cell but this remains controversial. The cells are separated from each other by irregular clefts, which vary in width among different tissues in a way that appears closely related to their functional needs. In the skin the clefts are about 4 nm

Fig 4–54.—Various types of capillary endothelium. In each case only parts of two cells are shown. The *arrows* point to interval between the two cells. **A,** "tight" junction between two endothelial cells of a skin capillary. **B,** wider junction between two endothelial cells in the liver. The cytoplasm of the cell on the left is very thin and is fenestrated. Some fenestrae (e.g., *1*) are closed by diaphragms; others (e.g., *2*) are not. The basement membrane is discontinuous. **C,** fenestrated endothelial cells of the renal glomerulus. These fenestrae allow the passage of large molecules such as proteins. The relatively thick continuous basement membrane is one of the structural features that prevent these molecules from entering the urine.

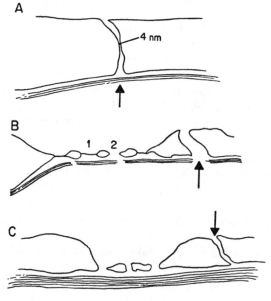

across at their narrowest. They permit free passage of water and small molecules but bar diffusion of large molecules such as proteins. In tissues in which exchange is rapid such as the liver and small intestine, the clefts are relatively wide and the endothelial cells are *fenestrated*, i.e., there are holes in the thin portions of the endothelial cytoplasm. The basement membranes are also thin and indistinct. The fenestrations have a diameter of about 50 nm. Some are sealed by a thin diaphragm; others are not (Fig 4–54). They are so numerous that the endothelium essentially offers no barrier to the passage of fluid and macromolecules (unless these are exceptionally large), and the interstitial fluid therefore contains relatively large amounts of protein. The endothelium serves only to the retain cells, platelets and the largest molecules. Water and dissolved substances pass freely out of the capillary, as water percolates through the soil of an irrigating ditch.

The capillaries of the renal glomerulus are also fenestrated and permit rapid filtration, but the basement membrane is thick and acts as a barrier that is partly responsible for preventing the passage of plasma proteins into the glomerular filtrate (see section on renal blood flow, Chapter 6).

Central nervous system capillaries are not fenestrated and the basement membranes are continuous. The clefts between adjacent endothelial cells are extremely narrow (1–2 nm) and many small molecules have difficulty in passing between them. This structural feature is in part responsible for the "blood-brain barrier" (see section on cerebral circulation).

THE INTERSTITIAL SPACE.—The properties of the interstitial space have an important bearing on the movement of water and solutes between blood and tissue. This is particularly so in areas supplied by fenestrated capillaries where the endothelial cells appear to have negligible exchange-regulating function. Unfortunately very little is known about the interstitial material. It appears to exist in two phases: a colloid-rich gel containing about 1% of hyaluronate and a colloid-poor fluid phase. The free-fluid phase appears to be contained in cylindric, tortuous channels, which can be partially demonstrated by tracer substances. The channels vary greatly in size under different experimental conditions. Presumably these variations influence the transport of substances through the space.

CAPILLARY FUNCTION

Two main processes occur across the capillary walls: diffusion and filtration-absorption.

DIFFUSION.—Diffusion (discussed in detail in Chapter 1) is by far the most important process for transferring nutrients and waste products between blood and tissues. Lipid-soluble substances such as oxygen and carbon dioxide diffuse rapidly down their concentration gradients across the whole surface of the lipophilic membranes. Water and water-soluble substances such as ions and glucose pass through the capillary wall at a slower rate, mainly through intercellular clefts and fenestrae. Macromolecules diffuse through the fenestrae but may also be ingested by endothelial cells and transported in vesicles from one surface of the cell to the other by a process termed *cyto-*

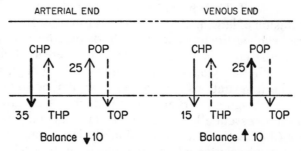

Fig 4–55.—Hydrostatic and osmotic forces governing filtration-absorption across capillary walls. At the arterial end, capillary hydrostatic pressure *(CHP)* exceeds plasma oncotic pressure *(POP)* and the pressure difference of 10 mm Hg causes filtration. At the venous end, plasma oncotic pressure exceeds capillary hydrostatic pressure and the pressure difference of 10 mm Hg causes absorption. Tissue hydrostatic *(THP)* and oncotic *(TOP)* pressures are considered to be small for the purposes of this diagram.

pempsis. The rate of diffusion of a substance is proportional to the area of the diffusing surface and this is controlled by the arterioles and precapillary sphincters. When previously closed capillaries open, the surface available for diffusion increases, diffusion distances decrease and concentration gradients rise. All these changes increase the diffusion rate.

FILTRATION-ABSORPTION.—Because the hydrostatic pressure within the capillaries is greater than is the hydrostatic pressure of the interstitial space, fluid tends to be filtered across capillary walls. It has been estimated that the volume filtered (exclusive of glomerular filtration) is about 20 L a day. The hydrostatic force causing fluid to leave the capillaries is offset by an osmotic force generated by the difference in protein concentration between intracapillary and interstitial fluids. This osmotic force causes absorption of about 18 L of fluid per day from the interstitial compartment. Thus filtration exceeds absorption by about 2 L per day, but the excess is returned to the blood via the lymphatic system. The interaction of transcapillary hydrostatic and osmotic forces (Fig 4–55), together with lymphatic function, ensures a remarkable constancy of the fluid partition between vascular and extravascular compartments.

HYDROSTATIC PRESSURES.—Pressure at the arterial end of the true capillaries averages 35 mm Hg. At the venous end it averages 15 mm Hg. These pressures may be quite variable depending on the activity of the arterioles and venules. The arterioles of many vascular beds exhibit vasomotion. When the arterioles are closed, capillary pressure falls to values close to venous pressure, i.e., 10–15 mm Hg. When they are dilated, capillary pressure rises to values that may be as high as 50 mm Hg. Precise measurements of capillary pressures are extremely difficult since they require cannulation of very minute vessels; the cannula may alter the flow and pressure in the vessel under study and so the values obtained may not be the same as those in the undisturbed state.

Tissue pressures are even more difficult to obtain. The consensus is that these pressures are usually small. Some authorities consider that the interstitial pressure may be subatmospheric in many tissues. Others believe that these "negative" pressures are artifacts owing to deficiencies in the methods used. In the following description of the balance of forces across the capillary wall, tissue pressure will be assumed to be negligible.

CAPILLARY PLASMA ONCOTIC PRESSURE.—The capillary wall is freely permeable to water and solutes but in general it does not permit passage of appreciable quantities of the plasma proteins that are present to the extent of 7 gm/100 ml in human plasma. The most abundant protein is albumen and this also has the smallest molecular weight. Because capillary walls are relatively impermeable to proteins, these substances exert an osmotic force, the *oncotic pressure,* which averages about 25 mm Hg. Part of the osmotic force of the plasma proteins is attributable to a Donnan effect (Chapter 1).

TISSUE ONCOTIC PRESSURE.—This pressure is extremely difficult to determine experimentally since all attempts to do so are bound to disturb the labile sol-gel condition of the interstitium. It has usually been considered to be negligible.

BALANCE OF FORCES ACROSS CAPILLARY WALL

Starling suggested that fluid was filtered out of the capillary at its arterial end because the forces favoring filtration (capillary hydrostatic pressure + tissue oncotic pressure) were greater than the forces favoring reabsorption (plasma oncotic pressure + tissue hydrostatic pressure). If tissue hydrostatic and oncotic pressures are assumed to be negligible, the driving force tending to filter fluid out of the arterial end of the capillary is $35 - 25 = 10$ mm Hg. The situation at the venous end of the capillary is reversed and there is a net force of $25 - 15 = 10$ mm Hg favoring reabsorption (see Fig 4–55). Because of the fluctuating state of arteriolar smooth muscle and the precapillary sphincters, it is probably extremely rare for such an accurate balance of forces to exist in any single capillary. Rather, some capillaries show net filtration, others net absorption. The overall effect, however, is that the volume of fluid filtered at the arterial end approximately equals that reabsorbed at the venous end of the vascular bed. Although Starling formulated this view several decades ago, its basic operation remains unquestioned; however, recent measurements indicate that some modification is needed. It is now clear that the capillaries on the venous side of most vascular beds have a surface area several times greater than have those on the arterial side. Moreover, the permeability of the "venous" capillaries is greater than those of the "arterial" capillaries. Both features would lead to a relative preponderance of reabsorption over filtration and, to explain the fact that there is actually a slight net filtration, it is necessary to postulate either a negative tissue pressure or a significant tissue oncotic pressure. These alternatives are currently being examined.

LYMPHATIC FUNCTION

The lymphatic system consists of a network of endothelial tubes. Their terminations are blind sacs lined by a low endothelium having overlapping cells with wide gaps between them. These gaps and the absence of a basement membrane permit free passage of interstitial fluid into the lymphatics. However, the overlapping endothelial cells act as "flap valves" that prevent flow in the reverse direction. The lymphatic terminals anastomose freely and join to form successively larger tributaries that have endothelial valves and a thin coat of smooth muscle cells. Lymphatic vessels convey lymph through aggregations of lymphoid tissue, the lymph nodes and eventually via one of two large lymphatic ducts into the blood vessels. The thoracic duct drains lymph from the entire body with the exception of the right side of the head and neck, right arm and right hemithorax, which are drained by the right lymphatic duct. The thoracic duct empties into the venous system at the junction of the left subclavian and internal jugular vein. The right lymphatic duct opens into the venous system at a corresponding position on the right.

The lymphatic system is responsible for returning a portion of the fluid filtered from the arterial ends of the capillaries into the blood. It also transports lymphocytes from the lymph nodes and large molecules, especially proteins, from the interstitial space. The continual withdrawal of protein from the interstitial fluid prevents its oncotic pressure from rising and so disturbing the vital filtration-absorption mechanism of the capillary wall. The lymphatic system also is a major route for absorption of fat, in the form of chylomicrons, from the intestine.

The pressure within the terminal lymphatics is 0-2 mm Hg. Fluid enters the lymphatics when interstitial fluid rises above this value. Lymph flow from muscles increases greatly during exercise because of greater capillary filtration and the squeezing of the interstitial fluid and lymphatics by contracting muscle. Respiratory pressure changes also affect lymph flow in the same way that they assist venous return. The lymphatic valves ensure that flow is unidirectional toward the large veins at the base of the neck. Active contractions of lymphatic smooth muscle and endothelial cells also promote lymph flow.

Thoracic duct lymph has a high protein content of 4-5 gm/100 ml, mainly derived from the hepatic and intestinal lymph that drains exclusively into this vessel. It will be recalled that the liver and intestinal capillary endothelia and basement membranes permit considerable protein leakage.

Because of the wide gaps between the endothelial cells of the terminal lymphatics, bacteria, blood cells and tissue cells can readily enter lymph vessels, which thus become important routes for the spread of infection and neoplasms. On the other hand, the lymphoid tissues are important components of the body's defenses because they can phagocytose bacteria and produce antibodies.

EDEMA

Edema is the excessive accumulation of fluid in tissue spaces. It occurs when there is an imbalance between the forces favoring filtration and absorption across the capillary wall or when lymphatic function is impaired.

Raised venous pressure, as in congestive heart failure (see below), increases capillary filtration and is a common cause of edema. The resistance of the venules is relatively low so that a rise in central venous pressure is effectively transmitted to the capillaries. Under normal circumstances a 10 mm Hg rise in venous pressure will increase capillary pressure by 8 mm Hg. Raised arterial pressure has much less effect on capillary pressure because of high arteriolar resistance. A rise of 10 mm Hg in arterial pressure produces only a 2-mm Hg increase in capillary pressure.

Reduced plasma oncotic pressure also causes edema when protein intake is severely reduced, as in malnutrition, or when severe protein losses occur, as in the *nephrotic syndrome* in which 20 gm protein per day may be lost in the urine.

Severe local edema may develop when the lymphatics are obstructed by cancer or chronic inflammation. A spectacular form of lymphatic edema, *elephantiasis,* occurs in certain tropical countries and is due to obstruction of lymphatics by the filarial parasite *Wuchereria bancrofti.*

Increased capillary permeability caused by bacterial or animal toxins or by histamine released during allergic responses may also produce edema.

In many cases the cause of edema is multifactorial. For example, in congestive heart failure there is not only an elevation of venous pressure but commonly a reduction in serum protein concentration owing to reduced food intake and impaired liver function. The poor peripheral circulation also causes increased capillary permeability. Additionally, sodium and water are abnormally retained, with a consequent expansion of extracellular fluid volume.

THE REGIONAL CIRCULATIONS

When the body is at rest, blood flow to the systemic organs and tissues is regulated to meet their metabolic demands. The major exceptions are the kidneys and under some circumstances the skin, which receive higher blood flows than required to sustain their metabolism. These higher flows are necessary for transport of waste products and heat generated by the body as a whole. During physiologic or pathologic stresses the distribution of cardiac output may be altered in order to assure adequacy of flow to the brain and heart. Under these conditions the remaining vascular beds may temporarily receive lower blood flows than are required for their functions. Generally speaking, the greatest variations in blood flow occur in skeletal muscle, skin, kidney, gastrointestinal tract and liver. Blood flow to the brain is relatively constant.

Determinants of Regional Blood Flow

Poiseuille's law (see section on hemodynamics), although an oversimplification, indicates that flow in a given vascular bed will be related directly to the arteriovenous pressure gradient across the bed and to the fourth power of the radius of the "resistance vessels" or arterioles. The other variables in Poiseuille's law, i.e., length of vessels and viscosity of blood, tend to be relatively constant in health and play no significant role in regulating regional blood flow.

Effects of changes in pressure gradient.—Mean arterial pressure is regulated within quite narrow limits by continuously operating baroreflex adjustments (see section on the arteries). Central venous pressure also is kept relatively constant by reflex mechanisms and by Starling's law of the heart. If central venous pressure rises, ventricular end-diastolic volume increases, and this leads to stronger myocardial contractions, which tend to reduce venous pressure toward normal levels. The arteriovenous pressure gradient across vascular beds therefore varies little under most physiologic conditions.

AUTOREGULATION.—Even when the pressure gradient across a vascular bed does change, blood flow may not alter significantly because of a process called *autoregulation*. Two major hypotheses have been proposed to explain this. The *metabolic* hypothesis states that the degree of contraction of resistance vessels is determined by the extracellular concentration of vasodilator metabolites. When blood flow is reduced, concentration of metabolites increases, arterioles dilate, vascular resistance falls and flow tends to return to its previous level. Conversely, an increase in blood flow "washes out" vasodilator metabolites, arterioles constrict, vascular resistance rises and flow returns toward control level. The *myogenic* hypothesis states that the degree of contraction of the resistance vessels is determined by their transmural pressure. When this pressure increases, the greater stretch of the smooth muscle cells evokes contraction (see section on vascular smooth muscle). When the pressure falls, the diminished stretch causes dilatation. Thus, alterations in arterial pressure are balanced by reciprocal changes in resistance and blood flow therefore tends to remain constant. It is possible, indeed likely, that both metabolic and myogenic mechanisms operate in most vascular beds but that the relative contribution of each varies from organ to organ. Metabolic autoregulation seems to be prominent in skeletal and cardiac muscle and in the brain. Myogenic autoregulation may be more prominent in intestine and liver. The kidney autoregulates exceptionally well but special mechanisms may be involved (Chapter 6).

Two aspects of autoregulation deserve emphasis. (1) Autoregulation is rarely complete so that flow alters when pressure changes, although to a much smaller degree than Poiseuille's law would predict, and (2) autoregulation occurs only over a limited pressure range, about 60–180 mm Hg.

EXTRAVASCULAR COMPRESSION.—Tissue pressure is normally low compared to intravascular pressures and does not significantly influence blood flow. Contracting muscle can, however, generate substantial pressures. Thus pressure in the subendocardial layer of the left ventricle may exceed aortic systolic pressure during ventricular contraction, and subendocardial blood flow may be halted or even reversed. Similarly, strong contractions of skeletal muscle may mechanically reduce blood flow in vessels that traverse it. Even smooth muscle may generate substantial pressures. For example, blood flow to segments of the intestine may be greatly reduced during strong contractions.

NEUROHUMORAL AND METABOLIC VASCULAR CONTROL.—The resistance vessels of most tissues are innervated by sympathetic nerves. Their effect depends on the density of innervation and on the number of alpha adrenotropic receptor sites, as well as on poorly understood extra- and intracellular environmental factors. The intestinal, splenic, hepatic, renal, cutaneous and resting skeletal muscle vascular beds are very strongly constricted by maximal sympathetic nerve stimulation, whereas cerebral and myocardial vessels show only minor constriction, which is readily overcome by local metabolites. As might be expected, the actions of circulating norepinephrine are qualitatively quite similar to sympathetic nerve stimulation, but plasma norepinephrine concentration is so low, except in severe stress, that its effects are small in comparison to the neural responses. In contrast, physiologic concentrations of epinephrine dilate skeletal muscle, coronary and intestinal vessels because they activate vascular beta adrenotropic receptors (Chapter 3). Higher epinephrine levels constrict these same vessels, presumably because alpha receptor activation predominates at these concentrations. The beta receptor activity of renal and cutaneous arterioles is negligible and these vessels are constricted by all doses of epinephrine. The parasympathetic nervous system innervates relatively few vascular beds including the salivary glands, pancreas, myocardium, external genitalia and parts of the brain.

Many other vasoactive substances may circulate in the blood. They include the vasoconstrictor peptides angiotensin and vasopressin, the vasodilator peptide bradykinin and several vasodilator gastrointestinal hormones (see section on the splanchnic circulation). Other vasoactive agents may enter the circulation in pathologic conditions. For example, histamine and serotonin may appear in the blood during allergic reactions.

The arterioles are strongly influenced by local chemical changes in the extracellular fluid that surrounds them. Reduction in oxygen tension, increase in carbon dioxide tension, increased potassium and hydrogen ion concentration and increased osmolarity usually cause vasodilatation. The relative roles of these factors will be discussed along with other vasodilator mechanisms when the individual vascular beds are considered.

It is obvious from the foregoing that control of resistance vessels involves

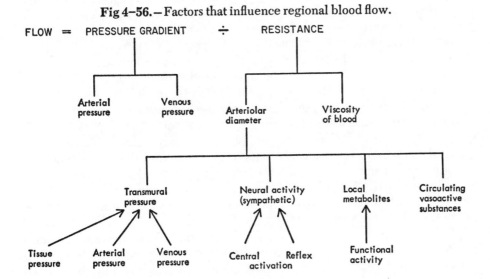

Fig 4–56.—Factors that influence regional blood flow.

interaction of physical, myogenic, neural, hormonal and chemical factors (Fig 4–56). Unfortunately for the student, the relative strength of these factors and the reactivity of the vessels vary from region to region so that the behavior of each vascular bed must be individually learned.

CIRCULATION IN SKELETAL MUSCLE

The average adult has 30 kg of skeletal muscle, which at rest consumes 50–60 ml of oxygen per minute (about 20% of the body's total oxygen consumption). Blood flow through resting muscle is 2–4 ml/100 gm per minute and about 6 ml of oxygen is extracted per 100 ml of blood flow per minute.

The chief characteristic of the circulation through skeletal muscle is its tremendous variability. The total blood flow through resting skeletal muscle is about 750 ml per minute or about 15% of the cardiac output. These values may rise during strenuous exercise to 20,000 ml per minute and 85% of the cardiac output.

As previously noted (Chapter 2), there are two types of skeletal muscle: white (or phasic) muscle constitutes 80–85% of the total, and the remainder is red (or tonic) muscle. White muscle is biochemically specialized for performance of rapid movements, red muscle for more prolonged contractions. These different types of muscle also show striking differences in blood supply and its control. Red muscle has a greater density of open capillaries and a resting blood flow that may be 10–20 times that of white muscle. Maximal flow is also higher in red muscle but only by a factor of 2 to 3. These differences are probably related to the more sustained contractions and greater sustained metabolic activity of red muscle.

REGULATION OF BLOOD FLOW TO RESTING MUSCLE

The blood vessels of resting skeletal muscle maintain considerable *myogenic tone*, i.e., their smooth muscle exhibits contraction even when deprived of extrinsic neural and chemical vasoconstrictor influences.

SYMPATHETIC NEURAL CONTROL.—In the resting intact animal, myogenic tone is enhanced by the continuous activity of the sympathetic nervous system. This is most clearly shown by the substantial, though transient, increase in muscle blood flow when the sympathetic nerves are cut (Fig 4–57). Similar effects can be shown in man by blocking the nerves with local anesthetic.

In the resting individual, less than 2 impulses per second pass along the sympathetic vasoconstrictor nerves to skeletal muscle vessels. Blood flow is reduced when this discharge rate is increased, and at rates of 10–15 per second it may fall almost to zero in white muscle. The arterioles in tonic muscle are less responsive, and maximal sympathetic stimulation will reduce blood flow to only 40–50% of control. The innervation of muscle veins is much less dense than that of the arteries and the veins are also less responsive to norepinephrine. Consequently, increased sympathetic nerve stimulation causes greater arteriolar than venular constriction and mean capillary pressure falls. This causes absorption of fluid from the extravascular spaces into the capillar-

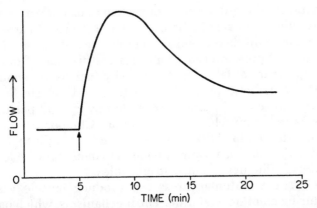

Fig 4–57.—Effect of cutting *(small arrow)* the sympathetic nerves supplying the blood vessels of skeletal muscle. Blood flow increases rapidly to several times the control value but then declines during the next few minutes because of autoregulatory influences.

ies. The reduction of flow into the veins, secondary to arteriolar constriction, also causes a passive recoil of the vein walls, which reduces the volume of the venous blood reservoir of this tissue. Normally, skeletal muscle contains 600–800 ml of blood or 12–16% of the entire blood volume, and sympathetic vasoconstrictor activation can mobilize 100–200 ml. Although most of this blood is expelled by passive recoil of the veins, a small volume is expelled by active sympathetic venoconstriction. Capillary surface area is initially reduced by sympathetic stimulation but the effect is relatively transient because it is soon counteracted by the accumulation of vasodilator metabolites.

SYMPATHETIC CHOLINERGIC VASODILATATION.—The skeletal muscle vessels of some species dilate when the sympathetic nerves are stimulated in the presence of a drug that blocks norepinephrine release. The dilatation is transient and fades within a minute or two despite continuing nerve stimulation. The response is blocked by pretreatment with atropine and must therefore be due to cholinergic vasodilator fibers in the sympathetic nerves. Cholinergic vasodilatation is readily demonstrated in dogs and cats by stimulation of an area in the lateral hypothalamus termed the "defense" area. Other components of the resulting cardiovascular response, sometimes called the "alarm" or "defense" reaction, include increased heart rate, cardiac output and systemic arterial pressure, and adrenergic vasoconstriction in gut, kidney and skin. This integrated response has been thought to be of value in preparing an animal for "fight or flight" reactions, although this has not been conclusively demonstrated. Species variation in the distribution of cholinergic vasodilator fibers is puzzling. These fibers have not been found in man, monkey, polecat, rat, badger and hare. The general significance of this vasodilator system is therefore obscure.

REFLEX CONTROL.—The blood vessels of skeletal muscle participate actively in circulatory reflexes. Changes in vascular tone are produced by variations in the discharge rate of adrenergic sympathetic vasoconstrictor fibers.

The sympathetic cholinergic vasodilator system does *not* participate in reflex responses. Stimulation of arterial baroreceptors by a rise in arterial pressure inhibits sympathetic discharge and the muscle vessels dilate. Conversely a reduction in arterial pressure causes a reflex constriction of muscle vessels. Stretch of low pressure atrial receptors, e.g., by increased venous return, reflexly inhibits sympathetic vasoconstrictor tone and muscle flow increases. Diminished venous return (caused for example by upright posture or hemorrhage) produces reflex muscle vasoconstriction. Chemoreceptor stimulation by reduction in arterial blood PO_2 or pH or by elevation of PCO_2 induces reflex vasoconstriction that may be powerful enough to overcome the local vasodilator effects of these influences on muscle arterioles.

EMOTIONAL STRESS.—Mental stress can produce muscle vasodilatation, particularly during emotional fainting. The mechanisms, which have not been fully worked out, involve inhibition of sympathetic constriction and increased secretion of epinephrine. A cholinergic vasodilator component also may be involved since atropine reduces the response.

HORMONAL CONTROL.—Infusion of norepinephrine into the arterial supply of skeletal muscle produces arteriolar and venous constriction owing to activation of the vascular alpha adrenotropic receptors. Blood flow is therefore reduced and blood content of the infused muscle declines. Capillary surface area is transiently reduced. Mean capillary pressure falls and fluid is absorbed from the interstitial space. These effects closely resemble those of sympathetic nerve stimulation. On the other hand, norepinephrine given *intravenously* may increase muscle blood flow because the rise in blood pressure may induce reflex vasodilatation, which overcomes the local constrictor response.

The effects of epinephrine are dose dependent. Intra-arterial infusions of physiologic concentrations stimulate the beta adrenotropic receptors and produce vasodilatation and increased blood flow. Larger doses produce constriction because the alpha constrictor response then outweighs the dilator effect. Epinephrine given intravenously usually increases muscle blood flow because of the combined effects of beta receptor stimulation and reflex vasodilatation.

Angiotensin, vasopressin and prostaglandin $F_{2\alpha}$ are potent constrictors, whereas bradykinin and the E prostaglandins are vasodilators.

Fig 4–58.—Reactive hyperemia *(RH)* of the skeletal muscle circulation following occlusions of the arterial supply for 5 and 20 seconds.

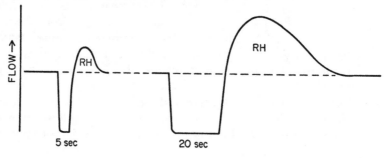

REACTIVE HYPEREMIA.—If skeletal muscle blood flow is arrested for a few seconds by clamping the supply artery, which is then released, blood flow rises transiently to a much higher level than in the preocclusion period. This "reactive hyperemia" subsides within seconds or minutes depending on duration of the preceding circulatory arrest (Fig 4–58). The mechanism of hyperemia probably involves both myogenic and metabolic factors. The fall in transmural pressure downstream from the occlusion reduces the myogenic stimulus to vasoconstriction. Circulatory arrest also allows accumulation of vasodilator metabolites, although there is no consensus as to which of the many proposed is mainly responsible.

EXERCISE

Severe exercise by trained athletes may increase the oxygen demand of exercising muscles more than 20-fold. This increased demand is met (1) by increased blood flow, (2) by increased oxygen extraction from the blood and (3) by changes in the oxygen dissociation curve that cause oxygen to be released more readily from hemoglobin. The overall cardiovascular adjustments to exercise are more fully discussed later in this chapter; only the *local* circulatory changes in skeletal muscle will be described in this section.

The increased muscle blood flow in exercise is brought about almost entirely by metabolic factors. Neural mechanisms play no significant role. This has most clearly been shown by the similarity of blood flow increases in sympathetically denervated and normal limbs during exercise. However, blood vessels of normally innervated *resting* muscle constrict during exercise. This implies that local factors associated with exercising muscle overcome the tendency toward neural vasoconstriction.

The metabolites responsible for exercise vasodilatation have been sought for decades but remain elusive. The venous effluent from exercising muscle contains higher concentrations of potassium, hydrogen and phosphate ions than does arterial blood. Interstitial oxygen tension of the exercising muscle is lower and carbon dioxide tension is increased. Adenosine, various nucleotides, Krebs cycle intermediates, lactate and histamine also are released, and local tissue osmolarity rises. All these agents have been proposed as the main cause of exercise hyperemia but no one factor completely reproduces all the effects observed. Probably a complex interaction of some or all of the substances named is responsible.

Exercise greatly increases the number of open capillaries in muscle. This increases the diffusion surface and diminishes diffusion distances for oxygen, carbon dioxide and the various metabolites. Mean capillary pressure also rises and is partly responsible for increased transudation of fluid from plasma into muscle interstitial space and for increased lymph flow. Transudation is aided by the rise in interstitial osmotic pressure caused by metabolic breakdown of large molecules into more numerous smaller molecules. Because of the increased capillary filtration, total plasma volume diminishes and hematocrit increases.

Most muscle groups are enclosed by tough fascial sheets. Therefore, contraction of muscles compresses the blood vessels within these compartments. This reduces arterial flow, but the resulting accumulation of metabolites relaxes arteriolar smooth muscle and enables greater flow during the succeeding period of skeletal muscle relaxation. Thus rhythmic exercise produces an overall increase in flow. Sustained skeletal muscular contraction causes a sustained reduction in flow, and fatigue and ischemic muscle pain may occur. Muscle contractions also compress the veins and, because of the valves, their contained blood is driven toward the heart. This is an important factor in maintaining adequate cardiac filling during exercise. Emptying of the muscle veins lowers their intraluminal pressure between contractions; the arteriovenous pressure gradient is thereby increased and a greater blood flow for a given degree of arteriolar dilatation can be achieved.

INTERMITTENT CLAUDICATION. — This term is applied to pain that develops when blood flow fails to increase sufficiently to meet the oxygen requirements of exercising muscle. The commonest cause is obstructive disease of the arteries. The pain is usually experienced in the calf but may also develop in the arch of the foot, the thigh or the buttock. The pain appears during exercise and is believed to be due to accumulation of an unknown metabolite, which stimulates sensory nerve endings in the ischemic muscle. The severity of the pain may make the patient stop exercising. During the enforced rest the concentration of the metabolites falls and the pain disappears.

CUTANEOUS CIRCULATION

The main function of the circulation through the skin is to participate in regulation of body temperature. The blood flow rate is therefore highly dependent on external climatic factors. When the ambient temperature is 25–30 C, a nude person of average weight has a skin blood flow of about 200 ml per minute, or 4% of the cardiac output. At very low external temperatures, flow may fall to 20 ml per minute and it may increase to as much as 3,000 ml per minute when the outside temperature is high. Thus skin, like muscle, has an extraordinary range of blood flow.

Organization of cutaneous blood vessels is well suited to thermoregulation. The arterioles break up into dense capillary networks beneath the epidermis and these networks drain into a capacious venous plexus. In some respects this plexus functions analogously to a car radiator. Blood from the interior of the body is circulated close to the external surface through a series of tubes that greatly increase the surface area available for radiation of heat. In some exposed parts of the skin such as the palms, fingers, soles, toes, ears and nose, arteriovenous anastomoses are present. When open, these vessels short-circuit the capillary beds, and the blood flow rate through the superficial venous plexus increases rapidly, thereby augmenting heat loss. When the arteriovenous anastomoses are closed, blood flow rate through the venous plexus is decreased and heat is conserved. The amount of blood in the skin greatly influences its insulating properties. When the vessels are dilated, its insulating power is reduced, facilitating heat loss. When they are constricted, insulation is increased and heat is conserved.

CONTROL OF SKIN BLOOD FLOW

The thermoregulatory function of the skin is mainly regulated by the sympathetic nervous system but in certain circumstances local chemical factors also play a role.

NEURAL CONTROL. — The skin vessels, particularly the arteriovenous anastomoses, are richly supplied by sympathetic adrenergic nerve fibers. Exposure to cold increases the frequency of sympathetic discharge, which produces vasoconstriction and reduces cutaneous blood flow. This effect is maximal at nerve impulse frequencies of about 10 Hz. Exposure to warmth produces sympathetic inhibition, vasodilatation and increased blood flow. At normal external temperatures there is a background sympathetic discharge of about one to two impulses per second that maintains a certain level of vasoconstriction or "neurogenic tone" in the arterioles and arteriovenous anastomoses. This tone is considerable in the hands and feet, and blood flow increases substantially in these areas when the nerves are cut or pharmacologically blocked. The tone is less marked in other skin areas where there is only a modest flow increase following sympathetic block.

Recent experiments have shown that stimulation of sympathetic nerves to the canine paw, in the presence of drugs that selectively block the constrictor response, causes arteriolar dilatation, which persists for some time after the stimulus is withdrawn. The substance released by this stimulation is neither acetylcholine nor histamine and its nature and significance in man are unknown.

It is commonplace that skin veins are constricted by cold and become dilated and prominent in hot weather. This is because cold causes increased sympathetic discharge and active venous constriction. At the same time the volume of the veins is passively reduced by diminished inflow from the constricted arterioles. Neurogenic venous tone is minimal at comfortable external temperatures, and the dilatation of the veins produced by external heat is mainly the passive result of increased inflow into the venous system caused by arteriolar dilatation.

Because of the large capacity of the cutaneous veins and their ability to undergo relatively large changes in volume, the skin is an important blood reservoir.

LOCAL EFFECTS OF TEMPERATURE CHANGES. — When an area of skin is warmed, local skin vessels dilate even if the sympathetic nerves have been previously blocked. This is due to the local effects of heat on vascular smooth muscle, to increased production of vasodilator metabolites and to axon reflexes.

Local cold produces vasoconstriction that is maximal at 10 C to 15 C. At lower temperatures the vasoconstriction is interrupted periodically by episodes of vasodilatation that particularly affect arteriovenous anastomoses. Its cause is obscure. One possibility is that the vessels become paralyzed by extreme cold and therefore dilate. The influx of warm blood restores contractile function and the vessels again constrict. The intermittent dilatation may be regarded as a protective mechanism that delays onset of skin necrosis.

REFLEX EFFECTS OF LOCAL HEATING AND COOLING. — If the circulation to a leg is obstructed, e.g., by means of an inflated blood pressure cuff, and the limb distal to the cuff is heated, vasodilatation occurs in other skin areas such as the hands. Since the circulation in the heated part is occluded, the vasodilatation elsewhere must be reflex. It is initiated by stimulation of "warm" thermoreceptors in the skin, and remote dilatation is due to inhibition of sympathetic constrictor tone. Similarly, cooling a limb activates "cold" receptors and induces reflex sympathetic vasoconstriction in other cutaneous areas.

When the circulation is intact, these reflex effects are enhanced by the effects of warmed or cooled blood reaching heat-regulating centers in the hypothalamus. These centers are sensitive to temperature changes of less than 0.5 C. Warming induces vasodilatation, and cooling produces vasoconstriction by appropriate modulation of sympathetic outflow to the cutaneous vessels.

OTHER REFLEX RESPONSES. — Skin vessels participate in baroreflexes but their responses are less striking than those induced by thermoregulatory requirements. Increased systemic arterial pressure causes reflex vasodilatation, and hypotension produces reflex vasoconstriction. These responses are respectively due to reduction and increase of sympathetic discharge. They are relatively weak and transient. Pallor of the skin is a classic physical sign of many types of shock (see section on abnormalities of cardiovascular function) but is only partly due to reflex neural vasoconstriction induced by hypotension. Increased amounts of circulating vasoconstrictors such as catecholamines, angiotensin and vasopressin also play a role. Skin vessels do not appear to participate very strongly in the circulatory response to chemoreceptor stimulation.

HORMONAL AND CHEMICAL CONTROL OF SKIN VESSELS. — Vasoconstrictors and vasodilators include the following:

1. *Catecholamines.* — Skin vessels possess only alpha receptors. Hence epinephrine and norepinephrine induce constriction, and isoproterenol is without effect.

2. *Other constrictors.* — Vasopressin and angiotensin are potent cutaneous vasoconstrictors.

3. *Cutaneous vasodilators.* — The skin vessels dilate in response to numerous drugs, e.g., nitrates, papaverine, histamine, nitroprusside. However, the vasodilator of most physiologic significance is the peptide bradykinin. Exposure of the body to high temperatures causes a greater increase in cutaneous blood flow than can be explained by inhibition of sympathetic tone. The extra flow often begins with onset of sweating, which is believed to cause the release of an enzyme that splits off bradykinin from a serum globulin. Although bradykinin has been found in skin perfusates during sweating, there is still controversy as to whether the active vasodilatation can entirely be ascribed to bradykinin release.

RESPONSES OF SKIN TO TRAUMA; THE TRIPLE RESPONSE. — If the skin is firmly stroked with a pointed object, a red line appears within a few seconds along the track of the point. This is soon followed by a red "flare," which spreads irregularly on either side of the line. In a few minutes a wheal develops along the track and replaces the red line. These three manifestations —

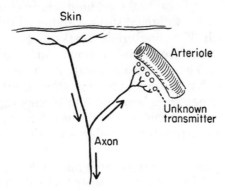

Fig 4–59.—Axon "reflex." Stimulation of the skin causes arteriolar dilatation (see text).

red line, flare and wheal—constitute the "triple response." The red line is due to dilatation of vessels induced by the mechanical stimulus. The flare is due to arteriolar dilatation and is abolished by prior chronic denervation of the sensory nerves to the skin but not by acute denervation. It is thought to be produced by an axon *reflex*. It is postulated that sensory stimuli that arise from the skin are conducted along the sensory nerve axon and then *antidromically* along a branch of the axon to arterioles, which then dilate (Fig 4–59). The transmitter is unknown. The wheal is due to increased capillary permeability, which causes local edema. Intradermal injections of histamine and certain other agents produce similar changes but the precise chemicals released by mechanical stimulation of the skin that are responsible for the flare and wheal are unknown.

EMOTIONAL RESPONSES.—Blood vessels of the skin, particularly those of the face, undergo a wide variety of responses to emotion. These responses are poorly understood and differ widely from person to person. Blushing is due to sympathetic inhibition and perhaps also to active vasodilator mechanisms including bradykinin release. Acute fear, anxiety or alarm often induces "cold feet" due to sympathetic vasoconstriction, which particularly affects peripheral areas including the face, hands and feet. The constriction can be very intense, particularly in women, and spasm of the digital arteries may become so severe that the fingers become dead white and numb (Raynaud's phenomenon). When the episode ends, a period of hyperemia ensues and the fingers then become hot, red, throbbing and painful. Similar responses are frequently induced by exposure to cold, and affected individuals appear to have an abnormal reactivity of the digital vessels to a variety of stimuli.

RENAL CIRCULATION

The gross and microscopic anatomy of the kidney is described in Chapter 6. Figure 4–60 shows the arrangement of renal vessels. The renal artery gives off a number of branches, the *interlobar* arteries, which pass between the lobes, or pyramids, of the kidney. At the junction between the cortex and the medulla, these vessels give rise to the *arcuate* arteries from which the *interlobular* arteries pass toward the renal surface giving off side branches termed *afferent arterioles*. Each afferent arteriole breaks up to form a group of capillaries, the *glomerulus*, which comes into intimate relationship with Bowman's capsule of the nephron. The glomerular capillaries open into the *efferent arteriole*, which is somewhat narrower than the afferent arteriole. The efferent arterioles of the cortical nephrons break up into another capillary network that surrounds the cortical parts of the nephron. The efferent arterioles of the juxtamedullary glomeruli also provide a similar network to the proximal and dis-

Fig 4–60.—Renal vasculature (see text). (After Pitts, R. F.: *Physiology of the Kidney and Body Fluids* [3d ed.; Chicago: Year Book Medical Publishers, Inc., 1974]. Used by permission.)

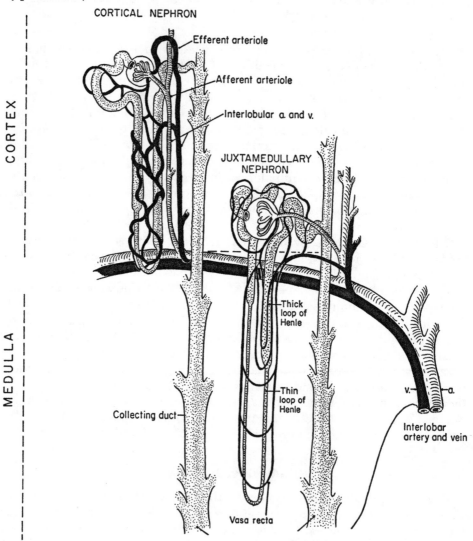

tal tubules but, in addition, send wide, thin-walled hairpin vessels, the vasa recta, into the medulla in close association with the loops of Henle. The capillary networks drain into interlobular veins and then successively into arcuate, interlobular and renal veins. The arteries and afferent arterioles receive a dense sympathetic innervation but the efferent arterioles are only sparsely innervated. The medullary vessels receive fewer sympathetic fibers than do those of the cortex.

The structure and arrangement of the renal vessels are intimately related to the functioning of the nephron (sections on kidney and tubular function, Chapter 6). The glomerular capillaries have a high intraluminal pressure, about 50 mm Hg, which favors capillary filtration. The capillaries themselves are fenestrated and provide little hindrance to bulk flow, but the thick basement membrane, together with structural adaptations of the capsular epithelium, largely prevents filtration of proteins. The vasa recta are essential components of the mechanism that enables the kidney to produce a concentrated urine. The *peritubular* capillary pressure is 15–20 mm Hg but the kidney has an unusually high interstitial pressure, about 15 mm Hg, so that transmural peritubular capillary pressure is very low, favoring reabsorption of fluid from the tubule.

RENAL BLOOD FLOW

The kidney receives one of the highest blood flow rates of any organ in the body. The kidneys together weigh only 300 gm but receive 20% of the entire cardiac output. Obviously, this huge flow is not required to meet the metabolic demands of the kidney but, rather, to be processed by the nephrons so that homeostasis of the "milieu intérieure" of the entire body is maintained.

The high blood flow rate relative to tissue metabolic needs causes renal arteriovenous oxygen difference to be very low, about 1.5 ml O_2/100 ml of blood compared to the average arteriovenous oxygen difference for the whole body of 5 ml per minute. The renal arteriovenous oxygen difference remains constant over a wide range of blood flows. In other words, oxygen consumption rises when blood flow rises and falls when blood flow falls. This is because the main determinant of oxygen consumption by the kidney is the process of active tubular transport. The higher the blood flow the greater the glomerular filtration rate and the amount of solute to be transported by the tubules. Similarly, lower blood flows reduce the amount of active transport and hence the rate of oxygen consumption.

Because renal blood flow is so high and because the body can tolerate short periods of reduced renal function without harm, the renal vasculature is strongly involved in emergency situations, e.g., severe exercise, hemorrhage, shock. Renal vasoconstriction then assists in maintenance of adequate systemic arterial pressure and so helps to sustain blood flow to heart, brain and exercising muscle (see section on integrated responses of the circulation).

INTRARENAL BLOOD FLOW DISTRIBUTION.—The cortex receives approximately 90% of the total renal flow, the outer medulla about 8% and the inner medulla 2%.

FACTORS REGULATING RENAL BLOOD FLOW

1. ARTERIAL PRESSURE.—The kidney shows a high degree of autoregulation. Changes in arterial pressure over the range 80–180 mm Hg cause very little change in renal blood flow. The glomerular filtration rate (GFR) also stays remarkably constant, which means that the resistance changes that maintain blood flow constant despite alterations in pressure must occur in the afferent arterioles. The ability to maintain a constant GFR protects the body from large fluctuations in water and solute excretion that would otherwise occur when arterial pressure changed. The two major hypotheses that have been proposed to account for renal blood flow autoregulation are discussed in Chapter 6 and are mentioned only briefly here. The *myogenic* hypothesis proposes that a rise in arterial pressure distends the afferent arterioles, which respond by constricting. Renal vascular resistance therefore rises and largely prevents any increase in renal blood flow, glomerular capillary pressure or GFR. The *tubular feedback* hypothesis proposes that a rise in arterial pressure causes a change in the pressure or composition of the tubular fluid, which leads to afferent arteriolar constriction possibly through some vasoactive intermediary. The proponents of these hypotheses have been debating their merits and weaknesses for decades but the issue remains unresolved.

2. NEUROHUMORAL CONTROL.—The renal circulation participates in baro- and chemoreflex adjustments of the circulation, although its responses are relatively small unless the reflexes are rather strongly engaged. Increased sympathetic nerve activity evoked by such factors as exercise, emotion, postural change and hemorrhage produces vasoconstriction and reduces renal blood flow. The afferent arterioles constrict more than do the efferent arterioles because of their relatively greater innervation and so glomerular filtration is reduced. The kidneys are normally a low-resistance pathway, and intense sympathetic constriction can substantially increase total systemic vascular resistance as well as local resistance. This serves to maintain arterial pressure and blood flow to less constricted regions. Sympathetic stimulation also causes renin release, which leads to angiotensin production and sodium retention, which assist in maintenance of arterial pressure and extracellular fluid volume.

The adrenotropic receptors of the kidney vessels are of the alpha type so that epinephrine and norepinephrine are strong renal vasoconstrictors, and beta receptor agonists such as isoproterenol are without effect.

The renal arterioles share with the gastrointestinal vessels the unusual property of dilating in response to use of dopamine. This response is believed to be mediated by "dopamine receptors," which are not blocked by propranolol and are therefore distinct from beta adrenotropic receptors. The pharmacologic actions of dopamine have led to its use in treatment of hypotensive states associated with low urinary output. Catecholamines such as norepinephrine tend to raise arterial pressure but at the expense of increased renal vasoconstriction and reduced urine flow. Dopamine, on the other hand, increases arterial pressure by its action on the heart and by constricting skin and muscle vessels. The renal vessels, however, are dilated and GFR and urinary output increase.

Angiotensin and vasopressin are potent renal vasoconstrictors, whereas acetylcholine, bradykinin, serotonin and some prostaglandins are renal vasodilators.

In addition to their direct effects on vascular smooth muscle, many constrictor agents tend to inhibit renin release. On the other hand, norepinephrine and sympathetic nerve stimulation increase renin release via a beta receptor mechanism. The beta adrenergic blocking agent propranolol is useful in management of hypertension. Its effect, particularly in patients with high blood renin concentrations, is partly attributable to inhibition of renin release.

CONTROL OF RENAL BLOOD FLOW DISTRIBUTION.—The renal medulla synthesizes a prostaglandin, PGE_2, a potent renal vasodilator that is broken down in the renal cortex. The dilator effect of PGE_2 is therefore relatively more pronounced in the medulla and inner cortex and this causes a redistribution of intrarenal blood flow favoring these regions. Both norepinephrine and angiotensin II release PGE_2, which reduces the constrictor action of these agents, especially in the inner cortical and medullary regions. Interestingly, inhibitors of prostaglandin synthesis also reduce basal renal blood flow, and this raises the possibility that continuous prostaglandin synthesis is important in establishing the basal flow level.

THE SPLANCHNIC CIRCULATION

The term "splanchnic circulation" refers to the circulation through the stomach, intestines, pancreas, liver and spleen. The arterial circuits to these organs are arranged in parallel but the venous blood from the gastrointestinal tract and spleen drains into a common vein, the portal vein, which subse-

Fig 4–61.—Diagram of the arrangement of the splanchnic vessels. The capillary beds of the spleen and gastrointestinal tract and the liver sinusoids are in parallel arrangement and are fed by the arteries shown. The liver sinusoids are also connected in series with other capillary beds via the portal vein.

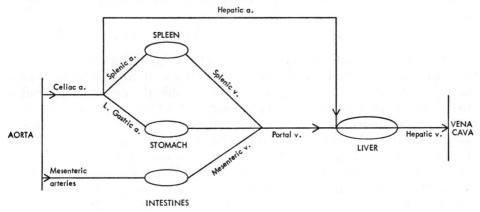

quently subdivides into progressively smaller branches that eventually empty their blood into the liver sinusoids (see section on the liver circulation). Thus the gastrointestinal circulation is connected in series with the liver sinusoids, and about 70% of the blood flowing into the liver is *venous* blood that has just passed through the gastrointestinal or splenic capillaries. The remaining 30% is delivered by the hepatic artery. After passing through the liver, blood is conveyed by the short wide hepatic veins into the inferior vena cava and thence to the right atrium (Fig 4–61).

The series connection between gastrointestinal capillaries and hepatic sinusoids means that substances absorbed from the intestine must first come into close relationship to liver cells before entering the general circulation. The liver can therefore regulate the entry of nutrients and drugs into the circulation and can metabolize and detoxify potentially harmful materials such as amines and ammonia if these are absorbed from the gut.

MEASUREMENT OF SPLANCHNIC BLOOD FLOW.—This is achieved in man by using the *Fick principle*. A nontoxic dye, such as indocyanine green, which is taken up only by the liver and otherwise does not leave the circulation, is infused into a vein until the arterial dye concentration becomes constant. A sample of blood is then withdrawn from a catheter previously inserted into the hepatic vein under x-ray guidance. Splanchnic blood flow (SBF) can then be calculated as follows:

$$\text{SBF (ml/min)} = \frac{\text{dye taken up by liver/minute (mg)}}{\text{arterial dye concentration} - \text{hepatic vein dye concentration (mg/ml)}}$$

The numerator of this fraction is easily obtained because the amount of dye taken up per minute by the liver must equal the known infusion rate of dye when the arterial dye concentration is constant. This method is applicable only to measurement of steady state flows but modifications are available for measuring fluctuating flows.

The splanchnic circulation is quantitatively the most important regional circulation. Its blood flow averages 1,200–1,500 ml per minute, about 25% of the resting cardiac output.

GASTROINTESTINAL CIRCULATION

The stomach and intestines receive blood from three major vessels: the celiac, superior mesenteric and inferior mesenteric arteries. The numerous branches of these vessels anastomose freely outside the gut forming arcades in the mesentery. Branches from these arcades penetrate the gut wall and contribute to an extensive submucosal plexus from which smaller vessels arise to supply the villi, glands and smooth muscle. The muscle also receives blood from the superficial vessels of the serosa. Each artery is accompanied by one or more veins. The copious anastomoses among the vessels in all layers of the gut wall ensure maintenance of flow during movements of the convoluted intestine. All the vessels are innervated by an adrenergic nerve plexus, which is denser in the arteries than in the corresponding veins.

Flow to the gastrointestinal tissues is not uniformly distributed. The mucosa receives about 70%, submucosa 7% and muscle 23% of the gastrointestinal flow.

The function of the gastrointestinal tract is to digest and absorb food. This requires the secretion of 5–8 L of digestive juices per day. This fluid is supplied by the gastrointestinal circulation, which is also responsible for supporting the metabolism of gastrointestinal cells. The mucosa particularly is a very metabolically active tissue that is constantly renewing its epithelial cells as well as elaborating enzymes and providing energy for active transport processes. In order to meet these requirements, the gastrointestinal capillary bed is unusually extensive, with a large surface area and a very high permeability. The capillaries are fenestrated and the basement membranes are thin and incomplete. As a result, a given pressure gradient will produce 20 times as much filtration of water across an intestinal capillary as across a skeletal muscle capillary.

Factors Influencing Gastrointestinal Circulation

1. ARTERIAL PRESSURE.—The intestinal vasculature autoregulates well. Increases or decreases in arterial pressure, respectively, induce arteriolar constriction or dilatation that not only prevent major alterations in flow but also significant changes in capillary pressure. This is particularly important for the highly permeable capillary beds of the stomach and intestine because edema could rapidly develop from slight capillary pressure increments.

2. VENOUS PRESSURE.—Elevation of venous pressure produces a "venous arteriolar response" in which the arterioles and precapillary sphincters constrict. This is probably a myogenic response, which also serves to maintain a near-constant capillary pressure.

3. INTRALUMINAL PRESSURE.—When the cat intestine is passively distended by relatively low pressures, intestinal blood flow is usually slightly increased, probably because of the release of vasoactive substances from the intestinal wall. Higher distention pressures transiently reduce flow but this quickly returns to normal or above normal values despite continuing distention. The intestinal arteriovenous oxygen difference narrows appreciably and it seems likely that the mucosal circulation is reduced or obliterated by compression against the tough submucosa and that blood flow continues at a rapid rate through dilated vessels in the muscle coat.

4. NEURAL CONTROL.—Stimulation of the mesenteric sympathetic nerves constricts the intestinal arterioles. When the stimulus rate is at the upper limit of the physiologic range (i.e., 10–15 Hz), intestinal blood flow may almost cease. In anesthetized cats and dogs, the constriction gradually wanes and blood flow returns toward prestimulation values despite continuing stimulation. This phenomenon is termed *vasoconstrictor escape.* When nerve stimulation is discontinued, a period of increased blood flow (hyperemia) occurs (Fig 4–62). The cause of the escape, which is also seen during infusions of norepinephrine and other vasoconstrictor drugs, has not been fully established. One hypothesis is that redistribution of blood flow occurs in the intes-

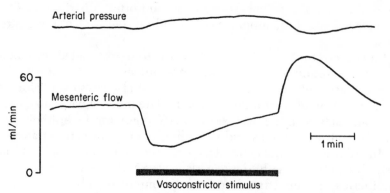

Fig 4–62.—Vasoconstrictor escape in the mesenteric circulation of a cat. Blood flow "escapes" to close to normal values despite a continuing vasoconstrictor stimulus. Note the poststimulus hyperemia.

tinal wall, with a gradually increasing flow in one area during the escape. The area of increased flow has been variously thought to be in the submucosa, deep mucosa or villus tips. Another hypothesis is that escape occurs because of a peculiarity of the excitation-contraction coupling mechanism of mesenteric arterial smooth muscle cells such that increasing sequestration or expulsion of free calcium ions occurs during the period of stimulation. This could be a direct effect of the constrictor or an indirect action mediated via an intermediary such as cAMP or a prostaglandin. The redistribution hypothesis cannot be the sole cause of escape because this also has been demonstrated in isolated mesenteric arterial strips and in the splenic and hepatic vascular beds of cats. Although the escape phenomenon is of considerable interest to vascular physiologists, its existence in man has not been established.

Sympathetic nerve stimulation causes a sustained reduction in intestinal volume owing to expulsion of blood from the intestine. Two mechanisms are involved. The first is a passive recoil of the veins caused by reduced inflow of blood from the constricted arterioles. The second is active venoconstriction caused by stimulation of the sympathetic fibers supplying the venous smooth muscle.

The parasympathetic nerves have no direct influence on the arterioles or veins of the intestine. Parasympathetic nerve fibers are found in association with the larger mesenteric arterial branches in some animals, but their significance is unknown. On the other hand, stimulation of the parasympathetic fibers of the vagus increases intestinal motility and luminal pressure. The intestinal wall becomes very pale during vigorous contractions, indicating that the blood vessels in the wall are compressed by the contracting intestinal muscle and by the raised intraluminal pressure. The resulting ischemia may be partly responsible for the pain of intestinal spasm.

5. HORMONES.—Norepinephrine produces effects resembling sympathetic nerve stimulation, i.e., arteriolar constriction, escape, venous constriction and reduced capillary filtration. The effect of epinephrine varies with dose. The smallest effective doses dilate arterioles, owing to activation of beta adreno-

tropic receptors, and intestinal blood flow increases. Larger doses produce a biphasic effect: initial constriction followed by dilatation. Still higher doses produce effects similar to those of norepinephrine owing to overwhelming alpha adrenotropic receptor stimulation. Beta adrenotropic receptors appear to be sparse in veins, whereas alpha receptors are relatively more prominent. Hence, all effective doses of epinephrine produce venous constriction. Dopamine, the biologic precursor of norepinephrine, increases mesenteric blood flow. This effect persists following beta adrenoreceptor blockade and is thought to be mediated by specific "dopamine receptors." The physiologic significance of these receptors is unknown.

The gastrointestinal tract produces numerous hormones. Many of these, including gastrin, secretin, cholecystokinin and vasoactive intestinal peptide (VIP), are intestinal vasodilators. Their physiologic circulatory role has not been defined, but probably they are responsible in part for the increased gastrointestinal blood flow during digestion and absorption of food.

6. OTHER VASOACTIVE SUBSTANCES. — Angiotensin reduces gastrointestinal blood flow, but a rapid "escape" occurs during continuous infusions. Vasopressin produces vasoconstriction without escape. Acetylcholine, prostaglandins of the E series, histamine and serotonin are gastrointestinal vasodilators and increase blood flow.

CIRCULATORY CHANGES ASSOCIATED WITH DIGESTION AND ABSORPTION. — Gastric mucosal blood flow increases markedly when acid secretion is increased before and after ingestion. This is discussed fully in Chapter 7. Intestinal blood flow also increases after food intake and remains elevated about 30% above normal for several hours. The mechanism of this increase is uncertain, but it is at least partly mediated by hormones. Oil placed in the duodenum of cats increases intestinal blood flow, probably owing to release of cholecystokinin, a mesenteric vasodilator. Secretin also dilates intestinal vessels. The elevated intestinal metabolism associated with secretion, absorption and motility may also increase blood flow via the accumulation of vasodilator metabolites.

SPLENIC CIRCULATION

The spleen is an important blood reservoir in some animals and can undergo large and rapid changes in size, thereby reducing or increasing circulating blood volume. In man, however, the spleen weighs only 150 gm, and splenic contractions could increase the normal blood volume of 5 L by a negligible 1 to 2%.

The main functions of the human spleen are to destroy aging red cells and to produce lymphocytes and monocytes, which are important components of the immune system that protects the body against infectious agents and their antigens.

The spleen is enclosed in a thick connective tissue capsule which sends trabeculae into the splenic substance, dividing it into many small intercommunicating compartments. These compartments contain two types of tissue, red pulp where red cells are broken down, and white pulp where lymphocytes and monocytes are produced. The splenic artery breaks up into branches, which pass along the trabeculae. Each branch passes through and supplies the white pulp and then enters the red pulp. Some authorities believe that blood is then poured directly into the tissue spaces of the red pulp and finds its way into a system of sinuses via gaps between the endothelial cells of these vessels. This is the "open circulation" theory. Others believe that the circulation is "closed," i.e., there is a continuous vascular endothelium from the arteries through the red pulp to the sinuses and then to the veins. In some animals red cells can be sequestered for some time in interstices of the pulp and in the sinuses before passing into the general circulation, but this appears not to be the case in man.

Little is known of the factors that regulate splenic blood flow in man, and the following account is derived mainly from experimental observations in animals.

Splenic blood flow averages 10 ml/min/kg body weight in anesthetized cats and dogs. Because of the highly variable blood content of the spleen in these animals, flows expressed as milliliters per gram of organ necessarily vary widely but average about 90 ml/min/100 gm.

Splenic arterial blood flow in anesthetized animals shows small oscillations with a period of 30–40 seconds. These oscillations may transiently become much larger following brief occlusion of the splenic artery but not of the splenic vein. This suggests that the oscillations are induced by rapid changes in arteriolar transmural pressure. Splenic venous flow also shows oscillations but these are a few seconds out of phase with arterial flow. Consequently there are periods when splenic inflow exceeds outflow and vice versa. As a result, splenic volume and weight also show periodic variations.

NEURAL CONTROL.—The splenic vessels have a dense sympathetic adrenergic innervation. Stimulation of these nerves constricts the arterioles and may produce substantial reductions in splenic blood flow. As in the intestine, vasoconstrictor escape occurs, but it is of smaller magnitude and slower onset. A moderate hyperemia develops and flow oscillations are strikingly enhanced when the stimulation is discontinued. The splenic arterioles do not receive a parasympathetic nerve supply.

HORMONAL CONTROL.—Norepinephrine produces effects similar to those of sympathetic nerve stimulation. Epinephrine causes vasodilatation in small doses and vasoconstriction in larger doses. Vasopressin produces vasoconstriction.

OTHER CHEMICAL FACTORS.—Isoproterenol, acetylcholine, histamine and prostaglandins E_1 and E_2 dilate, whereas angiotensin, serotonin and prostaglandin $F_{2\alpha}$ constrict, splenic arterioles.

THE SPLEEN AS A BLOOD RESERVOIR

The splenic capsule in cats and dogs contains much smooth muscle. The splenic reservoir can be emptied by contraction of the capsular muscle or of the splenic veins. Both mechanisms are usually employed, although capsular muscle appears to be somewhat more sensitive than veins to most excitatory stimuli. The splenic reservoir is depleted during a number of stresses including exercise, hemorrhage and asphyxia. The emptying of the spleen under these circumstances is accomplished by increased activity of the sympathetic nerves, although increased concentrations of circulating catecholamines play a subsidiary role. Splenic contraction can substantially increase the circulating blood volume in cats and dogs and, since the hematocrit of splenic blood is higher than the average hematocrit, there is an even greater relative increase in circulating red cell mass. These changes augment cardiac output and tissue oxygen delivery and are therefore helpful adaptations to the stresses mentioned. It must be re-emphasized however that the reservoir function of the spleen is negligible in man.

CIRCULATION IN THE LIVER

The liver is composed of a complex system of branching and anastomosing plates of liver cells separated by wide "leaky" capillaries termed sinusoids. The endothelial cells that line the sinusoids and their basement membranes have wide fenestrae, which allow passage of even large molecules. The interstitial fluid and lymph from the liver therefore have a very high protein content. Some of the endothelial cells, the Kupffer cells, are actively phagocytic and can ingest bacteria. They are also involved in the production of antibodies.

The sinusoids are fed by two streams of blood of different origin and composition and under different pressures. One stream is conveyed by the hepatic artery at systemic arterial pressure and delivers approximately one third of the total liver blood flow. The remaining two thirds consists of blood that has already flowed through the capillaries of the spleen or gastrointestinal tract and therefore has a much lower pressure, around 10 mm Hg, and a lower oxygen saturation (70–80%). These two streams become confluent at the junction of the terminal branches of the hepatic artery and portal vein at the origin of the sinusoids (Fig 4–63). After passing through the sinusoids, blood is returned to the heart via the hepatic veins and inferior vena cava.

Fig 4–63. — Arrangement of the microcirculatory vessels of the liver.

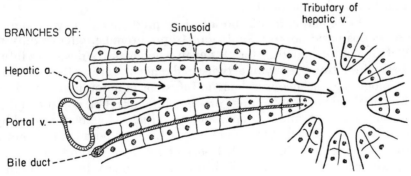

The liver receives about 25% of the cardiac output. Its veins contain about 400 ml of blood and it is an important blood reservoir.

FACTORS INFLUENCING HEPATIC BLOOD FLOW

1. PRESSURE. — Although hepatic arterial flow autoregulates well, the portal system does not. This is understandable when it is remembered that portal pressure influences two vascular beds: (1) the liver sinusoids and (2) the gastrointestinal and splenic vessels. It would be very difficult to make smooth muscle adjustments of the portal system that would simultaneously satisfy the requirements of all areas. Since a major function of the liver is to process blood from the gastrointestinal tract, it appears that to a considerable extent the portal system responds passively to changes in gastrointestinal flow, i.e., portal flow is largely determined by the tone of the gastrointestinal arterioles. In man, splenic blood flow is relatively small and so splenic arteriolar changes do not importantly affect portal flow.

2. NEUROHUMORAL CONTROL. — Plentiful sympathetic nerve terminals are present in the adventitiomedial coats of the hepatic artery and its branches down to and including the arterioles. The portal and hepatic veins and their radicals are less densely innervated. Stimulation of the hepatic sympathetic nerves constricts the arterioles and reduces hepatic blood flow, and this reduction is sustained in the dog but escapes in the cat. The type of constrictor response in man is unknown. Sympathetic stimulation causes passive reduction in portal flow and portal blood volume as a consequence of gastrointestinal vasoconstriction and reduced inflow into the portal system. Sustained active constriction of the portal and hepatic veins also occurs. These changes together can mobilize about 40% of the hepatic blood volume without materially altering venous resistance to flow. Sinusoidal pressures, and hence fluid filtration, do not change significantly.

The muscle coat of the hepatic veins is well developed in the dog and in some diving animals, and adrenergic nerves penetrate deeply into it. When these nerves are stimulated, the muscle contracts strongly, hepatic venous outflow is obstructed and considerable pooling of blood occurs in the splanchnic veins. This may account for the severe hepatic and portal congestion in dogs subjected to hemorrhagic or endotoxic shock (see section on abnormalities of cardiovascular function).

Norepinephrine and epinephrine infusions produce effects very similar to those of sympathetic nerve stimulation. Isoproterenol produces only mild vasodilatation of the hepatic arterioles even when massive doses are infused. Thus beta adrenotropic receptor activity in the hepatic arterioles is weak, and its physiologic significance is unknown. Infusions of norepinephrine directly into the portal veins of cats or dogs produce little effect on portal blood flow, but infusions into systemic veins produce an initial reduction followed by an increase. This indicates that the smooth muscle of the portal vein is not very sensitive to norepinephrine and that portal flow changes produced by intravenous administration of norepinephrine are secondary to changes in gastrointestinal flow.

REFLEX RESPONSES.—Blood flow to the liver is only weakly affected by baroreflex activity. Occlusion of the common carotid arteries in animal experiments produces relatively slight hepatic vasoconstriction, which escapes within a few minutes. Portal flow also is reduced but only to a small extent. Sympathetic tone to the hepatic vessels in the basal state is weak, and sympathetic nerve section produces little change in flow.

Hepatic flow decreases markedly during exercise, probably owing to a reflex increase in sympathetic nerve discharge.

3. METABOLIC CONTROL.—Metabolic responses are relatively weak. Reactive hyperemia does occur following a period of arterial occlusion but much less than in skeletal muscle and intestine. Hypoxia and hypercapnia cause only slight increases in hepatic blood flow.

In summary, the liver receives a large blood flow from two sources. Two thirds of the blood supply is venous blood derived from the gastrointestinal tract and spleen. The liver cannot regulate this flow and passively accepts what comes to it. Hepatic arterial blood flow is relatively constant and is only transiently and rather weakly influenced by changes in pressure or neural activity.

The liver is a metabolically active organ with varying metabolic needs. Increases in oxygen demand could be met by increasing oxygen extraction or by higher blood flow rates. The relative extent to which these mechanisms are used in different metabolic situations is unknown. The anatomical arrangement of the portal and hepatic arterial systems suggests that variations in the mixing of arterial and portal venous blood could regulate the oxygen content of the blood in the sinusoids, but the extent to which this occurs has not been determined.

CIRCULATION OF THE BRAIN

The brain has virtually no capacity for anaerobic metabolism and is highly dependent on an adequate delivery of oxygen by the circulation. Cerebral blood flow is remarkably constant at about 50 ml/100 gm per minute or 750 ml per minute for the average adult brain. Deprivation of flow for as little as 5 seconds may produce unconsciousness, and irreversible changes occur after 4–5 minutes. Changes in pH, PCO_2 and PO_2 of the internal milieu of the brain rapidly produce severe disturbances of neural excitability, and it is a function of the cerebral circulation to ensure that blood flow is regulated in order to maintain the constancy of these variables in the environment of the brain cells.

ANATOMY OF BRAIN CIRCULATION

Blood flow to the brain is delivered by the internal carotid and vertebral arteries. The right and left vertebral arteries enter the skull and unite on the ventral surface of the medulla oblongata to form the basilar artery. This vessel continues rostrally, giving branches to the brain stem, and ends by dividing to form the right and left posterior cerebral arteries.

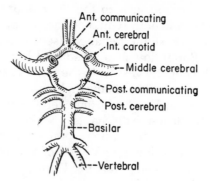

Fig 4–64.—Arterial circle of Willis.

Each internal carotid artery after entering the skull traverses the cavernous sinus and terminates by dividing into the anterior and middle cerebral arteries. The right and left anterior cerebral arteries are linked by a communicating artery. The right and left posterior cerebral branches of the basilar artery also are linked with the corresponding carotid system by communicating arteries. These anastomotic vessels form an arterial circle, the circle of Willis, at the base of the brain. The existence of this circle means that obstruction of a major artery such as the internal carotid can be compensated by inflow from the other arteries that join the circle (Fig 4–64).

The cerebral capillaries are drained by veins that open into a number of wide venous spaces or sinuses on the surface of the brain. These sinuses eventually leave the skull via the internal jugular veins, which pass down the neck alongside the common carotid arteries. The venous sinuses within the skull anastomose freely with the extracranial veins via communications that pass through the bone. The communications are numerous and substantial, as shown by the fact that ligation of the internal jugular veins in man causes no disability since blood can be readily directed through extracranial routes.

FACTORS REGULATING CEREBRAL BLOOD FLOW

1. PRESSURE.—The cerebral vascular bed shows pronounced autoregulation. Variations in arterial pressure over the range 60–160 mm Hg produce little change in blood flow. As in other vascular beds, cerebral autoregulation is probably due to a combination of myogenic and metabolic factors, although it recently has been suggested that the autonomic nerves play a more significant role than hitherto supposed. Certain animal experiments have suggested that the increase in vascular resistance that occurs in response to raised arterial pressure may be partly mediated by sympathetic nerves and that the vasodilator response to reduced pressure is partly mediated by parasympathetic nerves. It has been claimed that complete denervation of the cerebral vessels grossly impairs autoregulation. These findings await confirmation.

When arterial pressure falls below 60 mm Hg, autoregulation fails and flow tends to fall, but the cerebral hypoxia, hypercapnia and acidosis caused by reduced flow to the brain in the intact animal elicit a powerful "central nervous ischemic response." The sympathetic centers in the hypothalamus and brain stem discharge more frequently and produce cardiac stimulation and peripheral vasoconstriction, leading to a substantial rise in blood pressure that tends to restore cerebral blood flow at the expense of a reduction in blood flow in other areas such as kidney, gut, muscle and skin. This intense response is a "last ditch" effort that can be sustained for only a short period.

2. NEURAL CONTROL. — Histochemical methods have shown that the pial vessels on the brain surface receive both sympathetic constrictor and parasympathetic dilator fibers. Adrenergic innervation has also been observed in certain vessels within the brain. Sympathetic postganglionic fibers have their cell bodies in the superior cervical ganglion, whereas corresponding parasympathetic fibers have their cell bodies in the geniculate ganglion and are distributed by the seventh cranial (facial) nerve. Autonomic nerves have rarely been observed in the deep vessels of the cerebrum, cerebellum and spinal cord.

The role of the autonomic nerves in regulation of cerebral blood flow is controversial. They have little effect on resting vascular tone, and section of either the sympathetic or parasympathetic fibers to the brain produces only small and variable blood flow changes. Sympathetic nerve stimulation produces vasoconstriction and parasympathetic stimulation produces vasodilatation, but the responses are weak compared to those of most other vascular beds. Some investigators believe that sympathetic activation participates significantly in the rapid vasoconstrictor response to a rise in cerebral arterial pressure and to a fall in Pco_2. Similarly, the parasympathetic fibers carried by the facial nerve may participate in the rapid vasodilatation produced by a fall in cerebral perfusion pressure or by a rise in Pco_2.

3. LOCAL METABOLIC CONTROL. — Metabolic activity varies considerably in different parts of the brain. It is higher in gray matter, which mainly contains cell bodies, than in white matter, which consists mostly of nerve fibers. Correspondingly, average blood flow through the gray matter is five times higher than through white matter. Considerable variations occur. For example, gray matter flow in the inferior colliculus of cats was 1.74 ml/gram per minute compared to 0.74 ml/gram per minute in the olfactory cortex. Increased functional activity of localized areas of the brain is associated with increased blood flow in those areas. Thus during vigorous exercise of the right hand, blood flow is elevated in that portion of the left cerebral cortex that controls this activity. Blood flow to the brain increases dramatically during epileptiform seizures and falls during coma. The principal metabolic regulator appears to be the hydrogen ion concentration of the cerebrospinal fluid (CSF). This is dependent on the relative concentrations of CO_2 and bicarbonate ion. When CSF CO_2 rises, either as the result of increased metabolic production by brain tissue or because of elevated arterial Pco_2, CSF hydrogen ion concentration increases. This causes relaxation of the cerebral arterioles and increased blood flow. The CSF hydrogen ion concentration is also a potent regulator of respiration. Increased CSF hydrogen ion concentration stimulates ventilation, which tends to reduce Pco_2, whereas decreased CSF hydrogen ion concentration inhibits ventilation and tends to increase Pco_2. These effects on

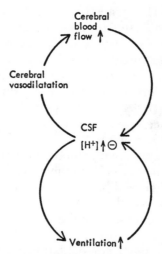

Fig 4–65.—Regulation of cerebral blood flow and ventilation by CSF hydrogen ion concentration. Increased CSF [H+] causes increased cerebral blood flow and stimulates ventilation, which restore CSF [H+] to normal levels. The symbol ⊖ indicates negative feedback control. Increased CSF [H+] may result from increased cerebral metabolism or elevated arterial P_{CO_2}.

cerebral blood flow and respiration ensure that CSF pH stays remarkably constant (Fig 4–65).

It is important to realize that a reduction in arterial pH caused by fixed acids has little dilator effect on cerebral blood flow because the hydrogen ion does not readily pass the blood-brain barrier. In fact, if the arterial hydrogen ion rises sufficiently to produce chemoreceptor stimulation of respiration, arterial P_{CO_2} may fall and cerebral vasoconstriction may result.

When arterial P_{CO_2} regulation is acutely impaired, cerebral blood flow changes substantially. Thus rapid CO_2 retention in respiratory disease is associated with increased cerebral blood flow, whereas excessive loss of CO_2 caused by acute hyperventilation may lower arterial P_{CO_2} levels to such a degree that intense vasoconstriction with dizziness and even syncope may occur. Chronic CO_2 retention, as in emphysema, does not alter cerebral blood flow because the kidneys compensate by retaining bicarbonate, and pH of the CSF is thereby kept within normal limits.

Reduction in arterial P_{O_2} produces little change in cerebral blood flow until the value falls below 50 mm Hg. Flow then increases. It is uncertain whether this is a direct effect of oxygen lack or due to concurrent production of lactic acid and an increase in tissue hydrogen ion concentration. Very high arterial oxygen tensions produce cerebral vasoconstriction but the precise mechanism is unclear.

PATHOLOGIC IMPLICATIONS

Recent development of methods that enable regional cerebral blood flow to be measured in man has yielded much important information concerning abnormal cerebral perfusion in disease. Many conditions result in a disparity between cerebral metabolic demands and the rate at which they can be met by the blood supply. Thus a blood flow rate within normal values may be inadequate if cerebral metabolism is increased. On the other hand, it may be excessive if cerebral metabolism is reduced. As in the myocardium (see section on the coronary circulation), it is the supply/demand *ratio* that is important. When this ratio falls, there is increased production of lactic acid, which causes cerebral vasodilatation. If lactic acid production is excessive, the brain vessels lose their power to autoregulate, and flow becomes susceptible to changes in arterial or intracranial pressure. The dangers of this situation are exemplified by the effects produced by cerebral edema, which may for example follow head injuries. The edema tends to reduce cerebral blood flow but the intense lactic acidosis causes vasodilatation and flow actually increases. This raises capillary pressure, which in association with capillary damage causes further edema, intracranial pressure elevation and further lactic acid production. A positive feedback loop is thus created, which ends in the patient's death unless the loop is interrupted by appropriate therapy. Edema of the brain can also be caused by increased permeability of the brain capillaries and may be responsible for the worsening symptoms and signs sometimes associated with brain tumor or hemorrhage.

Interesting changes in autoregulation occur in systemic arterial hypertension. The brain retains the ability to autoregulate but over the range 110–200 mm Hg compared to the normal autoregulatory range of 60–180 mm Hg. This means that brain blood flow will fall in hypertensive subjects when arterial pressure is acutely lowered below 110 mm Hg, and cerebral ischemia may therefore be produced by unduly vigorous hypotensive therapy.

CEREBROSPINAL FLUID

The student is urged to review the gross anatomy of the brain and its coverings as preparation for the following account.

Fig 4–66.—Major pathways of CSF flow.

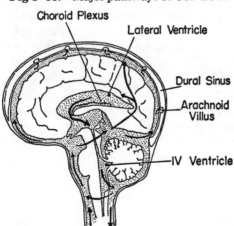

Choroid Plexus

Lateral Ventricle

Dural Sinus

Arachnoid Villus

IV Ventricle

The brain and spinal cord are surrounded by a clear fluid that also fills the cerebral ventricles and central canal of the spinal cord. This fluid, the CSF, is secreted by the *choroid plexuses* into the ventricles, leaves the ventricular system via the *foramina of Luschka and Magendie* and passes into the subarachnoid space. It is taken up again into the blood through the arachnoid villi, which protrude into the dural sinuses (Fig 4–66). The CSF circulates as follows: blood→choroid plexus→CSF→ventricles→subarachnoid space→arachnoid villi→blood. Pressure of the CSF is normally around 10 mm Hg when the body is horizontal.

Although the total volume of CSF is about 140 ml, it is turned over about six times a day so that the daily secretion of CSF is around 800 ml. The CSF and interstitial fluid of the brain provide the relatively constant chemical environment necessary for optimal transmission of signals among the billions of synapses. The CSF also provides a path for the secretion of certain waste products produced during neuronal metabolism. These substances are eventually returned to the blood by the arachnoid villi (the brain does not possess a lymphatic system). Other important functions of CSF are to provide buoyancy for the CNS and to cushion it during movements of the head.

Table 4–4 shows that the composition of the CSF resembles that of protein-free plasma, the main differences being lower K^+ and higher Cl^- concentrations in CSF.

TABLE 4–4.—COMPOSITION OF CSF

COMPONENT	CSF	PLASMA	CSF/PLASMA RATIO
°Na^+	147.0	150.0	0.98
°K^+	2.9	4.6	0.62
°Cl^-	113.0	99.0	1.14
°HCO_3^-	25.1	24.8	1.01
pH	7.33	7.41	—
†Protein	20.0	6,000.0	0.003
†Glucose	64.0	100.0	0.64
Concentrations in mEq/L(°) or mg/dl(†)			

A noteworthy feature of CSF is that its composition is kept remarkably constant by homeostatic mechanisms, which maintain pH and K^+, Mg^{2+}, Ca^{2+}, amino acid and catecholamine concentrations within narrow limits. Disturbances of homeostasis that allow CSF composition to fall outside these limits may have profound effects. For example, small changes in norepinephrine concentration produce numerous changes, including elevation of body temperature in some animals and reduced body temperature in others. Elevated glycine concentrations produce hypothermia, hypotension and motor incoordination. Changes in pH of CSF have important effects on the control of breathing (discussed in Chapter 5).

SECRETION OF CSF

The CSF is produced by the *choroid plexuses*—rich vascular plexuses that protrude into the ventricles. The endothelium of the choroidal capillaries is

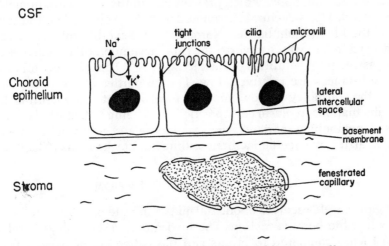

Fig 4–67.—The choroid epithelium and subjacent capillary.

separated from ventricular CSF by a specialized portion of ependyma, the choroid epithelium. The capillaries are fenestrated and porous and readily permit solute exchange. On the other hand, the apical portions of adjacent choroid epithelial cells are connected by tight junctions that act as barriers to prevent passage of proteins and other large molecules from blood into the CSF (Fig 4–67).

The CSF is mainly secreted by the choroidal epithelium, the primary process being active sodium secretion by a sodium pump in the brush border membrane. The net transfer of positive charge by the movement of Na^+ into the CSF is offset by the associated movement of Cl^- and HCO_3^- into, and the reciprocal movement of H^+ and K^+ out of the CSF. The movement of water into the CSF results from the osmotic gradient generated by local increases in extracellular solute concentration that occur as the result of active pumping of sodium out of the choroidal epithelial cells. The composition of the freshly secreted fluid is modified by the diffusion and transport of other solutes across the plexus. Moreover, as the CSF circulates, its composition is further modified by exchanges of solutes (1) with the brain across the pia and ependyma and (2) with the plasma across the arachnoid membrane.

The choroid plexuses also are involved in clearance of solutes from the CSF, e.g., drugs such as penicillin are actively transported out of the CSF across the plexus.

INTERSTITIAL FLUID OF THE BRAIN

The substance of the brain is very densely packed with neurons and glial cells, and the interstitial space is confined to narrow clefts about 20 nm wide between cells. These clefts communicate with each other and with the CSF so that the composition of interstitial fluid and CSF is very similar. It is of great importance to note that the endothelial cells of brain capillaries (with the exception of those of the choroid plexus and certain discrete areas of the brain) have tight junctions, which prevent free diffusion of certain molecules from the blood. This structural barrier and that of the choroid epithelium contribute to the blood-brain barrier. Normally, lipid-soluble substances, carbon dioxide, oxygen and water traverse the capillary endothelia relatively easily, but the passive permeabilities of Na^+, K^+, Cl^-, HCO_3^- and many other water-soluble substances are relatively low. The properties of the barrier help to conserve the constancy of the immediate environment of the neurons by preventing fluctuations in plasma composition from being transmitted to the CSF. Solutes of metabolic importance such as glucose and amino acids can rapidly cross the barrier by specific facilitated diffusion mechanisms (Chapter 1).

THE CORONARY CIRCULATION

The heart muscle receives about 200 ml per minute or 3–4% of the cardiac output when the body is resting. The myocardial vascular bed is relatively vasoconstricted compared to others and the extraction of oxygen from the blood passing through it is unusually high. Blood leaving the capillaries of left ventricular muscle contains only five volumes of oxygen/100 ml, a value lower than that of venous blood from any other part of the body. This fact has great physiologic significance since it means that increased myocardial oxygen demands can be met only to a minor extent by increased oxygen extraction. Furthermore, heart muscle has little capacity for anaerobic metabolism. Contractile force diminishes within seconds of coronary artery occlusion and irreversible damage occurs within minutes. Increased myocardial oxygen demands must therefore be met by a rapid increase in myocardial blood flow and it is hardly surprising, therefore, that the coronary circulation is primarily under metabolic control. The coronary resistance vessels are innervated by autonomic nerves but these play a subsidiary role in coronary regulation. Patients with transplanted, and therefore denervated, hearts have no demonstrable impairment of the coronary circulation.

ANATOMIC CONSIDERATIONS

The right and left main coronary arteries arise from the ascending aorta just above the valve cusps (Fig 4–68), and their distribution is rather variable. The left main vessel divides into anterior descending and circumflex branches, which supply the left ventricle, the anterior interventricular septum and part of the anterior free wall of the right ventricle. The right coronary artery passes downward in the AV groove and then curves around to the back of the heart. It supplies most of the right ventricle, about half of the diaphragmatic surface of the left ventricle and the posterior portion of the interventricular septum. The sinus node is supplied by the right coronary artery in 55% of hearts and by the left circumflex vessel in the remainder. The AV node and the AV bundle and its main branches are supplied by the right coronary artery in 90% of patients.

The major coronary arteries sit on the surface of the heart like a crown (corona = crown) and send numerous rather stout branches through the myo-

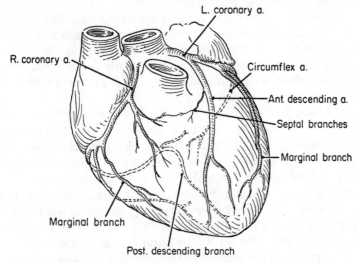

Fig 4–68. — Distribution of the major coronary arteries in man.

cardium toward the endocardial surface. These vessels divide frequently within the myocardium and supply a capillary network that provides an almost 1:1 ratio of capillaries to muscle fibers and is among the densest found in the body.

The terminations of the various coronary arteries and their branches are linked by small arterial anastomoses, but these usually are not capable of maintaining an adequate circulation if one of the major vessels is suddenly occluded. The coronary arteries are therefore termed "end-arteries."

After passing through the myocardial capillaries, blood flows into a number of coronary veins, most of which are tributaries of a large venous channel, the *coronary sinus,* which opens into the right atrium. The anterior wall of the right ventricle is drained by several *anterior cardiac veins,* which open separately into the right atrium. Parts of both ventricles, particularly the septal regions, drain via small *thebesian* veins directly into the ventricles. All the myocardial venous radicles communicate freely with one another.

The importance of a thorough knowledge of the normal anatomy of the coronary arteries and its variants has been underscored by recent developments in radiologic detection of obstructions in these vessels and in surgical management of these lesions.

CONTROL OF CORONARY CIRCULATION

PRESSURE GRADIENTS. — These can be rigorously studied only in controlled animal experiments. If a heart is arrested or fibrillated and its coronary arteries are perfused by a donor animal or from a reservoir, the effects of pressure gradients can be studied independently of other variables. The coronary arterioles autoregulate moderately well over the range 60–180 mm Hg. The relative role of myogenic and metabolic factors in producing autoregulation is unknown but, because of the great sensitivity of this vascular bed to metabolic changes, it seems likely that these play a prominent part. The autoregulatory response, as demonstrated under these conditions, is highly artificial. In the intact animal, changes in systolic arterial pressure alter the afterload of the left ventricle, thereby altering metabolic requirements for oxygen. These metabolic effects cause flow to increase when arterial pressure is raised and to decrease when pressure is lowered.

Myocardial blood flow is unique because it occurs mostly during diastole (see below). The "driving force" for coronary flow during this period is aortic diastolic pressure minus right atrial pressure. Because the coronary circulation is vasoconstricted at rest and has a large vasodilator capacity, the metabolic demands of the heart can usually be met by appropriate vasodilatation, even at relatively low diastolic pressures. If the coronary arteries are severely obstructed, however, the effective perfusion pressure is not aortic diastolic pressure but the pressure distal to the stenosis, which may be many mm Hg lower. The pressure gradient across the coronary capillaries may then be inadequate to supply the flow required even when maximum arteriolar vasodilatation is present, and serious myocardial disturbances may result.

EXTRAVASCULAR COMPRESSION FORCES.—With the exception of the larger arteries and veins, which sit on the surface of the heart, the coronary circulation is embedded in contracting myocardium. The contractions produce substantial increases in intramural pressure and may also deform the vessels passing through the muscle. However, even when the active effects of myocardial contractions on the vasculature are ignored, passive transmission of pressure through the ventricular wall considerably influences the distribution of blood flow within the ventricular muscle. Figure 4–69 is a cross section of the ventricles showing systolic and diastolic pressures in the ventricular cavities and coronary arteries in the open chest of a dog. During systole there is a substantial gradient across the left ventricular wall, 120 mm Hg inside and 0 mm Hg outside. Since the systolic coronary arterial pressure is about 120 mm Hg, there is no pressure gradient to drive flow in the subendocardium during

Fig 4–69.—Cross section of the ventricles showing systolic and diastolic transmural pressure gradients across their free walls.

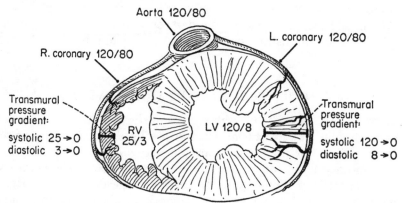

systole. However, the pressure difference between aortic systolic pressure and intramural ventricular pressure progressively increases from endocardium to epicardium so that substantial systolic flow can occur in the outer muscle layers. The pressure gradient across the ventricular wall during diastole is small, and an adequate driving force for flow exists in all layers. Thus the myocardium adjacent to the left ventricular cavity probably receives flow only during diastole, whereas the outer layers receive flow throughout the cardiac cycle. The situation is very different for the right ventricular free wall and the atrial myocardium. In these areas the intracavity systolic pressures are low relative to the coronary artery systolic pressure and flow occurs in all phases of the cardiac cycle.

PHASIC FLOW IN CORONARY ARTERIES.—The flow in a coronary artery can be measured from instant to instant throughout the cardiac cycle by means of electromagnetic or ultrasonic flowmeters. The flow pattern is complex and differs from that of other arteries (Fig 4–70). During isovolumic ventricular contraction, coronary flow in arteries supplying left ventricular muscle decelerates rapidly, occasionally to zero, because of the effects of extravascular compression just described. Flow tends to increase slightly during the early ejection phase because of rising aortic pressure. As aortic pressure falls from its peak, coronary flow also declines a little but then increases very rapidly during isovolumic relaxation when extravascular compression is greatly reduced. Flow declines slowly during the remainder of diastole along with the falling aortic pressure. When the ventricle is contracting strongly, compression of the intramural vessels may be so great that flow in the coronary arteries may be transiently reversed during systole, i.e., blood is squeezed out of the

Fig 4–70.—A, instantaneous flow in a coronary artery supplying left ventricular muscle (LCF). Aortic and left ventricular pressure (LVP) also are shown. Note that flow falls almost to zero during isovolumic contraction (IC), remains low during systole and increases rapidly during isovolumic relaxation (IR). B, flow pattern in an artery supplying right ventricular muscle (RCF) shows that systolic flow is relatively high because right ventricular pressure (RVP) is not sufficient to cause much impediment to flow.

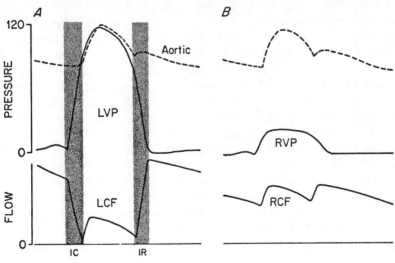

intraventricular arteries toward the aorta. Phasic flow in arteries supplying the right ventricle is similar to that of the left except that systolic flow is much higher because right ventricular systolic pressure is relatively low and a large perfusion gradient exists across the right ventricular wall even during systole.

Many factors may change the extravascular compression forces. These include hypoxia, autonomic nerve stimulation, circulating catecholamines and many drugs. Most of these factors also induce changes in myocardial metabolism and these frequently produce effects opposed to and more potent than the mechanical effects. Thus an increased ventricular rate might be expected to reduce coronary flow per minute since diastole is curtailed and the period of extravascular compression per minute is increased. In fact, coronary flow in conscious dogs increases as ventricular rate is increased from 50 to 250 beats per minute because increased myocardial metabolism induces coronary vasodilatation. This enables diastolic coronary flow to increase considerably despite the curtailment of diastolic time.

DISTRIBUTION OF CORONARY BLOOD FLOW.—Coronary flow per minute is evenly distributed throughout the left ventricular walls in anesthetized dogs, and mean basal values of approximately 70 ml/100 gm per minute have been obtained. Blood flow to the right ventricle is about 45 ml/100 gm per minute and atrial flow about 35 ml/100 gm per minute. The lower right ventricular and atrial flows presumably reflect the smaller tensions and consequently lower metabolic rates and oxygen demands. Recent evidence suggests that in *conscious* dogs subendocardial blood flow is about 40% higher than subepicardial flow.

Since total flow in the subendocardium is equal to or greater than flow in the subepicardium and since little or no blood flows through the subendocardium in systole, there must be a considerably higher diastolic flow in the subendocardium than in the subepicardium. The resistance of the subendocardial vasculature must therefore be lower. This relative vasodilatation is presumed to be metabolic in origin and attributable to the fact that the myocardial cells of the inner layer develop higher tensions than do the cells of the outer layer. They therefore consume more oxygen and produce more vasodilator metabolites. Furthermore, systolic obstruction to flow in the subendocardium may produce a reactive hyperemia in the subsequent diastole. Measurements of interstitial Po_2 have been somewhat variable but most authors have found subendocardial Po_2 to be lower than subepicardial Po_2. Nevertheless under basal conditions oxygen delivery appears to be adequate in all cell layers since products of anaerobic metabolism such as lactic acid are not normally detected in coronary venous blood. However, the subendocardium is much more vulnerable to ischemic injury than is the subepicardium. Subendocardial flow is dependent upon duration of diastole, aortic diastolic pressure and left ventricular end-diastolic pressure (shaded area in Fig 4–71). Factors that reduce diastolic time (e.g., tachycardia), lower arterial pressure (aortic insufficiency, coronary stenosis) or raise ventricular end-diastolic pressure tend to reduce coronary flow. This tendency can be offset by coronary arteriolar dilatation but, when this reaches its limit, the factors mentioned reduce subendocardial flow. The subepicardium is less vulnerable. It receives appreciable flow in

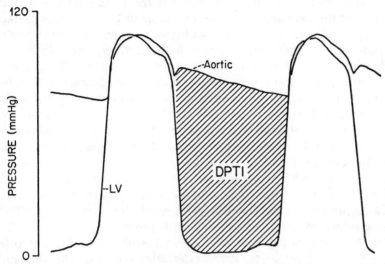

Fig 4–71.—The diastolic pressure time index *(DPTI)* is shown as the *shaded area* between the diastolic aortic and left ventricular *(LV)* instantaneous pressure curves.

systole, its vasculature has a higher basal resistance and probably a greater vasodilator reserve and its metabolic demands are less.

Neurohumoral Control of Coronary Flow

The coronary vessels are innervated by both sympathetic (adrenergic) and parasympathetic (cholinergic) fibers. The direct effects of stimulation of these fibers on the coronary vessels remain controversial despite much investigation.

There is no doubt that stimulation of the cardiac sympathetic nerves increases coronary blood flow in both anesthetized and conscious animals. However, this increase is associated with cardiac acceleration, increased myocardial force, elevated cardiac output and raised arterial pressure. The increased coronary flow is mainly due to the increased metabolism induced by these changes. However, the use of sensitive flowmeters has sometimes shown small increases or decreases in coronary flow that precede myocardial stimulation and are apparently due to a direct effect of the transmitter on coronary smooth muscle. After pretreatment with the beta adrenergic blocking agent propranolol, adrenergic nerve stimulation produces a small reduction in coronary flow. Infusions of norepinephrine directly into the coronary arteries have produced initial coronary vasodilatation in some studies and initial vasoconstriction in others. Whatever the initial response, it is always small and is followed by vasodilatation secondary to the metabolic effects of myocardial stimulation. After propranolol, which blocks the myocardial stimulation, intracoronary norepinephrine produces only a relatively slight vasoconstriction.

The direct effects of adrenergic nerve stimulation on the coronary arterioles are therefore unclear. Both vasodilatation and vasoconstriction have been observed. These effects are small compared to the metabolic vasodilatation that overwhelms the primary response, and their physiologic significance is doubtful. The results are compatible with the presence of weak activity of both alpha and beta adrenotropic receptors in the coronary arterioles.

When the experimental circumstances are carefully controlled, parasympathetic (vagus) stimulation produces coronary vasodilatation. However, under normal circumstances vagal stimulation slows heart rate and reduces arterial pressure, cardiac output and myocardial oxygen usage. The constrictor effect of reduced metabolism antagonizes the direct dilator effect of the neurotransmitter acetylcholine on the coronary arterioles and the net result is a reduction in coronary flow.

Angiotensin and vasopressin constrict the coronary arterioles and reduce coronary flow. The effect is transient and flow soon returns to normal values as a result of a number of factors, including the metabolic and physical effects of the increased arterial pressure that these agents produce.

Evidence is accumulating that prostaglandins play a significant role in coronary circulatory physiology. Coronary arteries synthesize prostaglandins and increased release of these agents occurs during coronary vasodilatation induced by hypoxia or ischemia. A recently discovered prostaglandin, *prostacyclin*, may be an important coronary flow regulator, since it is a potent vasodilator. It also inhibits platelet aggregation and may therefore have pathophysiologic significance.

EFFECTS OF BARO- AND CHEMOREFLEXES.—Elevation of systemic arterial pressure increases impulse traffic along the baroreceptor nerves and this induces bradycardia, reduced cardiac output and peripheral vasodilation (see section on arterial pressure). These effects reduce myocardial metabolism and coronary blood flow. Carotid sinus nerve stimulation has been used in an experimental approach to the management of angina pectoris. Electrodes were chronically implanted on a carotid sinus nerve and the patient activated these at the onset of angina. Although coronary flow decreased, myocardial oxygen demands fell disproportionately and the blood supply/oxygen demand ratio in the ischemic area of myocardium improved and the angina was aborted.

Decreases in systemic pressure evoke baroreflex responses that tend to increase myocardial oxygen demands and coronary flow. The relatively weak direct effects of sympathetic nerve activity on the coronary vessels probably do not significantly modify these metabolically induced effects.

Recent experiments in dogs suggest that chemoreceptor stimulation produces coronary vasodilatation by activating cholinergic parasympathetic vasodilator fibers and also by inhibiting preexisting sympathetic constrictor tone. It is not known whether this occurs in man.

METABOLIC CONTROL OF CORONARY FLOW

There is a strong correlation between myocardial metabolism and coronary blood flow. Those factors that increase metabolism, i.e., increased rate, force, "contractility," afterload or wall tension, also increase coronary flow. Decreases in these variables reduce flow. The close relationship between metab-

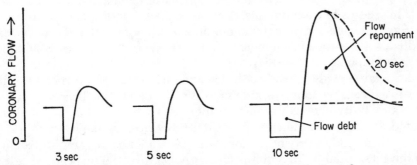

Fig 4–72.—Reactive hyperemia following mechanical occlusion of a coronary artery for 3, 5, 10 and 20 seconds. Peak hyperemic flow becomes maximal after 10 seconds' occlusion, but duration of hyperemic flow continues to increase after longer occlusions. Note that flow repayment exceeds flow debt.

olism and flow is related to the fact that the oxygen content of venous blood from resting myocardium is only 4–5 volumes/100 ml of blood, so that very little additional oxygen can be obtained by increasing extraction and increased oxygen demands can be met only by increased blood flow. This contrasts strongly with other tissues such as skeletal muscle and gut.

When a coronary artery is briefly occluded and then released, marked reactive hyperemia occurs. The "flow repayment" in the hyperemic phase always exceeds the "flow debt," often by a considerable margin (Fig 4–72). The magnitude of the hyperemia flow increases, within limits, with the duration of the preceding occlusion. This suggests that a vasodilator factor accumulates in ischemic tissue and that its final concentration is dependent on the period of occlusion. Many vasodilator substances are released from active myocardium. They include carbon dioxide, hydrogen ions, potassium ions, phosphate, adenosine and prostaglandins. None of these agents entirely explains reactive

Fig 4–73.—Possible scheme whereby coronary blood flow is linked to myocardial metabolism via adenosine (see text). (From Berne, R. M.: Physiol. Rev. 44:1, 1964. Used by permission.)

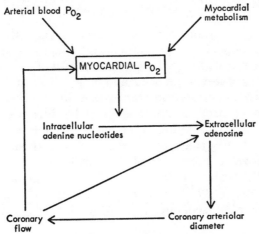

hyperemia. Coronary venous blood loses its vasodilator activity after oxygenation. This seems to exclude potassium, phosphate, prostaglandins and many other agents as the cause of the hyperemia and suggests that the responsible vasodilator is destroyed somewhere in its passage from the interstitial fluid to the venous blood.

Considerable evidence indicates that adenosine may be the metabolic link. This substance is a potent vasodilator, it is produced by the beating heart during basal activity and its production increases when cardiac metabolism is elevated. Its concentration in interstitial fluid rises markedly within a few seconds of coronary artery occlusion. Moreover, adenosine is rapidly removed from the interstitial fluid. Some is taken up again by the myocardial cell and reconverted to adenosine nucleotides, and the remainder passes into the capillaries where it is rapidly deaminated to form inosine, which is not vasoactive. The scheme illustrated in Figure 4–73 has been suggested. Increased cardiac activity tends to reduce myocardial oxygen tension, which causes a net increase in formation of adenosine from ATP. Adenosine diffuses into the interstitial fluid where it produces relaxation of the coronary arteriolar smooth muscle. Coronary blood flow is increased. This tends to remove adenosine from the interstitial space and to restore myocardial oxygen tension. Coronary occlusion also reduces myocardial oxygen tension, and by a similar chain of events coronary vasodilatation ensues. Certain facts are not fully explained by this hypothesis: (1) reactive hyperemia is not significantly prolonged by agents such as dipyridamole and lidoflazine, which are known to potentiate the vascular effects of adenosine, and (2) adenosine infused into the coronary arteries does not produce the magnitude of dilatation seen during severe hypoxia.

The metabolic link that so successfully causes coronary blood flow and hence oxygen delivery to be coupled with cardiac oxygen requirements may well include adenosine as a major participant. However, other known vasodilator metabolites probably contribute and it is possible that additional vasodilator agents remain to be discovered.

ANGINA PECTORIS

When myocardial oxygen demands exceed oxygen delivery, substances accumulate in the hypoxic myocardium that stimulate its sensory nerves. The most common manifestation of this stimulation is a crushing pain, *angina*, which is felt in the center of the chest and may radiate to the left arm, back or neck. The pain commonly occurs during exertion because of the increased oxygen requirement associated with tachycardia, increased contractility and elevated cardiac output. Another common setting for angina is severe emotional stress, which may cause tachycardia, increased cardiac output, hypertension and an elevated plasma catecholamine level.

The most frequent cause of angina pectoris is coronary atherosclerosis, with severe focal narrowing of the coronary arteries. As the stenosis develops, the tendency to myocardial ischemia is offset by dilatation of the coronary resistance vessels. This enables myocardial function to be normally maintained but at the expense of a reduced coronary vasodilator reserve. Symptoms occur when this reserve is inadequate to meet myocardial oxygen demands. At first, angina develops only during severe stress but, as the disease progresses and reserve becomes smaller, relatively mild stress may cause angina and eventually it may occur at rest.

The relationship between angina and stress is used in the procedure of exercise testing. A deliberate attempt, under carefully controlled conditions, is made to produce an imbalance between oxygen delivery by the coronary circulation and myocardial oxygen demands. The subject is exercised on a treadmill at progressively higher work loads. Supply-demand imbalance is indicated by the ECG appearance of 1 mm or more horizontal or down-sloping S–T segment depression. Angina may occur. When the work load is reduced, these effects promptly disappear. Disturbances of rhythm, e.g., ventricular ectopic beats, may be noted. This test is very useful in establishing the diagnosis of ischemic heart disease, in following its progress and in assessing its response to medical or surgical therapy.

Under certain circumstances, angina pectoris may be experienced even when coronary arteries are normal. It occurs, for example, in aortic stenosis in which myocardial oxygen requirements are raised by left ventricular hypertrophy and increased afterload. When aortic valvular insufficiency is present, aortic diastolic pressure is reduced and left ventricular end-diastolic pressure may be raised. These factors tend to reduce blood flow particularly to the subendocardial muscle, which receives most or all of its blood in diastole. If diastolic time is reduced by tachycardia, blood flow may be inadequate to supply the metabolic demands, and subendocardial ischemia and angina result.

Therapy for angina is directed toward reducing myocardial oxygen demands and increasing collateral blood flow. The drug of choice, nitroglycerin, reduces arterial pressure slightly and so reduces afterload. It also dilates veins and reduces central venous pressure (preload). The heart becomes smaller and hence requires a smaller expenditure of energy to produce a given pressure (see section on regulation of cardiac function; the Laplace law). It is also possible that nitroglycerin dilates collateral vessels and causes redistribution of blood flow within the myocardium, favoring the more vulnerable subendocardial layer. The beta adrenergic blocking drug propranolol also is of considerable value in angina. It slows the heart, reduces contractility and may reduce arterial pressure slightly. All these effects reduce myocardial oxygen demand. Propranolol also may cause a favorable redistribution of coronary flow, with a relative increase in subendocardial perfusion.

Nitroglycerin and propranolol are synergistic. Nitroglycerin reduces arterial pressure and evokes reflex tachycardia because of baroreflex sympathetic activation. These adverse effects are blocked by propranolol. On the other hand, propranolol tends to increase end-diastolic pressure and volume. These adverse effects are counteracted by nitroglycerin.

If medical treatment is ineffective, surgical treatment, usually saphenous vein bypass grafting, may be successful when the stenoses are relatively localized.

MYOCARDIAL INFARCTION

If a portion of the heart is inadequately perfused for more than a few minutes, muscle cells die. The dead area is called an *infarct* and usually is surrounded by a zone of edema and damaged cells.

If the infarction is extensive, the residual effective muscle mass may be inadequate to sustain an adequate circulation, and pump failure with pulmonary edema or cardiogenic shock (see section on abnormalities of cardiac function) may develop. The infarct may interfere with conduction of excitation over the heart and may alter the automaticity of potential pacemakers. Serious disturbances of rhythm including lethal ventricular fibrillation may then develop. If the AV node, AV bundle or bundle branches are infarcted, various degrees of heart block may ensue. If the papillary muscles are affected, mitral insufficiency may occur.

Medical treatment is directed toward improving oxygenation of the ischemic zone. Arrhythmias are corrected and aortic pressure is maintained with appropriate drugs. Oxygen is administered and the patient is sedated to reduce his myocardial oxygen requirements. In recent years sodium nitroprusside has proved useful in patients with severe infarcts requiring aggressive therapy. This vasodilator agent reduces the systolic afterload and also lowers central venous pressure, reducing the size of the heart. Myocardial oxygen demand therefore declines. Left ventricular end-diastolic pressure falls, pulmonary congestion is relieved and subendocardial blood flow improves.

If these measures fail, *counterpulsation* may be successful. Blood is withdrawn from the aorta during systole and reinfused in diastole using a special pump synchronized with the ECG. This technique reduces left ventricular afterload and oxygen consumption and increases the diastolic pressure gradient across the coronary circulation, thereby increasing diastolic coronary flow. In a variant of this method, a balloon is introduced into the thoracic aorta via the femoral artery. The balloon is inflated during diastole and deflated during systole.

THE PULMONARY CIRCULATION

The function of the pulmonary circulation is to transport oxygen and carbon dioxide between the tissues and gas-exchanging surfaces of the lungs. Lung vessels are very distensible and this permits them to accept maximal increases in cardiac output with very little rise in pulmonary arterial pressure. Because the resistance of the pulmonary circuit is low compared to systemic resistance, all pulmonary vascular pressures are less than corresponding systemic pressures. The low capillary pressure is one of the factors that prevent fluid from transuding into the alveolar spaces and reducing the ventilatory capacity of the lung.

The main pulmonary arterial trunk divides into right and left pulmonary arteries, which enter the corresponding lung and branch to supply the bronchopulmonary segments. These arteries and their subdivisions have much less smooth muscle than do systemic arteries, and the terminal arterioles may be devoid of muscle. Conventional descriptions indicate that blood passes through a network of wide capillary tubes around the alveoli into the thin walled pulmonary veins and then into the left atrium. However, recent animal studies have indicated that blood may flow as a thin sheet between endothelial surfaces joined by posts rather than through capillary tubes. The situation is comparable to a flood pouring through the space between floor and ceiling of a large parking structure, the internal supporting struts being analogous to the endothelial posts. Whatever its precise arrangement, the exchange area is huge, about 100 sq m, and the separation between blood and air is less than 1 μ. The junctions between endothelial cells are relatively tight and this helps to prevent fluid transudation into the alveoli.

The pulmonary vessels are capacious and the lung forms an important blood reservoir containing about 500 ml of blood, of which more than 50% is in the veins.

Pulmonary vascular smooth muscle is supplied by autonomic nerves but, in contrast to their role in the systemic circulation, they do not appear to participate significantly in pulmonary blood flow regulation.

DISTRIBUTION OF PULMONARY BLOOD FLOW

GRAVITATIONAL EFFECTS. — At the level of the lung hilus, mean pulmonary arterial pressure is about 15 mm Hg. In the upright posture, the lung apex is about 15 cm above and the lung base 15 cm below the hilus. Because of gravity, the effective arterial pressure at the apex will be reduced by 15 cm of blood, equivalent to 11 mm Hg. Similarly, effective arterial pressure at the lung base will be increased by 11 mm Hg. Intraluminal pressures in the arteries of the lung therefore vary with their location from 4 mm Hg at the apex to 26 mm Hg at the base. All the lung vessels are very compliant and their diameters increase as transmural pressure rises. The vessels at the lung base are therefore wider and offer a lower resistance to flow than do the apical vessels, and pulmonary flow in the upright position accordingly increases progressively from apex to base.

The vessels within the lung parenchyma are exposed to alveolar pressure. In the uppermost part of the lung (zone 1, Fig 4–74) mean arterial pressure is about 4 mm Hg but diastolic pressure may be subatmospheric. Capillary and venous pressures are also subatmospheric. All the vessels in this zone will therefore be collapsed in diastole by alveolar pressure and no flow will occur.

In the middle zone of the lung (zone 2; see Fig 4–74), pulmonary arterial pressure exceeds atmospheric pressure, but pressure in the pulmonary veins and in the venous ends of the capillaries tends to be subatmospheric and these vessels collapse. Pressure upstream from the collapsed vessels therefore rises to pulmonary arterial pressure, the collapsed segment is forced open and flow occurs. Because of the energy loss associated with flow, lateral pressure on the vascular walls falls and collapse recurs. "Fluttering" of the vessel walls is therefore produced. Flow in the midzone of the lung is thus independent of variations in pulmonary venous pressure in the subatmospheric range and is governed only by the arterial pressure-alveolar pressure difference.

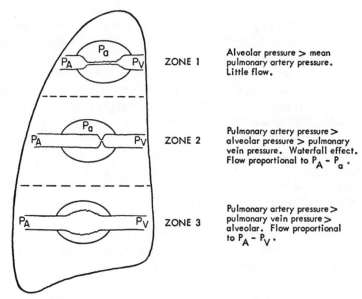

Fig 4–74.—Effects of gravity on the pulmonary circulation. P_A, P_V, P_a = pulmonary arterial, pulmonary venous, pulmonary alveolar pressures, respectively.

This situation has been termed a "waterfall effect" because flow over a waterfall is independent of the level of the stream below (pulmonary venous pressure) and is determined only by the height of the source above the lip of the fall (pulmonary artery pressure − alveolar pressure).

The vessels of the lower zone of the lung (zone 3; see Fig 4−74) have intravascular pressures that exceed alveolar pressure, and collapse does not occur. Flow through this zone is therefore governed by the pulmonary arteriovenous pressure gradient.

The foregoing description applies only to the distribution of pulmonary blood flow in the upright position. When the body is horizontal, a different distribution occurs because the effects of gravity are experienced in the anteroposterior axis rather than from apex to base. Furthermore, blood flow will be more evenly distributed because the distance from front to back is much smaller than the apex-base dimension of the lung.

In summary, the low pulmonary intravascular pressures make them extremely susceptible to gravitational effects and the transmural pressures are of much greater importance as determinants of flow than in the systemic vascular bed. In the upright position, the interaction of gravitational forces and intra- and extravascular pressures is such that blood flow increases progressively from the apex to the base of the lung.

These effects have an important bearing on the symptoms and signs produced by elevated left atrial pressure, which occurs, for example, when the left ventricle fails or when left atrial emptying is impaired, as in mitral stenosis. The high atrial pressure increases all pulmonary vascular pressures. Transmural pressures of the upper zone vessels are elevated, their diameters are increased and flow is greater than normal. Pressures in the lower zones are also increased and interstitial edema tends to occur. In some patients a protective pulmonary arterial and arteriolar constriction develops in the lower zones and the increased resistance reduces lower zone flow. Because flow in the upper zone is increased and flow in the lower zone is reduced, flow through the lung becomes more homogenous than is normal when the patient is upright. Passive enlargement of upper zone vessels and the narrowing of lower zone vessels are easily detected radiologically and are an important sign of raised left atrial pressure. Although pulmonary arterial vasoconstriction protects the lung from pulmonary edema, it does so only at the expense of greater pulmonary hypertension and right ventricular overload.

When a patient with high left atrial pressure lies down, vascular pressures increase in those parts of the lung that were above the hilus in the vertical position. Furthermore, systemic venous return is increased and the lung becomes more congested and less compliant. A greater effort is required to breathe and the patient experiences a sensation of breathlessness, which is relieved by resuming the upright position. (This symptom is termed orthopnea.) Exercise also causes elevated pulmonary vascular pressures and breathlessness (exertional dyspnea). If pulmonary capillary pressure rises to sufficiently high levels, transudation of fluid from plasma into the alveoli (acute pulmonary edema) develops. Edema formation is particularly prominent in the lower lobes because gravity increases capillary pressure, as described above.

EFFECTS OF BREATHING. — During inspiration pleural pressure becomes more negative and so the large intrapleural vessels expand and their resistance decreases. Pulmonary blood volume, which is mostly located in the pulmonary veins, may double. Inspiration also distends the alveoli and elongates the surrounding alveolar vessels, thereby increasing their resistance. Thus the effects of inspiration on the intrapleural and alveolar vessels are opposed and the change in total vascular resistance is negligible. The inspiratory reduction in intrathoracic pressure increases systemic venous return into the right side of the heart, and the stretching of the right ventricular muscle fibers increases right ventricular output. However, because the capacitance of the pulmonary vessels is increased, the elevated output of the right side of the heart is not immediately transmitted to the left side, and indeed left ventricular stroke volume may transiently fall at the beginning of inspiration.

During expiration the above changes are reversed, pulmonary blood volume diminishes and systemic venous return is impeded. During forced expiration, as occurs for example during severe coughing or straining to empty the bladder or rectum, venous return may be so reduced as to produce fainting.

CHEMICAL CONTROL OF PULMONARY BLOOD FLOW

HYPOXIA. — The pulmonary resistance vessels are constricted by reductions in alveolar PO_2. This response can be elicited by breathing less than 12% oxygen and is clearly of value in adjusting ventilation-perfusion relationships. Thus if a portion of the lung is underventilated, local alveolar oxygen tension

falls and arterioles constrict. This prevents needless perfusion of poorly ventilated areas and passage of deoxygenated blood into systemic circulation.
Arterial hypoxemia has a much greater effect on systemic than on pulmonary
circulation and produces no effects on the pulmonary vessels if alveolar Po_2 is
kept high. Excessive arterial CO_2 tension, e.g., in respiratory acidosis, or increased arterial hydrogen ion concentration, as in metabolic acidosis, also
produce pulmonary vasoconstriction.

NEURAL CONTROL OF PULMONARY VESSELS

Sympathetic nerve stimulation produces only a slight increase in pulmonary vascular resistance but may reduce pulmonary blood volume by constricting the larger pulmonary arteries and veins. Constriction of the pulmonary veins may facilitate left atrial filling. The relatively minor effect on pulmonary resistance is fortunate since it would obviously be inappropriate if a
large increase in resistance to right ventricular outflow occurred during episodes of increased sympathetic activity. Norepinephrine produces effects
that mimic those of sympathetic stimulation.

INTEGRATED RESPONSES OF THE CIRCULATION

In previous sections we have considered the individual functions of the
heart, arteries, capillaries and veins and also have reviewed the factors that
control the circulation through particular vascular beds. It is now time to consider how the function of the various parts is integrated into a smoothly functioning whole. Two examples have been chosen. In *exercise* the circulation
faces the problem of providing a huge blood flow to exercising muscle while
maintaining adequate flow to vital organs such as the brain. The adaptations
to sudden changes in *posture* involve rapid conpensatory adjustments to the
effects of gravity.

EXERCISE

Muscular exercise constitutes the strongest physiologic stress on the human
cardiovascular system. The degree of stress can be determined by measuring
the body's oxygen consumption. About 250 ml of oxygen per minute is consumed at rest and sedentary persons can increase this to about 2,500 ml per
minute during treadmill exercise against a grade. Champion athletes can
increase their oxygen consumption much more, to about 5,500 ml per minute.
The maximum oxygen consumption ($M\dot{V}o_2$) for any adult remains remarkably
constant when tested repeatedly and is an index of the maximal functional
ability of the circulation. It declines with age and prolonged bed rest and can
be increased by training.

The increased oxygen consumption of exercise occurs mainly in the exercising skeletal muscle and in the heart and respiratory muscles. It is made
possible by increased delivery of oxygen to these tissues and greater oxygen
extraction from the blood. The increased delivery is achieved by increased

cardiac output and by redistribution of blood flow. These changes demand integration of the responses of cardiac and vascular smooth muscle by appropriate modifications of neural, hormonal and chemical control.

CHANGES IN CARDIAC FUNCTION

Stroke volume, cardiac output and heart rate increase. Stroke volume becomes maximal at exercise levels requiring between 1 and 2 L per minute O_2 uptake. At these levels the diastolic volume of the heart does not alter appreciably and the increased stroke volume is due to a greater ejection fraction. As exercise becomes progressively more severe, further increases in cardiac output are achieved solely by an increase in heart rate. In a very fit individual who can increase his $M\dot{V}O_2$ to 5L per minute, stroke volume may rise from 100 to 150 ml per minute, heart rate from 50 to 200 beats per minute and cardiac output from 5 to 30 L per minute. It should be noted that the change in stroke volume alone would raise cardiac output by only 50% so that increased heart rate is the chief mechanism for increasing cardiac output in exercise. The maximal achievable heart rate declines with age and is about 220 minus the age in years.

CHANGES IN ARTERIAL PRESSURE

Systemic arterial systolic pressure invariably rises during exercise, sometimes to as high as 200 mm Hg, because of the augmented stroke volume and ejection rate. Diastolic pressure may also increase, but the increased diastolic runoff through the dilated arterioles of the exercising muscle causes the diastolic pressure increment to be smaller than the systolic pressure increment. When isolated muscle groups are strongly exercised, particularly if the exercise is isometric, the blood pressure rise may be disproportionately high. This is because the areas of vasodilation are smaller and reflex vasoconstriction occurs in a greater mass of resting muscle.

REDISTRIBUTION OF CARDIAC OUTPUT

Blood flow to exercising muscle increases enormously during maximal exercise from about 750 ml per minute to more than 20,000 ml per minute, and its share of the cardiac output may rise from 15% to 85%. This increase is due to profound arteriolar dilatation in the exercising muscle. At the same time, vasoconstriction occurs in the splanchnic and renal vascular beds, and their blood flows decline in proportion to the severity of exercise. At 100% $M\dot{V}O_2$ both flows may be reduced to about 30% of their resting values. This makes almost 2 L of blood per minute available to exercising muscle. Vasoconstriction in other areas may add another 500 ml per minute. The availability of this extra 2.5 L of blood enables a higher maximum level of work performance and $M\dot{V}O_2$ to be achieved. The percentage increment is inversely related to $M\dot{V}O_2$. Thus in a champion athlete with a very high $M\dot{V}O_2$ and a maximum cardiac output of 30 L per minute, 27 L would be delivered to exercising muscle. In the absence of the 2.5 L made available by cardiac output redistribution, only 24.5 L could be delivered and so $M\dot{V}O_2$ would fall by about 9%. On the other hand, in a patient with cardiac disease, a very low $M\dot{V}O_2$ and a maximum cardiac output of only 7 L per minute with 4 L per minute going to exercising muscle, the absence of blood flow redistribution would reduce muscle flow to 1.5 L per minute and $M\dot{V}O_2$ by 62%.

Vasoconstriction in renal, splanchnic and nonexercising muscle vascular beds maintains total vascular resistance at a higher level than it otherwise would be. This helps to maintain an adequate arterial pressure, essential for achieving high flow rates through the maximally dilated arterioles of the exercising muscle and for maintaining adequate cerebral and coronary blood flow.

The behavior of the skin blood vessels depends upon the severity of the exercise and the external temperature. When ambient temperatures are neutral, exercise usually causes cutaneous vasoconstriction. High environmental temperatures, however, demand increased skin blood flow, which reduces the ability to carry out prolonged exercise. Attempts to do so cause disproportionate tachycardia, and cardiac output may fall rather than rise owing to progressively falling stroke volume. This together with the intense vasodilatation in working muscle and skin does not allow arterial pressure to be maintained, and a falling pressure frequently heralds collapse.

MECHANISM OF CARDIOVASCULAR CHANGES

The increased heart rate, stroke volume and cardiac output and reduced splanchnic, renal and cutaneous blood flows suggest that increased sympathetic activity is the major cause of the observed changes. The main problem is to determine how this increased activity is engendered. The arterial pressure rises with the onset of exercise and it is therefore unlikely that baroreceptor reflexes are involved since they would produce sympathetic inhibition rather than activation. Reflex activation from mechanoreceptors in muscle and joints was suggested at one time, but recent evidence has excluded this as a major cause. It seems likely that the chief stimulus to sympathetic activation originates from the cerebral cortical areas responsible for initiation of exercise. The pathways involved are unknown but evidence in animals suggests that the hypothalamus may be an important relay station. Increased sympathetic activity accounts for all the observed effects of exercise on the circulation except for the vasodilatation in working muscle. At rest, sympathetic stimulation produces muscle vasoconstriction. Presumably the metabolic changes in active muscle protect its vessels from neural vasoconstriction. Animal experiments have shown that the sympathetic vasoconstriction is reduced by hypoxia and hyperkalemia, both of which occur in muscle during exercise.

The increased cardiac output in exercise can only be maintained if venous return is adequate. This is ensured by the massaging effect of the contracting muscle, by increased activity of the cardiac and respiratory pumps and by constriction of the large veins, particularly those of the splanchnic area and

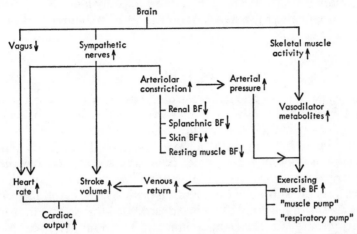

Fig 4–75.—Circulatory changes during muscular exercise. BF = blood flow.

skin. This venous constriction serves to increase circulating blood volume and to maintain an adequate filling pressure for the heart.

Figure 4–75 summarizes the cardiovascular responses discussed above.

INCREASED OXYGEN EXTRACTION

We have seen from the above that cardiac output in very fit individuals increases *sixfold* during maximum exertion and that a relatively small amount of extra blood is made available to exercising muscle by splanchnic, renal and cutaneous vasoconstriction. On the other hand, $M\dot{V}O_2$ in the same individuals may be 20 times greater than rest $\dot{V}O_2$. The difference is explained by the ability of working muscle to extract more than three times as much oxygen from the blood than can resting muscle. The increased extraction is aided by a shift to the right of the hemoglobin-oxygen dissociation curve in response to a fall in pH (from 7.4 to 7), and by a rise in temperature (from 37 to 40 C) in working muscle. An additional favorable feature is that hemoglobin concentration rises slightly as the result of increased filtration of plasma in working muscle. The arterial oxygen content therefore increases slightly (about 3%). When strenuous exercise is performed, the delivery of oxygen often fails to meet the immediate oxygen demands of the exercising muscle, which then begins to utilize anaerobic metabolism. After exercise, oxygen consumption may remain elevated for periods of a few minutes to two hours. The excess postexercise oxygen consumption is used to regenerate the depleted high energy phosphate stores and to oxidize the lactic acid produced in the hypoxic muscle. The ability of exercising muscle to use glycolytic (anaerobic) pathways considerably enhances the work capacity of muscle for relatively short periods of exercise. This is not the case for longer exercise periods because the amount of anaerobic metabolism is limited and provides a negligible *percentage* increase in work capacity under these circumstances. The oxygen demands of *prolonged* exercise must be almost entirely met during exercise. In practice the upper limit for steady-state exercise is about 4 L of oxygen per minute.

POSTURE

The principal effects of gravity on the circulation are due to its effects on the venous system and were covered in the section on the veins. During standing, transmural pressures in the veins below heart level increase, the veins dilate, blood tends to pool in them and venous return to the heart tends to diminish. Various compensatory measures offset these adverse effects. Pressure in the arterial baroreceptor areas falls, mainly as the direct result of postural change. On moving from the supine to the standing position, the carotid sinuses move from heart level to a position 10 inches or so above this so that transmural carotid sinus pressure falls by about 20 mm Hg. The diminished cardiac output caused by pooling of blood in the dependent veins also causes arterial pressure to fall. These effects on the carotid baroreceptors, together with effects on other mechanoreceptors in the heart and aortic arch, lead to increased sympathetic nerve discharge and peripheral arteriolar constriction. Total peripheral resistance is thereby increased and maintains arterial pressure at a near normal value despite reduced cardiac output. Venoconstriction also occurs and this reduces the degree of pooling and maintains an adequate central venous pressure. The activity of the cardiac sympathetic nerves also increases and heart rate goes up.

These compensatory mechanisms are often impaired during prolonged bed rest, febrile illnesses and hot weather. Fainting may then occur if posture is rapidly changed to the standing position. The importance of the sympathetic nervous system in postural adjustments is further shown in hypertensive patients being treated with drugs that block sympathetic activity. These patients become prone to *orthostatic hypotension,* i.e., arterial pressure falls abnormally when they stand.

Strong gravitational forces are frequently encountered in military aircraft flights and by astronauts during rocket flight. When these forces are directed toward the feet, blood tends to pool in the leg and abdominal veins. Fainting will occur unless a G-suit is worn, which applies counterpressure to the veins sufficient to prevent transmural pressures from rising to harmful levels.

ABNORMALITIES OF CARDIOVASCULAR FUNCTION

ARTERIAL HYPERTENSION

It is estimated that about 23,000,000 Americans have hypertension, which we may define somewhat arbitrarily as persistent elevation of diastolic arterial pressure above 90 mm Hg. Actuarial studies leave no doubt that high blood pressure shortens life because of its adverse effects on blood vessels and on cardiac, renal and cerebral function.

The pathophysiology of hypertension is an extremely complex, actively developing and controversial area. The following brief paragraphs are presented primarily to show what an important bearing a knowledge of the normal physiology of a controlled variable, arterial pressure, can have on our approach to a common disorder.

CAUSES OF HYPERTENSION

We have seen that mean arterial pressure is determined by cardiac output and total peripheral vascular resistance. In most patients with established hypertension, cardiac output is within normal limits. Thus the factors that maintain hypertension must be looked for among those that regulate resistance, i.e., in abnormalities of the neural, hormonal and chemical control of the arterioles.

Hypertension is not a single entity. There are many known causes, most of them rare, and only three will be listed as examples.

1. *Coarctation of the aorta.*—A congenital severe narrowing of the aorta increases vascular resistance and causes elevated blood pressure, particularly in parts of the body above the coarctation.

2. *Pheochromocytoma.*—A tumor of the adrenal medulla causes increased plasma catecholamine levels and hence increased cardiac output as well as peripheral resistance.

3. *Renovascular disease.*—Narrowing of the renal arteries causes increased renin production, elevated plasma concentrations of angiotensin and therefore increased peripheral resistance.

In the majority of hypertensive patients, no cause can be found and the condition is designated *essential* hypertension. This conceals our ignorance and it is already clear that persons with essential hypertension are not a homogeneous group with respect to the pathophysiology of their disorder.

SUGGESTED CAUSES OF ESSENTIAL HYPERTENSION

NEURAL MECHANISM.—In some patients plasma norepinephrine levels are elevated but not impressively. Nevertheless a significant reduction in blood pressure can be achieved in all hypertensive patients by drugs that interfere with sympathetic activity, and this implies at least a "permissive role" of the sympathetic nervous system. Denervation of the carotid and aortic baroreceptors in animals leads to hypertension, which indicates the importance of baroreceptors in maintaining normal pressure. Indeed in all hypertensive patients the baroreceptors appear to be "reset," i.e., they continue to regulate pressure but at an abnormally high level.

Another facet of neural control of blood pressure that has been exposed in recent years is the existence of catecholaminergic and serotoninergic interneurons in the brain stem that modulate baroreflexes. Abnormalities of function of these neurons have been shown in several animal models of hypertension but whether these are causal or are secondary to the hypertensive process is unknown. It is of interest that clonidine, an agent that stimulates brain stem alpha adrenergic receptors, is effective in lowering blood pressure in many patients.

RENIN-ANGIOTENSIN SYSTEM.—Over 40 years ago Goldblatt showed that hypertension could be induced in experimental animals by clamping a renal artery. Subsequently it was found that the hypertension was due to increased renin secretion leading to elevated plasma levels of angiotensin II. This octapeptide is a potent constrictor of vascular smooth muscle and also enhances sympathetic activity by several mechanisms, central and peripheral. In addition, it regulates aldosterone release from the adrenal cortex and so leads to increased extracellular fluid volume.

WATER AND ELECTROLYTE ABNORMALITIES.—These appear to play an important role. Most experimental types of hypertension are easier to produce, are more severe and are more resistant to therapy when the animals are salt-loaded. An important subgroup of hypertensive persons has an expanded extracellular fluid volume, and "overfilling" of the vascular tree seems to be the major cause of the elevated pressure.

ABNORMALITIES OF ARTERIAL WALL.—In established hypertension of whatever cause, arterial and arteriolar walls become thickened and their lumina become smaller. This structural change itself must account for some of the resistance increase. In some experimental models, increased amounts of sodium, calcium and water have been found in the arterioles. These changes may in part account for the hyperresponsiveness to vasoconstrictor stimuli that has been observed in hypertensive vessels.

GENETIC FACTORS.—A hereditary factor appears to be involved in many hypertensive patients. In the last few years a strain of rats that become spontaneously hypertensive has been developed. However, despite intense study the causes of hypertension in these animals remain unclear.

Although the various abnormalities that may lead to hypertension have been listed under separate headings, it should be emphasized that interactions are the rule rather than the exception. Thus abnormalities of adrenergic neural control, of the renin angiotensin system, of arterial wall electrolytes and of arteriolar thickness have all been found to occur together in the Goldblatt-type hypertensive model. Further, it must be stressed that the factors that trigger hypertension may not be the same as those responsible for its maintenance. Thus some authorities believe that human hypertension might be triggered by intermittent stress-induced cardiac output changes that, if frequently repeated, might lead to structural and functional changes that then maintain the abnormal state.

SUBGROUPS OF ESSENTIAL HYPERTENSION.—The pathophysiologic approach to hypertension discussed above has led to some useful groupings of hypertensive patients. Plasma renin levels have been particularly helpful, and low-, normal- and high-renin subgroups of hypertension have been recognized. The high-renin group tends to have abnormally high plasma catecholamine levels and many of these persons have renovascular abnormalities that should be carefully looked for, since a surgical cure might be possible. They respond well to propranolol, which blocks renin secretion as well as having beta adrenergic blocking actions elsewhere. The low-renin group rarely has elevated plasma catecholamines and responds poorly to propranolol. These patients have expanded extracellular and plasma volumes, and satisfactory reductions in pressure usually can be achieved by salt and water restriction and use of diuretics (see section on diuretics, Chapter 6).

CARDIAC FAILURE

Cardiac failure is said to be present when, despite an adequate filling pressure, the heart is unable to pump blood at a rate sufficient to meet the needs of the tissues. It must be distinguished from peripheral circulatory failure, in which the heart is at least initially capable of producing an adequate output but is unable to do so because of low venous filling pressure.

Cardiac failure occurs as the result of prolonged abnormal pressure or volume overloads or because of disease or injury of the myocardial cells.

PRESSURE OVERLOADS.—These may be produced by abnormally high vascular resistance, as in most types of hypertension, or by obstruction to outflow caused by stenosed valves. The precise mechanism whereby the heart adapts to pressure overloads is unknown. Obscure homeometric mechanisms appear to be involved initially and lead to myocardial hypertrophy (see section on regulation of cardiac function). Later, as these compensatory mechanisms fail, the heart dilates and the Starling mechanism enables cardiac function to be maintained. In contrast to volume overloads, pressure overloads are not associated with increased cardiac efficiency, and oxygen consumption increases *pari passu* with the increased pressure load.

VOLUME OVERLOADS.—These may occur as the result of abnormal "shunts." For example, in the congenital abnormality, *atrial septal defect,* blood may be shunted from the left to the right atrium through a large defect in the interatrial septum. The right ventricle may then have to pump up to three times as much blood as the left ventricle. Volume overloads may also occur because of incompetent valves. For example, in aortic valvular insufficiency, blood enters the left ventricle during diastole from the aorta as well as from the left atrium. Whatever the cause of the volume overload, the affected ventricle becomes dilated. This has two important consequences that depend on the operation of Starling's law and the law of Laplace. The initial length of the muscle fibers of the dilated ventricle is increased; this augments the force of contraction (Starling's law) and cardiac efficiency increases. This is partly offset by the effects of the Laplace relationship, which predicates that more tension is required to produce a given pressure within the ventricle when its diameter is increased.

When overloads persist, the ventricle hypertrophies. The number of sarcomeres increases and the mitochondria multiply. Hypertrophy is a useful adaptation since it spreads the increased load over a greater muscle mass. In some cases of severe aortic valve disease, the weight of the heart may increase to more than 300% of control values. There is a limit to the degree of hypertrophy that can occur, which appears to be set by the amount of DNA in each cell. This amount is fixed and does not increase as the cells hypertrophy. It can initiate the synthesis of only so much new structural material and, when the limit is reached, continuing overloads result in degeneration of myocardial cells and overgrowth of fibrous tissue. This is hastened by the fact that blood supply does not keep pace with muscle hypertrophy, and diffusion distances between blood and the interior of muscle cells increase.

There is evidence that sympathetic nervous activity is increased in heart failure. This shifts the cardiac function curve to the left and thus maintains cardiac function at a higher level than would otherwise be possible. Agents that interfere with sympathetic activation of the heart may induce or exacerbate cardiac failure. In established heart failure the norepinephrine content of the heart is usually greatly reduced, perhaps because of depletion produced by prolonged high-level sympathetic activity. These various adaptive mechanisms—the Starling effect, hypertrophy, and increased sympathetic activity—allow circulatory function to be maintained despite the presence of often severe mechanical defects or myocardial damage. However, the myocardial reserve is diminished so that abnormal signs or symptoms develop under stress and, eventually, as the disease progresses or the adaptive responses fail, abnormalities of function occur at rest.

HEMODYNAMIC CHANGES.—Pressure or volume loads often predominantly affect one cardiac chamber, although eventually functional impairment occurs throughout the heart. Thus when the left ventricle fails, pressures in the left atrium, pulmonary veins and capillaries rise and the lung becomes congested. Pulmonary arterial pressure also passively increases and an increased load is therefore placed on the right ventricle. This may be considerably aggravated in some conditions (e.g., certain cases of mitral stenosis) by pulmonary arteriolar constriction, which may develop via poorly understood mechanisms and lead to right ventricular failure with secondary distention of systemic veins, engorgement of the liver and peripheral edema.

When the symptoms and signs of cardiac failure are mainly due to congestion in the lungs, the term *left-heart failure* is frequently used. When the congestion is mainly in the systemic circuit, *right-heart failure* is said to be present.

ABNORMALITIES OF WATER AND ELECTROLYTE METABOLISM.—Cardiac failure is usually associated with retention of water and sodium chloride.

Fig 4–76.—Causes of water and salt retention in cardiac failure.

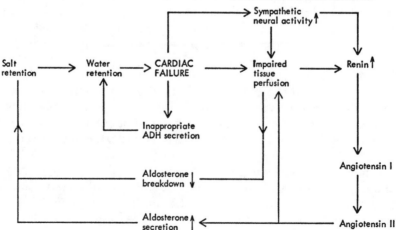

The mechanisms are complex and not fully understood but a currently accepted scheme is shown in Figure 4–76.

Heart failure leads to reduced renal perfusion pressure and a reduction in GFR. This causes increased secretion of renin and production of angiotensin I from renin substrate in the plasma. Angiotensin I is converted to angiotensin II by *converting enzyme*, which is present in many tissues, with particularly large amounts in the lung. Angiotensin II is a potent stimulus for aldosterone secretion from the adrenal cortex. Plasma aldosterone levels may be increased enormously in heart failure not only because of increased secretion but because circulatory impairment of the liver reduces aldosterone degradation. Aldosterone acts on the distal renal tubule to promote sodium retention and potassium excretion. Water is passively retained as a consequence of the sodium retention. Recent evidence suggests that lack of a factor that normally promotes urinary sodium excretion may contribute to sodium retention. This substance, "natriuretic hormone," has not yet been fully characterized.

CLINICAL FEATURES OF CARDIAC FAILURE

These can be readily understood on the basis of the disordered pathophysiology. Failure of the left side of the heart is characterized by dyspnea on exertion and later by orthopnea, paroxysmal nocturnal dyspnea and pulmonary edema. *Dyspnea* is the term applied to a subjective sensation of difficulty in breathing. It occurs during exertion in cardiac failure because the lung becomes congested secondary to the rise in pulmonary venous pressure that develops when left ventricular function is inadequate. The congestion makes the lung stiffer and the work of breathing is increased, leading to a feeling of breathlessness. When the patient is in the upright position, gravity reduces the venous pressure in the lung above heart level. This effect is lost when the patient lies down, and pulmonary congestion may then develop. The patient may therefore be able to breathe comfortably only when upright (orthopnea). This is particularly true at night when systemic edema that has developed during the day tends to be reabsorbed into the circulation. The patient may need to prop himself up with several pillows in order to breathe comfortably. If he slips down, he may be awakened by breathlessness (nocturnal dyspnea). If pulmonary capillary pressure becomes sufficiently high, fluid may transude from the plasma into the alveoli (pulmonary edema). The patient then becomes extremely breathless and frequently produces frothy sputum tinged with blood that has seeped through the congested and damaged lung capillaries.

Failure of the right side of the heart is characterized by the effects of raised systemic venous pressure and fluid retention. These include swelling of the liver, ascites and peripheral edema.

Treatment of heart failure includes drugs that improve myocardial force, e.g., digitalis, and agents that reduce excessive sodium and water loads (diuretics). When these agents are effectively administered, several liters of urine may be excreted, blood volume and venous pressures decline, the lung clears and cardiac output rises. Some patients may be refractory to these measures, in which case a recently developed therapy termed "afterload reduction" may be tried. A vasodilator agent such as sodium nitroprusside is continuously infused into a vein while left atrial pressure, systemic arterial pressure and cardiac output are carefully monitored. The drug reduces peripheral vascular resistance (afterload), and systolic ejection increases. Left atrial and left ventricular end-diastolic pressures (preload) also fall, in part due to the dilatation of veins that this agent produces, and so pulmonary congestion diminishes.

When these measures have proved effective, attention can be turned toward correction of the underlying cause, e.g., valve replacement, correction of shunts and treatment of hypertension.

SHOCK

Peripheral circulatory failure, or shock, is said to be present when cardiac output is insufficient to meet the functional demands of the tissue. This description is very similar to that of heart failure but there are important differences between the two. Cardiac failure usually is the result of increased pressure or volume overloads or of myocardial injury or disease; the circulating blood volume is high and cardiac filling pressures are increased. On the other hand, shock is frequently caused by noncardiac conditions (e.g., hemorrhage, trauma, toxemia) and circulating blood volume and cardiac filling pressures usually are low.

CAUSES OF SHOCK

Shock may be caused by (1) a reduction in blood volume, (2) an increased capacity of the systemic venous system or (3) myocardial damage. Frequently several causes may be operative. For example, the initiating factor in shock produced by severe bleeding is reduction of blood volume but, in later preterminal stages, inadequate perfusion pressure and reduced oxygen-carrying capacity may lead to impairment of myocardial function. The major causes of shock are as follows:

1. *Diminished blood volume:* hemorrhage, severe vomiting and/or diarrhea, water deprivation, burns.

2. *Increased vascular capacity:* bacterial toxins, anaphylactic shock.

3. *Myocardial damage:* myocardial infarction.

HEMODYNAMIC CHANGES

The changes produced by shock due to diminished circulating blood volume are summarized in Figure 4–77. Many of these are compensatory in character and are due to increased adrenergic activity. They include increased heart rate and peripheral vasoconstriction, particularly in the skin, kidney and gut. Brain and coronary circulations are spared and indeed the whole response may be looked upon as a means of maintaining adequate per-

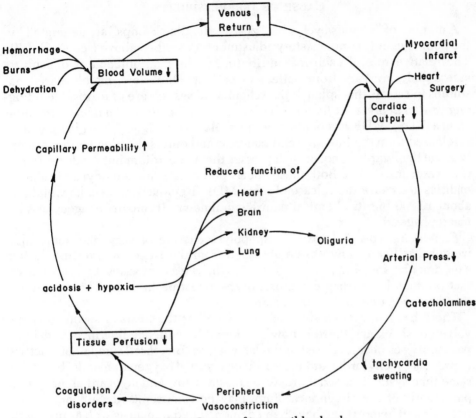

Fig 4-77.—Causes of irreversible shock.

fusion pressure and blood flow to the heart and brain. Unfortunately this can only be achieved at the expense of reduced perfusion of other organs, particularly muscle, skin, kidney and gut. Cutaneous vasoconstriction causes the striking pallor of the skin often seen in shock, and renal vasoconstriction causes the characteristic severe reduction in urine flow (oliguria). These changes are initially beneficial in maintaining arterial pressure, cerebral and coronary blood flows and blood volume but, when they are prolonged, tissue damage occurs and the patient becomes refractory to treatment (irreversible shock).

In certain types of shock, for example that initiated by gram-negative bacterial infection, the clinical picture is different. Toxins released by the bacteria produce vasodilatation and the pulses may be bounding and the skin flushed. Blood pressure may be normal and the patient's general condition may appear deceptively good, although urinary output is usually low. Cardiac output is often within or even above normal limits but nevertheless it is low relative to the increased metabolic demands produced by the infection. Tissue acidosis develops and the hemodynamic situation eventually changes to the low output, vasoconstrictive state described above.

CAUSE OF IRREVERSIBILITY

A number of "vicious cycles" or "positive feedback loops" are generated by the hemodynamic compensatory adjustments described above (see Fig 4–77). The relative role of each in contributing to the fatal outcome is difficult to assess and may vary from patient to patient. One view, mainly based on studies in cats, is that splanchnic ischemia causes release of toxins, including *myocardial depressant factor* (MDF) and *reticuloendothelial depressant substance* (RDS). MDF is a peptide of low molecular weight (500–1,000), which is released from the intestine and pancreas and enters the blood either directly or via the lymph. It not only depresses the myocardium but produces intestinal vasoconstriction. Both these effects perpetuate the hemodynamic abnormalities and cause the release of more MDF. RDS has a molecular weight of about 10,000 but its chemical nature is unknown. It impairs phagocytosis by macrophages.

Escherichia coli endotoxin, a lipopolysaccharide of very high molecular weight, has been shown to be absorbed from the ischemic intestine during experimental shock in animals. It causes the release of vasoactive substances that produce hemorrhagic necrosis of the intestine in dogs. This manifestation, however, does not occur in man.

There has been much discussion of whether the heart is secondarily involved in shock and there is now considerable evidence that myocardial impairment does occur, at least in the later stages. Increased sympathetic activity produces tachycardia and increased myocardial oxygen demands, but at the same time systemic arterial pressure is usually low and myocardial perfusion, particularly of the subendocardium, may be compromised. Subendocardial necrosis and hemorrhage have been demonstrated in man as well as in experimental animals in hemorrhagic shock. Myocardial function also is depressed by acidosis and possibly by specific myocardial depressant factors such as MDF, mentioned above.

The lung may be severely damaged in prolonged shock and the term "shock lung" is often applied. The alveolar epithelium and the lung capillaries are injured and fluid enters the alveoli. Portions of the lung collapse. Hypoxia develops and further impairs oxygen delivery to the poorly perfused systemic tissues.

SPECIFIC TYPES OF SHOCK

HEMORRHAGIC SHOCK.—Following a brisk hemorrhage of 10–30% of blood volume, systemic arterial and central venous pressures fall. Baroreceptor discharge from the aorta, carotid sinus and heart declines and this induces reflex sympathetic discharge, which produces peripheral vasoconstriction, especially in the kidney, skin and gut. The sweat glands also are sympathetically activated. Examination of a patient at this stage reveals pallor of the skin and mucous membranes, sweating, a rapid, thready pulse and low urinary output. The veins are collapsed. The skin feels cold and clammy. As explained, these sympathetic effects are compensatory. The cardiac effects and regional arteriolar constriction help to prevent drastic falls in arterial pressure. Capillary pressure falls and fluid moves into the circulation from the interstitial space. Other compensatory changes that tend to sustain blood volume are increased ADH secretion induced by reduced baroreceptor discharge from the atria and activation of the renin-angiotensin-aldosterone system. If hemorrhage continues and is inadequately treated, the various compensatory mechanisms prove ineffective and cardiac output falls, severe acidosis develops and the patient passes into an irreversible stage in which transfusions produce only temporary benefit.

GRAM-NEGATIVE SHOCK.—In about 25% of patients with gram-negative septicemia, shock develops. The manifestations vary widely among individuals but it is not uncommon to have bounding pulses and a warm skin in the early stages. Cardiac output may be within normal limits but is nevertheless inadequate for the raised metabolic demands. Mental aberrations are frequent and may reflect the imbalance of oxygen supply and demand in the brain. Myocardial failure may develop early in some patients, presumably due to the combined effects of toxins and adverse hemodynamic changes, and central venous pressure may then be elevated. In others, because of peripheral vasodilatation and pooling, venous pressure may be low. Eventually, because of inadequate blood supply, metabolic acidosis develops. Increasing sympathetic activity aggravates the acidosis by reducing regional perfusion and it also changes the clinical picture to that seen in late hemorrhagic shock, i.e., cold clammy skin, thready pulse and oliguria.

CARDIOGENIC SHOCK.—The most frequent cause of this condition, which kills approximately 100,000 Americans each year, is extensive myocardial infarction. Its features are similar to those of other types of shock except that there is a much more severe impairment of the cardiac pump. Thus, there are usually striking reductions in cardiac output, stroke volume and ejection fraction. The left ventricular function curve is shifted to the right and depressed (see Fig 4–30) and left ventricular filling pressure rises. Arterial pressure falls, urine output declines, the skin becomes cold and clammy and clouding of consciousness develops.

Cardiogenic shock used to be almost uniformly lethal but in recent years increased survival has occurred with the careful use of vasodilator drugs such as sodium nitroprusside. This agent causes arteriolar dilatation and reduces systemic vascular resistance (hence the "afterload" of the left ventricle). As a result, the ejection fraction and cardiac output increase. Arterial pressure commonly does not fall significantly because the reduced systemic vascular resistance is countered by the rise in cardiac output. Sodium nitroprusside also dilates systemic and pulmonary veins and thereby reduces cardiac filling pressure ("preload") and relieves pulmonary and systemic congestion.

BIBLIOGRAPHY

Abboud, F. M., et al.: Reflex control of the peripheral circulation, Prog. Cardiovasc. Dis. 18:371, 1976.

Baez, S.: Microcirculation, Ann. Rev. Physiol. 39:391, 1977.

Barger, A. C., and Herd, J. A.: The renal circulation, N. Engl. J. Med. 284:482, 1971.

Bergel, D. H.: *Cardiovascular Fluid Dynamics* (New York: Academic Press, 1972).

Berne, R. M., and Levy, M.: *Cardiovascular Physiology* (St. Louis: C. V. Mosby Co., 1977).

Bohr, D. F.: Vascular smooth muscle updated, Circ. Res. 32:665, 1973.

Braunwald, E. (ed.): *The Myocardium: Failure and Infarction* (New York: H. P. Publishing Co., Inc., 1974).

Braunwald, E., Ross, J., Jr., and Sonnenblick, E. H.: *Mechanisms of Contraction of the Normal and Failing Heart* (Boston: Little, Brown and Co., 1976).

Brody, M. J., and Zimmerman, B. G.: Peripheral circulation in arterial hypertension, Prog. Cardiovasc. Dis. 18:323, 1976.

Burton, A. C.: *Physiology and Biophysics of the Circulation* (Chicago: Year Book Medical Publishers, Inc., 1972).

Clement, D. L., and Shepherd, J. T.: Regulation of peripheral circulation during muscular exercise, Prog. Cardiovasc. Dis. 19:23, 1976.

Cumming, G.: The Pulmonary Circulation, in Guyton, A. C. (ed.): *MTP International Review of Science: Physiology* (Baltimore: University Park Press, 1974).

Feigenbaum, H.: *Echocardiography* (Philadelphia: Lea & Febiger, 1976).

Folkow, B., and Neil, E.: *Circulation* (New York: Oxford University Press, 1971).

Fowler, N. O. (ed.): *Diagnostic Methods in Cardiology* (Philadelphia: F. A. Davis Co., 1975).

Grant, R. P.: *Clinical Electrocardiography; the Spatial Vector Approach* (New York: McGraw-Hill Co., 1970).

Greenfield, A. D. M.: The Circulation Through the Skin, in Fenn, W. O., and Rahn, H. (eds.): *Handbook of Physiology* (Washington, D. C.: American Physiological Society, 1963).

Greenway, C. V., and Stark, R. D.: Hepatic vascular bed, Physiol. Rev. 51:23, 1971.

Guyton, A. C.: Venous Return, in Fenn. W. O., and Rahn, H. (eds.): *Handbook of Physiology* (Washington, D. C.: American Physiological Society, 1963).

Higgins, C. B., et al.: Parasympathetic control of the heart, Pharmacol. Rev. 25:119, 1973.

Hudlicka, O.: *Muscle Blood Flow* (Amsterdam: Swets & Zeitlinger B. V., 1973).

Ingvar, D. S., and Lassen, N. A. (eds.): *Brain Work: The Coupling of Function, Metabolism and Blood Flow in the Brain* (New York: Academic Press, 1974).

Interstitial fluid pressure and dynamics of lymph formation, Fed. Proc. 35:1861, 1976.

Kaley, G., and Altura, B. M. (eds.): *Microcirculation* (New York: Plenum Publishing Corp., 1977).

Langer, G. A., and Brady, A. J.: *The Mammalian Myocardium* (New York: John Wiley & Sons, Inc., 1974).

Laragh, J. H.: *Hypertension Manual; Mechanisms, Methods, Management* (New York: Yorke Medical Books, 1973).

Lassen, N. A.: Control of cerebral circulation in health and disease, Circ. Res. 34:749, 1974.

Lundgren, O., and Jodal, M.: Regional blood flow, Ann. Rev. Physiol. 37:395, 1975.

Marriott, H. J. L.: *Practical Electrocardiography* (Baltimore: Williams & Wilkins Co., 1972).

Maseri, A. (ed.): *Myocardial Blood Flow in Man, Methods and Significance in Coronary Disease* (Torino: Minerva Medica, 1972).

McDonald, D. A.: *Blood Flow in Arteries* (Baltimore: Williams & Wilkins Co., 1974).

Mellander, S., and Johansson, B.: Control of resistance, exchange and capacitance functions in the peripheral circulation, Pharmacol. Rev. 20:117, 1968.

Moir, T. W.: Subendocardial distribution of coronary blood flow and the effect of antianginal drugs, Circ. Res. 30:621, 1972.

Noble, D.: *The Initiation of the Heart Beat* (Oxford: Clarendon Press, 1973).

Onesti, G., Fernandes, M., and Kim, K. E.: *Regulation of the Blood Pressure by the Central Nervous System* (New York: Grune & Stratton, 1976).

Page, I. H.: Arterial hypertension in retrospect, Circ. Res. 34:133, 1974.

The Physiological Basis of Starling's Law of the Heart. Ciba Foundation Symposium (New York: American Elsevier Publishing Co., 1974).

Purves, M. J.: *The Physiology of the Cerebral Circulation* (London: Cambridge University Press, 1972).

Rapaport, S. I.: *Blood-Brain Barrier in Physiology and Medicine* (New York: Raven Press, 1976).

Rowell, L. B.: Human cardiovascular adjustments to exercise and thermal stress, Physiol. Rev. 54:75, 1974.

Rubio, R., and Berne, R. M.: Regulation of coronary blood flow, Prog. Cardiovasc. Dis. 18:105, 1975.

Shepherd, J. T.: *Physiology of the Circulation in Human Limbs in Health and Disease* (Philadelphia: W. B. Saunders Co., 1963).

Smith, E. E., et al.: Integrated mechanisms of cardiovascular response and control during exercise in the normal human, Prog. Cardiovasc. Dis. 18:421, 1976.

Somlyo, A. F., and Somlyo, A. V.: Vascular smooth muscle, Pharmacol. Rev. 20:197, 1968.

Stein, E.: *The Electrocardiogram* (Philadelphia: W. B. Saunders Co., 1976).

Stein, J. H.: The Renal Circulation, in Brenner, B. M., and Rector, F. C., Jr. (eds.): *The Kidney* (Philadelphia: W. B. Saunders Co., 1976).

Vascular reactivity in hypertension (Symposium): Fed. Proc. 33:121, 1974.

Vatner, S. F., and Pagani, M.: Cardiovascular adjustments to exercise, hemodynamics and mechanisms, Prog. Cardiovasc. Dis. 19:91, 1976.

Weissler, A. M.: *Non-Invasive Cardiology* (New York: Grune & Stratton, 1974).

West, J. B.: Respiratory Physiology (Baltimore: Williams & Wilkins Co., 1974).

Wood, E.: *The Veins* (Boston: Little, Brown & Co., 1965).

Zelis, R.: *The Peripheral Circulation* (New York: Grune & Stratton, 1975).

Zweifach, B. W.: Microcirculation, Ann. Rev. Physiol. 35:117, 1973.

Zweifach, B. W., and Fronek, A.: The interplay of central and peripheral factors in irreversible hemorrhagic shock, Prog. Cardiovasc. Dis. 18:147, 1975.

5

The Respiratory System

BRIAN J. WHIPP, Ph.D.

THE MECHANICS OF BREATHING

LIFE NECESSITATES exchange of energy between the environment and the organism. In a complex organism such as man, the nutrients essential for this energy exchange must be transported to the cells by highly specialized organ systems whose structures are exquisitely "designed" for the function they subserve.

The cells require a virtually continuous supply of both the nutritive elements of ingested food and oxygen. The predominant function of the lung* is to transport sufficient oxygen from the atmosphere into the blood so that the cardiovascular system is able to supply cellular oxygen requirements without undue stress. Also the lung must clear from the blood the consequent metabolic carbon dioxide in order to maintain the acid-base status of the blood within the relatively narrow limits compatible with life.

Performance of these functions involves (1) generation of negative, or subatmospheric, pressures within the lung, allowing atmospheric pressure to provide the motive force to propel air by convection into the gas exchange regions of the lung; (2) mixing the volume of inhaled air with the air already in the lung, consequently increasing its oxygen concentration and reducing its carbon dioxide concentration; (3) movement of oxygen into the blood, and carbon dioxide from the blood, by diffusion along the newly established concentration gradients and (4) the control system that "senses" whether specific chemical and physical features of the blood and cerebrospinal fluid, especially the oxygen and carbon dioxide levels, are appropriate for the body's current requirements.

In this chapter we will discuss the static and dynamic aspects of the mechanics of breathing. For convenience, special symbols used conventionally in pulmonary physiology are listed in Table 5-1.

---

*The convention of referring to both lungs as "the lung" is followed here.

244

TABLE 5-1.—SYMBOLS AND ABBREVIATIONS USED BY PULMONARY PHYSIOLOGISTS°

Special Symbols

Dash (–) above any symbol indicates a *mean* value
Dot (·) above any symbol indicates a *time derivative*

Gases

Primary Symbols

V = gas volume
$\dot{V}$ = gas volume/unit time
P = gas pressure
$\bar{P}$ = mean gas pressure
F = fractional concentration in dry gas phase
f = respiratory frequency (breaths/unit time)
D = diffusing capacity
R = respiratory exchange ratio
STPD = 0° C, 760 mm Hg, dry
BTPS = body temperature and pressure saturated with water vapor

Examples

V_A = volume of alveolar gas
$\dot{V}_{O_2}$ = O_2 consumption per minute
$P_{A_{O_2}}$ = alveolar O_2 pressure
$\bar{P}_{c_{O_2}}$ = mean capillary O_2 pressure
$F_{I_{O_2}}$ = fractional concentration of O_2 in inspired gas
D_{O_2} = diffusing capacity for O_2 (ml O_2/mm Hg per minute)
R = $\dot{V}_{CO_2}/\dot{V}_{O_2}$

Secondary Symbols

I = inspired gas
E = expired gas
A = alveolar gas
T = tidal gas
D = dead space gas
B = barometric

Examples

$F_{I_{CO_2}}$ = fractional concentration of CO_2 in inspired gas
$\dot{V}_E$ = volume of expired gas
$\dot{V}_A$ = alveolar ventilation per minute
V_T = tidal volume
V_D = volume of dead space gas
P_B = barometric pressure

Blood

Primary Symbols

Q = volume of blood
$\dot{Q}$ = volume flow of blood/unit time
C = concentration of gas in blood phase
S = % saturation of Hb with O_2 or CO

Examples

Q_c = volume of blood in pulmonary capillaries
$\dot{Q}_c$ = blood flow through pulmonary capillaries per minute
$C_{a_{O_2}}$ = ml O_2 in 100 ml arterial blood
$S\bar{v}_{O_2}$ = saturation of Hb with O_2 in mixed venous blood

Secondary Symbols

a = arterial blood

v = venous blood

c = capillary blood

Examples

$P_{a_{CO_2}}$ = partial pressure of CO_2 in arterial blood
$P\bar{v}_{O_2}$ = partial pressure of O_2 in mixed venous blood
$P_{c_{CO}}$ = partial pressure of CO in pulmonary capillary blood

Lung Volumes

V_T = volume of air inhaled or exhaled with each breath
VC = vital capacity = maximal volume that can be expired after maximal inspiration
IC = inspiratory capacity = maximal volume that can be inspired from resting expiratory level
IRV = inspiratory reserve volume = volume that can be inspired from end-tidal inspiration
ERV = expiratory reserve volume = maximal volume that can be expired from resting expiratory level
FRC = functional residual capacity = volume of gas in lungs at resting end-expiratory level
RV = residual volume = volume of gas in lungs at end of maximal expiration
TLC = total lung capacity = volume of gas in lungs at end of maximal inspiration

°Adapted from Comroe, J. H., Jr., et al.: *The Lung* (2d ed.; Chicago: Year Book Medical Publishers, Inc., 1962). (Based on Fed. Proc. 9:602, 1950.)

STATICS OF BREATHING

Functional Anatomy of the Lung

Functionally the lung may be conveniently considered as (1) airways that conduct gas to and from the gas exchange regions but are not themselves directly involved in arterializing the venous blood and (2) gas exchange regions.

After inspired air has passed through the nose and/or mouth, the pharynx and the larynx, it enters the trachea or the lung "proper."

The *trachea* in man is a single conducting airway approximately 10–11 cm long and 2 cm in mean diameter. It is supported by struts of U-shaped cartilage, the ends of which are located posteriorly and are banded together by smooth muscle. The trachea bifurcates into the right and left main *bronchi*, which themselves subsequently bifurcate into smaller bronchi and so on for a total of 11 branchings. A common feature of the bronchi is their cartilaginous support and helical bands of smooth muscle.

From the 12th branching through the 16th, the airways, each of progressively smaller diameter but progressively greater total cross-sectional area, are termed *bronchioles* rather than bronchi since they lack cartilage. The bronchioles, however, have strong helical bands of smooth muscle. The volume of air contained in the respiratory tract through the terminal bronchioles does not exchange gas with venous blood and, consequently, is effectively dead space with respect to gas exchange. This volume, the *anatomical dead space,* is about 150 ml* that in normal man.

The transition from the 16th to the 17th generation airway branching is of enormous functional significance. The terminal bronchioles are replaced by *respiratory bronchioles,* which after several subsequent branchings lead to *alveolar ducts, alveolar sacs* and finally the *alveoli* themselves. All these re-

Fig 5–1.—Representation of component structures of an acinus, i.e., lung regions distal to terminal bronchioles *(TB)*, which subserve gas exchange. *RB* = respiratory bronchioles; *AD* = alveolar ducts; *AS* = alveolar sacs.

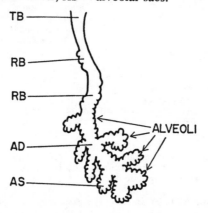

---

*A useful rule of thumb is that anatomical dead space in milliliters is about equal to the lean body weight in pounds.

gions that are capable of gas exchange are collectively defined as an *acinus* (Fig 5–1).

Thus a single airway with a cross-sectional area of about 2 sq cm leads, via 24 successive generations, to approximately 300 million alveoli with a total surface area of 70–90 sq m.

Gas exchange takes place across the *alveolar-capillary membrane,* a structure superbly suited to its function because of its huge surface and the fact that it is very thin, with an average thickness of about 1 μ.

In order to traverse the alveolar-capillary interface, an oxygen molecule must pass through the following layers:

1. A thin layer of fluid that lines the alveolus and has important surface-active properties that serve to stabilize alveoli of different sizes.

2. The alveolar epithelium, consisting predominantly of flat, thin *type I* cells and granular, more cuboidal type II cells. Type I cells, although less numerous than type II, account for more than 90% of the alveolar surface area and consequently are the predominant cells for gas exchange. Type II cells, characterized by high metabolic activity and numerous cytoplasmic inclusion bodies, are (almost certainly) the source of the surface tension-lowering material (alveolar surfactant) that is extruded into the alveoli to form the surface layer.

3. Alveolar and capillary-endothelial basement membranes with an intervening interstitial space, more prominent in some regions than in others (where most gas exchange occurs, the membranes appear to fuse into a single structure).

4. The capillary endothelium, not to be considered an "inert" structure, which subserves important biochemical functions such as transformation or inactivation of blood-borne materials such as amines, polypeptides and prostaglandins.

The lung has no inherent rhythmicity and no intrinsic motor system capable of causing ventilatory volume changes. It moves in response to forces transmitted from the chest wall via the *pleurae.*

The lung is covered with a membrane, the *visceral pleura;* the inner surface of the chest wall is covered with *parietal pleura.* The pleural interface or "space" between the visceral and parietal pleurae is not a cavity in the usual sense but rather is normally an air-free apposition of the pleural surfaces that holds the lung and chest wall together and causes them to move in unison. A very small quantity of fluid, possibly only a few milliliters, intervenes as a fine film between the pleural surfaces. This allows some sliding motion between the pleurae, for example, when the lung expands downward on contraction of the diaphragm.

FORCES INVOLVED IN BREATHING

1. INTRAPLEURAL PRESSURE.—The pressure in the intrapleural space is negative with respect to atmospheric pressure at the end of a normal, passive exhalation. This is due to the recoil properties of the lung and chest wall. The lung's elasticity causes it to attempt to retract to its intrinsic equilibrium vol-

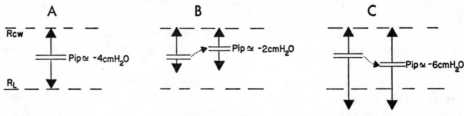

NORMAL CONDITION   REDUCED LUNG RECOIL   INCREASED LUNG RECOIL

Fig 5–2.—Balance of forces from chest wall recoil (R_{cw}) and lung recoil (R_L) that defines functional residual capacity (FRC) of lung—represented by two *solid horizontal lines*. **A** = normal condition. When lung recoil is reduced, as shown by *shorter arrow* in **B**, FRC increases. When lung recoil increases, as shown in **C**, FRC tends to decrease. P_{ip} = intrapleural pressure.

ume (effectively the gas-free lung). The elastic properties of the chest wall cause it to expand to adopt its own equilibrium position (which is at a volume of about 50–60% of the fully expanded state). When no volitional contractions of the muscles of breathing are applied to the chest wall, the combined lung and chest wall system adopt an equilibrium position in which the recoil forces of the individual structures are equal and opposite. The resultant lung volume is defined as the *functional residual capacity* (FRC). This is a crucially important volume for understanding lung function because, in conditions in which lung recoil is reduced or increased by disease, a new equilibrium position will be achieved, as shown in Figure 5–2.

Let us suppose that normally at FRC the equal and opposite forces acting on the intrapleural space lead to a value of intrapleural pressure (P_{ip}) of −4 cm H_2O relative to the atmosphere (see Fig 5–2, A). If the recoil force of the lung is reduced (see Fig 5–2, B), as in patients with pulmonary emphysema, a new equilibrium will be established, with the lung pulled to a higher volume because the force from the chest wall is less opposed by retraction of the lung. Consequently, FRC increases. But in addition, the sum of both retractive forces on the intrapleural space is now less; hence the intrapleural pressure is less negative than normal. The opposite will apply when the lung recoil force is increased, such as in patients with pulmonary fibrosis. It should be noted, however, that as the lung pulls the chest wall toward the new combined equilibrium position, the chest wall becomes farther removed from its own individual equilibrium configuration; hence its recoil is greater at the new FRC, and the combined forces cause FRC to be low and P_{ip} to be more negative than normal.

2. RESPIRATORY MUSCLES.—The forces that cause the lung and chest wall to expand above FRC are provided by the respiratory muscles. Inspiration is effected by three groups of muscles.

a. *Diaphragm.*—The diaphragm is innervated by the phrenic nerves, originating from cervical nerves 3–5. It is the predominant muscle of inspiration; when it contracts, it descends, causing its usual dome shape to be flattened.

b. *External intercostals.*—These are innervated by thoracic motor nerves 1–11. When they contract, they elevate the rib margins upward and outward.

Thus, acting in unison with the diaphragm, the external intercostals increase the size of the thorax in all three dimensions. Although quantitatively less important than the diaphragm at low levels of ventilation, the external intercostals become progressively more involved at higher ventilations.

c. *Accessory muscles.*—At high rates of ventilation, accessory muscles of breathing such as the scalenes, sternocleidomastoids and trapezius are called into play.

Expiration, on the other hand, is usually passive at low ventilatory rates, being effected by the recoil of the thorax back to its equilibrium configuration when the inspiratory muscles relax. However, at high rates of ventilation or in forced exhalations, expiratory muscles aid the spontaneous recoil. The expiratory muscles, *internal intercostals* and *abdominal muscles* cause the ribs to be lowered and abdominal pressure to increase, causing the diaphragm to be pushed up at a greater rate.

3. LUNG COMPLIANCE.—The greater the inspiratory muscle activity the greater the volume to which the chest wall is usually expanded. But as this causes the lung to be pulled farther and farther away from its equilibrium position, its recoil or retractive force increases. The intrapleural pressure consequently becomes more and more negative as the thoracic volume is progressively increased, up to the limits of expansion.

The distending pressure across the lung is the difference in pressures between the inside and outside of the lung, i.e., the difference between intrapleural pressure (P_{ip}) and alveolar pressure (P_{alv}). This difference, or lung distending pressure, is defined as *transpulmonary pressure* (P_{TP}); i.e.,

$$P_{TP} = P_{alv} - P_{ip}$$

It is important to recognize that it is both convenient and conceptually valid to consider the *distensibility* or *compliance* of the lung under static conditions. For example, lung compliance (C_L) is defined as the change in volume of the lung (ΔV) produced per unit distending pressure (ΔP_{TP}); i.e.,

$$C_L = \frac{\Delta V(\text{ml})}{\Delta P_{TP}\ (\text{cm H}_2\text{O})}$$

$$= \frac{\Delta V(\text{ml})}{\Delta(P_{alv} - P_{ip})\ (\text{cm H}_2\text{O})}$$

Consequently, only the volume change is considered and not the rate at which the volume change is attained. If the rate were considered in this measure of compliance, clearly the measurement would be influenced by factors such as the resistance of the airways for airflow, which, although an essential feature of lung function, would complicate the interpretation of this distensibility per se. Practically, therefore, to measure static lung compliance, the changes in lung volume, intrapleural pressure and alveolar pressures are needed.

The lung volume change can be readily measured with a simple spirometer.

As this compliance measure is a static one, i.e., made under conditions of no

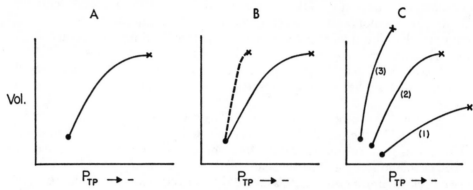

Fig 5–3.—**A**, change in lung volume induced by changes in transpulmonary pressure (P_{TP}) over the range from FRC (●) to TLC (x). The slope of the curve at any point represents lung compliance. **B**, comparison of lung pressure-volume curve with air (*solid line*) or fluid (*dashed line*) inflations. **C**, effects of increased (*1*) and decreased (*3*) lung recoil on normal (*2*) lung compliance curve.

airflow into or out of the lung, alveolar pressure must be the same as atmospheric pressure (i.e., our general reference pressure and considered to be zero, by convention). Alveolar pressure is therefore zero at points of zero airflow, i.e., at the end of inspiration or expiration of *any* volume (if it were not, air would flow into or out of the lung down the pressure gradient).

Thus static lung compliance is simplified to

$$C_L = \frac{\Delta V}{\Delta P_{ip}}$$

Although it is possible to introduce a needle or catheter into the intrapleural space and measure P_{ip} directly, the procedure is a potentially dangerous one, as the lung could be punctured or air could accidentally be introduced into the intrapleural space (pneumothorax, i.e., freeing the lung from its adhesive forces to the chest wall, allowing the lung to collapse from its intrinsic recoil). The P_{ip} is ordinarily estimated by measuring the pressure in the thoracic esophagus, which transmits pressure from the intrapleural space—especially changes in P_{ip}—with fairly good reliability (except in diseases that decrease the compliance or distensibility of the esophagus itself). This is usually done by inserting a long, narrow balloon on the end of a catheter (attached to a pressure-measuring device) into the lower third of the thoracic esophagus. The subject then takes a maximal inhalation to total lung capacity (TLC), and the change in volume is compared with the change of P_{ip}. Some air is then exhaled and new recordings are taken. This is repeated over the range of lung volumes (Fig 5–3, A). The slope of the resulting volume-pressure curve (i.e., $\Delta V/\Delta P$) at any point, therefore, gives the lung compliance at that volume.

Several features of this compliance curve are worth noting. The lung is most compliant, i.e., most distensible, in the normal range of tidal volume breathing. It becomes less compliant or more "stiff" at high lung volumes. For normal adults, a change of P_{ip} from −4 to −7 cm H_2O would induce a volume change (i.e., tidal volume) of approximately 600 ml:

$$C_L = \frac{\Delta V}{\Delta P}$$

$$= \frac{600 \text{ ml}}{(-4) - (-7) \text{ cm H}_2\text{O}}$$

$$= 200 \text{ ml/cm H}_2\text{O}$$

The same change of intrapleural pressure (i.e., 3 cm H_2O) nearer TLC would change lung volume by much less, e.g.,

$$C_L = \frac{\Delta V}{\Delta P}$$

$$= \frac{50 \text{ ml}}{(-20) - (-23) \text{ cm H}_2\text{O}}$$

$$= 17 \text{ ml/cm H}_2\text{O}$$

It is therefore usual to report the compliance of the lung at FRC.

Also, normal smaller subjects (e.g., children) have less change of volume of their (smaller) lungs than do normal subjects with larger lungs. When large size disparities are apparent, it is customary to normalize compliance to lung size in order that inferences may be drawn regarding the normality of the measured lung distensibility.

This normalized value is termed the *specific compliance,* i.e.,

$$\text{specific compliance} = \frac{\text{measured compliance}}{\text{lung volume at which measured}}$$

In Figure 5–3, B, an important characteristic of lung compliance is depicted. If the compliance curve is derived on an experimental animal's lung with normal air inflations, a curve similar to the solid line is obtained. If the curve is subsequently repeated with the lung filled with saline, the dashed curve is obtained. This demonstrates that when the normal air-fluid interface of the lung is replaced with a fluid-fluid interface, the interfacial forces operating on the lung are altered so that a much lower distending pressure is required to inflate the fluid-filled lung to the same volume.

Thus a large proportion of the total recoil of the lung is normally attributable to surface forces rather than the recoil being totally defined by the structural elastic elements of the lung (see below).

In Figure 5–3, C, typical compliance curves for patients with increased lung recoil, for example in pulmonary fibrosis (curve 1), and for patients with decreased lung recoil, such as patients with pulmonary emphysema (curve 3), are compared with the normal pattern (curve 2). Increased recoil causes reduced FRC (as previously described), a low compliance and reduced TLC. Decreased recoil leads to increased FRC, high compliance and high TLC. As chest wall function may be normal in all three conditions, it is predominantly the lung, and particularly its recoil, that normally limits further expansion of the thorax.

4. CAUSES OF LUNG RECOIL. — The lung tends to recoil to a smaller volume even after a maximal volitional exhalation (i.e., at *residual volume*). Two basic factors account for this retractive, or recoil, force.

a. *Lung elastic tissue.* — Elastic and collagen fibers are present in the alveolar walls and bronchial tree and, when distended, tend to return to equilibrium configuration.

b. *Surface forces.* — At an air-fluid interface, the forces between fluid molecules act in all directions except toward the air side. This tends to an imbalance in force distribution such that the surface of the fluid attempts to contract to a smaller and smaller surface area. This happens at the alveolar air-fluid interface and leads to surface forces that play an important role in recoil of the lung.

PULMONARY SURFACTANT. — The interfacial recoil forces in the lung would be markedly more powerful, and consequently the lung more difficult to ventilate, were it not for the presence in the alveolar lining film of material capable of reducing the surface tension (surfactant). Although the specific nature of pulmonary surfactant is still being debated, most evidence strongly indicates that it is a lipoprotein material (rich in the phospholipid dipalmitoyl lecithin) synthesized in type II alveolar cells and extruded onto the surface of the alveoli.

Pulmonary surfactant is formed late in fetal life. On occasion, especially in premature births,[*] insufficient pulmonary surfactant is available, or is inactive. This leads to the *neonatal respiratory distress* syndrome in which the lungs are stiff, with some alveoli collapsed and some likely to be filled with fluid.

Importantly, however, normal surfactant has the ability not just to reduce surface tension, thereby allowing easier lung distention, but, remarkably, to alter surface tension in proportion to the volume of the alveolus. Were it not for this feature, the lung with its hundreds of millions of alveoli of various sizes would be unstable. For example, as the alveolus can be modeled as a sphere with an open airway leading from it, its pressure *(P)* will be defined by the surface tension *(T)* and the radius *(r)* of the alveolus. Laplace's law predicts that the pressure in the alveolus will be described by the equation

$$\text{alveolar pressure} = \frac{2 \times \text{surface tension}}{\text{radius}}$$

$$P = \frac{2T}{r}$$

Thus if surface tension were to be equal throughout the lung, alveoli with the smallest radius would have the highest pressure. Consequently, the small alveoli would empty down the pressure gradient into larger ones. However, in reality the alveoli of different sizes are stable in the lung owing to the sur-

---

[*]It is, therefore, another example of Shakespeare's pervading genius that, from the adjectives available to modify "world" in the opening soliloquy of Richard III, he chose "breathing," i.e., "Deformed, unfinished, sent before my time into this breathing world, scarce half made up." — making us realize that, presumably owing to premature birth and an undeveloped surfactant system, Richard had to fight even to breathe following his birth.

face tension being reduced in proportion to the radius. This causes equal alveolar pressures in alveoli of different sizes. This remarkable process is thought to be caused by the surface-active molecules being more densely packed at the surface when alveoli are small; hence there is greater surface tension-lowering potential than in the larger alveoli where the packing is less dense.

REGIONAL VARIATIONS OF ALVEOLAR SIZE.—Up to this point we have considered the intrapleural pressure to be uniform throughout the lung. It is not. Careful measurements have clearly shown that intrapleural pressure at the apex of the lung is markedly more negative than at the base. In normal upright man, the pressure at the apex might be -8 cm H_2O compared to -2 at the base.

The cause of this vertical variation of intrapleural pressure relates to the total tendency of the lung to retract from the chest wall. The contribution from the weight of the lung being "pulled down" by gravitational forces is greater in the apex; i.e., there is more lung below it to be affected by gravity than at the base.

As a consequence of the variation in negative pressure in the intrapleural space, alveolar size will be greater in the region of most negative pressure. Apical alveoli tend to be larger in normal, upright man than do basal alveoli. In the supine position, much of this vertical gradient of alveolar size is lessened, as the gravitational effects on intrapleural pressure are reduced in this posture.

Normally, therefore, in the upright position, apical and basal alveoli are on different portions of their own normal compliance curve (Fig 5–4). For this reason, for a given change in intrapleural pressure, more air ventilates the *basal* alveoli than those in the apex (see Fig 5–4, A). Thus for normal breathing near FRC, apical alveoli have large volumes but small changes of volume,

Fig 5–4.—Regional variations of lung ventilation. **A,** during normal breathing, apical (*a*) alveoli ventilate less per unit volume than do basal (*b*) alveoli. **B,** near TLC, both *a* and *b* alveoli ventilate more evenly. **C,** near residual volume, *a* alveoli ventilate more than do *b* alveoli. ΔV = volume of lung; P_{ip} = intrapleural pressure.

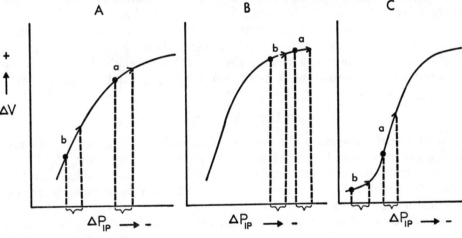

and basal alveoli have smaller volumes and larger changes of volume. Most of the normal inspirate, therefore, goes to the lower regions of the lung.

If we were to breathe near TLC, all lung regions would be on the flatter part of the compliance curve; hence they would ventilate more equally (see Fig 5–4, B).

If we were to breathe near the residual volume of the lung, where the lung becomes less distensible (or has low compliance), predominantly due to some airway closure in basal regions at low lung volumes, the normal pattern of distribution would be reversed. Most of the initial inspirate would go to the apex (see Fig 5–4, C). Therefore, depending on initial lung volume, the normal inspired volume is directed to different regions of the lung.

The patterns of regional ventilation for a maximal inhalation beginning at residual volume can therefore be summarized as follows:

Initially most of the inspirate is directed into the apex, with little ventilation of basal regions.

Both regions then tend to ventilate at the same rate, as both regions come onto the more linear portion of the compliance curve.

Then the basal regions ventilate better than do the apical regions, since the apical alveoli reach the flat part of their compliance curve.

Finally, both regions ventilate at about the same rate (but poorly), as both fall on the flat portion of their compliance curve.

On the subsequent exhalation, the reverse pattern is established.

5. REGIONAL DISTRIBUTION OF PULMONARY BLOOD FLOW.—Owing to the effect of gravity on blood in the lung, the distribution of pulmonary perfusion is such that blood flow is greatest in the basal lung regions in upright man. Blood flow is progressively reduced with height of the lung to where, at the apex, blood flow can be extremely poor, especially during the diastolic phase of the pulmonary artery pressure pulse (see section on abnormalities of cardiovascular function, Chapter 4).

It should therefore be recognized that both ventilation and perfusion are better at the base than at the apex for normal breathing close to the FRC of the lung. However, the vertical gradient of blood flow is markedly greater than the vertical gradient of ventilation. The ratio of ventilation to perfusion ($\dot{V}/\dot{Q}$) at the apex is therefore high and the $\dot{V}/\dot{Q}$ ratio at the base is *low*.

Normally at the *apex* there is a large alveolar volume, relatively low ventilation and very poor perfusion; therefore, ventilation per unit blood flow is high. At the *base* there is a smaller alveolar volume, relatively good ventilation and very high perfusion; therefore ventilation per unit blood flow is low.

LUNG VOLUMES

As previously described, the volume of air in the lung at the end of a normal, quiet exhalation is termed the functional residual capacity. The FRC is an important index of lung function, but other lung volumes and capacities are similarly useful. Important subdivisions of TLC (total lung capacity) are described in Figure 5–5.

As is apparent from Figure 5–5, all the volumes and capacities can be mea-

LUNG VOLUMES

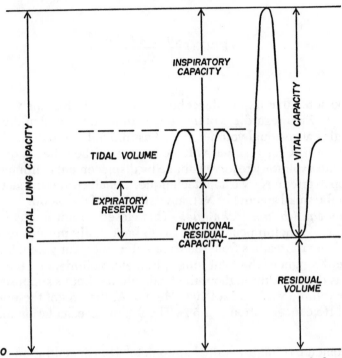

Fig 5–5.—Diagram of standard lung volumes and capacities, as might be determined by spirometry.

sured by simple displacement except for residual volume and FRC.

The displacement volumes are characteristically measured with a spirometer, in which movement of air to or from the lung causes changes in position of a carefully balanced cylindric bell, leading to recording of the volume changes on a volume-calibrated moving (usually rotating) chart.

To determine the FRC, different techniques are required:

1. *N_2 washout technique.*—Normally the lung contains approximately 80% N_2, which varies hardly at all during the breathing cycle. The actual N_2 concentration can be readily measured using a rapidly responding N_2 meter, now accessible to most pulmonary function laboratories. Using a valve system that allows the subject to inhale 100% O_2 and exhale into a spirometer, the test continues until all the N_2 that originally resided in the lung now resides in the spirometer, i.e., the N_2 has been "washed out" of the lung. In normal subjects this takes only a few minutes with normal, quiet breathing. In subjects with regions of the lung that ventilate poorly compared with other regions— i.e., subjects with maldistribution of ventilation—it may take substantially longer to completely wash out the N_2.

Let us imagine that after minutes of breathing 100% O_2 the subject has washed all the N_2 from his lung and has exhaled 40 L into the spirometer. The final mixed concentration of N_2 in the spirometer is found to be 5%.

The following computation is then made:

FRC volume $\times$ 80% N_2 = final spirometer volume $\times$ final spirometer % N_2

that is,

$$FRC = \frac{40 \text{ L} \times 0.05}{0.8}$$

$$= 2.5 \text{ L}$$

It should be noted that if the subject begins the test at FRC, the volume determined will be FRC regardless of the lung volume at the end of the test; if the subject begins at residual volume, the measured volume will be the residual volume. Once FRC or residual volume is known, the other is discerned by measuring the expiratory reserve volume by displacement spirometry.

By this open circuit N_2 washout technique, a small correction factor must be applied for the small amount of N_2 that washes out of the blood.

2. *Helium equilibrium technique.*—This closed circuit method requires a known volume of helium-containing gas to be initially present in a spirometer. The subject rebreathes from the spirometer sufficiently long for the He to equilibrate throughout the total lung-spirometer volume, i.e., the concentration of He is the same throughout the total volume. Let us suppose the initial spirometer volume was 2.5 L of 10% He gas. At the end of the equilibration period, the He concentration was 5%. The following calculation may then be readily made:

initial spirometer volume $\times$ initial spirometer He concentration

= final volume (lung and spirometer) $\times$ final mixed He
concentration i.e.,

2.5 L $\times$ 0.1 = (lung volume + 2.5 L) 0.05

$\therefore$ 2.5 L = initial lung volume

As this test is dependent on final lung volume at equilibrium, care must be taken to correct for any changes in lung volume from the original FRC position.

It should be noted that these techniques measure only the volume of gas in communication with the spirometer during the test. For example, the total gas in the lungs at FRC would be underestimated by these spirometric techniques if, in a diseased lung, some gas were "trapped" behind collapsed airways, as is commonly the case.

3. *Plethysmographic technique.*—All the gas in the thorax, whether in normal communication with the atmosphere or "trapped" behind collapsed airways, can be measured using an ingenious application of Boyle's law, which states that, if temperature is constant, there is a constant relationship between the pressure and volume of a given quantity of gas; i.e.,

$$PV = P_1 V_1$$

A subject is seated in an airtight box that allows him to breathe through a mouthpiece that is within the box. At the end of a normal exhalation (i.e., at FRC), alveolar pressure *(P)* is equal to atmospheric pressure, as no gas is flowing. A shutter then closes off the mouthpiece. The subject is requested to continue "breathing" against the closed shutter. During the next "breath," the thorax enlarges, creating a new thoracic gas volume by decompression, i.e., original $V + \Delta V$. This is measured from the pressure change within the plethysmograph. Also a new gas pressure (P_1) is measured between the shutter and the subject.

Thus, original alveolar pressure *(P)* × original thoracic gas volume *(V)* = new thoracic gas volume $(V + \Delta V)$ × new alveolar pressure P_1:

$$P \cdot V = P_1 \times (V + \Delta V)$$

As P, P_1 and ΔV are measured, we can readily determine V, the total thoracic gas volume at FRC.

DYNAMICS OF BREATHING

Gas flows down gradients of pressure. For gas flow in the lung the pressure gradient of concern is the difference between atmospheric and alveolar pres-

Fig 5–6.—**A,** volume (V_T), intrapleural pressure (P_{ip}) and airflow $(\dot{V})$ changes for normal resting and exercising breaths. The *dashed line* on the P_{ip} curve represents pressure needed to produce that lung volume under conditions of no airflow. The *shaded area* is the extra pressure required to generate air and tissue flow. **B,** dynamic inspiratory pressure-volume curve. *Stippled area* is static pressure-volume work. *Shaded area* is the dynamic component of inspiratory work equivalent to work required to generate air and tissue flow. I = inspiration; E = expiration.

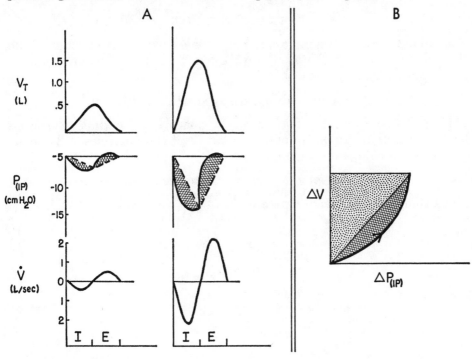

sure. As previously stated, alveolar pressure varies due to changes in the dimension of the thorax.

Inspiration leads to expansion of the thorax, causing alveolar pressure to fall until the end of inspiration when alveolar pressure again equals atmospheric pressure, i.e.,

$$\text{at FRC, } P_{atm} = P_{alv} \therefore \text{ no air flow}$$

$$\text{during inspiration, } P_{atm} > P_{alv} \therefore \text{ air flows into lungs}$$

Typical changes of volume, flow and intrapleural and alveolar pressures during a respiration cycle are given in Figure 5–6. Several features of this figure are worthy of consideration: under conditions of no airflow (i.e., under static conditions) changes in volume will be simply related to changes in intrapleural pressure (i.e., the static lung compliance relationship); however, during inspiration, additional force—as reflected by greater negativity of the intrapleural pressure change—is needed. The extra force is needed to overcome the flow resistances associated with breathing. These are (1) airway resistance to gas flow and (2) resistance to pulmonary tissue flow (i.e., tissue viscosity).

The difference between the static force requirement and the dynamic flow requirement of intrapleural pressure change, represented by the shaded area in Figure 5–6, is consequently greatest when air flow is greatest.

In addition, the maximum change in air flow during the breathing cycle effectively occurs at the time of maximum change of alveolar pressure.

Figure 5–6 also allows us to understand the basic concept of the work of breathing.

Work of Breathing

Breathing causes changes in thoracic volume owing to contraction of the skeletal muscles of ventilation. The work done by a muscle in moving an object is expressed by

$$\text{work} = \text{force} \times \text{distance moved, or force} \times \text{length}$$

This equation, however, is of no value to us since we cannot directly measure the force-length relationships of the various respiratory muscles in the intact animal. We can, however, compute work as

$$\text{change in intrapleural pressure} \times \text{change in lung volume}$$

This amounts to the same thing, since

$$P \cdot V = \frac{\text{force}}{\text{area}} \times \text{vol}$$

$$= \frac{\text{force}}{\text{length}^2} \times \text{length}^3$$

$$= \text{force} \times \text{length}$$

$$= \text{work}$$

Thus, by measuring the changes in intrapleural pressure and change in lung volume, work of breathing can be computed as shown in Figure 5–6. The total area (stippled and shaded) between the ΔV and ΔP axes is the total work of breathing. The stippled area is, of course, the work done to overcome the static recoil forces of the lung over that volume. The shaded area represents the flow-resistive work, of which airway resistance normally accounts for about 80% of the total and viscous tissue resistance for 20%. In subjects with high breathing resistance, the dynamic part of the loop is markedly increased, since the work of breathing is high.

AIRWAY RESISTANCE

Resistance to airflow is analogous to electric resistance as defined by Ohm's law, i.e.,

$$\text{resistance} = \frac{\text{potential difference}}{\text{current flow}}$$

For the airways the potential difference is the difference between alveolar and atmospheric pressure, and flow is the flow rate of the air. Thus,

$$\text{airway resistance} = \frac{P_{alv} - P_{atm}}{\text{flow rate}}$$

We are able to estimate alveolar pressure fairly accurately using an esophageal balloon, as the alveolar pressure is the sum of the intrinsic recoil pressure of the lung at that volume and the applied intrapleural pressure, i.e.,

$$P_{alv} = P_{recoil} + P_{ip}$$

As recoil is derived from the static pressure-volume relationship of the lung at the given volume and P_{ip} is measured directly during a (usually) forced exhalation, P_{alv} can be determined. Airflow can be readily measured; therefore, airway (and tissue) resistance can be estimated. A body plethysmograph also is often used to determine airway resistance specifically; the reader is referred to any standard textbook of pulmonary mechanics for this technique.

Typical values for airway resistance normally range between 0.5 and 2 cm H_2O/L per second. Although the cross-sectional area of a bronchiole is small compared with a large bronchus, the total cross-sectional area of all the small airways is large compared with the large airways. Consequently, the total resistance to airflow of the small airways is small compared with the large ones. In fact, only about 20% of the airway resistance is related to airways smaller than 2 mm. Unfortunately, therefore, considerable increases in small airway resistance must occur before alterations in total airway resistance can be discerned with the usual tests.

Several factors should be borne in mind.

1. Airway resistance is not constant in a given subject but, rather, varies with lung volume. As the airways are affected by intrapleural pressure, so their cross-sectional area varies with this surrounding pressure. At high lung

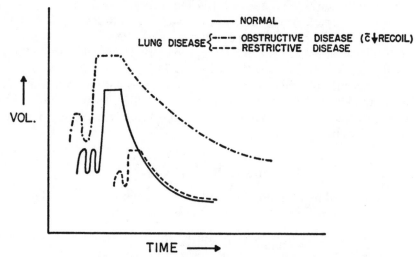

Fig 5–7.—Representation of typical forced expiratory spirograms (following a maximal inhalation) in the three conditions shown. Note that maximum airflow is reduced in both obstructive and restrictive lung disease. But when observed at equivalent lung volume, only the obstructive disease has low airflow.

volumes, the airways are distended; at low lung volumes or during forced expiratory efforts when intrapleural pressure becomes positive, airways can undergo *dynamic compression* and their resistance to flow becomes relatively high as a consequence.

2. Contraction of bronchial smooth muscle or plugging of the bronchi with mucus also increases airway resistance.

3. Recoil pressure of the lung also becomes greater at high volumes and relatively low at low lung volumes. Thus despite efforts at maximal airflow in both cases, peak flow is high at high lung volume (i.e., recoil high, resistance

Fig 5–8.—Maximum expiratory flow-volume curve in three states. Note that, normally, flow accelerates to maximum and thereafter decreases relatively linearly with decreasing lung volume. Airway obstructive disease (associated with low lung recoil) results in low expired flows both in absolute terms and relative to lung volume. In contrast, restrictive lung disease results in low flow in absolute terms but normal or slightly high flows when corrected for lung volume.

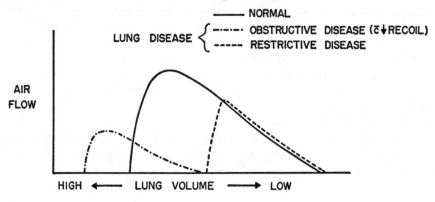

relatively low) and low at low lung volume (i.e., recoil lower and resistance higher). If a subject exhales maximally from TLC to residual volume into a simple displacement spirometer, a volume versus time plot is described. The slope of this relationship (i.e., flow) is highest at high lung volume and becomes progressively less as volume decreases (Fig 5–7). It should be noted that in a disease associated with reduced lung recoil, such as pulmonary emphysema, the maximal rate of airflow is low *despite* high lung volume, whereas a restrictive lung disease results in low maximum airflow *due to* the reduced lung volume. Also if expired air flow is plotted against expired (or lung) volume, a characteristic flow-volume curve is attained, as shown in Figure 5–8. Peak flows of 10 L per second are not uncommon in normal subjects and the maximum flow decreases as volume decreases. The flow-volume curves of patients with two kinds of lung disease also are shown for comparison.

In the condition with reduced lung recoil, the flows are low despite a large lung volume. Also, diseases associated with airways constriction or narrowing, e.g., chronic bronchitis or bronchial asthma, lead to low flow at these lung volumes. However, in a disease with increased lung recoil such as pulmonary fibrosis, the absolute peak flow is low owing to inability to distend the lung to high volume, but the flow at a given (lower) lung volume can be normal, or even slightly higher owing to increased recoil.

4. Turbulent airflow requires greater driving pressure to generate a given airflow than if the flow were laminar.

For laminar flow, the flow $\dot{V}$ is proportional to the geometry of the tube (or airway) through which the air flows. This is given by the Poiseuille equation:

$$\dot{V} = \frac{\Delta P \pi r^4}{8l\eta}$$

where ΔP is the driving pressure, r the radius, l the length of the tube and η the coefficient of viscosity (assuming straight, unbranched tubes).

Another way of considering this equation is

$$P = \frac{8l\eta}{\pi r^4} \cdot \dot{V}$$

$$= k_1 \dot{V}$$

where k_1 is resistance.
But for turbulent flow the driving pressure required is given by

$$\Delta P = k_1 \dot{V} + k_2 \dot{V}^2$$

where k_1 is the laminar portion of the resistance and k_2 is the turbulent portion of the resistance. Thus turbulence, which can occur at very high flow rates, at branchings or across a constricted region of airways, requires greater pressure to generate a given rate of air flow. The general feature of whether flow will be turbulent or laminar is given by the Reynolds number (Re). This is a dimensionless number:

$$Re = \frac{v \cdot 2r \cdot d}{\eta}$$

where v is the mean linear velocity, r the tube radius, d the gas density and η the viscosity.

When Re > 2,000, turbulence usually occurs. It should be noted, however, that Re is proportional to the gas density and inversely to the viscosity. Hence, reducing the density of a gas, e.g., by breathing high concentrations of helium instead of nitrogen, makes it less likely that flow will be turbulent. It is also apparent from the equation that where mean linear velocity is low, such as in the small airways, turbulence is less likely to occur than in regions such as the trachea where linear velocity is high.

VENTILATION AND ALVEOLAR GAS PRESSURES

The essential function of the lung is to oxygenate the blood and remove CO_2 at rates proportional to the body's metabolic requirements. The gas transfer between lung and blood occurs down gradients of pressure for each gas, which are established by cyclic ventilation of the alveolar gas with atmospheric air. A very brief statement of some of the physical laws that define the behavior of gases in the lung, as elsewhere, is warranted.

1. *Avogadro's law.* — Equal volumes of gas at the same pressure and temperature contain the same number of molecules. Alternatively, equal numbers of molecules of different gases at the same temperature and pressure occupy the same volume.

One mole of an ideal gas at standard temperature and pressure contains 6.02×10^{23} molecules, known as Avogadro's number, and occupies a volume of 22.4 L. Although CO_2 does not quite conform as an ideal gas, 1 M of CO_2 occupies 22.26 L at standard temperature and pressure.

2. *Boyle's law.* — At constant temperature the volume of a gas is inversely related to its pressure. If, for example, a given quantity of gas at pressure P and volume V is compressed, so that its new pressure and volume are P_1 and V_1, then:

$$PV = P_1V_1$$

This law is clearly of great importance in understanding movement of gas in and out of the lungs.

3. *Charles's law.* — At constant pressure the volume of a gas is directly proportional to its absolute temperature (i.e., the Kelvin scale).

Thus, considering the volumes $(V$ and $V_1)$ of a gas at two different temperatures $(T$ and $T_1)$, then:

$$\frac{V}{V_1} = \frac{T}{T_1}$$

This law is important in relating gas volumes that are commonly expressed at different temperatures. Ventilation is normally referred to body temperature, i.e., 37 C or 310 K, whereas volumes of CO_2 expired are commonly expressed at standard temperature, i.e., 0 C or 273 K.

General gas law. — This combines the three laws given above in a single statement of the behavior of an ideal gas:

$$PV = nRT$$

where P is gas pressure, V its volume, T its temperature (degrees K), n the number of moles of gas and R the gas constant, the value of which depends on the units in which the variables are expressed. If, for example, P is in atmospheres, V is in liters, and T is in degrees K, R has a value of 0.082 L · atmosphere/M · degree.

Graham's law.—The rate of diffusion of a gas through a mixture of gases is dependent upon its molecular weight. Light molecules travel faster and therefore diffuse more rapidly. In fact, the diffusion rate of a gas is inversely proportional to the square root of its molecular weight. This tells us that O_2 will diffuse faster than CO_2 in a gas phase, since

$$\frac{D_{O_2}}{D_{CO_2}} = \frac{\sqrt{MW_{CO_2}}}{\sqrt{MW_{O_2}}} = \frac{\sqrt{44}}{\sqrt{32}} = \frac{6.6}{5.6} = 1.18$$

or O_2 diffuses approximately 18% faster than CO_2 in the gas phase.

GAS PARTIAL PRESSURE

The total atmospheric, or barometric, pressure (PB) at sea level is about 760 mm Hg; in other words, the pressure exerted by the sum of all gases in atmospheric air is 760 mm Hg.

However, atmospheric air, which we will assume to be dry for this example, is a mixture containing 20.93% O_2, 0.04% CO_2 and 79.03% N_2.[*] Dalton's law of partial pressures states that the pressure exerted by a mixture of gases is equal to the sum of the individual (i.e., partial) pressures exerted by each gas. Thus $P_B = P_{O_2} + P_{CO_2} + P_{N_2}$. For example, of the total 760 mm Hg dry atmospheric pressure, O_2 would exert a pressure of

$$\frac{20.93}{100} \times 760 = 159 \text{ mm Hg}$$

i.e., the partial pressure of a gas is equal to the fractional concentration (F) of that gas multiplied by the total pressure. Thus, for a gas x

$$P_x = F_x \cdot P_B$$

For dry atmospheric air

$$P_{O_2} = \frac{20.93}{100} \times 760 = 159 \text{ mm Hg}$$

$$P_{CO_2} = \frac{0.04}{100} \times 760 = 0.3 \text{ mm Hg} \quad \text{(this pressure is so small that physiologists commonly assume it to be zero)}$$

$$P_{N_2} = \frac{79.03}{100} \times 760 = 601 \text{ mm Hg}$$

---

[*]The other "inert" gases in atmospheric air such as argon, neon and krypton are normally present in such small proportions (<1%) that their concentrations are normally included in the N_2 value. Consequently, N_2 refers to the "inert" gas factor.

$$\text{Total } P_B = \frac{100}{100} \times 760 = 760 \text{ mm Hg}$$

In the above example we considered the air to be dry. Normally, atmospheric air contains water vapor, and consequently the partial pressure of H_2O must proportionally reduce the pressures exerted by the other gases, i.e.,

$$P_B = P_{O_2} + P_{CO_2} + P_{N_2} + P_{H_2O}$$

When atmospheric air is inhaled, it becomes saturated with H_2O at body temperature by the time it has traversed the trachea. The water vapor pressure is not constant but varies with temperature, as shown to the nearest mm Hg in Table 5–2. It should be noted that at normal body temperature of 37 C, P_{H_2O} is 47 mm Hg. This means that the total pressure exerted by O_2, CO_2 and N_2 is now only $760 - 47 = 713$. The P_{O_2} of moist tracheal air is therefore

$$\frac{20.93}{100} \times 713 = 149 \text{ mm Hg}$$

ALVEOLAR GAS PRESSURES.—As O_2 is absorbed into pulmonary capillary blood and CO_2 evolves into alveolar gas, the P_{O_2} of alveolar gas ($P_{A_{O_2}}$) is less than that of saturated atmospheric air. Similarly, P_{CO_2} in alveolar gas ($P_{A_{CO_2}}$) is higher than in atmospheric air. For example, the fractional concentration of CO_2 in alveolar gas (expressed on a dry basis) can be readily determined, using a rapidly responding CO_2 analyzer, and has a normal value of approximately 5.6%. Consequently,

$$P_{A_{CO_2}} = F_{A_{CO_2}} \times (P_B - 47)$$

$$= \frac{5.6}{100} \times 713$$

$$= 40 \text{ mm Hg}$$

Alveolar P_{O_2}, calculated by a similar technique, is approximately 100 mm Hg. It should be noticed that the change in P_{O_2} between moist inspired air and alveolar gas is about 49 mm Hg (i.e., $149 - 100$), whereas the change in P_{CO_2} is only 40 mm Hg. This apparent discrepancy is due to the fact that at rest, on a normal Western diet, the body's metabolic processes release less CO_2 than it uses O_2 by volume.

These alveolar values, however, should be considered the average of the

TABLE 5-2.—WATER
VAPOR PRESSURES

TEMPERATURE (C)	P_{H_2O} (mm Hg)
30	32
32	36
34	40
36	45
37	47
38	50
40	55

TABLE 5–3.—PARTIAL PRESSURES OF RESPIRATORY GASES
(AT SEA LEVEL)—mm Hg

	DRY AMBIENT AIR	TRACHEAL (INSPIRED) AIR—37 C	ALVEOLAR AIR	ARTERIAL BLOOD	RESTING MIXED VENOUS BLOOD
P_{O_2}	160	150	100	90	40
P_{CO_2}	0.3	0.3	40	40	46
P_{N_2}°	≈600	≈563	≈573	≈583	≈583
P_{H_2O}	0	47	47	47	47
P_{total}	760	760	760	760	716

°P_{N_2} usually represents partial pressure of *all* inert gases in this context.

alveolar gas during inspiratory and expiratory cycles. During inhalation, alveolar P_{O_2} increases owing to the addition of atmospheric air and decreases during exhalation owing to the continued extraction of O_2 into the blood when no further atmospheric air is being brought to the alveoli. Thus both alveolar P_{O_2} and P_{CO_2} fluctuate throughout the respiratory cycle, O_2 being highest at the end of inspiration and lowest just prior to addition of atmospheric air into the alveoli on the next breath. CO_2 obviously fluctuates in an opposite direction to the O_2, being lowest at the end of inhalation.

The normal mean alveolar P_{O_2} and P_{CO_2} (Table 5–3) are therefore defined by the rate of ventilation of the alveoli with fresh atmospheric air and the rate of metabolism.

It is clear that the amount of CO_2 produced per minute ($\dot{V}_{CO_2}$) is equal to the amount cleared from the alveoli by exhalation minus the amount taken in during inhalation, i.e.,

$$\dot{V}_{CO_2} = \frac{\text{expired alveolar volume} \times \text{alveolar } CO_2 \text{ concentration}}{\text{minute}}$$

$$- \frac{\text{inspired alveolar volume} \times \text{inspired } CO_2 \text{ concentration}}{\text{minute}}$$

or

$$\dot{V}_{CO_2} = \dot{V}_{A_{exp}} \cdot F_{A_{CO_2}} - \dot{V}_{A_{insp}} \cdot F_{I_{CO_2}},$$

but, as $F_{I_{CO_2}}$ can be considered to be effectively zero, the minus term in this equation drops out, leaving:

$$\dot{V}_{CO_2} = \dot{V}_A \cdot F_{A_{CO_2}}$$

or

$$F_{A_{CO_2}} = \frac{\dot{V}_{CO_2}}{\dot{V}_A}$$

Thus, when related to partial pressures,

$$\frac{P_{A_{CO_2}}}{P_B - 47} = \frac{\dot{V}_{CO_2}}{\dot{V}_A}$$

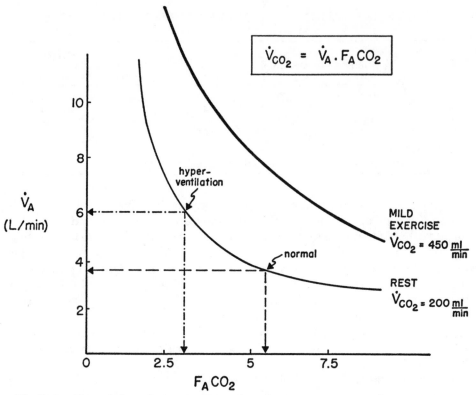

Fig 5–9.—Hyperbolic relationship between alveolar ventilation ($\dot{V}_A$) and alveolar CO_2 concentration (FA_{CO_2}), which defines the rate of CO_2 production ($\dot{V}_{CO_2}$). Thus a given $\dot{V}_{CO_2}$ can be attained by an infinite combination of $\dot{V}_A$ and FA_{CO_2}. Increased $\dot{V}_{CO_2}$, such as seen during muscular exercise, shifts the $\dot{V}_{CO_2}$ isopleth upward, as shown.

This is a vitally important relationship, as it shows that, at any given rate of metabolic CO_2 production, the steady-state value for PA_{CO_2} is inversely related to the rate of alveolar ventilation. This hyperbolic relationship is graphically described in Figure 5–9. It should be noted that a given $\dot{V}_{CO_2}$ can be achieved by a high $\dot{V}_A$ operating at a low PA_{CO_2} or by a low $\dot{V}_A$ with a high PA_{CO_2}.

Figure 5–9 also shows that if metabolic rate is higher, for example in an exercising subject, the same relationship holds, but now the $\dot{V}_{CO_2}$ hyperbola (also called a $\dot{V}_{CO_2}$ isopleth to stress the constant value of $\dot{V}_{CO_2}$ throughout the curve) is shifted upward.

A similar relationship holds for alveolar P_{O_2}, i.e.,

$$FI_{O_2} \frac{FA_{N_2}{}^*}{FI_{N_2}} - FA_{O_2} = \frac{\dot{V}_{O_2}}{\dot{V}_A}$$

In such equations as $FA_{CO_2} = \dot{V}_{CO_2}/\dot{V}_A$ great care must be taken with the condi-

---

*The term FA_{N_2}/FI_{N_2} is the correction factor that is to be applied, resulting from the fact that normally inspired ventilation is greater than expired ventilation because of the imbalance of $\dot{V}_{CO_2}$ and $\dot{V}_{O_2}$ described above.

tions of the measurement. For example, $\dot{V}CO_2$ and $\dot{V}O_2$ are usually expressed as volumes at standard temperature and pressure, dry (STPD), whereas $\dot{V}A$ is usually expressed at body temperature and pressure, saturated (BTPS). A correction term is therefore usually applied to relate the quantities to the same conditions. For example, if $\dot{V}A$ is measured BTPS and $\dot{V}CO_2$ is measured STPD in the equation,

$$\dot{V}CO_2 = \dot{V}A \cdot FA_{CO_2}$$

then $\dot{V}CO_2$ must be corrected by $(273 + 37)/273$ to correct to body temperature (i.e., 37 C) and by $760/(760 - 47)$ to correct to saturated gas volume. The correction for $\dot{V}CO_2$ (BTPS) from $\dot{V}CO_2$ (STPD) is therefore

$$\frac{273 + 37}{273} \times \frac{760}{760 - 47} = 1.21$$

or

$$1.21 \, \dot{V}CO_2 \, (\text{STPD}) = \dot{V}A \, (\text{BTPS}) \cdot FA_{CO_2}$$

If expressed as partial pressure,

$$1.21 \, \dot{V}CO_2 = \frac{\dot{V}A \cdot PA_{CO_2}}{PB - 47}$$

hence,

$$PA_{CO_2} = 863 \cdot \frac{\dot{V}CO_2(\text{STPD})}{\dot{V}A(\text{BTPS})}$$

The relationship $FA_{CO_2} = \dot{V}CO_2/\dot{V}A$ allows us to make an important distinction between the two terms, *hyperpnea* and *hyperventilation*, which are often, but sometimes wrongly, used interchangeably.

Hyperpnea refers to any increase in ventilation; hyperventilation is an increase in ventilation out of proportion to the increase in metabolic $\dot{V}CO_2$ and results in a low arterial PCO_2. Thus during moderate muscular exercise $\dot{V}A$ can be appreciably elevated but the increase is usually in proportion to $\dot{V}CO_2$; hence PA_{CO_2} and arterial PCO_2 remain constant and the condition is one of *hyperpnea*, not hyperventilation. The increased ventilation associated with emotional stresses is usually out of proportion to the increase in $\dot{V}CO_2$; hence arterial PCO_2 and PA_{CO_2} is reduced and the condition is one of *hyperventilation*.

Conversely, reduced ventilation is categorized as *hypopnea* whereas a reduction of $\dot{V}A$ greater than the reduction in $\dot{V}CO_2$ leads to increased PA_{CO_2} and the condition is one of *hypoventilation*.

It should be noted that it is possible for a subject to be both hyperpneic and hypoventilating at the same time. If a subject exercises so that both $\dot{V}CO_2$ and $\dot{V}A$ are increased, but the increase in $\dot{V}A$ is insufficient for the rate of CO_2 production, then $\dot{V}CO_2/\dot{V}A$ will be increased causing PA_{CO_2} and arterial PCO_2 to be increased, i.e., hypoventilation. It must be pointed out, however, that ruthless adherence to the convention that, if PA_{CO_2} and arterial PCO_2 are low, the subject *is* hyperventilating can lead to incongruities, as in the following example. A subject volitionally increases $\dot{V}A$, causing a low PA_{CO_2}. This condition often leads to apnea (i.e., cessation of breathing) until CO_2 and O_2 levels return to values that cause respiratory stimulation. In the apneic period the subject had not been breathing at all but, because his CO_2 was low, could be considered to have been hyperventilating. Clearly, observation of the subject shows that he *was not* hyperventilating during the apnea, but recognition of his alveolar and arterial gas partial pressures acknowledges that he must *have been* hyperventilating.

DEAD SPACE VENTILATION

Of the volume of air inhaled per breath, i.e., the tidal volume (VT), a portion reaches the alveoli to effect gas exchange, whereas another portion remains in the conducting airways, or anatomical dead space. The volume of air entering the alveoli per breath (VA) = $VT - VD_{anat}$. The alveolar ventilation per minute is therefore

$$\dot{V}A = VT \cdot f - VD_{(anat)} \cdot f$$

or

$$\dot{V}A = \dot{V}E - \dot{V}D_{(anat)}$$

Thus, the effectiveness of the total inspired volume for gas exchange depends on the size of the dead space.

If $\dot{V}E$ = 8,000 ml per minute, f = 10 per minute and $VD_{(anat)}$ = 150 ml,

$$\dot{V}A = \dot{V}E - \dot{V}D_{(anat)}$$
$$= 8,000 - (150 \times 10)$$
$$= 6,500 \text{ ml/min}$$

But if the same ventilation were achieved with a larger dead space, perhaps 300 ml,

$$\dot{V}A = 8,000 - (300 \times 10)$$
$$= 5,000 \text{ ml/min}$$

thereby causing a lower alveolar ventilation for the same total ventilation.

Similarly, the pattern of breathing affects alveolar ventilation. A minute ventilation of 8 L per minute can be achieved by an infinite number of combinations of VT and f.

If $\dot{V}E$ = 8 L/min, VT = 800 ml, f = 10 and $VD_{(anat)}$ = 200 ml,

$$\dot{V}A = 8,000 \text{ ml/min} - (200 \times 10)$$
$$= 6,000 \text{ ml/min}$$

but if the subject achieved his 8 L per minute total ventilation by breathing with VT = 400 ml and f = 20 with the same dead space, then:

$$\dot{V}A = 8,000 \text{ ml/min} - (200 \times 20)$$

$$= 4,000 \text{ ml/min}$$

Consequently, the effective or alveolar ventilation is reduced, although both $\dot{V}E$ and VD were the same. It should be noted here that we have considered anatomical dead space to be the same at two different tidal volumes in order to stress the extreme importance of the pattern of breathing on the efficiency of ventilation. In reality, anatomical dead space tends to increase as tidal volume increases because, at larger lung volumes, intrapleural pressure is more negative, thereby increasing the transluminal pressure. This leads to the intrapulmonary airways becoming more distended, hence increasing the dead space.

MEASUREMENT OF ANATOMICAL DEAD SPACE

If the lung were constructed such that during expiration the first gas from the lung contained exclusively dead space air, followed by an abrupt transition where alveolar air would be the exclusive expirate (i.e., no mixing of alveolar and dead space gas), we could simply measure the volume exhaled until the abrupt increase in CO_2 was seen and this would be the volume of the conducting airways (Fig 5–10). In reality, although the initial part of the expirate is derived exclusively from the dead space and the final part exclusively from the alveoli, there is an intermediate region in which some alveoli empty before all the dead space is cleared. Hence, expired PCO_2 has the shape shown in Figure 5–10.

We see from Figure 5–10 that, if we draw a line on the plot of exhaled PCO_2 versus volume such that area B, B[1], C equals C, D[1], D, the volume exhaled between A and B[1] will be the anatomical dead space. This is because the tech-

Fig 5–10.–Plot of PCO_2 against exhaled volume during a single exhalation. During $A{\rightarrow}B$, only dead space air leaves the mouth. During $B{\rightarrow}C{\rightarrow}D$, a mixture of dead space and alveolar gas is exhaled and, during $D{\rightarrow}E$, alveolar gas is exhaled. See text for discussion of constructed areas $B{\rightarrow}B'{\rightarrow}C$, and $C{\rightarrow}D'{\rightarrow}D$. Anatomical dead space is equivalent to volume exhaled between A and B'.

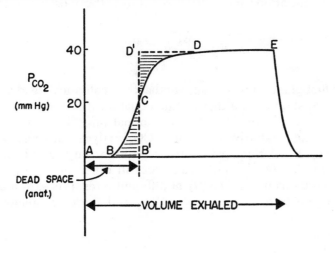

nique effectively allows us to consider the nature of the expiratory pattern were the transition from dead space to alveolar gas in the expirate abrupt and instantaneously complete.

The technique can be complicated, however, by a steep increase in the alveolar plateau phase of the expirate, as in some patients with airway disease, rather than its being virtually flat, as in normal (resting) subjects.

Another technique for measuring the anatomical dead space is the *CO_2 proportion technique.* This is based on the fact that the CO_2 in mixed expired air is less than that in an alveolar sample because it has been diluted with the volume of atmospheric air contained in the dead space just prior to exhalation. The more disparate these concentrations, the greater the fraction of dead space in the expirate.

We know that the volume of CO_2 exhaled in a breath is

$$\text{V}_{CO_2} = \text{V}_\text{T} \cdot \text{F}_{E_{CO_2}}$$

but also

$$\text{V}_{CO_2} = \text{V}_\text{A} \cdot \text{F}_{A_{CO_2}}$$

$$\therefore \text{V}_\text{T} \cdot \text{F}_{E}CO_2 = \text{V}_\text{A} \cdot \text{F}_{A_{CO_2}}$$

but

$$\text{V}_\text{A} = \text{V}_\text{T} - \text{V}_\text{D}$$

therefore by substitution,

$$\text{V}_\text{T} \cdot \text{F}_{E}CO_2 = \text{V}_\text{T} \cdot \text{F}_A CO_2 - \text{V}_\text{D} \cdot \text{F}_{A_{CO_2}}$$

or

$$\frac{\text{V}_\text{D}}{\text{V}_\text{T}} = \frac{\text{F}_{A_{CO_2}} - \text{F}_{E_{CO_2}}}{\text{F}_{A_{CO_2}}}$$

and, as multiplying top and bottom of the equation by $\text{P}_\text{B} - 47$ transforms the fractional concentrations into partial pressures, the most usual form of this relationship is

$$\frac{\text{V}_\text{D}}{\text{V}_\text{T}} = \frac{\text{P}_{A_{CO_2}} - \text{P}_{E_{CO_2}}}{\text{P}_{A_{CO_2}}}$$

Although at first glance it would appear that the anatomical dead space is easy to measure, we should note that in this equation (the Bohr equation) $\text{P}_{A_{CO_2}}$ refers to the mean alveolar CO_2. In the normal lung this can be reasonably determined using a rapidly responding CO_2 analyzer, such as was used in Figure 5–10. But in patients with airway (or pulmonary vasculature) disease, large variations in alveolar P_{CO_2} can occur in different regions of the lung and, as these regions usually empty at different rates, it is extremely difficult to determine the true mean alveolar P_{CO_2} and therefore anatomical dead space.

MEASURING PHYSIOLOGIC DEAD SPACE

The only ventilation that is effective in gas exchange is the ventilation of perfused alveoli. Measurement of the physiologic dead space (the sum of the anatomical dead space and volume of the ventilated but not perfused alveoli, i.e., the alveolar dead space) is therefore often more important than that of the anatomical dead space. To do this we make use of a modification of the Bohr equation, again using the CO_2 proportion concept described above: (1) Mixed expired CO_2 concentration is lower than that of the perfused alveoli because it has been diluted, both by atmospheric air in the anatomical dead space and by the atmospheric air (virtually) that came from the nonperfused alveoli; (2) the more disparate these values, the greater the proportion of total expirate that must originate in the physiologic dead space; (3) although we are unable to determine the P_{CO_2} of only the perfused alveoli from expired gas measurements, we know that the perfused alveoli are in equilibrium for CO_2 with arterial blood (except in conditions of significant right-to-left shunts); (4) therefore, we can determine the P_{CO_2} of only the perfused alveoli from measurement of arterial blood P_{CO_2}.

Hence in the equation

$$\frac{V_D}{V_T} = \frac{P_{A_{CO_2}} - P_{E_{CO_2}}}{P_{A_{CO_2}}}$$

if we replace $P_{A_{CO_2}}$ by arterial P_{CO_2} ($P_{a_{CO_2}}$), then the dead space we compute is that of the total ineffective ventilation space

$$\therefore \frac{V_{D(physiol)}}{V_T} = \frac{P_{a_{CO_2}} - P_{E_{CO_2}}}{P_{a_{CO_2}}}$$

In normal subjects the physiologic dead space is approximately equal to the anatomical dead space and accounts for about 25–30% of the tidal volume at rest. In patients with lung diseases such as pulmonary vascular occlusive disease or in obstructive lung disease in which alveolar septa and associated blood vessels are destroyed, the physiologic dead space can be appreciably larger than the anatomical dead space, e.g.,

1. Normal subject with a tidal volume of 600 ml

$$\frac{V_D}{V_T} = \frac{P_{a_{CO_2}} - P_{E_{CO_2}}}{P_{a_{CO_2}}}$$

$$= \frac{40 \text{ mm Hg} - 28 \text{ mm Hg}}{40 \text{ mm Hg}}$$

$$\therefore = 30\%$$

2. Subject with lung disease leading to increased alveolar dead space, also with a tidal volume of 600 ml

$$\frac{V_D}{V_T} = \frac{Pa_{CO_2} - PE_{CO_2}}{Pa_{CO_2}}$$

$$= \frac{40 \text{ mm Hg} - 20 \text{ mm Hg}}{40 \text{ mm Hg}}$$

$$\therefore = 50\%$$

DIFFUSION

During inspiration, gas moves down the tracheobronchial tree by convection. When the gas reaches the level of the alveolar ducts, the total cross-sectional area of the airways has become progressively so great that the linear velocity of the gas has been reduced to an extremely low value. In fact, gas movement from the alveolar duct to the surface of the alveolus is probably accomplished almost entirely by diffusion.

Although diffusion would be an ineffective process for appropriate rates of gas transfer were the diffusion pathway long, the average alveolar diameter is only of the order of 100 μ, and so diffusion equilibrium throughout the alveolus takes place rapidly. In fact, Comroe has estimated that, even if the diffusion distance were as large as 0.5 mm, diffusion of the gas in the alveolar phase would be 80% complete within 0.002 second and, as the blood flowing through the lung is exposed to the gas exchange surface of the pulmonary capillary bed for 0.75–1.0 second, it is clear that alveolar gas diffusion does not normally limit gas transfer into the blood. However, in a disease such as pulmonary emphysema, the diffusion distance can be much larger because some alveolar septa are destroyed and larger single air sacs are formed where numerous alveoli were originally located. Hence in these abnormal units, diffusion within the air space may impair efficiency of oxygenation of the blood, although it must be reiterated that this is *not* normally limiting in any sense.

The most essential component of the gas transfer is, however, the uptake of O_2 across the alveolar-capillary membrane into the blood and CO_2 evolution in the reverse direction.

The rate of gas exchange, i.e., volume per unit time, from alveolus to blood depends on the following:

1. DIFFUSION CONSTANT $(d)^*$ OF THE GAS.—This depends not only on the rate of diffusion in the gas phase but also on solubility of the gas in the liquid phase, i.e.,

$$d \propto \frac{\text{solubility (S)}}{\text{molecular weight (MW)}}$$

Although CO_2 diffuses about 20% more slowly than O_2 in the gas phase, it is approximately 24 times more soluble in water than is O_2, i.e., the sol-

---

*Lower case d is used to symbolize this diffusion constant. The upper case D will be reserved for the variable "diffusing capacity" of the lung.

ubility* of CO_2 is 0.59, whereas O_2 has a solubility of 0.024. The relative rates of gas transfer across the alveolar-capillary membrane are therefore conventionally presented as

$$\frac{\text{rate of } CO_2 \text{ transfer}}{\text{rate of } O_2 \text{ transfer}} = \frac{S_{CO_2}}{S_{O_2}} \times \frac{\sqrt{MW_{O_2}}}{\sqrt{MW_{CO_2}}}$$

$$= \frac{0.59}{0.024} \times \frac{5.6}{6.6}$$

$$= \frac{20}{1}$$

2. GEOMETRY OF ALVEOLAR-CAPILLARY INTERFACE. — The greater the surface area for diffusion the greater the net rate of gas transfer. For the lung this is the area (A) of the alveolar surface in contact with functioning pulmonary capillaries. The diffusion rate is, however, inversely related to the diffusion distance (T).

Thus the geometric factor affecting diffusion is A/T. It should be remembered that the total diffusion pathway for O_2 is as follows: (a) alveolar diffusion, (b) diffusion through alveolar surface lining (surfactant) layer, (c) diffusion through alveolar-capillary membrane, (d) diffusion through plasma, (e) diffusion through red cell (erythrocyte) membrane, (f) diffusion through intracellular fluid of the red cell and, finally, (g) chemical combination with hemoglobin.

3. RATE OF GAS TRANSFER. — This also depends on the pressure gradient across the capillary bed.

Thus, combining (1), (2) and (3), we see that the net rate of gas transfer or volume per unit time (V) is given by

$$\dot{V} \propto d \times \frac{A}{T} \times (P_1 - P_2)$$

UPTAKE OF O_2 INTO BLOOD

As mixed venous blood flows from the pulmonary artery into the beginning of the pulmonary capillary bed, it has a normal PO_2 at rest of about 40 mm Hg. It is brought into close apposition with alveolar air, which has a PO_2 of 100 mm Hg or more. A gradient of approximately 60 mm Hg is therefore established for PO_2; consequently, O_2 is taken up into the pulmonary capillary blood until it is in equilibrium with the alveolar air. This equilibrium normally takes 0.25 to 0.3 second. But as pulmonary capillary blood normally takes an average 0.8 second to traverse the capillary bed (at rest), there is usually ample time available for diffusion equilibrium.

The pattern of increase in pulmonary capillary PO_2 from its mixed venous

---

*Solubility in this case is expressed as milliliters of gas dissolved per milliliter of fluid at a pressure of 1 atm.

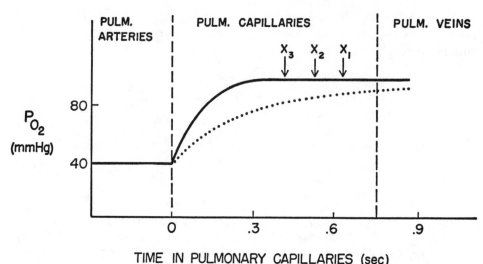

Fig 5–11.—Time course of pulmonary capillary P_{O_2} profile. Note that normally (*solid line*) the blood is fully saturated within about 0.3 second after entering pulmonary capillary. X_1, X_2 and X_3 represent three progressively greater cardiac outputs, i.e., reducing capillary transit time. In diseases that increase markedly the alveolar-capillary path length, the rate of P_{O_2} increase in the pulmonary capillary bed is reduced (*dotted line*). In extreme cases the blood may leave lung capillaries before time for equilibrium with alveolar P_{O_2}.

value to its equilibrium with alveolar air is complex; the instantaneous pressure gradient for P_{O_2}, which defines the instantaneous rate of O_2 transfer, is itself dependent upon the characteristics of O_2 reaction with red cell hemoglobin (see below). However, the general time course of pulmonary capillary P_{O_2} is given by the solid curve in Figure 5–11. It should be noted that the initial rate of uptake is rapid owing to large P_{O_2} gradient from alveolus to capillary. As the gradient is progressively reduced, the rate of transfer decreases until P_{O_2} reaches equilibrium; that unit of blood will then take up no more O_2.

If cardiac output is increased, such as during muscular exercise, the transit time through the pulmonary capillary bed will be shortened. However, if cardiac output is doubled, for example, capillary transit time is not proportionally halved, as the lung has great capacity to recruit previously nonperfused, or poorly perfused, capillaries and has some capacity to distend the already perfused capillaries. This increases the pulmonary capillary blood volume. The reduction in transit time that does occur still does not normally reach the levels at which O_2 transfer is impaired, as shown by the symbols X_1, X_2 and X_3, which represent three progressively higher work rates (see Fig 5–11).

Insufficient time for O_2 equilibrium can, however, impair O_2 transport across the alveolar-capillary bed under at least three kinds of conditions.

1. *Increased diffusion path length.* — If the alveoli are partially filled by some exudate, such as the proteinaceous material that is extruded into the alveolar space in the disease pulmonary alveolar proteinosis, or by pulmonary

edema fluid, the path length for O_2 diffusion can be appreciably increased. This leads to a slower rate of rise of pulmonary capillary Po_2 toward its equilibrium value with alveolar Po_2. Under these conditions the time for equilibrium may be increased from the normal 0.3 to 0.6 second or even, in extreme cases, to values greater than the 0.8 second normally available for diffusion equilibrium. When the time needed for O_2 equilibrium exceeds the capillary transit time, clearly the pulmonary venous blood and therefore arterial blood will have low Po_2 and be hypoxemic. But even if there is just sufficient time for equilibrium at rest, increases in cardiac output leading to decreased pulmonary capillary transit time will tend to cause hypoxemia, which becomes progressively greater as cardiac output is progressively increased.

2. *Reduction of functioning pulmonary capillary bed.* — Any condition leading to fewer functioning pulmonary capillaries, such as blockade of pulmonary vessels with emboli, will cause the mean transit time of blood in the capillary bed to be decreased, i.e., the same cardiac output, or pulmonary blood flow, will traverse a smaller volume in less time. Thus, the residence time of the blood in the pulmonary capillaries approaches that needed for equilibrium. Exercise can lead to insufficient time for equilibrium in such cases and arterial hypoxemia ensues. It should be noted that the actual structure of the diffusion path length may be unaltered in those regions through which the diffusion occurs; rather, the hypoxemia simply results from lack of sufficient time in the pulmonary capillary bed for diffusion equilibrium.

3. *Reduced alveolar Po_2.* — When the driving pressure for O_2 transfer is reduced, e.g., low alveolar Po_2 due to a sojourn at high altitude or when a subject is given a low O_2 concentration to breathe, the rise time of pulmonary capillary Po_2 will be slowed. Again, this can lead to the pulmonary blood leaving the capillary bed before it has come into equilibrium with the alveolar gas, especially during exercise.

The most striking condition for diffusion impairment of O_2 transfer is therefore a combination of a low alveolar Po_2 and exercise in a subject with pulmonary disease such as those mentioned above.

In contrast, CO_2 retention owing to diffusion impairment is rarely a problem clinically even though the impairment is sufficient to cause severe hypoxemia.

DIFFUSING CAPACITY (OR PULMONARY GAS TRANSFER INDEX)

The pulmonary diffusing capacity (DL) is defined as the volume of gas taken up per unit time ($\dot{V}$) divided by the pressure gradient for that gas across the alveolar-capillary interface $(P_1 - P_2)$, i.e.,

$$DL = \frac{\dot{V}}{P_1 - P_2}$$

If we recall that the general formula for gas transfer across the lung is

$$\dot{V} = d \cdot \frac{A}{T} \cdot (P_1 - P_2)$$

then it is apparent that D_L really takes into account not only the specific diffusing characteristics of the gas but also the area and thickness of the intervening diffusing pathway.

For O_2 the diffusing capacity,

$$D_{L_{O_2}} = \frac{\dot{V}_{O_2}}{P_{A_{O_2}} - P_{\bar{c}_{O_2}}}$$

We can readily measure O_2 uptake in a given time and also measure, or estimate closely, mean alveolar P_{O_2}. Mean capillary P_{O_2} ($P_{\bar{c}_{O_2}}$), however, is extremely difficult to measure, as it represents an integrated mean value of the instantaneous P_{O_2} of capillary blood as it equilibrates. As stated above, this has a complex time course that is influenced by the O_2 in solution and by its reaction and binding with hemoglobin of the red cells.

To circumvent this inherent difficulty with O_2 measurement, the diffusing capacity for carbon monoxide (CO) is usually determined. Carbon monoxide is typically used because of the following properties:

1. CO follows the same pulmonary transfer route as O_2 and binds to hemoglobin on the same site.

2. Except in areas where CO pollutes the air or in subjects who smoke tobacco, the CO concentration in venous blood entering the pulmonary capillary bed is effectively zero.

3. The affinity of CO for hemoglobin is about 210 times greater than that of O_2. This means that if small quantities* are administered to a subject, the CO in the capillary blood will be preferentially bound to the hemoglobin; consequently the P_{CO} in the capillary blood will be effectively zero. Hence,

$$D_{L_{CO}} = \frac{\dot{V}_{CO}(ml/min)}{P_{A_{CO}}\ (mm\ Hg)}$$

4. $D_{L_{O_2}}$ can be readily determined from $D_{L_{CO}}$ by referring to the solubilities and molecular weights of the gases:

$$\frac{D_{L_{O_2}}}{D_{L_{CO}}} = \frac{Sol.\ O_2}{Sol.\ CO} \times \frac{MW_{CO}}{MW_{O_2}}$$

$$= \frac{0.024}{0.018} \times \frac{28}{32}$$

$$= 1.23,\ i.e.,$$

$$D_{L_{O_2}} = 1.23 \cdot D_{L_{CO}}$$

Several tests are used to determine $D_{L_{CO}}$, but a common procedure is to inhale a gas mixture containing a low concentration ($\sim$0.3%) CO and then hold the breath for 10 seconds. While the breath is being held, CO enters the blood

---

*As CO can be a lethal gas in somewhat higher concentrations (it binds to the transport site of O_2 on the hemoglobin and impairs O_2 delivery to the tissues), only a very small amount is inhaled in these tests; e.g., 0.3% CO.

and obviously the more CO that enters in this time the greater the diffusion capacity for the gas.

The volume of CO taken up by the blood during breath-holding is computed ($\dot{V}CO$) and the mean alveolar CO during the time is measured. Hence DL_{CO} is computed. The normal value for DL_{CO} by this test is about 25 ml per minute per mm Hg in resting subjects.

FACTORS AFFECTING DL_{CO}

In addition to the thickness and area of the surface, several other factors affect DL_{CO}:

1. HEMOGLOBIN (Hb) CONCENTRATION.—The greater the Hb concentration, the greater is the number of sites for combination with CO per unit blood volume and thus the greater the rate of CO uptake. Hence, polycythemia leads to high DL_{CO}, whereas anemia leads to low DL_{CO}.

2. PULMONARY BLOOD VOLUME.—Increased volume of blood in the capillary bed increases the effective surface area for diffusion, especially if there is recruitment of pulmonary capillaries, and makes more Hb molecules available for uptake of CO. Thus exercise or recumbency increases DL_{CO} by increasing pulmonary blood volume.

The DL_{CO}, however, does not increase simply due to increases in pulmonary blood flow per se. This is because so few of the available binding sites for CO on the Hb are filled during this test. Consequently, so many are still available for binding that it does not alter the rate of CO uptake if new red cells replace the original ones in the pulmonary capillary bed with increased blood flow.[*] Therefore, the uptake rate of CO would be maintained for a long period even though blood flow were to stop. Most conditions in which pulmonary blood flow increases, however, lead to a simultaneous increase in pulmonary blood volume, or alteration of the perfusion pattern, so that DL_{CO} is affected, but not by flow per se.

3. BODY SIZE.—The DL_{CO} increases with body size, presumably owing to the larger pulmonary surface area available for diffusion.

4. ALTERED ALVEOLAR PO_2.—If O_2 concentration is high, there is greater competition for Hb-binding sites. This leads to reduced DL_{CO}; similarly, low PO_2 increases DL_{CO}. As the membrane component of DL_{CO} is presumably unaltered by these variations in PO_2, it is possible to perform DL_{CO} tests at various levels of PO_2 and determine the membrane and the Hb reaction component.

For example, the total resistance to diffusion of the gas is made up of the sum of the membrane resistance and the resistance offered to the Hb binding. Hence

$$\frac{1}{DL} = \frac{1}{D_m} + \frac{1}{\theta \cdot Vc}$$

where DL is the total diffusing capacity, D_m is the membrane component (this

---

[*]This is not true for O_2.

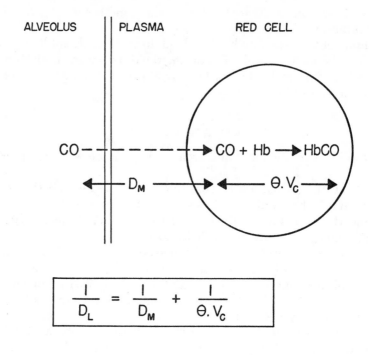

$$\frac{1}{D_L} = \frac{1}{D_M} + \frac{1}{\theta \cdot V_c}$$

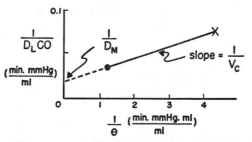

Fig 5–12.—Components of diffusing capacity $(D_{L_{CO}})$. D_M is the membrane component and includes both alveolar-capillary and red cell membranes. V_c is the volume of blood in the capillary bed and θ is the rate at which 1 ml of blood will combine with CO. As θ varies predictably with P_{O_2}, if $D_{L_{CO}}$ is determined at various levels of θ, e.g., room air ($\bullet$) and 100% O_2 breathing (X), the membrane component (D_M) and the capillary volume (V_c) can be estimated as shown.

includes alveolar-capillary and red cell membranes), Vc is the volume of blood in the pulmonary capillaries and θ is the reaction rate coefficient for the chemical combination of Hb with the gas. If $D_{L_{CO}}$ is measured at various levels of P_{O_2} (i.e., altering the reaction rate coefficient (θ) then plotting $1/D_L$ against $1/\theta$ results in a straight line relationship (Fig 5–12) with a slope of $1/V_c$ (hence we can determine pulmonary capillary blood volume) and intercept at $1/D_M$. This technique is therefore a refinement of the commonly used $D_{L_{CO}}$ tests, as it allows us to determine whether or not reduced $D_{L_{CO}}$ is due predominantly to impaired membrane diffusion.

GAS TRANSPORT IN THE BLOOD

The respiratory gases are transported in the blood either in simple physical solution in the aqueous phases, as is the case for N_2, or in both physical solution and chemical combination with other molecules, as is the case for O_2 and CO_2.

OXYGEN TRANSPORT

1. DISSOLVED OXYGEN. — The amount of O_2 transported in solution in plasma is defined by Henry's law, which states that the quantity of gas that can dissolve in a fluid is equal to the product of the gas partial pressure and the solubility coefficient (α), i.e.,

$$\frac{\text{volume } O_2}{\text{volume plasma}} = Po_2 \cdot \alpha$$

α for O_2 in blood at 38 C is 0.023 ml O_2/ml of blood per atmosphere pressure. Therefore, per mm Hg Po_2, 1 ml blood can transport $0.023 \div 760 = 0.00003$ ml O_2 in physical solution. Accordingly, 100 ml of blood at a normal arterial Po_2 of 100 mm Hg will transport 0.3 ml O_2. Similarly, mixed venous blood, which has a Po_2 of about 40 mm Hg at rest, will carry $0.003 \times 40 = 0.12$ ml dissolved O_2/100 ml of blood. It should be noticed that the relationship is linear so that if a subject breathes 100% oxygen and his arterial Po_2 is 600 mm Hg, the amount transported will be proportionately increased to $0.3 \times 6 = 1.8$ ml O_2/100 ml.

It is clear that the quantity of O_2 transported in physical solution is not capable of supplying any appreciable proportion of the body's metabolic O_2 requirements (about 250 ml per minute at rest). In fact, if dissolved O_2 were the only transport form for O_2, an arterial Po_2 of 1,667 mm Hg would be required to provide the resting metabolic requirements if the cardiac output were normal and all the O_2 were extracted by the tissues so that mixed venous Po_2 were zero.[*] A simple calculation shows this:

$$\dot{V}o_2 = \dot{Q}(Ca_{O_2} - C\bar{v}_{O_2})$$

where Ca_{O_2} and $C\bar{v}_{O_2}$ are the O_2 contents of arterial and mixed venous blood. Substituting given values, we have

$$\dot{V}o_2 = 5{,}000 \text{ ml/min} ([0.00003 \times 1{,}667] \text{ ml/ml} - [0.00003 \times 0] \text{ ml/ml})$$

or

$$= 5{,}000 \, (0.05)$$

$$= 250 \text{ ml/min}$$

Alternatively, if the arterial Po_2 were normal at 100 mm Hg, the blood would contain 0.3 ml/100 ml or 3 ml/L of dissolved O_2. In order to supply 250 ml per minute to the tissues at this arterial O_2 content (again assuming mixed

---

[*]This, of course, is an impossible situation.

venous P_{O_2} to be zero), the cardiac output ($\dot{Q}$) would have to be at least 83 L per minute:

$$250 \text{ ml/min} = \dot{Q} \text{ L/min} \times (3 \text{ ml/L} - 0 \text{ ml/L})$$

that is,

$$\dot{Q} = \frac{250 \text{ ml/min}}{3 \text{ ml/L}} = 83 \text{ L/min}^*$$

2. Oxygen in chemical combination with hemoglobin.—The Hb of red blood cells has the remarkable property of reversibly combining with large quantities of O_2 in loose chemical combination. Thus the interaction is one of oxygenation, not oxidation.

The Hb molecule consists of four iron-containing heme groups, each of which is combined with the protein globin. Each globin molecule is made up of four polypeptide chains, which, in normal adult Hb (Hb A), consists of two α chains (each with 141 amino acids) and two β chains (each with 146 amino acids). These amino acid chains are folded into a complex three-dimensional structure, within which each heme group is located such that its iron molecule is readily accessible to O_2. Each Hb molecule might therefore be considered as four identical Hb subunits. The conjunction of the subunits modifies the function of each individual subunit as described below.

Each subunit contains only about 0.335% iron but it is this iron in the ferrous form (Fe^{2+}) that confers the ability for loose and reversible O_2 combination with Hb. Certain drugs can oxidize the iron in Hb to the ferric (Fe^{3+}) form. The resulting compound, methemoglobin, no longer posesses the property of loose, reversible combination with oxygen. Drugs that produce methemoglobin are therefore dangerous to life if taken in sufficient quantity.

As each heme group contains one molecule of iron and each iron molecule can combine with one molecule of O_2, the correct general equation is given as

$$Hb_4 + 4O_2 \rightleftarrows Hb_4(O_2)_4$$

although the simpler statement (for each O_2 reaction) $Hb + O_2 \rightleftarrows HbO_2$ is commonly used. As the iron in each heme group comprises 0.335% of the Hb subunit and the atomic weight of iron is 56, the molecular weight of an individual subunit is

$$56 \div 0.335/100 = 16,700 \text{ gm}$$

As one iron molecule can combine with one O_2 molecule (or 1 M Hb with 1 M O_2), 16,700 gm per Hb subunit can maximally combine with 22,400 ml O_2. Therefore, 1 gm Hb can combine with

$$\frac{22,400}{16,700} = 1.34 \text{ ml } O_2 \dagger$$

---

*A highly trained athlete exercising maximally has a cardiac output of about 30 L per minute.

†This reflects the value commonly given for the molecular weight of Hb. Under physiologic conditions in the blood, the effective Hb molecular weight may be somewhat lower. The combining capacity of 1 gm of Hb may be as high as 1.39 under physiologic conditions.

Each red cell contains approximately 30×10^{-12} gm Hb or 250–300 million Hb molecules. It is conventional, however, to consider units of 100 ml of blood, which normally contains about 15 gm Hb/100 ml of blood. As each gram of Hb has the capacity to combine with 1.34 ml O_2, the capacity of blood Hb for O_2 transport is

$$15 \times 1.34 = \frac{20.1 \text{ ml } O_2}{100 \text{ ml}}$$

The amount of O_2 that actually combines with Hb is dependent both on the Po_2 of the blood and certain physicochemical properties of the blood such as its Pco_2, [H^+], temperature and concentration of certain organic phosphates.

Oxygen-Hemoglobin (HbO₂) Dissociation Curve

The quantity of O_2 bound to hemoglobin does not increase linearly with increasing Po_2 but, rather, has a characteristic sigmoid relationship (Fig 5–13). This curve, with its steep slope between 10 and 50 mm Hg Po_2 and its shallow slope above approximately 70 mm Hg, is of enormous physiologic import. It is apparent, even from cursory observation of the curve, that the blood will retain a high proportion of its capacity to transport O_2 despite moderate decreases in arterial Po_2 (induced, for example, by ascent to altitude or by lung disease).

When arterial blood reaches the tissue capillary bed, large quantities of O_2 are unloaded from the blood as capillary Po_2 falls, resulting in a mixed venous Po_2 of 40 mm Hg or less. Similarly, as mixed venous blood reaches the lung, relatively large amounts of O_2 will be loaded into the blood as Po_2 begins to increase above the mixed venous value.

It is important to recognize that this sigmoid relationship between Po_2 and O_2 content of Hb is a feature of its four-subunit structure. The characteristic

Fig 5–13.—Sigmoid relationship between O_2 content and O_2 partial pressure of blood. The % O_2 saturation of hemoglobin (HbO₂) is given on right-hand scale. Normal resting values for arterial and venous blood at sea level are shown.

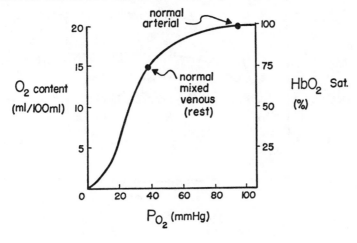

sigmoid shape is due to interaction between individual Hb subunits (heme-heme interaction) such that the presence of O_2 on one heme group affects the affinity for O_2 binding at the other sites. And as, with progressively increasing Po_2, the four O_2 binding sites become occupied sequentially, the successive bindings cause the content or "affinity" curve to become sigmoid.

A structure such as myoglobin, which is identical to a single Hb subunit and therefore has a molecular weight of 16,700, has a hyperbolic rather than a sigmoid dissociation curve and consequently does not release its bound O_2 until Po_2 becomes very low. In fact, myoglobin is still 50% saturated with O_2 despite Po_2 being less than 10 mm Hg, whereas Hb is normally 50% saturated when Po_2 falls only to about 27 mm Hg. Therefore, at a Po_2 of 10 mm Hg, which is close to the physiologic minimum attainable from exercising muscle, the Hb saturation with O_2 is only 10%, whereas myoglobin still binds 50% of its potential maximum O_2.

O_2 CAPACITY, SATURATION AND CONTENT

The O_2 capacity is the maximal amount of O_2 that will combine with Hb at high Po_2. Thus the capacity will vary with the Hb concentration of the blood. At a normal 15 gm Hb/100 ml, the O_2 capacity is

$$15 \times 1.34 \text{ ml } O_2/\text{gm Hb} = 20.1 \text{ ml}/100 \text{ ml}$$

The O_2 saturation (S_{O_2}) is the percentage of the O_2 capacity that is occupied with O_2. Thus

Fig 5–14.—Effects of anemia and polycythemia on O_2 content and Hb saturation relationships with O_2 partial pressure (Po_2). Note that O_2 content curve reaches values appropriate for Hb content of the blood. In contrast, the HbO_2 saturation always reaches 100% at high Po_2. *Dashed line* represents the curve if 33% of Hb were bound with CO, i.e., providing the same approximate functional Hb value as the 10 gm/100 ml anemic condition. Note that HbCO curve is displaced to left of anemic curve, evidencing impaired O_2 release.

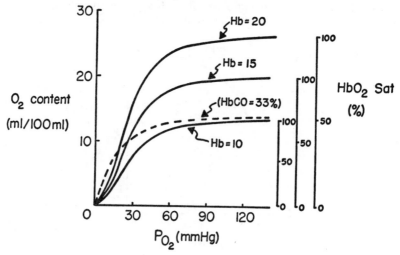

$$S_{O_2} = \frac{O_2 \text{ bound to Hb} \times 100}{O_2 \text{ capacity}}$$

It should be noted that the O_2 capacity varies with the concentration of Hb but the S_{O_2} does not. The O_2 content of the blood (C_{O_2}) is therefore the sum of the O_2 in physical solution and the amount bound to Hb, i.e.,

$$C_{O_2} = (PO_2 \times \alpha) + (Hb \times S_{O_2} \times 1.34)$$

where α is the solubility coefficient of O_2 in blood, Hb is the hemoglobin concentration (gm per 100 ml), S_{O_2} is the HbO_2 saturation (%) and 1.34 is the combining capacity for O_2 of 1 gm Hb.

Figure 5–14 shows the effect of anemia (low Hb concentration) and polycythemia (high Hb concentration) on the HbO_2 dissociation curve plotted as O_2 content and also as percentage saturation. Clearly the greater the Hb concentration, the greater will be the content of O_2 per 100 ml of blood at any given PO_2. Hence the sigmoid curve of HbO_2 dissociation will asymptote at values appropriate for Hb concentration. On the other hand, the same fraction of total available Hb sites will be occupied by O_2 at a given PO_2 regardless of Hb concentration. Hence the O_2 saturation curve will be the same at various Hb concentrations, whereas the O_2 content curve will not.

FACTORS AFFECTING HbO_2 DISSOCIATION CURVE

Up to this point we have considered the HbO_2 dissociation curve as though it were in a constant position relative to PO_2. In reality, the position of the curve varies with temperature, pH, PCO_2 and concentration of certain organic phosphate compounds (e.g., 2,3-diphosphoglycerate [2,3-DPG]). Thus a whole "family" of curves may be constructed relating S_{O_2} to PO_2, as these conditions vary (Fig 5–15).

Fig 5–15.—Effects of PCO_2, H^+ and temperature on position of HbO_2 saturation curve. As shown, increases in PCO_2, H^+ or t degrees shift the curve to the right, thereby aiding O_2 unloading. Decreases in PCO_2, H^+ or t degrees shift the curve to the left, thereby aiding O_2 loading onto Hb.

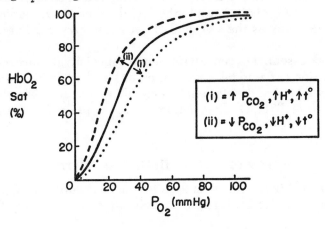

TABLE 5-4.—VALUES ON "NORMAL" HbO_2
DISSOCIATION CURVE°

Po_2 (mm Hg)	So_2 OF HB (%)	CONDITION
95	97	"Normal" arterial blood
60	90	Hypoxic arterial blood (saturation still high)
40	75	Mixed venous blood (at rest)
27	50	Po_2 for 50% saturation (i.e., P_{50})

°pH = 7.4; Pco_2 = 40 mm Hg; t = 37C.

1. *Temperature.*—As the temperature of blood rises, the curve shifts to the right (see Fig 5-15). Thus during muscular exercise the blood unloads O_2 at a higher Po_2, a condition clearly beneficial for the increased O_2 requirements of the exercising tissue. Hypothermia on the other hand shifts the curve to the left, thus retaining O_2 bound to the Hb until a lower Po_2. (The redness of the ears in cold weather attests to the high saturation of capillary blood at these low temperatures—although other factors are also involved.)

2. *2,3-Diphosphoglycerate.*—The 2,3-DPG is an end-product of red cell metabolism, and its concentration is increased under chronic conditions of low Po_2, anemia or low cardiac output. This causes the HbO_2 dissociation curve to be shifted to the right (i.e., decreased O_2 affinity), thus assisting the unloading of O_2 at the tissues. Although other phosphate compounds such as ATP also affect the affinity of O_2 binding, their concentrations within the red cell are much lower than that of 2,3-DPG, hence the special importance of this phosphate. Storage of blood (in a blood bank for example) leads to decreased 2,3-DPG concentrations, hence an increased affinity for O_2 binding to Hb reflected in a left shift of the HbO_2 curve. This should be considered prior to transfusion of blood.

3. *pH and Pco_2 (Bohr effect).*—Increased Pco_2 and/or increased [H+] (i.e., decreased pH) leads to a rightward shift of the curve. Decreased Pco_2 and/or [H+] (i.e., increased pH) shifts the curve to the left. The effect of CO_2 on the HbO_2 dissociation curve is largely, but not entirely, due to its H+-forming properties. Thus at the tissue level, O_2 unloading is aided (right shift) by concomitant H+ and CO_2 evolution into the blood, whereas O_2 loading is aided (left shift) in the lung as the CO_2 and H+ of pulmonary capillary blood decrease.

It is therefore essential in considering—and especially in constructing—the HbO_2 dissociation curve to take account of, or standardize, the conditions in the blood. It will benefit the student to remember a few approximate key locations on the "normal" HbO_2 dissociation curve, i.e., pH = 7.4; Pco_2 = 40 mm Hg; t = 37 C (Table 5-4).

EFFECT OF CO ON HbO_2 DISSOCIATION

The gas CO combines with Hb at the site usually occupied by O_2. However, the affinity (M) of Hb for CO is approximately 210 times that of O_2. Thus in the presence of a mixture of both O_2 and CO (such as would be the case in an area with exhaust fumes from an automobile, or gas inhaled during cigarette smoking), the proportion of HbCO to HbO_2 in the blood would be given by

$$\frac{[\text{HbCO}]}{[\text{HbO}_2]} = M\,\frac{\text{Pco}}{\text{Po}_2}$$

As M has a value of about 210, this means that, at a normal arterial Po_2 of 100 mm Hg, over 50% of the Hb would be bound with CO at an arterial Pco of just 0.5 mm Hg. At Pco of 1 mm Hg, the Hb is virtually completely saturated with CO. Small quantities of CO in inspired air therefore appreciably affect O_2 transport capacity of the blood.

The CO binding with Hb also affects the shape of the HbO_2 dissociation curve (see Fig. 5–14), shifting the curve to the left, thereby impairing the unloading of O_2 from Hb.

TRANSPORT OF CO_2

The CO_2 is transported in the blood in four forms: (1) as dissolved CO_2, (2) as carbonic acid (H_2CO_3), (3) as bicarbonate ions (HCO_3^-) and (4) combined with amino groups of Hb and plasma protein (carbamino compounds).

1. PHYSICALLY DISSOLVED CO_2. — The concentration of dissolved CO_2 in blood is defined by Henry's law, i.e., CO_2 concentration = Pco_2 × solubility coefficient (α) for CO_2. In the case of CO_2 the solubility (α) is conveniently expressed in millimoles per liter per units of mm Hg, where α = 0.03 mM/L/mm Hg for plasma at normal body temperature. Therefore, at a normal arterial Pco_2 of 40 mm Hg, the amount of CO_2 transported in the plasma as dissolved CO_2 is

$$40 \times 0.03 = 1.2 \text{ mM/L}$$

or about 5% of total CO_2 transported in arterial blood. Mixed venous blood at rest has a Pco_2 of 46 mm Hg. Therefore, the amount of dissolved CO_2 is

$$46 \times 0.03 = 1.38 \text{ mM/L}$$

As the total resting venous to arterial difference for CO_2 (in all forms) is only about 2 mM/L (Table 5–5), it is clear that dissolved CO_2 exchange accounts for about 10% of total CO_2 exchanged at the lung.

2. CARBONIC ACID. — CO_2, which is evolved from tissue metabolism, combines with H_2O to form carbonic acid (H_2CO_3), i.e.,

$$CO_2 + H_2O = H_2CO_3$$

However, only about 0.001% of the CO_2 molecules are transported in the hydrated H_2CO_3 form. Thus, only a negligibly small quantity of CO_2 (0.0017 mM/L at Pco_2 = 40 mm Hg) is transported as carbonic acid.

3. BICARBONATE ION. — The H_2CO_3 formed in the reaction of H_2O and CO_2 readily dissociates, as follows:

$$H_2CO_3 \rightleftharpoons H^+ + HCO_3^-$$

TABLE 5–5.—NORMAL VALUES FOR BLOOD OF
CO$_2$ TRANSPORT[*]

CO$_2$ TRANSPORT	ARTERIAL BLOOD	MIXED VENOUS BLOOD	ARTERIAL/VENOUS DIFFERENCE
Plasma (mM/L)			
Dissolved CO$_2$	1.2	1.4	+0.2
Carbonic acid	0.0017	0.0020	+0.0003
Bicarbonate	24.4	26.2	+1.8
Carbamino CO$_2$	–	–	–
Total	25.6	27.6	+2.0
Red cell of 1 L blood			
Dissolved CO$_2$	0.44	0.51	+0.07
Bicarbonate	5.88	5.92	+0.04
Carbamino CO$_2$	1.10	1.70	+0.60
Plasma fraction of 1 L blood			
Dissolved CO$_2$	0.66	0.76	+0.10
Bicarbonate	13.42	14.41	+0.99
Whole blood			
pH	7.40	7.36	−0.04
Pco$_2$ (mm Hg)	40.0	46.0	+6.0
Total CO$_2$ (mM/L)	21.5	23.3	+1.8
(vol %)	48.0	52.0	+4.0

[*]Modified from Nunn, J. F.: *Applied Respiratory Physiology with Special Reference to Anaesthesia* (London: Butterworth, 1969).

The complete reaction is therefore given by

$$\overset{\text{(i)}}{} \quad \overset{\text{(ii)}}{}$$
$$CO_2 + H_2O \rightleftharpoons H_2CO_3 \rightleftharpoons H^+ + HCO_3^-$$

Reaction (i) proceeds very slowly in the plasma. However, because of the presence of the enzyme carbonic anhydrase, the rate of formation of HCO$_3^-$ within red cells is much more rapid. The hydrogen ions formed in this reaction are buffered by Hb, and the HCO$_3^-$ diffuses through the red cell membrane into the plasma to re-establish equilibrium conditions (Fig 5–16). Therefore, plasma HCO$_3^-$ increases when CO$_2$ is added to the blood, predominantly because of reactions within the red cell.

As the red cell membrane is relatively impermeable to cations, it is plasma anions, mainly Cl$^-$, that move into the red cell to retain electric neutrality (chloride shift, or Hamburger effect). These translocations are governed by the Gibbs-Donnan equilibrium:

$$\frac{HCO_3^- \text{ in plasma}}{HCO_3^- \text{ in red cell}} = \frac{Cl^- \text{ in plasma}}{Cl^- \text{ in red cell}}$$

Water also moves into the red cell to maintain osmotic equilibrium and this accounts for the slight swelling of the red cell as it goes from systemic arterial to systemic venous conditions. The opposite reactions occur in the pulmonary capillaries as CO$_2$ leaves the blood.

Approximately 90% of the total CO_2 transported in arterial blood is carried as HCO_3^-, which has a concentration of 24 mEq/L of plasma or 21.5 mEq/L of blood (note in Table 5–5 that more of the HCO_3^- is transported in the plasma than in the red cell fraction of the blood).

4. CARBAMINO TRANSPORT.—As described above, of the total CO_2 transported by arterial blood, 5% is in physical solution and 90% is carried as HCO_3^-. The other 5% is transported in combination with amino groups of Hb and, to a much lesser extent, of plasma proteins.

The general equation may be represented as

$$R = NH_2 + CO_2 \rightarrow R - NH\ COO^- + H^+$$

However, as reduced Hb can combine with more CO_2 than can oxygenated Hb, oxygenation in the lung assists CO_2 unloading, whereas deoxygenation in the peripheral tissues augments CO_2 loading onto Hb. Therefore, of the total CO_2 transported in arterial blood, 5% is dissolved, 5% is as carbamino compounds (predominantly Hb) and 90% is HCO_3^-. But of the CO_2 transferred by

Fig 5–16.—Blood transport of CO_2, released by tissue metabolism. CO_2 is transported in both plasma and red cells. CO_2 (and O_2) follows the opposite route during gas exchange in the lung. See text for further discussion. $PrCO_2 = CO_2$ combined with plasma proteins.

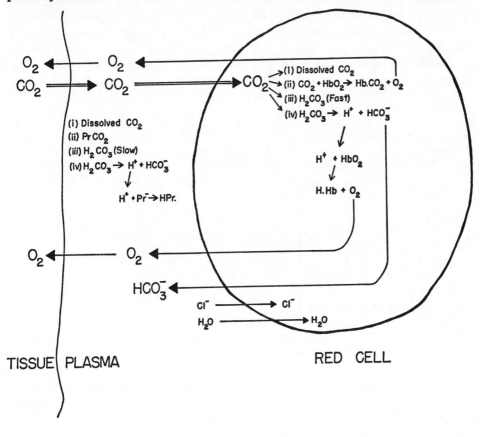

gas exchange, about 10% is from dissolved CO_2, 30% from carbamino CO_2 and 60% from HCO_3^-.

"TRUE" PLASMA AND "SEPARATED" PLASMA

The role of the red cells and Hb in CO_2 transport is exemplified by the well-known observation that, if plasma is separated from the red cells by centrifugation and is then equilibrated with P_{CO_2} at various levels, the slope of this "separated plasma" CO_2 content to P_{CO_2} curve is much less steep than if the equilibration with CO_2 were to occur before separating the red cells. This latter "true" plasma relationship of CO_2 content to P_{CO_2} reflects predominantly the additional CO_2, which is transported as plasma bicarbonate but which originated in reactions within the red cell.

CO_2 DISSOCIATION CURVE

Within the physiologic range, the CO_2 dissociation "curve," i.e., the plot of CO_2 content (ml CO_2/100 ml of blood) against P_{CO_2} (mm Hg), is usually linear, although over a wider range of P_{CO_2} the relation is curvilinear (see Fig 5–16). Consequently, increases in P_{CO_2} due to hypoventilation lead to proportional increases in the CO_2 content, and decreased P_{CO_2} due to hyperventilation leads to proportional decreases in CO_2 content, within the physiologic range. The CO_2 dissociation curve therefore differs from the O_2 dissociation curve in this regard.

In addition, as reduced Hb can bind more CO_2 – in proportion to the degree of reduction – than can oxygenated Hb, a family of curves is apparent in such a plot (Fig 5–17).

Fig 5–17.—Relationship between blood content of CO_2 and its partial pressure over a range of P_{CO_2} from 0–90 mm Hg. In the physiologic range (e.g., 35–55 mm Hg), the curve is relatively linear. Note that reduced blood (*dashed line*) can transport more CO_2 at a given P_{CO_2} than can oxygenated blood (*solid line*).

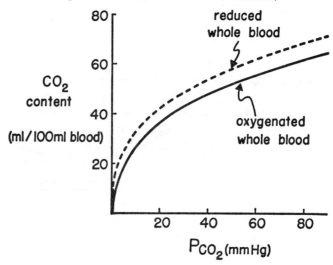

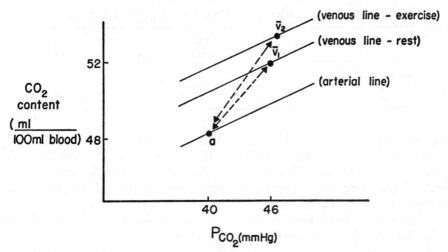

Fig 5–18.—Representation of relatively linear portions of CO_2 dissociation curves for normal arterial and mixed venous blood. As blood traverses the pulmonary-capillary bed, it shifts to the more oxygenated arterial line (*dashed arrows*).

This ability of degree of oxygenation of Hb to modify the quantity of CO_2 transported in the blood at a given P_{CO_2} is termed the Haldane effect and is clearly beneficial to gas exchange. (About 70% of the Haldane effect in CO_2 exchange is derived from the changes in carbamino transport with the degree of oxygenation.) Consequently, as blood flows through the pulmonary capillary bed, the relationship of C_{CO_2} to P_{CO_2} shifts downward from the venous line to the arterial line, thereby making the "effective" CO_2 dissociation curve steeper, as shown by the dashed arrows in Figure 5–18.

Some useful values for arterial and venous CO_2 and O_2, applicable to the normal resting individual, are presented in Table 5–6.

MECHANISMS OF HYPOXEMIA AND HYPERCAPNIA

In man breathing normal air at sea level, arterial hypoxemia, i.e., reduced arterial O_2 partial pressure, can result from four different causes, or any combination of the four. These are (1) alveolar hypoventilation, (2) diffusion abnormality, (3) right-to-left shunt and (4) ventilation-perfusion ($\dot{V}_A/\dot{Q}$) inequality.

ALVEOLAR HYPOVENTILATION

The fundamental relationship between the rate of metabolic O_2 utilization ($\dot{V}_{O_2}$) and alveolar ventilation ($\dot{V}_A$) defines the level of alveolar P_{O_2}, i.e.,

$$\dot{V}_{O_2} = \dot{V}_A(F_{I_{O_2}} - F_{A_{O_2}})$$

or

$$863\frac{\dot{V}_{O_2}(\text{STPD})}{\dot{V}_A(\text{BTPS})} = P_{I_{O_2}} - P_{A_{O_2}}$$

TABLE 5-6.—RESTING GAS EXCHANGE VALUES

BLOOD	C_{CO_2} (VOL %)	PCO_2 (mm Hg)	C_{O_2} (VOL %)	PO_2 (mm Hg)
Mixed venous blood	52	46	15	40
Arterial blood	48	40	20	100
Difference	4	6	5	60

It is clear therefore that if alveolar ventilation is inappropriately low for the current metabolic O_2 requirements, the difference between inspired and alveolar PO_2 will increase; this means that alveolar PO_2 will fall. Consequently, arterial PO_2 is reduced by hypoventilation. The difference between alveolar and arterial $(A - a) PO_2$, however, remains unchanged.

Similarly, if alveolar ventilation is too low for the rate of metabolic CO_2 production, PA_{CO_2} and Pa_{CO_2} will rise, as described by

$$PA_{CO_2} = 863 \frac{\dot{V}CO_2 \, (\text{STPD})}{\dot{V}A \, (\text{BTPS})}$$

Thus alveolar ventilation leads to reduced arterial PO_2 (hypoxemia) and increased arterial PCO_2 (hypercapnia).

The fall of Pa_{O_2} is related to rise of Pa_{CO_2} by the gas exchange ratio (R) as shown by the following equation:

$$R = \frac{\dot{V}CO_2}{\dot{V}O_2} \simeq \frac{\dot{V}A \cdot FA_{CO_2}}{\dot{V}A(FI_{O_2} - FA_{O_2})^*}$$

which reduces to

$$R \simeq \frac{PA_{CO_2}}{PI_{O_2} - PA_{O_2}}$$

Thus when $R = 1$, the change in PA_{CO_2} and PA_{O_2} is the same; the change in Pa_{CO_2} is therefore equal to the change in Pa_{O_2}. However, as R normally is about 0.8 at rest, for each 10 mm Hg decrease in Pa_{O_2} due to hypoventilation, Pa_{CO_2} will normally increase by about 8 mm Hg.

It is important to recognize that alveolar hypoventilation can occur although the lung itself is normal such as in diseases affecting the medullary respiratory center, or respiratory neuromuscular diseases. Hypoventilation also is seen, however, in severe obstructive airways disease associated with high work of breathing.

DIFFUSION ABNORMALITY

As described in the section on diffusion, impaired diffusion from alveolus to pulmonary capillary blood can lead to arterial hypoxemia, which is commonly

---

*For clarity the normal small difference between inspired and expired alveolar ventilation is disregarded here.

exacerbated by high cardiac output states such as muscular exercise. As O_2 is loaded into the alveolus at an appropriate rate in this condition but fails to be loaded adequately into the blood, $(A-a)PO_2$ will be widened.

The CO_2 retention from diffusion impairment is rare. Any increase in Pa_{CO_2} that might ensue is normally corrected by ventilatory control mechanisms, which are exquisitely sensitive to Pa_{CO_2}. In contrast, ventilatory stimulation from low arterial PO_2 only becomes appreciable after Pa_{O_2} falls below approximately 60 mm Hg (see section on control of ventilation). Thus for moderate hypoxemia from diffusion impairment, Pa_{O_2} is decreased, $(A-a)PO_2$ is widened and Pa_{CO_2} can be normal. With more severe hypoxemia (i.e., leading to hypoxic stimulation of ventilation), Pa_{O_2} is low, $(A-a)PO_2$ is widened but now Pa_{CO_2} is low.

Right-to-Left Shunt

In normal subjects the bronchial circulation (to the larger airways) and thebesian vessels (to the myocardium) drain their venous blood into the arterialized blood, "downstream" of the pulmonary capillary bed. This venous blood consequently reduces mixed arterial blood PO_2 (Pa_{O_2}) to values less than pulmonary end-capillary blood. However, as this normal right-to-left shunt is only a small percentage of the cardiac output, this reduction in Pa_{O_2} is only a few mm Hg.

Cardiac and/or pulmonary disease can lead to marked increases in the fraction of the cardiac output that does not undergo gas exchange in the lung. Atelectic (collapsed) regions of the lung that still have pulmonary blood flow contribute to intrapulmonary right-to-left shunts, as do abnormal pulmonary artery-pulmonary venous fistulas. Intracardiac right-to-left shunts can also develop, such as through atrial or ventricular septal defects, if associated with abnormally high pressures in the right side of the heart. The reduction in arterial PO_2 from right-to-left shunts depends on several factors, e.g., (1) the shunt fraction of the cardiac output ($\dot{Q}s/\dot{Q}t$), (2) mixed venous O_2 content and (3) HbO_2 dissociation curve.

With a few assumptions, such as that pulmonary end-capillary blood has a uniform composition and that all blood that is "shunted" has a mixed venous O_2 content, the shunt fraction of cardiac output can be readily determined.

We know that total arterial O_2 flowing (per unit time) is made up of a pulmonary capillary component and a shunt component:

cardiac output $\times$ Ca_{O_2} = (pulmonary flow $\times$ end-capillary O_2 content)

$$+ \text{(shunt flow} \times C\bar{v}_{O_2})$$

$$\therefore \dot{Q}t \cdot Ca_{O_2} = (\dot{Q}_{pulm} \cdot Cc'_{O_2}) + (\dot{Q}s \cdot C\bar{v}_{O_2})$$

but

$$\dot{Q}_{pulm} = \dot{Q}t - \dot{Q}s$$

$$\therefore \dot{Q}t \cdot Ca_{O_2} = (\dot{Q}t \cdot Cc'_{O_2}) - (\dot{Q}s \cdot Cc'_{O_2}) + (\dot{Q}s \cdot C\bar{v}_{O_2})$$

or

$$\frac{\dot{Q}s}{\dot{Q}t} = \frac{Cc'_{O_2} - Ca_{O_2}}{Cc'_{O_2} - C\bar{v}_{O_2}}$$

In this equation the pulmonary end-capillary O_2 content (Cc'_{O_2}) is determined by assuming that end-capillary P_{O_2} is equal to the "ideal" alveolar P_{O_2} (PA_{O_2}), where PA_{O_2} is that alveolar P_{O_2} that would be seen if the lung exchanged gases ideally. (Clearly the true mean alveolar P_{O_2} would be extremely difficult to measure in a lung with maldistribution of $\dot{V}A/\dot{Q}$, for example.)

PA_{O_2} can be determined from the following equation:

$$PA_{O_2} = PI_{O_2} - \frac{Pa_{CO_2}}{R} + \left[Pa_{CO_2} \cdot FI_{O_2} \cdot \frac{(I - R)}{R} \right]^{*}$$

or, in simplified form, can be estimated as

$$PA_{O_2} = PI_{O_2} - \frac{Pa_{CO_2}}{R}$$

As CO_2 in the shunted blood is higher than in pulmonary end-capillary blood, right-to-left shunt also tends toward CO_2 retention. This is rarely observed, however, owing to the normally small venous to arterial P_{CO_2} difference (about 6 mm Hg) compared with O_2 (about 60 mm Hg) and also because the mechanisms of ventilatory control normally respond to increased P_{CO_2} to return it to normal. As stated earlier, if P_{O_2} falls appreciably owing to the right-to-left shunt, hypoxemia will stimulate ventilation by its effect on the peripheral chemoreceptors and can lead to reduced Pa_{CO_2}. Thus, moderate right-to-left shunt leads to reduced Pa_{O_2}, widened $(A-a)P_{O_2}$ but relatively normal Pa_{CO_2}. Severe right-to-left shunts cause markedly reduced Pa_{O_2}, wide $(A-a)P_{O_2}$, and also reduced Pa_{CO_2}.

$\dot{V}A/\dot{Q}$ INEQUALITY

Under conditions in which overall, or average, $\dot{V}A$ is approximately equal to overall, or average, $\dot{Q}$ (i.e., a normal average $\dot{V}A/\dot{Q}$), the lung may still have regions of high $\dot{V}A/\dot{Q}$, normal $\dot{V}A/\dot{Q}$, and low $\dot{V}A/\dot{Q}$ interspersed. The consequence of such a distribution is that blood from the high $\dot{V}A/\dot{Q}$ region will reflect hyperventilation, with high P_{O_2} and low P_{CO_2}; that blood from the normal $\dot{V}A/\dot{Q}$ regions will have normal values for P_{O_2} and P_{CO_2}; and that blood from the low $\dot{V}A/\dot{Q}$ region will reflect hypoventilation, with low P_{O_2} and high P_{CO_2}. The overall arterial P_{O_2} and P_{CO_2} will therefore result from averaging the total contents of the gas from each "stream" in proportion to the blood flow from each region.

However, the blood dissociation curve for O_2 is nonlinear. The result is that low $\dot{V}A/\dot{Q}$ regions lead to both low P_{O_2} and low O_2 content, whereas high

---

*The term in the square brackets normally accounts only for a few mm Hg (and is zero when R = 1) and so is often neglected in estimating PA_{O_2}. It should always be included, however, for precise analysis.

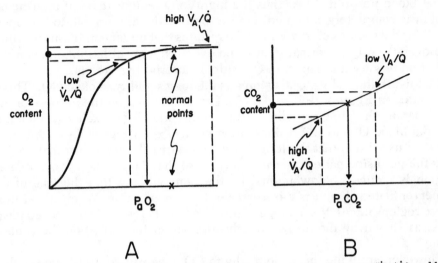

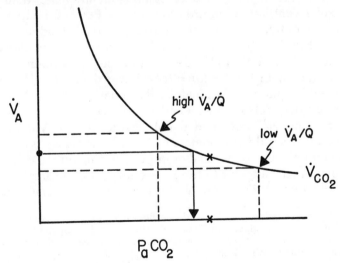

Fig 5–19. – **A,** effects of regions of high ventilation to perfusion ratio ($\dot{V}_A/\dot{Q}$) and low $\dot{V}_A/\dot{Q}$ in the lung on mean arterial O_2 content (●) and PO_2 (shown by *arrowhead on abscissa*). **B,** effects of regions of high $\dot{V}_A/\dot{Q}$ and low $\dot{V}_A/\dot{Q}$ in the lung on mean arterial CO_2 content (●) and PCO_2 (*arrowhead on abscissa*). x = normal value. For further discussion see text.

$\dot{V}_A/\dot{Q}$ regions lead to high PO_2, with only small increases of O_2 content above the normal value (because the O_2 dissociation curve is relatively flat in this range). Mixing blood from low $\dot{V}_A/\dot{Q}$ regions with blood of high $\dot{V}_A/\dot{Q}$ regions will therefore result in an average PO_2 that is "weighted" toward the low $\dot{V}_A/\dot{Q}$ blood value (Fig 5–19). The actual value for the mean PO_2 will depend on the volumes of blood from each "region" that stream into the mixed ar-

Fig 5–20. – Effects of regions of high ventilation to perfusion ratio ($\dot{V}_A/\dot{Q}$) and low $\dot{V}_A/\dot{Q}$ on mean alveolar PCO_2 (*arrowhead on abscissa*) and alveolar ventilation (●). x = normal values. See text for further discussion.

terial blood per unit time. Thus the high $\dot{V}_A/\dot{Q}$ regions (even if their hemoglobin is completely saturated) tend to be functionally unable to "compensate" for the regions of low $\dot{V}_A/\dot{Q}$. The result is that under conditions in which the overall $\dot{V}_A/\dot{Q}$ is normal (that is, no overall hypoventilation), uneven distribution of $\dot{V}_A$ with respect to $\dot{Q}$ results in arterial hypoxemia.

Maldistribution of $\dot{V}_A/\dot{Q}$ also impairs the lung's ability to clear CO_2. This is illustrated in Figures 5–19 and 5–20. Let us suppose, for the purpose of this explanation, that lung initially operates with "ideal" gas exchange such that arterial blood O_2 and CO_2 values are at points "x" in Figure 5–19, A and B. Now let us assume that the lung develops maldistribution of ventilation (but that the perfusion pattern remains normal) so that Pa_{O_2} of the hypoventilated region is reduced below normal by the same amount that the Pa_{O_2} of the hyperventilated region is raised above the normal value. When blood from these regions mixes, the average of the O_2 contents results in a reduced mean Pa_{O_2}, as shown by the arrows which originate at the dot (·) on the content scale.*

However, applying the same model to CO_2 (see Fig. 5–19,B), we see that, because of the relatively linear CO_2 dissociation curve,* the resulting mixed arterial P_{CO_2}, as shown by the arrow, is at or close to normal. But it is important to consider how the lung with such maldistribution of ventilation effects such changes of P_{CO_2}. The hyperventilated and the hypoventilated lung regions in our model are shown on the graphic representation (see Fig 5–20) of the equation

$$\dot{V}_{CO_2} = \dot{V}_A \cdot F_{A_{CO_2}}$$

or

$$\dot{V}_{CO_2} = \dot{V}_A \cdot \frac{P_{A_{CO_2}}}{P_B - 47}$$

It should be noticed that the increment of alveolar ventilation required to maintain P_{CO_2} of the hyperventilated region is much greater than the decrement of alveolar ventilation required to maintain P_{CO_2} of the hypoventilated region. Consequently, although the mixed arterial blood can be normal (points "x" in Fig. 5–20) in such a condition, the total alveolar ventilation must be increased (symbol • in Fig 5–20). A further consequence is that mean alveolar P_{CO_2} (arrow in Fig 5–20) will be lower than if the lung exchanged gas in an "ideal" manner.† Thus maldistribution of $\dot{V}_A/\dot{Q}$ results not only in a widened $(A-a)P_{O_2}$ but also in a widened $(a-A)P_{CO_2}$.

The pattern of arterial blood and alveolar gas tensions in $\dot{V}_A/\dot{Q}$ maldistribution is such that with mild or moderate maldistribution, Pa_{O_2} is low, $(A-a)$ P_{O_2} is widened and Pa_{CO_2} can be normal or low depending on the degree of ventilatory stimulation consequent to the hypoxemia (or other factors such as regional lung distortion). In severe $\dot{V}_A/\dot{Q}$ impairment, associated with severe

---

*Effects such as those induced by Haldane and Bohr effects are neglected in this example.

†For simplicity, we have neglected in Figure 5–20 the effect of "weighting" the mean alveolar P_{CO_2} further toward the high $\dot{V}_A/\dot{Q}$ value, as would actually be the case in this example because of the unequal ventilation of each region.

TABLE 5-7.—ALVEOLAR AND ARTERIAL BLOOD GAS
TENSIONS TYPICALLY ASSOCIATED WITH
CAUSES OF HYPOXEMIA

CAUSE OF HYPOXEMIA	Pa_{O_2}	$(A-a)Po_2$	Pa_{CO_2}	$(a-A)Pco_2$
Alveolar hypoventilation	↓	Normal	↑	Normal
Diffusion impairment	↓	↑	Normal or ↓	Normal or ↑
Right-to-left shunt	↓	↑	Normal or ↓	↑
Maldistribution of $\dot{V}A/\dot{Q}$	↓	↑	Normal, ↓ or ↑	↑

airways obstruction, hypoventilation can ensue, owing to the increased work
of breathing, and cause increased Pa_{CO_2}. This, of course, reduces Pa_{O_2} even
more.

It should be noted that of the four independent causes of arterial hypoxemia
at sea level, only hypoventilation is inevitably associated with CO_2 retention
and it is also the only cause that does not lead to increased $(A-a)Po_2$ (Table
5–7). The $(A-a)Po_2$ might, therefore, be considered the best single indicator
of the efficiency of the lung for gas exchange, as (1) diffusion impairment,
(2) right-to-left shunts and (3) maldistribution of $\dot{V}A/\dot{Q}$ all widen the gradient.

CONTROL OF VENTILATION

The inherent rhythmicity of breathing is determined by specialized clusters of neurons ("centers") in the brain stem. The centers in the medulla are
absolutely essential for maintenance of respiratory rhythmicity; section
through the most caudal region of the medulla leads to apnea (cessation of
breathing). The centers in the pons modify breathing by influencing the
intrinsic pattern of discharge established by the medullary respiratory neurons.

MEDULLARY CENTER

The cyclic pattern of inspiration and expiration results from the sequential
firing of groups of inspiratory and expiratory neurons within the medullary
reticular formation. When the inspiratory group fires, inspiration is effected by
phrenic and intercostal nerve discharge and at the same time the expiratory
cells are inhibited. Expiratory nerve activity then builds up and inhibits the
firing of the inspiratory neurons as inspiratory activity fades, thereby leading
to expiration. The functional separation of inspiratory and expiratory neurons
has led to the terms inspiratory and expiratory "centers." This classification,
however, is functional rather than anatomical, since inspiratory and expiratory
neurons are interspersed in each region (Fig 5–21). Regions with predominantly inspiratory neurons are located bilaterally in the more ventral aspects

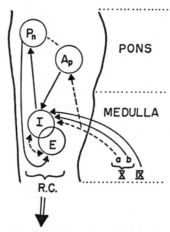

Fig 5–21.—Schematic representation of organization of brain stem respiratory neurons. P_n = pneumotaxic center, A_p = apneustic center; I and E = inspiratory and expiratory regions, respectively, of the medullary respiratory center *(RC)*. IX = input from the glossopharyngeal nerve, i.e., carotid bodies; X = vagal input from *(a)* pulmonary stretch receptors and *(b)* aortic bodies. To avoid further complexity, influences of inputs from cranial nerves IX and X on the medullary expiratory neurons have been omitted. *Solid arrows* = stimulatory influences; *dashed arrows* = inhibitory influences; *double arrow* = efferent motor pathways to muscles of breathing.

of the medullary reticular formation. Those with predominantly expiratory neurons are located bilaterally somewhat caudal and dorsal to the inspiratory center.

PONTINE CENTERS

The coordinated pattern of normal breathing involves the influences of pontine respiratory neurons on the medullary center.

1. *Apneustic center.*—This is situated in the middle and caudal pons. Sectioning the brain stem at the caudal pons results in *apneustic* breathing (i.e., large inspiratory gasps or even inspiratory "spasm") if the vagi also have been sectioned. With vagi intact, such sectioning does not lead to apneustic breathing but rather a more erratic pattern than normal. Thus the influence of the apneustic center on the medullary center is restrained by vagal afferents (predominantly stretch receptors from the lung) and also by neural information from the "pneumotaxic center."

2. *Pneumotaxic center.*—This is located in the upper pons. Stimulation of this region leads to accelerated respiration. It is thought to function by receiving impulses from the medullary inspiratory center, leading to stimulation of pneumotaxic efferents to the medullary expiratory center and inhibition of the apneustic center, which also functions to end inspiration. When this influence is sufficient, expiration begins.

When we speak of a brain stem respiratory "center," we are really referring to a highly complex interrelationship of nerve cells in both medulla and pons, the above being a highly simplified account of the general characteristics.

REGULATION OF RESPIRATORY CENTER ACTIVITY

The inherent respiratory pattern is modified from moment to moment by neural inputs to the medullary center from (1) the cerebral cortex, (2) limbic system and hypothalamus, (3) central chemosensitive regions, (4) peripheral chemoreceptors, (5) the lung, (6) proprioceptors and (7) other influences such as pain.

1. CORTICAL INFLUENCES.—Respiration can be altered volitionally. However, any such increase in ventilation that is out of proportion to metabolic CO_2 production leads to hypocapnia. Similarly, volitional reductions of ventilation, of which breath-holding is the extreme case, leads to hypercapnia and hypoxia.

2. LIMBIC AND HYPOTHALAMIC INFLUENCES.—The limbic system and the hypothalamus have important roles in affective behavior, e.g., anger, rage, fear, which often have respiratory manifestations. Although the responses described above for cortical stimulation to ventilation apply in general to these influences, both metabolic (e.g., muscular) and cardiovascular stimulation also occur in association with these behavioral changes.

3. CENTRAL CHEMOSENSITIVE REGIONS.—Chemosensitive regions are found bilaterally on the ventral-lateral aspects of the medulla in an area approximately bounded by the pontine-medullary junction, the pyramidal tracts and cranial nerve roots 7–11 (although other regions such as the floor of the fourth ventricle also demonstrate chemosensitivity). These areas are deliberately termed regions of chemosensitivity rather than chemoreceptors, as no morphologically discrete chemoreceptor cells have been demonstrated. Application of acidic materials to these regions, however, causes immediate ventilatory stimulation, whereas alkaline solutions or procaine depress ventilation.

These ventrolateral regions can be considered to subserve chemoreceptor function and appear to be functionally distinct from the deeper-lying respiratory center in the medullary reticular formation (although there is by no means unanimity on this point among investigators in the field).

As the chemosensitive regions are located on the surface of the medulla, they are exposed to CSF, which has very low buffering capacity. They are separated from the blood by the blood-brain barrier, which is relatively impermeable to charged molecules such as H^+ and HCO_3^- but which readily allows transfer of CO_2.

Mechanism of stimulation.—The medullary chemosensitive regions should be considered H^+ "receptors." Increased CSF H^+ augments and decreased H^+ reduces ventilation. These regions also are indirectly affected by CO_2 because of its H^+-forming action. For example, if arterial blood P_{CO_2} increases, CO_2 will enter the CSF and increase $[H^+]$, which stimulates $\dot{V}_E$. Hypoxia per se does not stimulate these central regions. In fact, in the absence of aortic and carotid bodies (the hypoxic ventilatory chemoreceptors), ventilation is depressed at low P_{O_2}. Hypoxia can affect ventilation indirectly by a central action. At very low P_{O_2}, cerebral tissue anaerobiosis can ensue, leading to CSF lactic acidosis, which affects ventilation via H^+.

4. PERIPHERAL CHEMORECEPTORS.—A chemoreceptor has the ability to sense changes in the chemical environment of its fluid milieu and encode this information into patterns of neural (or humoral) discharge to influence some function.

The peripheral chemoreceptors that affect ventilation are carotid and aortic bodies.

a. CAROTID BODIES.—These are discrete structures near the bifurcation of the common carotid into the internal and external carotid arteries. They transmit information to the medullary respiratory center via the carotid sinus nerve (nerve of Hering) and thence via the glossopharyngeal nerve (cranial nerve IX), causing increased ventilation from both increased tidal volume and breathing frequency. Hering's nerve, however, also contains efferent fibers to the carotid bodies, which can modulate the effect of a given carotid body stimulus on the afferent neural discharge pattern. In addition, sympathetic nerves to the carotid bodies from the superior cervical ganglion have an important role in modifying blood flow to these organs.

The carotid bodies have a very high metabolic rate per gram of tissue, with an O_2 uptake of about 10 ml/100 gm per minute.* However, its blood flow is probably the greatest of any organ in the body per gram of tissue, being about 2,000 ml/100 gm per minute. Hence under normal conditions its arteriovenous O_2 difference is so extremely small that the metabolic needs of the carotid bodies can be subserved predominantly by dissolved O_2, with little HbO_2 desaturation.

Physiologic stimuli to carotid bodies.—The carotid bodies are stimulated by reduced Po_2, by increased Pco_2 and H^+, by decreased carotid blood flow (e.g., by sympathetic vasoconstriction of the carotid arterioles, or hemorrhagic hypotension), by high blood temperature and by chemicals such as sodium cyanide and lobeline. When more than one of these stimuli are present simultaneously, they interact in a complex manner. Possibly the most important interaction among stimuli is that hypoxia potentiates the effects of Pco_2 or H^+ and vice versa. This will be discussed in the section describing the ventilatory response to these stimuli.

b. AORTIC BODIES.—These have less discrete localization than have carotid bodies, being scattered around the arch of the aorta, especially between the arch and pulmonary artery, and also between the subclavian and carotid arteries.

The aortic bodies are responsive to the same stimuli as are carotid bodies but transmit information to the medullary respiratory center via the vagus nerve (cranial nerve X), leading to increased ventilation from increases in both tidal volume and breathing frequency. The relative importance of aortic bodies varies with the species involved but is generally appreciably less than that of carotid bodies. In cat and dog, the aortic bodies may account for 10–20% of peripheral chemoreceptor-induced hyperpnea, whereas in the rabbit they have little or no influence on ventilation. In man, the aortic bodies

---

*To provide a frame of reference, this value approximates the highest recorded O_2 uptake for a highly trained athlete exercising to exhaustion.

appear to have no influence on ventilation or, if they do, it is trivially small compared with carotid bodies.

5. THE LUNG.—Neural information from lung receptors is transmitted to the brain stem in large part via vagal afferents. It is convenient to consider the lung reflexes in three categories: (1) stretch receptor, (2) irritant receptor and (3) type J (juxtapulmonary-capillary) receptor.

a. STRETCH RECEPTOR.—These reflexes are important in establishing the pattern of breathing. Inflation of the lung leads to inhibition of inspiratory activity (Hering-Breuer reflex). This inhibition is abolished by cervical vagotomy, which consequently leads to deeper and slower breathing. The reflex originates in slowly adapting pulmonary stretch receptors located in airway (predominantly bronchial) smooth muscle. Although important in some animals, the inflation reflex tends to be relatively weak within the normal tidal volume range in man.

b. IRRITANT RECEPTORS.—Inhalation of irritant materials such as noxious gases or dust leads to stimulation of rapidly adapting irritant receptors in the epithelium of the large, predominantly extra-pulmonary airways. These receptors can also be stimulated mechanically, resulting in hyperpnea, bronchoconstriction and often coughing. Both pulmonary stretch and irritant reflex information is transmitted via myelinated vagal afferents.

c. TYPE J RECEPTORS.—Sensory (presumably) nerve endings of small, unmyelinated vagal afferents have been demonstrated, although somewhat sparsely distributed, in the alveolar-capillary interstitial space. They are thought to be stimulated by interstitial distortion, such as at very low lung volumes, or congestion, and also by small pulmonary emboli. Type J receptors are often demonstrated experimentally by injection of nonphysiologic chemical agents (e.g., phenyldiguanide), which stimulate them.

The J reflex is characterized by rapid, shallow breathing, bradycardia and hypotension. These effects are abolished by bilateral cervical vagotomy.

6. PROPRIOCEPTORS.—Large myelinated afferents from muscle spindles and articular proprioceptors can cause hyperpnea when appropriate electric stimuli are applied. Transmission along these nerve pathways of impulses initiated by the movements of muscles and joints may account for part of the hyperpnea of muscular exercise, although the issue has not been resolved to any certain degree. Stimulation of smaller unmyelinated fibers ("C" afferents) from muscle are also thought to be involved.

7. OTHER INFLUENCES.—Pain can provide potent stimulation of ventilation, leading to marked hypocapnia and alkalosis. Increased temperature stimulates $\dot{V}E$ via skin, carotid body and hypothalamic mediation. Decreased temperature has the opposite effect, except when shivering ensues. Shivering usually leads to hyperpnea because of the increased metabolic rate and, possibly, the shivering motion. Increases in arterial blood pressure tend to reduce ventilation (whereas decreases in pressure stimulate ventilation). A large part of this effect is due to the reduced peripheral chemoreceptor discharge caused by the increased pressure, but part of the response is probably owing to baroreceptor (aortic and carotid sinus) stimulation.

Control of ventilation, therefore, involves complex interactions of afferents

TABLE 5–8.—SIMPLIFIED SCHEME OF NEURAL AND HUMORAL STIMULI TO VENTIL.

Phrenic nerves
Intercostal nerves

Medullary-pontine respiratory "center"

	Cortex: motor efferents	Subcortex: hypothalamus, limbic system	Peripheral nerves: sensory afferents—muscle spindles, articular proprioceptors, pain afferents	Lung (vagal) afferents: Irritant receptors (embedded in epithelium of predominantly larger airways and larynx)	Stretch receptors (embedded in bronchial smooth muscle of predominantly medium-sized airways)	"j" receptors (free nerve endings in the alveolar capillary interstitium)	Peripheral chemoreceptors: aortic and carotid bodies	Central "chemoreceptors": ventrolateral surfaces of medulla
SITE								
STIMULI	Volition	Emotion: fear, etc.	Movement, pain, etc.	Local mechanical or chemical irritation	Lung inflation	Distortion of interstitium (edema, emboli or mechanical distortion, etc) or chemical "irritation"	Low Pa_{O_2}, High Pa_{CO_2} → potentiation; High arterial [H+] → additive effect; Low PO_2 → potentiation	High CSF [H+] (High CSF Pco_2?)
SPECIAL REMARKS	All exclusive neural stimulation to ventilation leads to hypocapnia and acute respiratory alkalosis. (I.e., increased ventilation without a simultaneous increase in metabolic rate must lead to depletion of the body's CO_2 stores. Hence, the reduced Pa_{CO_2} and $PCSF\ CO_2$ and the acute arterial and CSF alkalosis.)			Adapt rapidly to stimulation; reflex effects: cough or tachypnea, bronchoconstriction; myelinated afferent fibers.	Adapt slowly to inflation; reflex effects: inhibition of inspiration, mediate Hering-Breuer inflation reflex; myelinated afferent fibers.	Stimulation can lead to rapid, shallow breathing (with intense stimulation, apnea results), also hypotension and bradycardia; nonmyelinated afferent fibers.	Low Pa_{O_2} not an effective stimulus to breathing until $Pa_{O_2} \approx 60$mm Hg (at normal Pa_{CO_2}). Even less effective if Pa_{CO_2} and/or [H+] is low.	Low Pa_{O_2} and $PCSF\ O_2$ should be regarded as depressant to central nervous system. CO_2 equilibrates rapidly between blood and CSF. [H+] and [HCO3-] equilibrate much more slowly.

from numerous receptor sites in the body (Table 5–8). Accordingly, care must always be taken when attempting to interpret the responses to "a" stimulus, especially when the ventilatory response itself alters the intensity of other stimuli that affect breathing.

VENTILATORY RESPONSES TO HYPOXIA

Inhalation of an hypoxic gas mixture, in normal man, leads to an increase in ventilation. However, as a consequence of this ventilatory stimulation, arterial Pco_2 is lowered. Therefore, the response is not simply to hypoxia but rather to hypoxia with concomitant hypocapnia. And, since normal subjects vary in their responsiveness to CO_2, it is difficult to discern the specific chemoreceptor-induced response under these conditions. It is common, therefore, to add CO_2 to the inspired gas, as appropriate, to maintain constancy of alveolar and arterial Pco_2 during the hypoxic test.

When performed at several different levels of PA_{O_2} a hypoxic ventilatory response curve is constructed (Fig 5–22), the shape of which is hyperbolic. It is clear from this relationship that, at normal spontaneous levels of Pco_2, hypoxia is a poor stimulus to ventilation until Pa_{O_2} reaches about 60 mm Hg. As Po_2 falls below this value, ventilation increases more and more rapidly. When PA_{CO_2} is maintained at 40 mm Hg during this test, $\dot{V}E$ begins to increase at a somewhat higher PA_{O_2}.

Fig 5–22.—Effect of arterial Po_2 on ventilation. Note hyperbolic relationship. At progressively higher arterial Pco_2, arterial hypoxia becomes progressively a more potent stimulus to breathing.

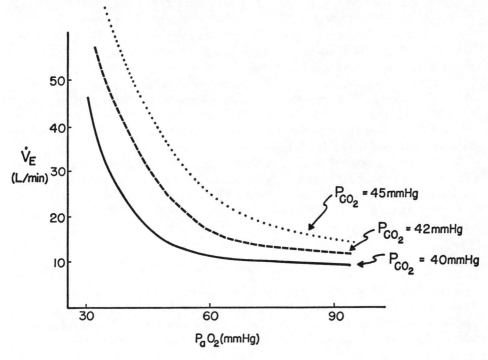

If the test is repeated, with P_{CO_2} (and/or H^+) again maintained constant throughout but at an increased level, hypoxia becomes a more potent ventilatory stimulus. The higher the P_{CO_2} during the test the more potent a stimulus hypoxia becomes.

When P_{CO_2} is lower than normal throughout the test, the ventilatory response to hypoxia is lower than normal and does not become significant until low levels of P_{O_2} are reached.

The peripheral chemoreceptors are the exclusive mediators of this ventilatory response to hypoxia but there is a wide variation in responsiveness even in ostensibly normal subjects. Chronic hypoxia, such as in man born, or resident for decades, at high altitudes markedly attenuates peripheral chemosensitivity to hypoxia.

Another useful, and very specific, test of peripheral chemosensitivity is to induce hypoxemia by inhalation of a hypoxic gas mixture (great care should be taken to ensure that the subject is not already hypoxemic before this is done). The hypoxic drive is then abruptly abolished by inhalation of 100% O_2 for two or more breaths. The $\dot{V}E$ then decreases (by an amount that is an index of the hypoxic drive at that P_{O_2}) after a short delay of one or two breaths, i.e., the transit time to the peripheral chemoreceptors. As the maximum decrease in $\dot{V}E$ typically occurs before the altered CO_2 (owing to this fall in $\dot{V}E$) can cause a central modifying effect, the test is quite specific as an index of hypoxic ventilatory sensitivity.

Four kinds of hypoxia can influence the peripheral chemoreceptors.

1. Hypoxic hypoxia, i.e., reduced arterial P_{O_2}. The ventilatory responses were described above.

2. Anemic hypoxia, i.e., reduced HbO_2 despite possibly normal Pa_{O_2}, e.g., owing to anemia or effects of CO. Ventilation is typically not stimulated by this mechanism until very low HbO_2 saturation (possibly resulting in metabolic acidosis). This is good evidence that the specific stimulus to $\dot{V}E$ is reduced Pa_{O_2} rather than HbO_2 saturation.

3. Stagnant hypoxia, consequent to reduced blood flow. Stagnant hypoxia of the peripheral chemoreceptors leads to increased ventilation. It has been suggested that this stimulation may result from reduced flow, washing out less of the organ's metabolic end-products such as CO_2 and H^+, hence a form of "autostimulation" of the carotid bodies.

4. Histotoxic hypoxia, i.e., inhibition of the enzymes of the tissue respiratory electron transport chain. This limits, or even abolishes, oxidative phosphorylation in these cells. Injection of sodium cyanide, which inhibits cytochrome oxidase, into the common carotid artery is often used to demonstrate ventilatory and cardiovascular responses to carotid body stimulation.

VENTILATORY RESPONSES TO CO_2

Inhalation of a gas mixture rich in CO_2 causes ventilatory stimulation. However, unlike the hyperbolic response curve to reduced P_{O_2}, the CO_2 response curve, which is constructed by plotted $\dot{V}E$ in the steady state at several elevated levels of PA_{CO_2} or Pa_{CO_2} is linear. In normal man this steady-state re-

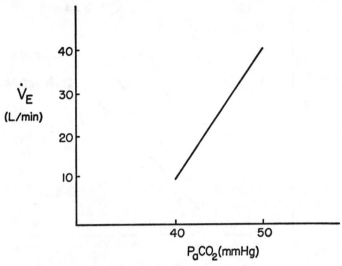

Fig 5–23.–Effect of arterial P_{CO_2} on ventilation. Note linearity over the range of P_{CO_2}.

sponse curve usually has a slope of about 3 L per minute per mm Hg (Fig 5–23) or 1.5 L per minute per mm Hg per square meter of body surface area.

The CO_2-induced hyperpnea is mediated by both peripheral and central chemoreceptors, but at normal arterial P_{O_2} the central component accounts for ≈80% of the response. Thus the absence of peripheral chemoreceptors, which abolishes ventilatory response to hypoxia in man, does not appreciably impair responsiveness to CO_2 at normal Pa_{O_2}.

Fig 5–24.–Effect of simultaneous metabolic acidosis or alkalosis on ventilatory response to arterial P_{CO_2}. Note that these conditions shift the CO_2 response curve, as shown, without appreciable change of slope.

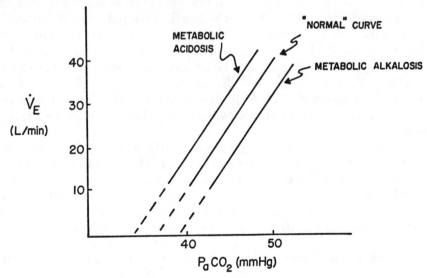

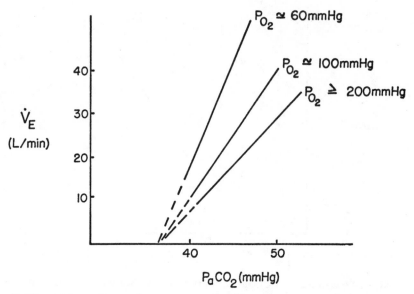

Fig 5–25.—Effect of simultaneous arterial hypoxia on ventilatory response to CO_2. Note that slope of CO_2 response curve is increased in proportion to degree of hypoxia. The zero ventilation intercept is not appreciably affected by the changes in PO_2.

Carbon dioxide stimulates ventilation largely through its H^+-forming properties. It is not surprising, therefore, that altered H^+ in arterial blood and CSF, induced by metabolic acidosis, provides additive stimulation to ventilation. This is shown in Figure 5–24, where the CO_2 response curve, associated with metabolic acidosis, is shifted to the left of the control curve but has the same slope, i.e., the increment of $\dot{V}E$ induced by the metabolic acidosis is constant at all levels of PCO_2, hence the additive effects of the stimuli. Metabolic alkalosis, by the same token, shifts the curve to the right without a change of slope. Thus, although $\dot{V}E$ is higher than normal at any given level of PCO_2 when there is concomitant metabolic acidosis (and lower when there is metabolic alkalosis), if we consider the change of $\dot{V}E$ to be induced by a change in PA_{CO_2} (i.e., $\Delta\dot{V}E/\Delta PCO_2$ to be an index of "sensitivity" to CO_2), the sensitivity is the same in nese conditions. It is useful therefore to consider that metabolic acidosis and alkalosis change the zero ventilation intercept of the CO_2 response curve (sometimes called a "threshold") but do not affect the "sensitivity."

In contrast, simultaneous hypoxia or hyperoxia changes the sensitivity to CO_2 but does not appreciably affect this threshold (Fig 5–25). Hypoxia therefore potentiates the effect of CO_2, and the peripheral chemoreceptors are the site of this potentiating effect. It has been shown that this characteristic "fan" of CO_2 response curves at various levels of Pa_{O_2} can be demonstrated in the total impulse traffic up Hering's nerve as well as in single afferent fibers of the nerve.

We should not, however, be led to believe that, in conditions associated with simultaneous hypoxia and hypercapnia, $\dot{V}E$ will always be markedly high. This occurs when such a test is performed on normal man. Patients with severe airways obstructive disease or primary alveolar hypoventilation syndrome, for example, can be both hypoxic and hypercapnic at rest. When a CO_2 response test is performed on such subjects, they almost invariably demonstrate low ventilatory sensitivity to CO_2.

Finally, it should always be remembered that these tests only reflect the overall ventilatory response to the stimulus. They are not specific tests of chemoreceptor responsiveness. For example, chemoreceptor responsiveness could be normal and yet the ventilatory response to CO_2 be low if high airway resistance, or respiratory neuromuscular disease, were to provide mechanical impairment to ventilation.

ROLE OF VENTILATION IN ACID-BASE REGULATION

Carbon dioxide in the blood is part of an equilibrium reaction involving both pH and HCO_3^-, i.e.,

$$CO_2 + H_2O \overset{*}{\rightleftharpoons} H_2CO_3 \rightleftharpoons H^+ + HCO_3^-$$

where the asterisk represents the reaction that is catalyzed by the enzyme carbonic anhydrase. Hence from the law of mass action, the dissociation constant K of the reaction

$$H_2CO_3 \rightleftharpoons H^+ + HCO_3^-$$

is given by

$$K = \frac{(H^+) \times (HCO_3^-)}{(H_2CO_3)}$$

However, as H_2CO_3 is (1) present in the blood in such small concentrations and (2) its concentration is proportional to that of dissolved CO_2, we can also write

$$K' \text{ (note: value different from } K \text{ above)} = \frac{(H^+) \times (HCO_3^-)}{(CO_2)}$$

If we now take logarithms of each side

$$\log K' = \log H^+ + \log \frac{(HCO_3^-)}{(CO_2)}$$

or

$$-\log H^+ = -\log K + \log \frac{(HCO_3^-)}{(CO_2)}$$

but as $pH = -\log H^+$ and $pK' = -\log K'$ by definition, then:

$$pH = pK' + \log \frac{HCO_3^-}{\alpha P_{CO_2}}$$

where α is the solubility coefficient, relating P_{CO_2} in mm Hg to CO_2 in mM/L, having a value in plasma at 38 C of 0.03. This means that pH will be defined by the ratio of HCO_3^- to αP_{CO_2}. Any normal ratio of HCO_3^- to αP_{CO_2} will give a normal pH regardless of whether both HCO_3^- and αP_{CO_2} are normal, both are proportionally low or both are proportionally high.

The normal ratio of HCO_3^- to αP_{CO_2} is 20:1, i.e., normal plasma HCO_3^- is 24 mM per L and P_{CO_2} is 40 mm Hg or $0.03 \times 40 = 1.2$ mM per L,

$$\frac{HCO_3^-}{\alpha P_{CO_2}} = \frac{24}{1.2} = \frac{20}{1}$$

However, as $pK' = 6.1$ in plasma:

$$\begin{aligned} pH &= 6.1 + \log 20 \\ &= 6.1 + 1.3 \\ &= 7.4 \end{aligned}$$

A major function of ventilation is to control Pa_{CO_2} in proportion to the current HCO_3^- level so that pH remains at or close to normal.

If HCO_3^- is acutely decreased, e.g., by sudden increase in concentration of lactic acid, this acid-base condition is termed *acute metabolic acidosis*. Ventilation is then rapidly stimulated by the low pH, predominantly by peripheral chemoreceptor stimulation, resulting in a decrease in Pa_{CO_2} until the ratio $HCO_3^-/\alpha P_{CO_2}$ is returned to normal, hence normal pH. The sequence in such a condition can be characterized as follows:

1. ACUTE METABOLIC ACIDOSIS. — HCO_3^- decreased, pH decreased but initially no ventilatory increase, hence no P_{CO_2} decrease. The magnitude of the task is shown by the fact that the body produces 20,000–25,000 mM CO_2 per day (compared with the 50–100 mM of fixed acids produced daily and eliminated by the kidneys).

2. METABOLIC ACIDOSIS WITH PARTIAL RESPIRATORY COMPENSATION. — HCO_3^- decreased, pH still low but not as low as (1), due to $\dot{V}E$ increasing, causing Pa_{CO_2} to fall.

3. METABOLIC ACIDOSIS WITH FULL RESPIRATORY COMPENSATION. — [*] HCO_3^- decreased, but pH is now normal due to $\dot{V}E$ increasing to a level at which αP_{CO_2} is decreased in proportion to reduced HCO_3^-.

Similarly, if HCO_3^- is acutely increased, the sequence will be (a) *acute metabolic alkalosis:* HCO_3^- high, pH high, $\dot{V}E$ initially unchanged, hence no Pa_{CO_2} increase; (b) *metabolic alkalosis with partial respiratory compensation:* HCO_3^- high, pH still high, but not as high as in (1) as $\dot{V}E$ decreases, causing Pa_{CO_2} to rise; and (c) *metabolic alkalosis with full respiratory compensation:* HCO_3^- high, but pH now normal, as $\dot{V}E$ is decreased to a level such that αP_{CO_2} is increased in proportion to the increased HCO_3^-.

Ventilatory control therefore occupies a central role in acid-base regulation consequent to metabolic disturbances.

However, respiratory control mechanisms also can be the *cause* of respiratory acid-base disturbances. If, for example, Pa_{CO_2} were allowed to rise due

---

[*]The sequence is being suggested in this section. We are not asserting that fully compensated acid-base conditions always occur in practice.

to ventilation being inappropriately low for CO_2 production rate, the ratio $HCO_3^-/\alpha PCO_2$ would be low and hence pH would be low. This condition, *acute respiratory acidosis*, is compensated for by retention of bicarbonate, a process that may take several days to complete.

Similarly, acute reduction in Pa_{CO_2} consequent to hypoxemia, for example, induces *acute respiratory alkalosis*. Renal mechanisms compensate by increased excretion of HCO_3^-, thereby reducing plasma HCO_3^-.

EXAMPLES OF INTEGRATIVE ASPECTS OF RESPIRATORY CONTROL

1. RESPONSES TO ACUTE METABOLIC ACIDOSIS.—The sequence described above for respiratory compensation for metabolic acidosis involves rather complex interactions. For example, if metabolic acidosis is induced, pH of the arterial blood will be low, as stated, but pH of the CSF will rise. The reason is that the blood-brain barrier is relatively impermeable to H^+, especially with acute changes. Thus initially the blood is acidic but the CSF is not. However, as VE is increased owing to H^+ stimulation of the peripheral chemoreceptors, Pa_{CO_2} is reduced. This causes CO_2 to diffuse out of the CSF into the blood until a new CO_2 equilibrium is established. The resulting acute respiratory alkalosis in the CSF will restrain the full ventilatory effect of the arterial acidosis. As pH of the CSF is returned to normal by reduction of CSF HCO_3^-, possibly as a result of active transport, this restraining influence of the CSF alkalosis is removed. VE increases further, therefore, in this phase, as the full effect of the arterial blood metabolic acidosis is "allowed" to be manifest.

2. ASCENT TO HIGH ALTITUDE.—When a normal resident at sea level ascends to high altitude, hypoxic stimulation of the peripheral chemoreceptors leads to hyperventilation. The low CO_2 in the arterial blood also causes a low PCO_2 in the CSF. The full ventilatory effects of the hypoxemia are restrained by alkalosis of both blood and CSF. As HCO_3^- in the blood and CSF are lowered as compensatory mechanisms for the acute respiratory alkalosis, the pHs return toward normal. This relieves part of the "restraint" on the full ventilatory response to hypoxemia. Consequently, ventilation continues to increase over a period of weeks following the ascent.

Subjects born, or resident for decades, at high altitude have little ventilatory response to hypoxia owing to attenuated peripheral chemoreceptor hypoxic sensitivity.

3. HYPERVENTILATION PRIOR TO UNDERWATER SWIMMING.—It is well known that the duration of a breath-hold can be prolonged if it is preceded by a period of hyperventilation. This is due to the washout of large quantities of CO_2 from body stores and to reduction of PCO_2 and H^+ in both blood and CSF to levels where there is little or no ventilatory stimulating effect. The amount of additional O_2 loaded into the blood, however, is small due to the relatively flat slope of the HbO_2 dissociation curve in this region. But Pa_{O_2} will be increased from the hyperventilation. Such hyperventilation prior to underwater swimming is a dangerous procedure, as it is possible for the reduced O_2 transfer to the brain to cause loss of consciousness before appreciable ventilatory stimulation occurs.

The lack of ventilatory stimulation is due to the fact that, while the subject is underwater, the metabolic CO_2 gets loaded into the reduced CO_2 stores and, thus, P_{CO_2} builds up only slowly toward levels where ventilatory stimulation occurs. Metabolic O_2 uptake, however, can markedly reduce the O_2 content and P_{O_2} of the blood. But under the conditions of low P_{CO_2}, the hypoxemia does not provide significant ventilatory stimulation of the peripheral chemoreceptors until P_{O_2} is at very low levels, in fact, possibly at levels at which some central ventilatory depressant effects may be becoming manifest.

The O_2 transfer to the brain is impaired owing to at least three important factors: (1) reduced cerebral blood flow consequent to hypocapnia; (2) reduced O_2 content of the blood resulting from metabolic O_2 consumption; and (3) a left shift of the HbO_2 dissociation curve, i.e., O_2 dissociates from Hb less readily because of the reduced P_{CO_2} and H^+ in the blood.

Although some developing metabolic acidosis may complicate this simplified account, it is apparent that the reduced O_2 transfer to the brain in the face of little or no ventilatory stimulation, under these conditions, provides a potentially lethal combination.

RESPIRATION DURING MUSCULAR EXERCISE. — Muscular exercise can be considered an energetic process in which the chemical energy of ingested food is transformed into the mechanical energy of muscular contraction and work. However, the energy liberated in the catabolism of food cannot be used directly for the energetic requirements of contracting muscle. Rather, ATP is an obligatory intermediary in this process, with creatine phosphate serving a major storage function, preserving intramuscular ATP concentration close to normal, resting values during exercise.

Oxygen has an essential role in cellular energetics in allowing ATP to be formed aerobically within the mitochondria. This results in CO_2 and H_2O as readily excretable end-products. When O_2 availability to the contracting muscle is impaired, or is available but not used,* ATP continues to be formed in anaerobic reactions within the cytoplasm and lactic acid is formed as an additional end-product.

Aerobic metabolism. — Carbohydrate (e.g., glucose): $C_6H_{12}O_6 + 6\ O_2 \rightarrow$ $6\ CO_2 + 6\ H_2O + 38\ ATP$ or (2) fatty acids (e.g., palmitic acid): $C_{16}H_{32}O_2 + 23O_2 \rightarrow 16\ CO_2 + 16\ H_2O + 130\ ATP$.

Anaerobic metabolism. — Carbohydrate (e.g., glucose): $C_6H_{12}O_6 \rightarrow 2\ C_3H_6O_3$ (i.e., 2 lactic acid molecules) $+ 2\ ATP$. The lactic acid formed in these anaerobic reactions is virtually completely dissociated at the pH of body fluids, being predominantly buffered by the bicarbonate system.

In the steady state of moderately intense dynamic exercise† i.e., work rates associated with no significant increase in blood lactate – minute ventilation increases linearly with increasing $\dot{V}_{O_2}$ (Fig 5–26) with the ratio $\dot{V}_E/\dot{V}_{O_2}$, the ventilatory equivalent for O_2, having a value of about 25. Similarly $\dot{V}_E$ increases linearly with increasing $\dot{V}_{CO_2}$ (see Fig 5–26). But as the gas exchange

---

*For example, a muscle fiber with relatively few mitochondria and/or relatively low oxidative enzyme capacity might be expected to have a maximal rate of O_2 utilization, lower than the average of all contracting fibers in the muscle, despite presumably similar O_2 delivery.

†We shall not consider isometric, straining exercise here, as it is commonly associated with volitional apnea rather than with spontaneous breathing.

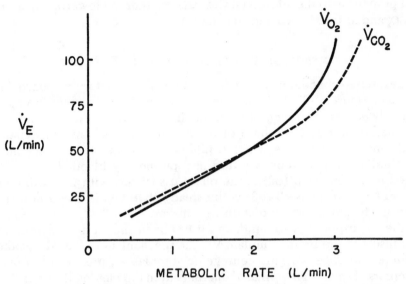

Fig 5–26.—Ventilatory response to increased metabolic rate during exercise. Note the relatively linear relationship of $\dot{V}_E$ and $\dot{V}_{O_2}$ at the lower work rates that engender metabolic acidosis. The $\dot{V}_E$–$\dot{V}_{CO_2}$ relationship tends to be "linear" over a wider range of metabolic rate than for O_2. See text for discussion.

ratio R (i.e., $\dot{V}_{CO_2}/\dot{V}_{O_2}$) is typically less than 1 at these work rates, $\dot{V}_E/\dot{V}_{CO_2}$ is about 28. It is important to recognize that, although $\dot{V}_E/\dot{V}_{O_2}$ and $\dot{V}_E/\dot{V}_{CO_2}$ are not appreciably different at these different exercise levels, the value of both functions is lower than at rest. This is because the physiologic dead space fraction of each breath is less during exercise; i.e., the lung is more efficient at subserving gas exchange during exercise than at rest. Consequently, both minute ventilation and alveolar ventilation increase linearly with increasing metabolic rate. Mean alveolar P_{O_2} and P_{CO_2} are thereby regulated at, or close to, resting levels in normal man at these work rates, although P_{CO_2} is the more precisely regulated variable. Mean arterial blood gas and acid-base status are also normal in such moderate exercise.

At slightly higher work rates, a metabolic (lactic) acidosis ensues and stimulates ventilation via the carotid bodies. $\dot{V}_E$, and therefore $\dot{V}_A$, increases at a faster rate than does $\dot{V}_{O_2}$, causing alveolar P_{O_2} to rise. However, as described above, the buffering of the lactic acid by, predominantly, $NaHCO_3$ results in an evolution of additional CO_2. Thus, the proportional increase in $\dot{V}_{CO_2}$ closely approximates the increase in $\dot{V}_E$ and $\dot{V}_A$ so that alveolar P_{CO_2} remains stable despite the increasing alveolar P_{O_2}. This response pattern is typically so precise that the increase in $\dot{V}_E/\dot{V}_{O_2}$ and $P_{A_{O_2}}$ without a change in $\dot{V}_E/\dot{V}_{CO_2}$ can be used to estimate, noninvasively, the work rate at which appreciable lactic acid begins to increase in arterial blood [i.e., the anaerobic threshold (Θ_{an})].

At even higher work rates, usually associated with arterial blood lactates greater than 4–5 mEq/L and decreasing arterial pH, ventilation is stimulated out of proportion to the additional CO_2 release, thereby lowering mean alveo-lar P_{CO_2} and further increasing alveolar P_{O_2}.

Control of Ventilation during Exercise

For exercise of moderate severity, the typical ventilatory response pattern is given in Figure 5–27. It is characterized by an abrupt increase in $\dot{V}_E$, which commonly occurs at, or even within, the first breath of exercise if the work is performed from rest. This phase 1 ($\phi1$) response occurs before any change in mixed venous gas tensions, which would reflect the increased tissue metabo-lism. Cardiac output, venous return and pulmonary blood flow are all in-creased in $\phi1$. Phase 2 ($\phi2$) is the phase associated with increasing mixed venous P_{CO_2}, and phase 3 ($\phi3$) is the steady state where blood flow, mixed venous gas tensions and ventilation are constant.

As the $\phi1$ response is so rapid, and it has been shown that afferent nerves from muscle to the brain stimulate $\dot{V}_E$, the mechanism of the $\phi1$ response is considered neurogenic. This neurogenic stimulation, presumably from pro-prioceptors of small unmyelinated afferents from the moving limbs, is thought to be sustained during the exercise, the slower $\phi2$ resulting from superim-posed blood-borne stimuli to the arterial and presumably central chemore-ceptors.

Certain observations are in conflict with this control scheme.

1. Arterial and alveolar P_{CO_2} do not decrease systematically in $\phi1$, as would be the case for simple neurogenic hyperventilation—although decreas-es can on occasion be seen, especially in excitable subjects or knowledgeable physiologists who act as subjects.

2. The $\phi1$ response does not typically occur if a work increment is imposed

Fig 5–27.—Schematic representation of time course of ventilation following on-set of dynamic muscular exercise of moderate intensity. Ventilation initially increases within a breath or so of beginning work ($\phi1$), then after a short delay increases slowly ($\phi2$) to its steady state ($\phi3$). Similar responses (in opposite direction) can be observed at the off-transient.

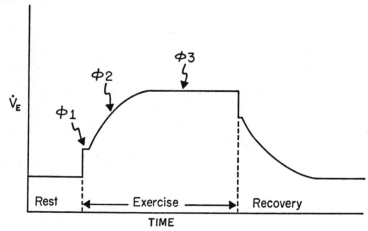

upon prior work, despite requiring the recruitment of muscle units and equal increments of metabolic rate. Rather, a slower pattern of increase is observed.

3. The $\phi 1$ response can be abolished by prior hyperventilation and markedly reduced arterial P_{CO_2}.

4. The only known ventilatory chemoreceptors are located on the arterial side of the circulation and in the brain stem: arterial blood gas and acid-base status are normal during moderate exercise.

Although it is difficult to conceive how the $\phi 1$ response can be mediated other than by neurogenic mechanisms, a signal proportional to the increased CO_2 delivery rate to the lung by the increased blood flow in $\phi 1$ may also be involved and thereby account for the constancy of P_{CO_2}.

At higher work rates associated with moderately increased blood lactate (i.e., $\simeq 3-4$ mEq/L), $\dot{V}E$ receives further acidic stimulation, mediated by the carotid bodies. However, as pH can be essentially normal owing to compensatory hyperventilation, it is hard to isolate a stimulus in the arterial blood under conditions of normal pH and P_{O_2} with *low* P_{CO_2}. Only at very high work rates is there demonstrable stimulus to $\dot{V}E$ in the arterial blood as pH falls owing to only partial respiratory compensation for the acidosis but, consequently, Pa_{CO_2} is driven even lower, thereby restraining the full effect of the low pH on stimulating $\dot{V}E$.

The function of the respiratory system, to provide adequate oxygenation of the blood and to maintain arterial acid-base balance, is clearly maintained over a wide range of work rates by a control system that precisely matches ventilation with the metabolic and acid-base requirements of the body. The mechanisms of this mediation await elucidation.

BIBLIOGRAPHY

Adamson, J. W., and Finch, C. A.: Hemoglobin function, oxygen affinity and erythropoietin, Ann. Rev. Physiol. 37:351, 1975.

Benesch, R., and Benesch, R. E.: Intracellular organic phosphates as regulators of oxygen release by haemoglobin, Nature 221:618, 1969.

Comroe, J.H., Jr., et al.: *The Lung: Clinical Physiology and Pulmonary Function Tests* (Chicago: Year Book Medical Publishers, Inc., 1962).

Cotes, J. E., and Hall, A. M.: The Transfer Factor for the Lung; Normal Values in Adults, in Archangeli, P. (ed.): *Introduction to the Definition of Normal Values for Respiratory Function in Man* (Alghero: Panminerva Medica, 1970).

Cunningham, D. J. C.: The control system regulating breathing in man, Quart. Rev. Biophys. 6: 433, 1974.

Farhi, L. E.: Gas Stores of the Body, in Fenn, W. O., and Rahn, H. (eds.): *Handbook of Physiology* (Washington, D. C.: American Physiological Society, 1964).

Farhi, L. E.: Ventilation-Perfusion Relationship and Its Role in Alveolar Gas Exchange, in Caro, C. A. (ed.): *Advances in Respiratory Physiology* (London: Arnold Publishers, Ltd., 1966).

Finch, C. A., and Lenfant, C.: Oxygen transport in man, N. Engl. J. Med. 286:407, 1972.

Filley, G. F., et al.: CO uptake and pulmonary diffusing capacity in normal subjects at rest and during exercise, J. Clin. Invest. 33:530, 1954.

Forster, R. E.: Exchange of gases between alveolar air and pulmonary capillary blood: Pulmonary diffusing capacity, Physiol. Rev. 37:391, 1957.

Forster, R. E.: Interpretations of Measurements of Pulmonary Diffusing Capacity, in Fenn, W. O., and Rahn, H. (eds.): *Handbook of Physiology* (Washington, D. C.: American Physiological Society, 1965).

Fry, D. L., and Hyatt, R. E.: Pulmonary mechanics, a unified analysis of the relationship between pressure, volume and gasflow in the lungs of normal and diseased human subjects, Am. J. Med. 29:672, 1960.

Guz, A.: Regulation of respiration in man, Ann. Rev. Physiol. 37:303, 1975.

Hyatt, R. E., and Black, L. F.: The flow-volume curve: A current perspective, Am. Rev. Resp. Dis. 107:191, 1973.

Macklem, P. T., and Mead, J.: The physiological basis of common pulmonary function tests, Arch. Environ. Health 14:5, 1967.

Mead, J., et al.: Significance of the relationship between lung recoil and maximum expiratory flow, J. Appl. Physiol. 22:95, 1967.

Mitchell, R. A.: Cerebrospinal Fluid and the Regulation of Respiration, in Caro, C. G. (ed.): Advances in Respiratory Physiology (London: Arnold Publishers, Ltd., 1966).

Mitchell, R. A., and Berger, A. J.: Neural regulation of respiration, Ann. Rev. Resp. Dis. 111:206, 1975.

Nadel, J. A., et al.: Early diagnosis of chronic pulmonary vascular obstruction: Value of pulmonary function test, Am. J. Med. 44:16, 1968.

Nunn, J. F.: Applied Respiratory Physiology with Special Reference to Anesthesia (London: Butterworth, 1969).

Otis, A. B.: Quantitative Relationships in Steady-State Gas Exchange, in Fenn, W. O., and Rahn, H. (eds.): Handbook of Physiology (Washington, D.C.: American Physiological Society, 1964).

Otis, A. B.: The Work of Breathing, in Fenn, W. O., and Rahn, H. (eds.): Handbook of Physiology (Washington, D. C.: American Physiological Society, 1964).

Perutz, M. F.: Haemoglobin; the molecular lung, N. Sci. Sci. J. 50:676, 1971.

Radford, E. P., Jr.: The Physics of Gases, in Fenn, W. O., and Rahn, H. (eds.): Handbook of Physiology (Washington, D. C.: American Physiological Society, 1964).

Rahn, H., and Farhi, L. W.: Ventilation, Perfusion and Gas Exchange—the $\dot{V}_A/\dot{Q}$ Concept, in Fenn, W. O., and Rahn, H. (eds.): Handbook of Physiology (Washington, D. C.: American Physiological Society, 1964).

Rahn, H., and Fenn, W. O.: A Gradual Analysis of the Respiratory Gas Exchange: The O_2-CO_2 Diagram (Washington, D. C.: American Physiological Society, 1956).

Riley, R. L., and Cournand, A.: "Ideal" alveolar air and the analysis of ventilation relationships in the lungs, J. Appl. Physiol. 1:825, 1949.

Rossier, P. H., and Buhlmann, A.: The respiratory dead space, Physiol. Rev. 35:860, 1955.

Roughton, F. J. W., and Forster, R. E.: Relative importance of diffusion and chemical reaction rates in determining rate of exchange gas in the human lung, with special reference to true diffusing capacity of pulmonary membrane and volume of blood in the lung capillaries, J. Appl. Physiol. 11:290, 1957.

Rupp, J. F.: Applied Respiratory Physiology with Special Reference to Anaesthesia (London: Butterworth, 1969).

Severinghaus, J. W., and Stupfel, M.: Alveolar dead space as an index of distribution of blood flow in pulmonary capillaries. J. Appl. Physiol. 10:335, 1957.

Torrance, R. W.: Prolegomena, in Torrance, R. W. (ed.): Arterial Chemoreceptors (Oxford: Blackwell Scientific Publishers, 1968).

Turner, J. M., et al.: Elasticity of human lungs in relation to age, J. Appl. Physiol. 25:664, 1968.

West, J. B.: Causes of carbon dioxide retention in lung disease, N. Engl. J. Med. 284:1232, 1971.

West, J. B.: Ventilation/Perfusion and Gas Exchange (Oxford: Blackwell Scientific Publishers, 1970).

Widdicombe, J. G.: Reflex Control of Breathing, in Widdicombe, J. G. (ed.): Respiratory Physiology, M. T. P. International Review of Science—Physiology Series, vol. 2 (Baltimore: University Park Press, 1974).

6

Physiology of the Kidney

RAYMOND G. SCHULTZE, M.D.

THE CELLS of the human body, as in all other vertebrates, are bathed in a salt solution the composition of which is maintained with remarkable constancy as we move from environment to environment or from stress to stress. Claude Barnard, the noted French physiologist, observed toward the end of the last century that the success of mammals with their highly specialized organ systems was totally dependent on the preservation of a constant internal environment. The kidneys and the several control mechanisms that alter their function are primarily responsible for this complex task since, by regulating the excretion of water and solutes, significant changes in volume and composition of body fluids are prevented.

In the process of acquiring and metabolizing food, many chemical compounds enter the body fluids that cannot be utilized by the organism and must be eliminated from the body. For instance, herbivores ingest large amounts of potassium that must be excreted rapidly because elevations of plasma potassium may cause paralysis of both skeletal and heart muscle. On the other hand, severe potassium deficiency may also result in muscular weakness and paralysis and may lead to death. Thus excretion of this ion by the kidneys must be closely regulated. In other vertebrate species, large amounts of sodium are ingested that, if retained, would lead to intolerable increases in extracellular fluid volume and possible death from the accumulation of fluid in the lung. Conversely, loss of sodium in excess of intake may lead to extracellular fluid volume depletion, decreased blood pressure and ultimately shock and death. Thus sodium excretion by the kidney also must be precisely regulated.

In the following sections, the composition and volume of body fluids and the role of the kidneys and their controlling mechanisms in maintaining the constancy of the internal environment will be described.

BODY FLUIDS

The major constituent of the body is water. In young adult males, water accounts for 60% of body weight and in young adult females 50%. The differ-

313

ence is due to the greater amount of adipose tissue in females. Fat cells contain far less water than do muscle cells.

The body water can be divided into two major compartments. The *intracellular* compartment contains about two-thirds (30–40% of body weight) of total body water and the *extracellular* compartment one-third (20% of body weight). The extracellular compartment can be further divided into three subcompartments: (1) interstitial fluid, which fills the spaces between cells—about 15% of body weight; (2) plasma, 4.5% of body weight; and (3) small amounts of fluid in specialized compartments enclosed by epithelium, e.g., cerebrospinal, intraocular, pleural, pericardial and peritoneal fluids—0.5% of body weight.

Thus in a male weighing 70 kg, total body water is about 42 L. The intracellular fluid has a volume of about 28 L, the extracellular fluid 14 L. Plasma volume is about 3 L and interstitial fluid volume about 10.5 L. The special compartments contain less than 500 ml.

The composition of both intra- and extracellular fluid varies among different cell types and different tissues. However, these variations are relatively slight and for the purposes of this section can be neglected. Concentrations of the major constituents of idealized intracellular and interstitial fluids and of plasma are shown in Table 6–1. The reason for the great differences in composition between extra- and intracellular fluids have been fully discussed in Chapter 1.

The chief difference in composition between plasma and interstitial fluid is the much greater concentration of protein in plasma. In addition, small differences (about 3%) occur between the anion and cation concentrations of interstitial fluid and plasma because of the Donnan effect.

The osmotically active solutes in each of the major body fluid compartments (intracellular, interstitial and plasma) determine the volume of that compartment. The ability of a solution to cause osmotic effects depends on the number of particles it contains and is measured in *osmoles*. Thus the osmolality of a solution containing a given *non-ionizable* solute is equal to the molar concentration of the solute in solution. For example, 1 M of glucose or 180 gm dissolved in 1 L of water yields a solution with an osmolality of 1, and 180 gm of glucose is 1 Osm. The osmolality of a solution containing an *ionizable* solute is equal to the molar concentration of the solute in solution times the

TABLE 6–1.—CONCENTRATIONS (mEq/L)
OF SOME IMPORTANT CONSTITUENTS OF
MAJOR FLUID COMPARTMENTS

CONSTITUENT	INTRACELLULAR	INTERSTITIAL	PLASMA
Na^+	10	136.0	140.0
K^+	145	3.8	4.0
Ca^{2+}	4	2.4	5.0
Mg^{2+}	30	1.5	2.0
Cl^-	4	105.0	100.0
HCO_3^-	10	28.0	27.0
Protein	60	4.0	16.0

number of particles per molecule obtained by ionization. Thus 1 M of NaCl or 58.5 gm dissolved in 1 L of water yields a solution with an osmolality of 2, and 29.25 gm NaCl is 1 Osm. (It is assumed that NaCl ionizes completely to yield 2 particles per molecule.) Glucose, 180 gm, and NaCl, 29.25 gm, yield the same number of particles.

The osmole is an inconveniently large measure for physiologic studies and it is customary to express the osmolality of a physiologic solution as the number of milliosmoles per kilogram of water in the solution (one milliosmole = 1/1000 Osm).

The osmolality of a solution may be measured by determining its freezing point. The freezing point of water is reduced when it contains dissolved solutes, and the reduction is proportional to the concentration (or more accurately, the chemical activity) of the solute particles. The osmolality of intracellular and interstitial fluid is normally about 290 mOsm/kg H_2O. Plasma osmolality is about 1 mOsm higher because of the protein it contains. The higher osmolality of plasma is sufficient to offset the outwardly directed hydrostatic pressure of blood within capillaries and prevents loss of plasma fluid into the interstitium.

In clinical practice it is common to substitute the term *tonicity* for osmolality when referring to solutions. A solution is *isotonic* (isosmotic) if a normal cell does not change volume when exposed to it. A solution that causes the cell to swell is termed *hypotonic* (hyposmotic), and a solution that causes shrinkage of the cell is *hypertonic* (hyperosmotic). A 0.92% solution of NaCl is isotonic. More dilute solutions are hypotonic since, when a cell is exposed to it, there will be an osmotic movement of water into the cell, causing it to swell. Conversely, solutions more concentrated than 0.92% produce an osmotic movement of water out of the cell, causing it to shrink.

REGULATION OF VOLUME AND OSMOLALITY OF BODY FLUIDS

Although dietary intake of water and salt varies widely, the volume and composition of body fluids remain remarkably constant. However, dangerous variations can occur in disease, for example when large amounts of electrolytes are lost during severe vomiting and/or diarrhea. Severe environmental conditions such as lack of drinking water in the desert or on the ocean after shipwreck can also cause major and potentially lethal changes. In clinical medicine and surgery, large volumes of electrolyte solutions are often given intravenously and may lead to water and electrolyte imbalance. It is therefore important for the student to understand the effects of water and electrolyte deficits and supplements on the composition and volume of body fluid compartments. A few illustrative examples follow. In these examples, sodium is considered as essentially an extracellular ion, and potassium an intracellular ion.

EXAMPLE 1.—What would be the effect of drinking 1 L of water on the volume and osmolalities of intra- and extracellular fluids and on their sodium and potassium concentration? Let us assume the following initial values: volume

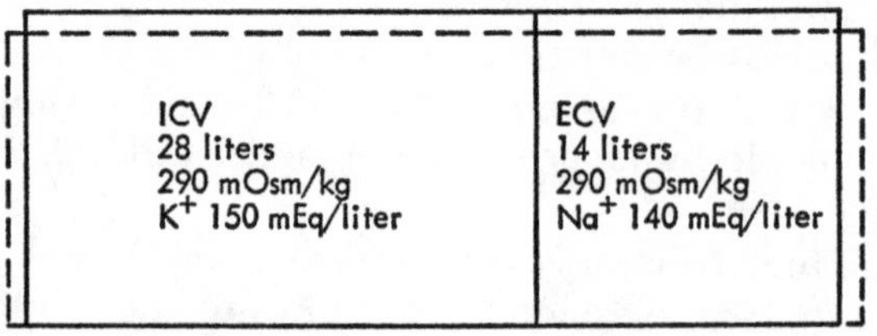

```
┌ ─ ─ ─ ─ ─ ─ ─ ─ ┬ ─ ─ ─ ─ ─ ─ ─ ─ ┐
│                 │                 │
│  ICV            │  ECV            │
│  28 liters      │  14 liters      │
│  290 mOsm/kg    │  290 mOsm/kg    │
│  K⁺ 150 mEq/liter│  Na⁺ 140 mEq/liter│
│                 │                 │
│                 │                 │
└─────────────────┴─────────────────┘
```

Fig 6–1.–Effect of infusing 1 L of water on the body fluids is shown by *dashed line*. *ICV* = volume of intracellular water; *ECV* = volume of extracellular water.

of intracellular water (ICV) = 28 L and extracellular water (ECV) = 14 L; osmolality = 290 mOsm/kg, intracellular K^+ = 150 mEq/L and extracellular Na^+ = 140 mEq/L.

When water is drunk, the initial effect is to dilute the extracellular fluid. Water will then move osmotically into the intracellular fluid until equilibrium is reached. Figure 6–1 (introduced by Darrow) is helpful in analyzing many water and electrolyte problems. Osmolality is shown on the vertical axis, volume on the horizontal axis. In the control situation we have the following:

Intracellular	*Extracellular*
28 L H_2O	14 L H_2O
290 mOsm/kg	290 mOsm/kg
K^+ = 150 mEq/L	Na^+ = 140 mEq/L

Total body water = 42 L
Total osmotically active solute = 42 × 290 = 12,180 mOsm
After drinking 1 L of water
Total body water = 42 + 1 = 43 L
Total solute = 12,180 mOsm
Osmolality = 12,180/43 = 283 mOsm/kg
Expansion of ECV = 14/42 × 1 = 1/3 L
Expansion of ICV = 28/42 × 1 = 2/3 L

Final ECF Na^+ concentration = $140 \times \dfrac{14}{14\frac{1}{3}}$ = 137 mEq/L

Final ICF K^+ concentration = $150 \times \dfrac{28}{28\frac{2}{3}}$ = 147 mEq/L

The volume and osmotic changes after drinking 1 L of water are represented by the dashed line in Figure 6–1. Thus, the effect of drinking water is to reduce the osmolality of the body fluids, increase the volume of extra- *and* intracellular fluid, reduce extracellular Na^+ concentration and reduce intracellular K^+ concentration. This situation normally is rapidly corrected by renal excretion of water.

EXAMPLE 2.—What would be the effect of infusing 2 L of isotonic saline on ECV and ICV and on Na^+ and K^+ concentrations of these fluids?

In this case the isotonic saline would increase the volume of extracellular fluid by 2 L (Fig 6–2). There would be no change in ICV or osmolalities or in intracellular K^+ or extracellular Na^+ concentrations. This situation would be rapidly corrected by renal excretion of NaCl and water.

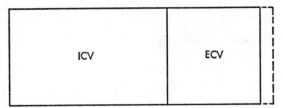

Fig 6–2.—Effect of infusing 2 L of saline on body fluid compartments is shown by the *dashed line* (see text).

EXAMPLE 3.—What would be the effect of rapidly absorbing 17.5 gm of ingested NaCl? Values are as follows:

17.5 gm NaCl = 300 mEq NaCl = 600 mOsm
Initial osmotically active solute = 42 × 290 = 12,180 mOsm
Final osmotically active solute = 12,180 + 600 = 12,780 mOsm
Final osmolality = 12,780/42 = 304 mOsm/kg
Initial ECF solute content = 14 × 290 = 4,060 mOsm
Final ECF solute content = 4,060 + 600 = 4660 mOsm
Final ECV = $\dfrac{4,660}{304}$ = 15.3 L
Final ICV = 42 − 15.3 = 26.7 L
Initial ECF Na$^+$ content = 14 × 140 = 1,960 mEq
Final ECF Na$^+$ content = 1,960 + 300 = 2,260 mEq
Final ECF Na$^+$ concentration = $\dfrac{2,260}{15.3}$ = 148 mEq/L
Final ICF K$^+$ concentration = $\dfrac{28 \times 150}{26.7}$ = 157 mEq/L

Thus the effect of salt ingestion is to increase the osmolality of intra- and extracellular fluid (Fig 6–3). Solute concentration is increased in both compartments. Extracellular Na$^+$ and intracellular K$^+$ rise. This situation is corrected by renal excretion of excess NaCl.

In the above examples and in many clinical situations, disturbances of body fluids originate in the extracellular compartment. Changes in extracellular volume are reflected primarily by cardiovascular disturbances. Thus small reductions in plasma fluid *volume* may cause *orthostatic hypotension*, i.e., a drop in arterial blood pressure on standing. More severe losses may cause peripheral circulatory failure, with low blood pressure and cardiac output. On

Fig 6–3.—Effect of rapidly absorbing 17.5 gm of NaCl is shown by the *dashed line* (see text).

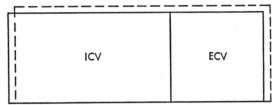

the other hand, increased interstitial and plasma volume may cause hypertension and edema. Changes in extracellular *osmolality* cause changes in intracellular volume associated most importantly with neurologic disorders, which may terminate in convulsions, coma or death.

The serum sodium concentration may be used as an index to assess disturbances of total body water. Its normal range is 135–145 mEq/L. *Hypernatremia*, i.e., a serum sodium concentration above 145 mEq/L, is indicative of increased extracellular osmolality and is usually associated with water deficit (dehydration). *Hyponatremia*, i.e., a serum sodium concentration below 135 mEq/L, usually suggests a relative or absolute gain of water (overhydration). In other cases, hyponatremia may indicate a relative or absolute loss of sodium. However, in most clinical situations, hyponatremia more accurately reflects changes in body water content than it does total body sodium. Thus serious sodium deficits may exist when the serum sodium concentration is within the normal range and there may be excess total body sodium in the presence of hyponatremia. Furthermore, extracellular fluid sodium concentration may be reduced in the presence of large amounts of another solute in the extracellular fluid, e.g., glucose. The added solute, by increasing the osmolality of the extracellular fluid, will cause water to shift from the intracellular compartment and dilute the sodium in the extracellular fluid.

Another difficulty is encountered when the nonaqueous phase of plasma is increased, e.g., by high concentrations of lipids. In these cases the concentration of sodium per liter of plasma may be low, whereas the concentration of sodium per liter of plasma *water* may be normal.

EXAMPLE 4.—A young man now weighing 70 kg has suffered severe water deprivation. His serum sodium is 170 mEq/L. How much water should he receive to restore his serum sodium to a normal level of 140 mEq/L?

Let us assume that 50% of this man's current body weight is water. Then

$$\text{current total body water (TBW}_1) = 50\% \times 70 = 35 \text{ L}$$

If we give him water, this will dilute both the intra- and extracellular compartments to the same degree. Thus

$$\text{TBW}_1 \times \text{Na} = \text{TBW}_2 \times \text{Na}_2$$

where TBW_1 and Na are current total body water and serum Na concentrations and TBW_2 and Na_2 are the final values. Therefore

$$35 \times 170 = \text{TBW}_2 \times 140$$

$$\therefore \text{TBW}_2 = 42.5$$

The patient needs 42.5 − 35 = 7.5 L of water to restore his total body water to normal.

BRIEF OVERVIEW OF KIDNEY FUNCTION

The primary functions of the kidneys are excretory. They must rid the body of water, sodium, potassium and some other inorganic substances that are

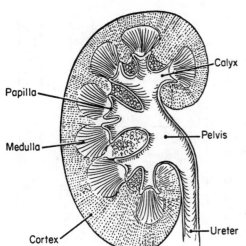

Fig 6–4.—Longitudinal section of the kidney.

ingested in excess of need. The kidneys also excrete solutes that are the end-products of protein and nucleic acid metabolism, urea, creatinine, water and nonvolatile acids.

Although there are two kidneys anatomically, they function physiologically as one. Each kidney weighs about 150 gm and is shaped like a common kidney bean. The kidneys lie in the retroperitoneal space between the first and fourth lumbar vertebrae. When observed after being cut longitudinally, the organ is seen to be divided into an outer, finely granular shell about 1 cm thick and an inner group of pyramidal-like structures that penetrate a hollow envelope, the renal *pelvis* (Fig 6–4). The outer shell is called the *cortex*, the inner portion the *medulla*. The tips of the pyramidal structures are called *papillae*. The renal pelvis is a collecting system with many extensions called *calyces*. It empties into the ureter, which serves as a conduit for urine to flow to the bladder.

Each of the kidneys consists of about 1,300,000 separate units called *nephrons*, consisting of a *glomerulus* and a *tubule* (Fig 6–5). The glomerulus is a tuft of capillaries arranged in parallel loops between two arterioles, *afferent* and *efferent*. The tuft of capillaries is surrounded by an envelope of epithelial cells called *Bowman's capsule*. This envelope is composed of two layers of epithelia which enclose a space that is contiguous with the tubule. The tubule forms the next segment of the nephron, is hollow, and averages about 30 mm in length. It may be divided anatomically into a *proximal segment*, the first portion of which is convoluted and the second portion straight; a *loop*, which has a *thin descending limb* and a shorter *thin ascending limb* followed by a *thick ascending segment;* and a convoluted *distal segment*. Distal tubules join together to form *collecting tubules*, which in turn terminate in the papillae.

The kidneys are very well supplied with blood. Between 20 and 25% of the cardiac output flows through the kidneys, and thus the normal renal blood flow is about 1,200 ml, of which 55% or about 660 ml per minute is plasma.

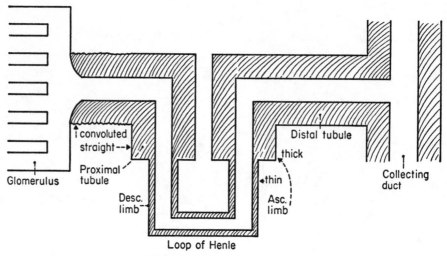

Fig 6-5. — Diagram of the nephron.

The capillaries of the glomerulus are highly permeable to water and to substances with a molecular weight of less than 10,000 daltons. Of the plasma flowing through the glomerular capillaries, 16–20% is filtered into Bowman's capsule and from there flows into the tubule. The hydrostatic pressure of blood inside the capillary supplies the energy for filtration. Quantitatively the rate of glomerular filtration (GFR) is about 125 ml per minute or 180 L per day. The entire extracellular fluid compartment can be filtered 13 times a day. The 180 L of filtrate contain 1,300 gm NaCl, 420 gm $NaHCO_3$, 55 gm KCl, 52 gm $CaCl_2$, 8 gm Na_2HPO_4, 36 gm $MgCl_2$ and 180 gm glucose. Since on a normal diet the urine contains roughly 10 gm NaCl, no $NaHCO_3$, 8 gm KCl, 0.5 gm $CaCl_2$, 1 gm NaH_2PO_4, and 0.6 gm $MgCl_2$ in about 2 L of water, it is obvious that large amounts of solute and water must be resorbed (Table 6-2).

The reabsorptive process occurs along the entire length of the tubule and is both selective and regulated. In terms of fluid alone, about 65% of the filtered

TABLE 6-2.—QUANTITY (gm) OF
SUBSTANCES FILTERED, ABSORBED AND
EXCRETED DAILY BY NORMAL MAN ON
AVERAGE DIET

SUBSTANCE	FILTERED	ABSORBED	EXCRETED
NaCl	1,300.0	1,290.0	10.0
$NaHCO_3$	420.0	420.0	0.0
KCl	55.0	47.0	8.0
$CaCl_2$	52.0	51.5	0.5
$MgCl_2$	36.0	35.5	0.5
Na_2HPO_4	8.0	7.0	1.0
Glucose	180.0	180.0	0.0
Creatinine	1.5	0.0	1.5
Urea	54.0	21.0	33.0
Uric Acid	9.0	8.3	0.7
H_2O	180.1	178.1	2.1

water is absorbed by the proximal tubule, 15% by the loop, 10% by the distal tubule and the remainder by the collecting duct. The absorption of sodium differs to some extent from that of the filtered water. Approximately 65% of sodium is reabsorbed in the proximal tubule where the reabsorbed fluid has the same sodium concentration as the filtrate. In the ascending limb of the loop, however, the sodium concentration in the reabsorbed fluid is higher than in the filtrate so that, by the end of the loop, only 5–10% of the filtered sodium remains to be absorbed by the distal tubule and collecting duct. Nearly all of the filtered bicarbonate and glucose is reabsorbed in the proximal tubule. The movement of urea is much more complex since it is partially reabsorbed in the proximal tubule and collecting duct, but enters the filtrate in the loop. Filtered creatinine passes through the entire tubule and into the urine without reabsorption.

As noted above, urea and creatinine are end-products of metabolism, and the excretion rate for these substances must equal the rate at which they are produced. The kidneys accomplish this task in a passive manner, as follows. Since the plasma concentration of these substances is proportional to their rate of production, the rate at which they are filtered through the glomerular capillary wall into the tubule is also proportional to the rate of production. Thus when much urea is produced because protein intake is elevated, plasma urea concentrations increase and the filtered load of urea increases. Since under normal circumstances the fraction of filtered urea excreted remains constant at about 0.6, as filtered load increases the rate of excretion increases. If protein intake is low and urea production falls, the converse occurs. Thus for creatinine and urea, urinary excretion is regulated by production and a specific mechanism to control excretion is not needed.

On the other hand, if the amount of water and sodium lost in the urine exceeds the amount ingested, the effect may be disastrous. Conversely, excretion of too little of these substances leads to their accumulation in excess of need and may also have serious consequences. Homeostasis is achieved by mechanisms that control excretion of each of these substances by changing the rate at which they are absorbed from the renal tubules. The kidney in these instances acts as an end-organ for the control systems.

The loss of some filtered substances represents a danger to the body. Glucose and amino acids are filtered and, if lost in the urine, represent the loss of both energy and essential building blocks for the body. Thus virtually the entire filtered load of these substances is recovered from filtrate.

RENAL BLOOD FLOW AND GLOMERULAR FILTRATION

As indicated previously, the kidney receives a large fraction of the cardiac output, and the blood flow per gram of tissue is substantial. However, it is not uniformly distributed. The cortex receives more than 90% of blood supplied through the renal artery, the outer medulla about 8% and the papillae the rest. In terms of flow per gram of tissue, flow through the cortex is 4–5 ml per minute per gram; through the outer medulla, 1.5 ml per minute per gram and through the inner medulla, 0.2 ml per minute per gram.

Renal blood flow (RBF) is related directly to the drop in blood pressure (ΔP) across the organ (primarily determined by arterial pressure since, normally, venous pressure is quite constant) and inversely to the total resistance to flow. The major resistances are the afferent arteriole (Raa) and efferent arteriole (Rea). Thus

$$RBF = K \frac{\Delta P}{Raa + Rea} \tag{1}$$

where K is a constant.

Blood flow through each glomerulus is a direct function of arterial pressure and an inverse function of the sum of the afferent and efferent arteriolar resistances. Furthermore, changes in efferent arteriolar resistance can alter the hydrostatic pressure in the glomerular capillary, but only at the expense of altering RBF. Thus when efferent arteriolar resistance decreases, glomerular blood flow increases and glomerular capillary hydrostatic pressure decreases. If efferent arteriolar resistance increases, glomerular flow decreases. Consequently the pressure drop across the afferent arteriole diminishes, resulting in a higher glomerular capillary pressure.

AUTOREGULATION OF RENAL BLOOD FLOW

One characteristic of RBF is that it remains relatively constant over a wide range of perfusion pressures. Below a mean arterial pressure of 80 mm Hg, blood flow varies directly with pressure but, between this value and a mean blood pressure of 200 mg Hg, blood flow increases only slightly with each increment in pressure because renal vascular resistance increases. The ability of the kidney to maintain a constant blood flow in the face of varying perfusion pressures is termed *autoregulation* (see section on regional circulations, Chapter 4). The most widely accepted explanation is that renal vascular resistance increases as perfusion pressure increases because of contraction of the afferent arteriole in response to stretch. Substantial evidence supports this view, the *myogenic* hypothesis. When perfusion pressure of the kidney is suddenly increased, there is a transitory increase in RBF, followed by increased resistance, leading to a short period of decreased blood flow (overshoot), after which flow returns to control values and the vascular resistance stabilizes at a higher level. The fact that this sequence of events occurs over a period of 10–15 seconds suggests strongly that autoregulation is a local phenomenon and may be a function of the muscle layer of the afferent arterioles.

The second major theory offered to explain autoregulation involves participation of the juxtaglomerular apparatus (see page 333). According to this theory, it is glomerular filtration that is closely regulated and RBF is only secondarily controlled. It is suggested that when the filtration rate in a single nephron increases, a larger amount of filtrate reaches the *macula densa* located at the end of the thick ascending limb of the loop of Henle. The increased filtrate present at this site is detected in an as yet poorly understood manner, and the macula densa signals specialized cells in the walls of the afferent and efferent arterioles to release *renin,* a proteolytic enzyme. Renin acts on a plasma globulin, *renin substrate,* leading to formation of the vasoconstrictor polypeptide *angiotensin,* which constricts the glomerular afferent arterioles and increases resistance to blood flow. This action also reduces glomerular capillary hydrostatic pressure and returns the filtration rate of the nephron to normal. The major support for this hypothesis has come from the observation that solutions containing concentrations of sodium equal to that of plasma cause cessation of filtration in the same nephron when injected retrograde from the distal convoluted tubule to the macula densa. It has also been shown that, if the loop of Henle is perfused from the proximal tubule at varying rates, glomerular filtration of that nephron will vary inversely with the rate of perfusion. That is, increases in the perfusion rate decrease GFR, whereas reductions in perfusion rate increase GFR. These data imply that the macula densa may play a central role in regulating the single nephron GFR.

Against this intellectually gratifying hypothesis stands the observation that the rate at which autoregulation occurs is faster than the transit time of filtrate through the proximal tubule and loop of Henle. Furthermore, experimentally reducing the GFR by decreasing perfusion pressure leads to an increase in renin release measured in renal vein blood or systemically. This observation is opposite to what would have been predicted by the macula densa hypothesis for control of RBF or GFR.

Autoregulation of blood flow in the face of variations in renal perfusion pressure does not prevent changes in RBF by other physiologic mechanisms. The sympathetic nervous system has a profound effect. Stimulation of the renal nerves leads to increased vasoconstriction and decreased RBF, whereas anatomical or chemical denervation leads to increased RBF. Many hormones including epinephrine, norepinephrine, angiotensin II and the prostaglandins also alter RBF, but their physiologic role in the day-to-day control of the renal circulation has not been clearly defined.

MECHANISM OF GLOMERULAR FILTRATION

As blood flows through the glomerular capillary tuft, about one fifth of the plasma water is filtered through the capillary wall into the surrounding space. The energy for this process is the net hydrostatic pressure across the capillary wall generated by the heart. The major opposing force is the oncotic pressure of the plasma, the osmotic pressure contributed by the large molecules of protein that are not filtered.

The barrier through which filtration takes place consists of three layers (Fig 6–6). The first of these is the endothelial lining of the capillaries. The endothelial cells contain many round, window-like openings that expose the underlying basement membrane, so that this layer of the barrier is quite porous.

The next barrier is the basement membrane, which appears to consist of amorphous material that is dense in the middle and less dense on either side. Functionally, the dense middle layer acts as a barrier to particles in plasma of a diameter of 10 nm or greater. The layer immediately beneath the endotheli-

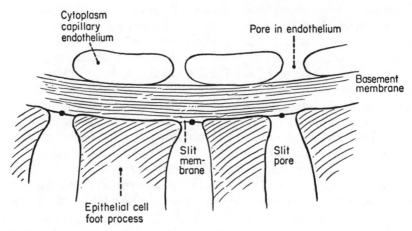

Fig 6–6.—Glomerular barrier to filtration. The endothelium contains many punched-out holes that expose basement membrane. The foot process of the epithelium converges with the other side. A thin membrane traverses the slit between endothelial foot processes.

um resembles a loose gel, and particles up to 14 nm in diameter may traverse it.

The epithelial cells that cover the outer loose layer of the basement membrane have long cystoplasmic projections called *foot processes.* The foot processes of one cell interdigitate with those of other cells and are separated from one another by a space of about 24 nm, called the *filtration slit.* Most electron micrographic studies have demonstrated that there is a thin membrane, appropriately called the *slit membrane,* crossing this space.

Molecules of several types (dextrans, polyvinylpyrolidone [PVP], proteins) have been used to study the filtration barrier presented by the glomerular capillary wall. It has been found that molecules that are not electrically charged pass more easily through the barrier than those that are. For instance, uncharged dextran and PVP molecules 3.5 nm in diameter are filtered about half as easily as are very small molecules. However, charged protein molecules of this size are barely filtered at all.

Inulin, a polyfructosan, with a molecular weight of about 5,000 and an effective molecular radius of 1.4 nm, is completely filtered, i.e., the concentration in plasma water is the same as the concentration in the ultrafiltrate in Bowman's space. The filtrate concentration of small protein molecules such as lysozyme (MW 14,600, effective radius 1.9 nm) and myoglobin (MW 16,900, effective radius 1.9 nm) is only 80% of their concentration in plasma water. Only 4% of hemoglobin (MW 68,000, effective radius 3.2 nm) is filtered and only 0.3% of serum albumin (MW 69,000, effective radius 3.6 nm) is filtered. Thus the normal glomerular capillary wall effectively prevents loss of serum proteins into the tubule and permits maintenance of an oncotic pressure gradient between plasma and interstitial fluid.

A diseased glomerulus, however, may allow relatively large amounts of albumin to be filtered and ultimately lost in the urine because the integrity of the filtration barrier has been disrupted. When the urinary excretion of albumin exceeds 0.05 gm/kg per day, the ability of the liver to replace that loss is usually exceeded and plasma albumin concentration falls. The consequent reduction in plasma oncotic pressure results in loss of plasma water (along with its dissolved salts) from the vascular space into the interstitial space. The excess interstitial fluid is termed edema (see section on the microcirculation, Chapter 4), and the syndrome of excessive protein excretion, decreased plasma albumin concentration and edema formation is called the *nephrotic syndrome.*

The behavior of the filtration barrier has been likened in *model* systems to a membrane with pores 7.5–10 nm in diameter and 40–60 nm in length covering 5% of the membrane surface. Since the total capillary surface is thought to be about 8,000 sq cm, up to 400 sq cm should be available for filtration. The experimental observation that the hydraulic permeability of glomerular capillaries is 50–300 times greater than that of capillaries in other parts of the body is consistent with this formulation. However, recent experiments have shown that fluid moves across the filtration barrier not only by bulk flow (the substance is carried in a stream), but also by diffusion (the substance moves through a solution saturating the barrier down its chemical activity gradient). Thus the pore hypothesis may not be necessary to explain the dynamics of filtration.

Recently, considerable information about the process of glomerular filtration has been obtained directly from experiments carried out in rats with glomeruli on their kidney surfaces. This fortunate anatomical arrangement has permitted direct measurements of hydrostatic pressure in the glomerular capillaries and in Bowman's space (Fig 6–7).

Fig 6–7.—Hydrostatic and oncotic pressures in glomerular capillary and in Bowman's space. P_{GC} = hydrostatic pressure glomerular capillary; P_T= hydrostatic pressure tubule; π_{GC} = oncotic pressure glomerular capillary; π_T = oncotic pressure tubule; P_{UF} = ultrafiltration pressure. As filtrate moves from the afferent to the efferent end of the capillary, filtrate moves across the basement membrane until oncotic pressure increases to the point where the ultrafiltration pressure is 0. Then filtration stops.

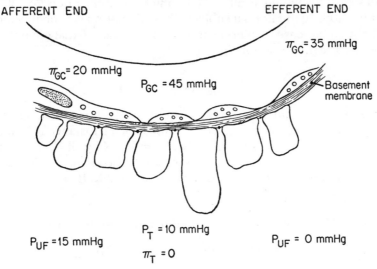

As fluid enters the glomerular capillary from the afferent arteriole, hydrostatic pressure is 45 mm Hg and colloid osmotic pressure 20 mm Hg. The pressure in Bowman's capsule is 10 mm Hg and, since there is no protein in the filtrate, the colloid osmotic pressure of the filtrate is zero. Thus at the beginning of the capillary, the force favoring filtration is 15 mm Hg. As fluid leaves the capillary, the protein concentration and oncotic pressure of the remaining plasma increase. Since hydrostatic pressure changes very little along the length of the capillary, filtration ceases when the colloid osmotic pressure reaches about 35 mm Hg. It is not known where along the capillary length the colloid osmotic pressure reaches this value but, if it happens by the midpoint of the capillary, only half of the available glomerular surface would be required for filtration.

It has also been shown in the rat that the rate of formation of glomerular filtrate can be increased by increasing the rate of plasma flow through the glomerulus. With each increment in flow a greater amount of fluid must be filtered before osmotic pressure rises to equal hydrostatic pressure and equilibrium will be reached closer to the efferent end of the capillary. Hence, the effective filtration surface will be greater at higher flow rates. Ultimately at a very high rate of plasma flow, filtration pressure equilibrium is not achieved and the filtration rate would no longer rise with increases in renal plasma flow, but only with increases of glomerular capillary pressure.

These measurements from the rat may not be representative of the situation in other species. Similar micropuncture measurements in dogs suggest that filtration pressure equilibrium is not normally reached within the glomerular capillary and, therefore, the rate of filtration will not be blood flow dependent, as it is in the rat. The observation that certain experimental maneuvers can increase RBF in these animals, without increasing GFR, supports this view. It is not known if man resembles the rat or the dog.

Clinical observations have shown that in patients with damaged hearts and reduced cardiac output, the GFR is normal or nearly so despite a reduction in renal plasma flow. Renal vascular resistance is very high and the fraction of renal plasma flow which is filtered is much greater—often 25–30% compared to 16–20% in the normal subject. This adaptation is probably mediated primarily by an increase in efferent arteriole resistance so that glomerular capillary hydrostatic pressure is maintained.

MEASUREMENT OF GFR

In order to understand how the GFR can be measured in man, it is first necessary to discuss *renal clearance*. The clearance of a substance X is defined as the *volume* of plasma that is completely cleared of X by the kidneys per unit time. We can calculate this clearance if we know the amount of X excreted in the urine per minute and the plasma concentration of X. If

U_X = concentration of X in urine (mg/ml)

V = volume of urine produced per minute (ml)

P_X = concentration of X in plasma (mg/ml)

then the amount of X excreted in the urine per minute = U_xV (mg) and the volume of plasma that would be completely cleared of U_xV mg of X per minute equals

$$\frac{U_xV}{P_x} \text{ (ml)} \tag{2}$$

Two facts require emphasis: (1) each substance excreted by the kidneys has its own clearance and (2) the clearance is the *virtual* volume of plasma cleared of a substance. This is because substances are *not* usually completely cleared in one passage through the kidney; some remain in the blood leaving the kidneys. The clearance calculation, however, tells us how much plasma would be required to deliver the amount of X appearing in the urine per unit time, *assuming* that the volume of plasma could pass *all* its content of X into the urine during a single passage through the kidney.

Different substances are handled in different ways by the kidney. Urea, for example, is filtered through the glomerulus but is partially reabsorbed in the pars convoluta of the proximal tubule, secreted in the pars recta of the same segment and by the thin ascending limb of Henle's loop and partially reabsorbed in the collecting tubule. Its calculated clearance of about 75 ml per minute represents the net effect of all these processes. Glucose is also filtered through the glomerulus but is almost *completely* reabsorbed, so that normally the glucose concentration in the urine is nearly zero. When plasma glucose concentrations are normal, glucose clearance is nearly zero.

Some substances, such as inulin, are filtered through the glomerulus but are neither reabsorbed nor secreted by the tubule. Thus the amount of inulin that appears per minute in the urine is the same as the amount filtered per minute through the glomerulus. Moreover, the concentration of inulin in the filtrate is the same as its concentration in the plasma. From first principles: amount of inulin in glomerular filtrate per minute (mg) = volume of glomerular filtrate produced per minute (GFR, ml/min) × concentration of inulin in filtrate (mg/ml). Thus,

$$\text{GFR} = \frac{\text{amount of inulin in filtrate/min}}{\text{concentration of inulin in filtrate}} \tag{3}$$

Since all of the filtered inulin appears in the urine,

$$\text{GFR} = \frac{\text{amount of inulin in urine/min}}{\text{concentration of inulin in plasma}}$$

$$= \text{inulin clearance}$$

Therefore, GFR can be estimated by measuring the clearance of a substance that is neither secreted nor absorbed by the tubule. In carrying out these measurements, inulin is infused intravenously until a constant plasma concentration is obtained (P_{In}). Excreted urine is collected over a precisely measured period of time, and the volume of urine excreted per minute (V) is calculated.

Then

$$GFR = \text{inulin clearance} = \frac{U_{In}V}{P_{In}} \tag{4}$$

Clinical application of this technique is cumbersome, and in clinical practice GFR can be estimated from the endogenous *creatinine clearance.*

Creatinine is an end-product of creatine metabolism. The amount of creatinine produced daily is a function of the body's muscle mass rather than of diet and, thus, plasma levels are fairly constant. The human kidney secretes creatinine and therefore its clearance would be expected to exceed simultaneous values obtained with inulin. However, there are other chemicals in plasma that are measured by the chemical methods used to estimate plasma creatinine. The resulting falsely high value for plasma creatinine offsets the error in the estimate of GFR caused by the creatinine secretory process. Creatinine clearance varies with surface area. Normal values for individuals with a "standard" surface area of 1.73 sq m are 110–150 ml per minute in males and 105–132 ml per minute in females. These values are quite close to the normal values for inulin clearance.

In contrast to creatinine and inulin, certain substances are actively secreted into the tubules and these substances have clearances that exceed GFR. If there were a substance that was completely removed from all of the plasma passing through the kidney, its clearance would equal renal plasma flow. Para-aminohippuric acid (PAH) comes closest to meeting this criterion. When its plasma concentration is low, most PAH that escapes filtration through the glomerulus is secreted by the proximal tubule. Thus, PAH clearance measures the plasma flow that has passed through the glomeruli and peritubular capillaries. The PAH clearance is also termed the *effective renal plasma flow* (ERPF) and represents about 85% of the *total* renal plasma flow. The other 15% perfuses nonsecreting portions of the kidney. Thus *total renal plasma flow* = ERPF × 100/85, and total RBF = ERPF × 100/85 × blood volume/ plasma volume. The RBF in normal adults is 1–1.5 L per minute when measured by this technique.

A more precise measurement of renal plasma flow may be obtained by sampling arterial and renal vein blood and directly measuring the concentration of any substance excreted by the kidney. Thus,

$$RPF = \frac{U_x V}{Ra_x - Rv_x}$$

where $U_x V$ = amount of the substance excreted per minute (equal to concentration of X in the urine $[U_x]$ multiplied by the volume of urine $[V]$ excreted per minute), and Ra_x and Rv_x = plasma concentrations of X in arterial and renal vein blood.

Alterations in GFR are commonly encountered in clinical medicine. Following infection of the pharynx or tonsils by bacteria of a particular species (*B hemolytic streptococcus*), the glomeruli may become damaged. The capillaries become clogged with cells and the products of coagulation of blood and the GFR per nephron is markedly reduced. As a result, ingested water and solute are retained and the volume of the extracellular fluid compartment increases. A marked increase in extracellular fluid volume may lead to elevated arterial blood pressure and permanent damage to the blood vessels. Fortunately, glomerular damage of this type usually reverses after a period of days, and glomerular filtration returns to normal.

Other chronic disease processes may decrease the glomerular filtration at a slow rate. When this happens, control mechanisms for the various solutes found in the blood alter tubular function so that excessive quantities will not be retained. Urea and creatinine are exceptions, since their levels in plasma rise because the excretion of both is a direct function of the amount filtered. Thus, when a reduction in filtration rate occurs, the filtered load decreases, as does the quantity excreted. As production continues at a constant rate, plasma urea and creatinine levels rise until the amounts excreted once more match production. A new steady state has been reached, but with higher levels of urea and creatinine in the plasma.

A MORE DETAILED VIEW OF TUBULAR FUNCTION

Before entering into a discussion of how the kidney maintains constancy of the internal environment, we shall examine the function of the tubules, which have the task of resorbing almost all of the 180 L of filtrate formed each day. Clearly the resorption of the components of the filtrate is selective and, if a precise balance between dietary intake and excretion is to be maintained, resorption must also be carefully controlled. The resorptive process occurs along the entire length of the nephron as well as in the collecting duct. In each segment of the tubule, characteristics of the reabsorptive process are somewhat different.

PROXIMAL TUBULE

The proximal tubule is joined to the epithelium of Bowman's capsule. It is approximately 14 mm in length and for two thirds of this distance it follows a convoluted course (pars convoluta), but the final third is straight (pars recta). The cells of the epithelium of the proximal tubule contain many mitochondria and are considered metabolically active. The luminal membrane of these cells is characterized by a layer of microvilli, approximately 1μ in length. The layer is very dense in the convoluted portion, less so in the straight portion.

The cells of the epithelium are joined to one another over a short distance along their lateral walls near the luminal border. This area has been called the "tight junction," but functional studies have demonstrated that it is actually porous since water and electrolytes move through it. Beneath the tight junction, the cells are separated from one another, and the narrow space between them is termed the lateral intercellular space (Fig 6–8). The cells lining the first third of the tubule are not simply cylindric but have extensive projections not unlike those of the glomerular epithelium. The projections of one cell interdigitate widely with those of adjacent cells, forming extensive lateral intercellular spaces. In the middle third of the proximal tubule, the number of

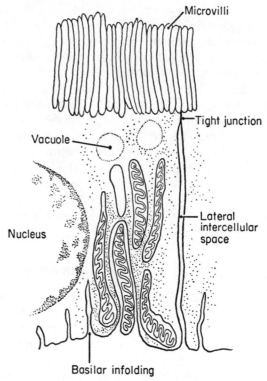

Microvilli

Tight junction

Vacuole

Lateral
intercellular
space

Nucleus

Basilar infolding

Fig 6–8.—Proximal tubular cell.

projections per cell decreases, and finally in the last third the cells have a shape closely resembling a cylinder. The number of mitochondria and micro-bodies declines as the distance from the glomerulus increases. The anatomi-cal changes correlate with a decrease in transport capacity along the tubule.

The proximal tubule is surrounded by a generous capillary network. In the outer cortex this network is derived in large part from the efferent arterioles of its own glomerulus. In the inner cortex the capillaries surrounding the proximal tubule may originate from the efferent arterioles of neighboring glomeruli.

The major force for filtrate reabsorption by both convoluted and straight segments of the proximal tubule is the *active reabsorption of sodium.* Water follows passively, leaving the tubular fluid isosmotic with plasma. Evidence for active sodium transport is derived from the observation that the proximal tubule can lower the concentration of sodium in fluid, isolated in its lumen between two columns of oil, to 20–30 mM less than the fluid bathing the peri-tubular surface. To accomplish this, sodium must be moved against its chemi-cal gradient and, therefore, its transport must be active. The conclusion is supported by the observation that ouabain, known to inhibit active sodium transport in other tissues, reduces sodium absorption in the proximal tubule.

The process of sodium absorption begins with the movement of the ion down its electrochemical gradient from the tubular lumen across the luminal membrane of the proximal tubular cell (Fig 6–9). The interior of the cell is electrically negative (about 70 mv) compared to the lumen, and the intracellular sodium concentration is about 100 mM less than the filtrate sodium concentration. In order to maintain this gradient, sodium that leaks into the cell must be pumped out. This is accomplished along the lateral walls of these cells, which are known to contain large amounts of Mg-dependent, Na–K activated ATPase, generally believed to be the active sodium pump of most tissues, including the kidneys. The sodium is pumped into the lateral intercellular spaces where its concentration increases enough to generate an osmotic gradient that provides the force for the movement of water from the lumen into the lateral intercellular spaces either through the tight junction between the apices of the cells or through the cell itself. Under optimal conditions for absorption, the absorbed filtrate in the lateral intercellular space diffuses rapidly through the tubular basement membrane and is picked up by a peritubular capillary and thus returned to the circulation. The rate of movement of the reabsorbed fluid in the capillary is determined by the Starling forces existing across the capillary epithelium. Potassium, calcium and magnesium are reabsorbed in parallel with sodium in the proximal tubule. The precise nature of the resorption process for these ions is unknown. Hydrogen ion is *secreted* by the proximal convoluted tubule. The major reabsorptive and secretory processes of the early proximal tubule are summarized in Figure 6–10.

By the time the filtrate enters the second half of the pars convoluta and the

Fig 6–9.—Movement of sodium chloride and water in proximal tubule.

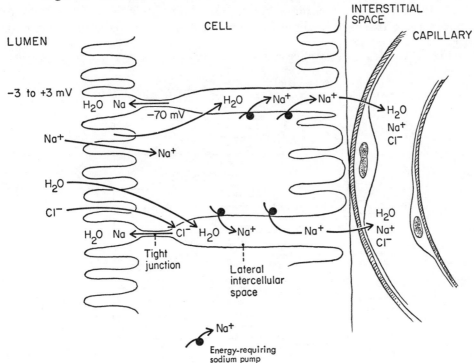

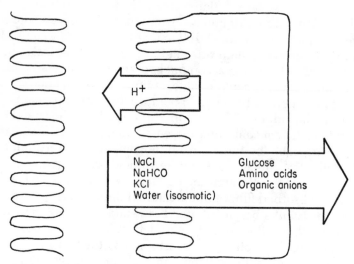

Fig 6–10.—Major reabsorptive and secretory processes in the early proximal tubule.

pars recta, it is markedly altered in composition. Most of the bicarbonate has been reabsorbed so that its concentration in the filtrate is much less than in plasma. Since the resorption in chloride is relatively slow, this anion is now present in a concentration greater than that of plasma. Many organic solutes such as glucose, amino acids, lactate and acetate have been completely reabsorbed.

Sodium absorption continues in the straight portion of the proximal tubule, but at a rate that is only about one third of that in the convoluted segment.

Fig 6–11.—Major reabsorptive and secretory processes in pars recta of the proximal tubule.

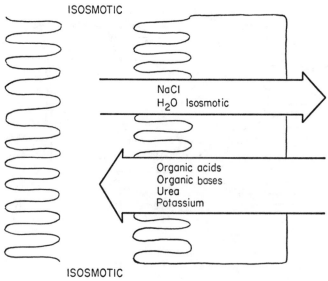

Chloride is passively absorbed with sodium, and the rate of bicarbonate absorption is much slower than in the early proximal tubule. Ordinarily, glucose and amino acids are not present in significant concentrations here but, when they are, they also are absorbed slowly. Four secretory processes have been found in the straight segment. The most important is the organic anion transport mechanism, the details of which are discussed in the section on regulation of extracellular volume and sodium balance. A similar, but less active system secretes organic bases, and recent experimental evidence suggests that the pars recta may secrete both urea and potassium (Fig 6–11).

LOOP OF HENLE

The next segment of the nephron is the loop of Henle. The descending portion begins as a thin extension of the pars recta of the proximal tubule. It has a very flat, uninteresting epithelium, and studies have indicated that it is as functionally uninteresting as its structure implies. The descending segment of the loop follows a course into the medulla toward a papilla and then turns around in a hairpin fashion and heads back toward the cortex as the *thin ascending* limb. The depth to which the loop descends depends on the location of the originating glomerulus. Those nephrons with glomeruli in the outer cortex have loops that descend only into the outer third of the medulla. Nephrons with glomeruli in the juxtamedullary region (about 15% of all nephrons) send their loops deep into the medulla, and many reach the papilla.

As the thin portion of the loop ascends, the epithelium suddenly changes. The cells become larger and more complex and the lumen has a greater diameter. This segment is called the *thick ascending* limb. The thick segment follows a direct path to the originating glomerulus of the nephron. Here it joins the distal convoluted tubule in the notch between the afferent and efferent arterioles. At the point of contact with the blood vessels, the epithelial cells of the ascending limb become high, columnar and densely packed and form a plaque called the *macula densa*. The macula densa abuts specialized cells in the walls of the afferent and efferent arterioles. The entire complex is called the *juxtaglomerular apparatus* and is responsible for synthesis and release of renin from the kidney.

The *thin descending* portion of the loop of Henle is inactive from a transport point of view. It is permeable to water and, therefore, the filtrate within it is in osmotic equilibrium with the interstitial fluid of the adjacent medulla. Considerable disagreement exists as to whether or not the thin descending limb is also permeable to sodium chloride and urea, but most evidence suggests that it is relatively impermeable to these solutes.

The membranes of the cells that make up both sections of the *ascending* limb have characteristics that differ entirely from those of the descending portion in that they are relatively water impermeable. The cells of the thick ascending limb are capable of transporting solute. Recent microperfusion studies have demonstrated that the ion species that is actively transported is *chloride* rather than sodium. Sodium and potassium are absorbed passively.

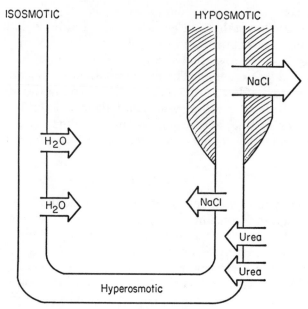

ISOSMOTIC   HYPOSMOTIC

NaCl

H₂O

H₂O   NaCl

Urea

Urea

Hyperosmotic

Fig 6–12.—Major solute and water movement in loop of Henle.

The reabsorbed fluid is hyperosmotic and therefore the filtrate becomes hypotonic as it flows through this segment. The filtrate entering the distal tubule represents 20% of the filtered water and about 10% of the filtered sodium (Fig 6–12).

DISTAL TUBULE AND COLLECTING DUCTS

The distal convoluted tubule is only a third as long as the proximal convoluted portion and is anatomically much less complex. It joins with other distal tubules to form the collecting ducts, which in turn descend again through the medulla to empty into larger ducts in the papillae. Ducts from the papillae empty into the collecting system.

Sodium chloride and water absorption continue in both the distal tubule and collecting duct. The distal tubule reabsorbs about half of the sodium reaching it, or about 5% of the filtered load. Reabsorption of water in the distal tubule lags behind that of sodium in some species (monkey, rabbit, dog) but, in the rat, water absorption under conditions necessitating water conservation is sufficient to raise the osmolality of the filtrate to that of plasma. The collecting duct is capable of absorbing almost all the remaining sodium and water when physiologic circumstances make it desirable to do so.

The distal tubule plays a primary role in secretion of potassium. At least half of the potassium excreted in the urine enters the filtrate as it passes the distal half of the distal tubule. Under certain circumstances, more is added in the collecting duct. If potassium must be conserved, these two segments are capable of potassium reabsorption.

Hydrogen ion is secreted by both the distal tubule and collecting duct. The concentration of hydrogen ions in urine within the collecting duct may be as much as 1,000 times greater than in plasma.

WATER BALANCE

Under steady-state conditions, the body maintains an accurate water balance, i.e., the amount of water ingested or formed by the body equals the amount lost from the lung, skin and gut plus the amount of water excreted. An example of a normal daily water balance is shown in Table 6–3.

Since losses through the skin and lungs depend more on physical activity and environmental temperature, only the acquisition of water and its urinary excretion can be controlled. Acquisition of water is regulated by the *thirst mechanism.* The system that leads to reduced water excretion is the concentrating mechanism of the kidney. The chief components of the latter are the medullary structures of the kidney and the central nervous system (CNS), which is responsible for releasing antidiuretic hormone.

These two systems maintain water balance so that the osmolality of body fluids remains constant and their volume remains adequate. They must deal with two major problems: (1) dilution of body solutes by excessive ingestion (or administration) of water and (2) hyperconcentration of solute in body fluids because of a decrease in total body water.

ADH plays a vital role as a messenger between the CNS and the kidney. The hormone is an octapeptide synthesized in neurons of the supraoptic and paraventricular nuclei of the hypothalamus and then transported along their axons to the posterior pituitary gland. The hormone is released by a variety of stimuli including an increase in osmolality of the interstitial fluid of the brain, a decrease in blood volume, stimuli from baroreceptors in both pulmonary and systemic circulation, and angiotensin II. Secretion is generally inhibited by dilution of the solute in body fluids and by an increase in the volume of body fluids, especially in the major systemic arteries. These stimuli frequently are additive but may also be competitive. The net result, ADH release or inhibition of release, depends on the strength of the various stimuli.

Antidiuretic hormone acts on the collecting tubule of the kidney where it increases water permeability by activating receptors in the contraluminal

TABLE 6–3.—NORMAL DAILY
WATER BALANCE

WATER IN (ML)		WATER OUT (ML)	
Beverages	1,000	Insensible loss	850
Food	800	(lungs, skin)	
Metabolic°	300	Sweat	50
Total	2,100	Feces	100
		Urine	1,100
		Total	2,100

°Derived from oxidation of hydrogen-containing foods, e.g., fats and carbohydrates.

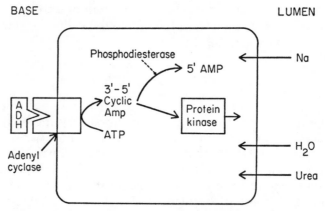

Fig 6–13.—Action of ADH (antidiuretic hormone) on a cell of collecting tubule.

membrane (Fig 6–13). (In the rat, it may also act on the last half of the distal tubule.) This leads to stimulation of adenylate cyclase and to accumulation of cAMP. Cyclic AMP acts on a protein kinase, which in turn increases the permeability of the luminal membrane to water by a factor of seven. Permeability of the membrane to urea and sodium also increases but to a lesser degree.

EFFECTS OF EXCESSIVE WATER INTAKE

When a large volume of water is ingested without solute, the concentration of solute in both the intra- and extracellular fluid compartments is reduced and, thus, osmolality falls. This change is detected by neurons of the supraoptic and paraventricular nuclei in the hypothalamus, which act as *osmoreceptors*. The exact mechanism of detection is not known. Possibly the neurons swell as extracellular fluid osmolarity falls and water enters them to keep intracellular osmolality equal to extracellular osmolality. In any case, dilution of the extracellular fluid causes secretion of ADH to be inhibited. In the absence of ADH, permeability of the collecting ducts to water is markedly reduced and urine, which has been made hypotonic in the ascending limb of Henle's loop by reabsorption of NaCl without water, remains dilute as it passes through the collecting duct on its way to the bladder. Further sodium and potassium absorption by the collecting duct may reduce urine osmolality to as little as 70 mOsm/L.

Conceptually, it is possible to divide a volume of dilute urine into two separate parts. One part will contain all of the solute that has been excreted in a concentration equal to that of plasma. This volume can be calculated by using the clearance formula, since the osmolar clearance (C_{osm}) is equal to the virtual volume of plasma cleared of all solute. Thus

$$C_{osm} = \frac{U_{osm}\,V}{P_{osm}} \qquad (5)$$

where U_{osm} is urine osmolality, V the volume of urine excreted per minute and P_{osm} the plasma osmolality.

The second part of the original urine volume will be water free of solute. By convention, this volume is called *free water*, and C_{H_2O} is used as a notation. Thus

$$V = C_{H_2O} + C_{osm}$$
$$C_{H_2O} = V - C_{osm}$$

where C_{H_2O} and C_{osm} are both expressed in milliliters per minute. The free water clearance is simply the difference between the actual volume of urine excreted and the hypothetical volume represented by C_{osm}. When urine is hypotonic, C_{H_2O} is always positive. When urine is isotonic, C_{H_2O} is zero and when the urine is concentrated, C_{H_2O} is negative (free water has been absorbed). Thus excessive water intake is followed by excretion of free water, leading to restoration of normal plasma and intracellular osmolality.

EFFECTS OF WATER LOSS OR DEPRIVATION

If water is lost in excess of solute, such as in heavy sweating, osmolality of body water increases. The osmoreceptors detect the change and cause release of ADH from the posterior pituitary. ADH circulates and acts on the basal and lateral membranes of the cells of the collecting tubules, as described above. As previously indicated, the fluid that enters the cortical collecting tubule from the distal tubule has an osmolality of about 150 mOsm/L. The ADH-induced increase in water permeability of the collecting tubule allows the filtrate to come into osmotic equilibrium with plasma in the cortex and then with the interstitial fluid of the medulla. Since the medullary interstitial fluid becomes progressively more hypertonic as the distance from the cortical-medullary junction to the papilla increases (see below), water is absorbed down an osmotic concentration gradient as it flows through the collecting duct. Thus the remaining solute in the filtrate is progressively concentrated in smaller and smaller volumes of water. The maximum urine osmolality achievable is equal to the osmolality of the interstitial fluid at the tip of the papilla. In man under extreme stress, this value is about 1,200–1,500 mOsm/L. This process conserves water by recovering more than 99% of the water filtered. However, it is important to emphasize that the kidney cannot restore fluid lost from the body. This can only be accomplished by ingestion of water stimulated by thirst.

URINARY CONCENTRATING MECHANISM

Details of the mechanism involved in the concentration of solute in urine have been the subject of considerable research and a brief description of the process will be given here. As noted above, the interstitial fluid of the medulla becomes progressively hyperosmotic from the cortical-medullary junction to the papillae. If a number of species are considered, the width of the medulla relative to the cortex correlates directly with the maximum osmolality achievable when the animal is deprived of water. Desert rats, which can excrete urine with an osmolality of 5,000 mOsm/L, have papillae 11 times longer than the thickness of the cortex. In man, the equivalent ratio is three to one.

Maintenance of medullary hypertonicity is a function of the complex anatomy of this region of the kidney and of the special characteristics of the membranes of the tubules that traverse it (Fig 6–14). The outer medulla contains thin descending limbs of the loops of Henle from both cortical and juxtamedullary nephrons. The descending thin limbs of the cortical nephrons reverse direction and quickly become thick ascending limbs. Only the juxtamedullary nephrons send descending limbs into the inner medulla. The descending thin limb turns in the papilla to become the thin ascending limb, which returns the tubular fluid to the outer medulla. At the juncture of the inner and outer medulla, the thin limb becomes the thick ascending limb, which returns to the glomerulus where the nephron originated. The vascular supply of the medulla consists of straight vessels (vasa recta), which originate from the efferent arterioles of juxtamedullary glomeruli and enter and leave this region in the form of a hairpin loop. The two sides of the vascular loop are interconnected at several levels. Collecting ducts traverse the medulla from the cortex to papilla.

The relationship between these structures is not haphazard, but highly or-

Fig 6–14.–Movement of salt, water and urea across loop of Henle and collecting duct.

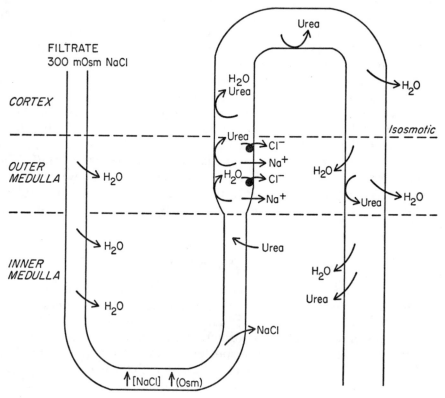

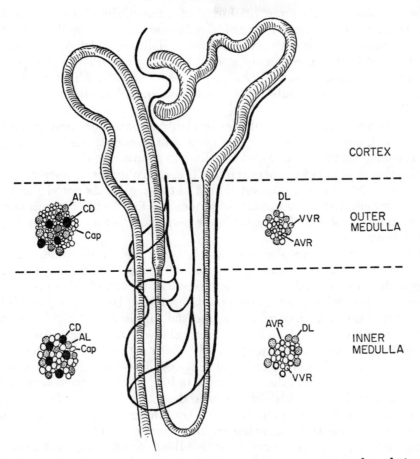

Fig 6–15.—Juxtamedullary nephron. The small inserts represent the relationship among the various structures. *AL* = ascending limb; *CD* = collecting duct; *Cap* = capillaries; *DL* = descending limb; *AVR* = arterial vasa recta; *VVR* = venous vasa recta.

ganized (Fig 6–15). In the outer medulla, vascular bundles are formed, which are made up of a central core of arterial and venous vasa recta, the former predominating. In a ring around the central core are thin descending limbs of the loop of Henle and additional venous vasa recta. Both the thin and thick parts of the ascending limb and the collecting ducts form bundles that are interspersed between the bundles containing the thin descending limbs. In the inner zone of the medulla, several thin ascending limbs surround a collecting duct and in the outer medulla thick ascending limbs surround them. Capillaries, which form between ascending and descending limbs of the vasa recta, surround the collecting ducts and adjacent ascending limbs of Henle's loops. In this anatomical arrangement, vessels or tubules containing fluid flowing toward the papilla are in close proximity to vessels or tubules that contain fluid flowing in the opposite direction. Thus, anatomically, the conditions for a countercurrent mechanism are met, i.e., fluid flowing in one direction always is near fluid flowing in the opposite direction.

The osmolality of the interstitial fluid of the medulla of all mammals increases progressively from the cortical-medullary junction to the papilla. In
the outer medulla, most of the increase in osmolality is contributed by sodium
chloride, but in the inner medulla NaCl and urea contribute equally to the
increase in interstitial osmolality over plasma osmolality. About one third of
the total increase in osmolality occurs in the outer, and two thirds in the inner
medulla.

Maintenance of a high medullary osmolality depends on three major processes. First, there must be an active process to accumulate solute. This process is confined to the thick ascending limb as it traverses the outer medulla,
and it is this segment of the nephron that is responsible for the increased concentration of NaCl in the interstitial fluid of this region. As noted above, the
ascending limb actively transports chloride (see Fig 6–14). Sodium but not
water follows passively. The result is a hyperosmotic interstitium and hyposmotic filtrate.

The second process results in accumulation of urea in the inner medulla.
Urea is reabsorbed from the medullary collecting duct into the inner medulla
along with water when ADH is present (see Fig. 6–14). Urea enters the thin
ascending (and perhaps the thin descending) limb of Henle's loop and is carried through the thick ascending limb, distal tubule and cortical and outer
medullary collecting duct, all of which are relatively impermeable to urea, to
the inner medullary collecting duct, where it may recycle by entering the
medullary interstitium or be excreted. The result of the anatomical arrangements of the medulla and the urea permeabilities of the various segments of
the nephron is the accumulation of urea in the inner medulla. Of interest, the
greater the need to concentrate the urine (and therefore the more ADH present) the higher the urea concentration in the inner medulla.

The third major important component of the concentrating mechanism are
the vasa recta, which supply the medulla with blood. In most tissues the pattern of blood flow is designed to carry away solutes. In the medulla of the kidney, the uncontrolled loss of solute would soon reduce the osmolality of the
interstitial fluid to that of the rest of the body and the concentrating function
of the medulla would be lost. To prevent this, the blood supply to the medulla
has evolved a unique anatomical structure. The efferent arterioles from juxtamedullary glomeruli enter the medulla perpendicularly and soon break up
into several descending vasa recta. At various levels in the medulla, vasa recta
leave the bundle and divide into a capillary network surrounding the collecting ducts and thick ascending limbs and then form the ascending vasa recta.
Descending and ascending vasa recta travel adjacent to one another in the
vascular bundles (see Fig 6–15). Plasma leaving juxtamedullary glomeruli
enters the medulla with an osmolality only slightly greater than that of plasma
elsewhere because of the increased protein concentration (the result of filtration). As the vasa recta descend into the hyperosmotic medulla, water moves
across the capillary wall into the interstitium. However, there is also a
chemical gradient for NaCl and urea from interstitital fluid to plasma. Since
the vessel wall is permeable to both of these substances, they move down
their gradients into the plasma.

On the ascending side, the conditions are reversed. The hypertonic fluid in the vessel is moving through areas where the interstitial fluid has a lower osmolality and a lesser concentration of NaCl and urea. Hence, water enters the vasa recta, and urea and NaCl leave to enter the interstitium. Furthermore, the protein concentration of the efferent arteriolar plasma was elevated owing to filtration in the glomerulus. The protein was further concentrated in the descending vasa recta. Thus in the ascending vasa recta, the high colloid osmotic pressure increases the uptake of water. The overall effect of the vasa recta is to keep NaCl and urea trapped in the inner medulla and to carry away the water that has been reabsorbed into the medulla from the collecting duct.

The processes so far discussed produce a hyperosmotic interstitium and a hyposmotic filtrate. How then does the kidney produce concentrated urine? The fluid entering the distal tubule is quite dilute. In the presence of ADH, water is absorbed down its osmotic gradient so that by the end of the *cortical* collecting duct the filtrate osmolality equals that of plasma. As the tubular fluid passes through the *medullary* collecting tubule, water moves across the duct epithelium into the hyperosmotic medullary interstitial fluid down an osmotic gradient. In the absence of ADH, the amount of fluid absorbed is very small and the tubular fluid remains hypotonic. In the presence of ADH, the permeability of the duct epithelium to water is very high. When sufficient ADH is present, the osmolality of the urine at the tip of the papilla is equal to that of the medullary interstitial fluid at that point. Furthermore, since ADH also increases the permeability of the medullary collecting duct to urea as well as water, urine urea concentration is the same as that of the interstitial fluid.

Quantitatively, the action of the concentrating mechanism on filtrate is measured by calculating the free water clearance (see above). Since the osmolality of concentrated urine is greater than that of plasma, C_{H_2O} must be negative. The clearance $-C_{H_2O}$ represents the water absorbed to increase the osmolar concentration of urine to levels greater than plasma osmolality.

Obviously, the concentrating mechanism for urine is incapable of restoring total body water to normal. It can only conserve what fluid the body has by concentrating the obligatory daily solute excretory load into the smallest possible urine volume. Human beings ingesting a normal diet have an excretory solute load of about 600 mOsm per day. If the maximum concentration that can be achieved is 1,200 mOsm/L, the minimum achievable urine volume is 500 ml. Fortunately, modern man has been able to provide himself with ready access to water so that the ability to concentrate urine maximally is not a dominant factor in survival.

Patients with severe chronic renal failure have an impaired concentrating mechanism and excrete urine with the osmolality of plasma. They have a urine volume of about 2 L per day and have no difficulty remaining in water balance. However, patients with pituitary diabetes insipidus who are without ADH excrete a markedly dilute urine (the osmolality is about the same as that of the fluid leaving the thick ascending limb) because their collecting tubules are relatively impermeable to water. These patients excrete large volumes of urine (15 L per day) and are in danger of becoming dehydrated if access to water is limited.

The thirst mechanism is responsible for the acquisition of water and can be stimulated by an increase in plasma osmolality and by a reduction in the effective arterial volume. Detectors in the hypothalamus can detect a change of less than 6 mOsm/L. Thirst initially abates when 80% of the water deficit has been ingested. Within a short time, however, the drive for water recurs until water volume is restored. The renin-angiotensin system also appears to be involved in the control of water acquisition; animal experiments have demonstrated that angiotensin II directly stimulates thirst.

CLINICAL WATER ABNORMALITIES

Dehydration is usually reflected in an elevation in body water osmolality or, since NaCl is a primary contributor to extracellular fluid osmolality, dehydration is present when the plasma sodium concentration is elevated. A heat-acclimatized individual in a very hot environment loses water by convection through the skin and by evaporation of hyposmolar sweat from the skin. In this setting his urine is small in volume and high in osmolality.

Clinically, the degree to which urine is concentrated is determined by measuring the specific gravity of the urine. A concentrated urine has a specific gravity of up to 1.035, whereas urine with the osmolality of plasma has a specific gravity of about 1.010. Dilute urine may have a specific gravity as low as 1.002. Dehydrated individuals should have a urine specific gravity of 1.020 or more.

Clinical states involving simple water overload are unusual. On occasion, damage to the brain has led to a drive for massive water ingestion. If the rate of water intake exceeds the maximum rate of water excretion (1,000 – 1,500 ml per hour), body fluids will become hyposmolar and plasma sodium concentration will fall. Reduction in extracellular fluid osmolality leads to reduced intracellular osmolality because fluid enters the cells. If this occurs suddenly, impaired function of the neurons of the brain may lead to confusion, stupor and eventually coma.

REGULATION OF EXTRACELLULAR VOLUME
AND SODIUM BALANCE

As noted previously, the distribution of water between the intra- and extracellular fluid compartments is determined by the amount of solute in each. Since sodium and its attendant anion are the major extracellular solutes, control of the body content of sodium determines extracellular fluid volume. Over the past three decades, several elements of the control mechanism for maintenance of sodium balance have been described and it is difficult to overemphasize their importance. For instance, if ingested sodium is retained in excess of need, the size of the extracellular fluid compartment will increase, leading to an increase in blood pressure and accumulation of fluid in the interstitial space. Interstitial fluid accumulation is especially dangerous when it occurs in the lung and interferes with oxygenation of hemoglobin. Conversely, loss of sodium in excess of intake and the subsequent decrease in extracellular fluid volume lead to decreased blood pressure and failure to perfuse the peripheral tissues adequately. Thus it is imperative that sodium balance be accurately maintained.

Sodium chloride is found in virtually all diets, and people of many cultures habitually use it to season food. The average dietary intake varies from population to population, person to person and from day to day in the same individual. The amount of salt ingested by each person induces variations in the volume of the extracellular fluid compartment. These variations will be minimized if external sodium balance (sodium intake = sodium excretion via all routes) is accurately maintained on a day to day basis. Although some ingested sodium may be lost through the skin in sweat or in gastrointestinal secretion, the kidneys are responsible for controlled excretion of this mineral. More than 95% of ingested sodium is excreted in the urine. The control mechanisms for sodium excretion act to regulate reabsorption of filtered sodium. For instance, if dietary salt intake is very low and losses through sweat exceed intake, the kidney must reabsorb all of the filtered sodium. Conversely, if intake is excessive, tubular absorption must be inhibited and a greater fraction of the filtered sodium must be allowed to pass into the urine.

Sodium excretion is not an end in itself but, rather, a function of the mechanism responsible for controlling extracellular fluid volume or some part of it. As noted, the extracellular fluid space may be divided into the interstitial fluid compartment and the intravascular compartment. Although these compartments are closely related, the intravascular compartment, and especially that component in the arterial tree, is of greatest importance.

The cardiac output is distributed under pressure to all tissues of the body. The flow through each organ is determined not only by the pressure gradient across its vessels, but also by the fraction of total body vascular resistance that its vascular bed provides. As the need for more oxygen develops or when end-products of metabolism begin to accumulate, the arterioles of the affected organ dilate, resistance is reduced and blood flow increases. Although it cannot be measured, the arterial system to the level of the distal arterioles must be anatomically larger after the vascular bed of the organ is dilated than in its previous state. The volume of blood within the anatomical vascular space must increase to maintain blood pressure. In a manner not understood, the body appears to be able to measure the size of its anatomical arterial volume, judge the availability of blood required to keep it full and under optimal pressure, and then regulate sodium excretion and to some extent water excretion according to that measurement. When anatomical arterial volume increases, salt and water are retained to ensure that it will remain full and pressurized. When anatomical volume is reduced and an excess blood volume results, salt and water are excreted until blood volume and arterial volume again have a normal relationship.

An excellent example of the renal manifestations of a rapidly changing arterial volume came from the study of individuals with large AV fistulas produced by trauma. When the fistulas were closed, salt and water excretion increased. The change in renal function was associated with a rise in diastolic pressure because the rate of emptying of the arterial tree decreased and its blood volume exceeded the physiologically appropriate volume. The kidney responded to what must have been interpreted by the body as a surfeit of arterial blood volume. When the fistulas were reopened, the kidney retained water and salt in order to increase blood volume.

There are other less dramatic experiments that expose the presence of a control system of this nature. When a subject is immersed in a thermo-indifferent bath, blood is redistributed from the extremities to the central venous system because of the external hydrostatic pressure. As a result, cardiac output is increased according to Starling's law of the heart and arterial volume expands. Natriuresis and diuresis ensue, which restores arterial blood volume to a physiologically appropriate size.

When an individual stands quietly, blood pools in the veins of the lower extremities and the return of blood to the heart falls. This results in decreased cardiac output, a fall in arterial blood volume and decreased pressure in the arterial system. The pressure drop is immediately compensated by the sympathetic nervous system but, again in a manner not understood, the reduced arterial blood volume is detected and the kidneys are instructed to conserve salt and water until the appropriate arterial volume is restored.

Several sites have been suggested for the receptors that might play a role in this control system. They include the baroreceptors in the cardiac atria, carotid sinuses and aortic arch. Additional receptors in the pulmonary artery, both ventricles and the great veins in the thorax have been described, but their physiologic importance is unknown.

It is possible that each of these detectors is involved in the control mechanism. The signals generated at the various receptor sites are integrated in several centers of the brain located beneath the floor of the third ventricle. Here also are found the supraoptic nuclei and paraventricular nuclei that are involved in synthesis and secretion of ADH. These nuclei respond not only to changes in extracellular fluid osmolality, but also to changes in blood volume. Thus in response to hemorrhage, ADH is released immediately, and levels remain high until the deficits are corrected. Posterior to these nuclei is the vasomotor center, which contributes to control of peripheral resistance through the autonomic nervous system. In an adjacent area the thirst center is located. Thirst is stimulated both by increases in extracellular fluid osmolality and decreases in blood volume. In some animals a salt appetite center has been described that operates in a manner analogous to the thirst center. In rats, a pair of nuclei that participate directly in the control of sodium excretion has been found in the posterior aspect of this region. Together, these nuclei may be considered to form a control center for maintenance of the effective arterial volume.

Some signals generated by effective arterial volume sensors act without going through the CNS. Decreased blood flow to the kidney, perhaps as a result of decreased effective arterial volume, may lead to a decrease in tangential stretch of the afferent arterioles in the kidney and release of the enzyme renin. Renin in turn initiates a series of events resulting in sodium retention and preservation of effective arterial volume. Increased perfusion of the kidney, as might be induced by increased blood volume, reduces renin release, and that ultimately allows an increase in salt and water excretion.

Once the volume control center determines body requirements for sodium and water, renal function must be altered appropriately. The regulation of water excretion has been discussed earlier. This section will deal with the mechanisms that control sodium.

Historically, modulation of sodium excretion by changes in the GFR was the first mechanism considered. As originally envisioned, the GFR increased slightly with an increase in blood volume and decreased when the opposite occurred. Thus, if the absolute rate of sodium reabsorbed by the tubule was fixed, any positive increment in filtered load of sodium would be excreted. Similarly, when sodium had to be conserved, the GFR and filtered sodium load would fall so that the tubule could reabsorb a larger fraction of the sodium presented to it.

It has been demonstrated subsequently, however, that the reabsorptive rate for sodium by the tubule is not fixed but varies with the filtered load of sodium. For example, the normal filtration rate of the rat nephron is about 36 nl per minute, and therefore the filtered load of sodium per nephron is 5.11 pEq per minute. About two thirds of the sodium is absorbed by the first 60% of the proximal tubule. If the filtration rate increases to 48 nl per minute, the filtered load will be 6.81 pEq per minute, and about 4.54 pEq per minute will be absorbed by the initial 60% of the proximal tubule. Similarly, if single nephron GFR falls to 30 nl per minute, the filtered load of sodium will be 4.26, and 2.84 pEq would be absorbed. The changes in proximal tubular reabsorptive rate with changes in the rate of formation of glomerular filtrate are a manifestation of a phenomenon called *glomerular tubular balance,* which may be defined as follows: For any given steady state of the effective arterial volume, the fraction of filtered plasma reabsorbed at any point along the proximal tubule remains relatively constant despite variations in single nephron GFR.

To a large extent, the same holds true for the distal nephron where an increase in filtrate delivery out of the proximal tubule brings about an increase in reabsorption. The process has been termed *tubulo-tubular balance.* Thus, variations in GFR are not effective in controlling the rate of sodium excretion because alterations in tubular reabsorptive rate prevent the full effect of the changes in filtered load from being expressed as changes in sodium excretion. It is apparent, therefore, that for regulation of sodium excretion to occur, net tubular transport of this ion must be stimulated or inhibited.

REGULATION OF TUBULAR REABSORPTION OF SODIUM

Several factors influence the rate of sodium reabsorption by the renal tubule. They include aldosterone, Starling forces (hydrostatic and osmotic pressures) operating on the peritubular capillaries and possibly *natriuretic hormone.*

ALDOSTERONE. — This mineralocorticoid is synthesized and secreted by the adrenal cortex, acts to stimulate sodium transport by the tubule and is a major factor in control of sodium excretion. The primary stimulus for aldosterone secretion is angiotensin II, a peptide derived from a circulating alpha-2 globulin (see below).

Renin is secreted by modified smooth muscle cells, the juxtaglomerular or type J cells located in the walls of both the *afferent* and *efferent* arterioles of the glomerulus. The renin-secreting cells are closely associated with the cells of the *macula densa* and constitute the *juxtaglomerular apparatus,* as previously described. Numerous sympathetic nerve terminals are found in this region.

The regulation of renin secretion has not been fully worked out. Secretion is stimulated by reduction in arterial pressure or blood volume and by chronic sodium depletion. Secretion of renin is inhibited by an elevated blood pressure and by expansion of the extracellular fluid volume. At least three hypotheses have been advanced to explain the variations in renin secretion.

1. *Baroreceptor hypothesis.*—Reduction in afferent arteriolar transmural pressure, e.g., resulting from a fall in systemic arterial pressure, decreases the "stretch" of the type J cells and stimulates renin release.

2. *Macula densa hypothesis.*—An increase in the sodium load passing the macula densa stimulates renin release.

3. *Adrenergic hypothesis.*—Increased sympathetic nerve activity engendered, for example, by a fall in arterial or atrial pressure may cause renin release in addition to constricting the glomerular arterioles. The receptor involved in renin release appears to be of the *beta* type, possibly *beta-1.*

Whatever the precise mechanisms of renin release, this enzyme acts on the circulating alpha-2 globulin, "renin substrate," and splits off a decapeptide, angiotensin I. Two additional amino acids are subsequently split off, mainly in the lung, by *converting enzyme* to yield the octapeptide angiotensin II, which is biologically active. Angiotensin II stimulates *aldosterone* production and secretion by the cells of the zona glomerulosa of the adrenal cortex. Aldosterone, in turn, stimulates reabsorption of sodium in the collecting tubule and perhaps in the distal end of the distal tubule in a regulatory fashion. That is, increasing plasma aldosterone levels progressively increase sodium reabsorption up to a maximum. There does not seem to be a regulatory action of aldosterone on other parts of the nephron. Thus, aldosterone regulates the transport of only 5% of the filtered load of sodium and is without a regulatory effect on the proximal tubule and Henle's loop, which reabsorb 95% of the filtered sodium. However, since 5% of the filtered sodium is equivalent to about 80 gm NaCl per day or 10–20 times the normal intake, this hormone is potentially of great importance in regulating salt balance.

When sodium ingestion declines or when sodium losses exceed intake, renin secretion is stimulated, and arterial angiotensin aldosterone concentrations increase. By stimulating the reabsorption of sodium, aldosterone acts to preserve the salt already in the body. In contrast, when sodium excretion is required, the secretion rate of aldosterone is very low, owing to the absence of angiotensin II, and recovery of filtered sodium from the collecting tubule is less avid.

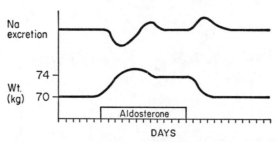

Na
excretion

74 —
Wt.
(kg) 70 —

Aldosterone

DAYS

Fig 6–16.—Response of a normal person to administration of aldosterone. The hormone causes sodium retention, and water is retained to maintain normal osmolality. Weight increases. By the fourth day, Na excretion increases (escape) and weight stabilizes. Administration of the hormone is stopped, Na excretion increases and weight returns to normal.

Aldosterone cannot be the only regulator of tubular sodium transport, however. If this hormone is administered in large amounts to animals, sodium retention does not persist indefinitely. Rather, after a period of marked sodium retention, sodium excretion increases and external balance is once more achieved despite the persistent sodium-retaining stimulus (Fig 6–16). This phenomenon has been termed "mineralocorticoid escape" and is due to the operation of other determinants of sodium excretion.

PERITUBULAR STARLING CAPILLARY FORCES.—Evidence for participation of other mechanisms in the control of tubular sodium transport has been accumulating swiftly in the last decade. It has been observed that the rate of resorption of filtrate by the proximal tubule is limited by the rate at which the resorbed fluid is carried away by the peritubular capillaries. When the resorbate is not picked up rapidly enough, it leaks back into the tubular lumen between the cells of the tubular epithelium (Figs 6–17 and 6–18). Thus forces which govern the movement of fluid across capillary mem-

Fig 6–17.—Resorbate moves easily from intercellular space to capillary. High plasma oncotic pressure (OP) and decreased capillary hydrostatic pressure (HP) favor capillary removal of fluid. There is little back diffusion into the lumen (thin arrow).

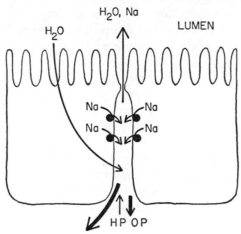

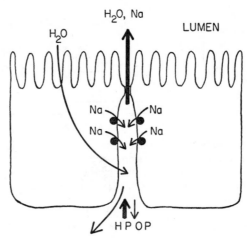

Fig 6–18.—Resorbate is not as well transported because of reduced plasma oncotic pressure *(OP)* and increased hydrostatic pressure *(HP)*. Back-leak into the lumen is markedly increased *(thick arrow)*.

branes operate in the kidney to partially control sodium reabsorption. For instance, if enough salt and water are ingested to increase extracellular fluid volume, an increase in hydrostatic pressure in the peritubular capillaries might follow. Furthermore, the increased plasma volume would lead to dilution of plasma proteins and reduction in peritubular capillary oncotic pressure. Both these events would slow capillary absorption of reabsorbed filtrate and ultimately limit net proximal tubular transport. The escape from the sodium-retaining effects of mineralocorticoids probably involves both these physiologic changes. Suppression of proximal sodium absorption leads to increased delivery of sodium and water to the distal nephron where part of the increment is absorbed by the distal tubule and the collecting duct, but a fraction escapes and is excreted. It is unknown to what extent distal nephron transport may be affected by changes in rates of capillary removal of reabsorbed filtrate.

When extracellular fluid volume is reduced, changes in the opposite direction occur. Hydrostatic pressure in the peritubular capillaries declines and oncotic pressure of the plasma is increased because of the increased concentration of protein in the plasma and because of an increase in the fraction of plasma water filtered at the glomerulus. The combination of decreased capillary hydrostatic pressure and increased plasma oncotic pressure assures efficient removal of the reabsorbed filtrate and thus decreases back diffusion from peritubular space to tubular lumen. Proximal tubular transport mechanisms become maximally effective under these conditions.

NATRIURETIC HORMONE.—In recent years considerable evidence has accumulated that there is another hormone that controls the rate of sodium excretion by the kidney. This hormone is thought to be natriuretic (increases salt excretion), and several investigators have suggested that its action is on the distal portion of the nephron (thick ascending limb, distal tubule and perhaps the collecting duct). Various assay systems have been used to demonstrate the substance in plasma and urine of volume-expanded animals or in uremic animals that must excrete a large fraction of their filtered sodium. The source and nature of the proposed hormone are not known and, in fact, some investigators doubt that it exists.

In order to understand the operation of the volume control mechanism as a whole, let us consider the response to hemorrhage (Fig 6–19). As blood is lost, intravascular volume falls and baroreceptors in the arterial tree and atria are stimulated. The vasomotor center responds by increasing sympathetic nervous system tone, and catecholamines are released from the adrenal gland. In response, the small arterioles of all organs but the heart and brain constrict. Peripheral resistance to blood flow increases, and arterial volume is maintained since rapid runoff into the capillary bed is prevented. The reduced renal perfusion and increased beta receptor stimulation cause a marked increase in renin secretion by the kidney. The resultant increase in plasma angiotensin II stimulates thirst, ADH secretion and aldosterone secretion. Thus angiotensin not only stimulates conservation of water but also its acquisition. The increased resistance to blood flow through the kidney may also reduce GFR, but to a lesser degree than reduction in blood flow. This disparity results in an increased filtration fraction and, therefore, increased peritubular capillary oncotic pressure. This change increases the efficiency of the tubular transport system, leading to increased resorption of filtered sodium and water. Although we cannot be certain because a sensitive assay is not available, it is

Fig 6–19.—Response of the organism to a decrease in blood volume. See text for details.

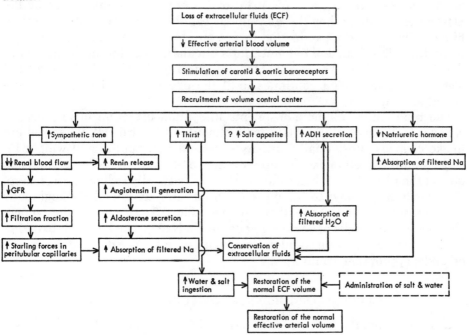

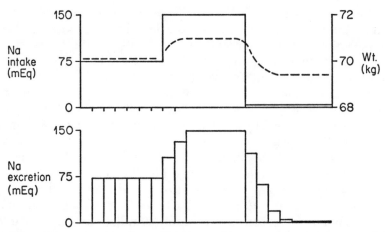

Fig 6–20.—Changes in weight and urinary Na excretion in response to changes in dietary Na.

reasonable to assume that secretion of natriuretic hormone is inhibited during hemorrhage. The result of all these processes is conservation by the renal tubules of filtered salt and water to the greatest extent possible and acquisition of water for re-expansion. In some species there may be an increase in salt appetite due to a mechanism entirely analogous to the thirst mechanism.

When extracellular fluid volume is expanded secondary to intake of much salt and water, the process is reversed. The sympathetic nervous system is inhibited, secretion of renin is suppressed and thirst is absent. ADH secretion stops and aldosterone secretion is reduced. Hydrostatic pressure in peritubular capillaries increases and plasma proteins are diluted, thus reducing the Starling forces that allow the peritubular capillaries to carry off reabsorbed filtrate. This leads to a reduction in net tubular sodium transport. Furthermore, there is substantial evidence for a circulating natriuretic hormone under these conditions. The result of these events is an increased rate of salt and water excretion, which continues until the effective arterial volume is restored to normal.

The time course of the adjustments in sodium excretion in response to changes in dietary sodium intake is illustrated in Figure 6–20. When sodium intake changes substantially, urinary excretion lags behind and weight increases until excretion catches up with intake. Usually the weight gain is no greater than 1–1.5 kg. When sodium intake is drastically reduced to near zero, it may take 2 to 3 days to bring excretion into line with intake. Weight falls but, after several days, it usually is not more than 0.5–1 kg below the average weight on a salt intake of 68 mEq per day (4 gm NaCl). The kidney is extremely efficient in conserving sodium. Less than 1–2 mEq per day may be excreted under appropriate conditions.

CLINICAL DISTURBANCES OF SODIUM BALANCE

Clinically important disorders of sodium balance may be divided into two general groups: (1) those characterized by excessive accumulation of salt and water and (2) those in which salt (and water) are lost from the body inappropriately.

As noted above, aldosterone controls sodium absorption by the collecting tubule. In the absence of aldosterone, the rate of transport of sodium by this structure is very slow and a large fraction of the sodium reaching the duct is lost in the urine. If dietary sodium is not sufficient to restore the losses, the extracellular fluid compartment shrinks, blood pressure falls, vital organs are not adequately perfused and death soon follows. This is not a disease of the kidney but rather of the adrenal gland. Salt loss occurs through the kidney because the organ has not received the "instruction" to conserve this ion. The disease process is reversed by administration of aldosterone or other salt-retaining adrenocortical steroids.

Some patients have diseases characterized by kidney damage that limits the ability of the tubule to reabsorb filtered sodium. Patients with chronic kidney infections may have this problem. They must be treated with NaCl supplements in the diet, since administration of sodium-retaining hormones is ineffective.

Diseases that cause increased total body sodium and water are commonly seen in the practice of medicine. At times, salt retention is the result of a disease of the kidney. Acute glomerulonephritis is characterized by inflammation of the capillaries in the glomeruli. When the inflammatory process is severe enough, the rate of filtrate formation may be markedly reduced and, as a result, salt and water excretion is impaired. If ingestion of these substances continues, they accumulate as extracellular fluid because they cannot be excreted.

Glomerular capillaries may become permanently damaged and, as a result, albumin in the plasma may cross the glomerular basement membrane and be lost in the urine. If the rate of albumin production is less than the rate of loss, the albumin concentration of plasma falls and plasma colloid osmotic pressure declines. Fluid that leaks from capillaries into the interstitial space normally is returned to the vascular space by the force represented by plasma colloid osmotic pressure. Obviously, as this force falls, the fluid remains in the interstitial space and the intravascular volume declines. The decrease in intravascular volume is detected and the kidney is instructed to increase tubular salt and water absorption and preserve the remaining extracellular fluid. Newly acquired salt and water is also conserved until the arterial plasma volume is large enough to satisfy the needs of the body.

Fluid accumulation may occur when the kidney is not diseased but another organ is. Diseases of the heart often lead to reduced cardiac output and the volume of blood that enters the arterial tree may then be inadequate to meet organ needs. Under these conditions, the volume control center responds as it would to hemorrhage. The kidney is "instructed" to conserve salt and water by the hormones aldosterone and ADH. Both the intravascular and interstitial compartments of the extracellular fluid increase. If the heart is not too badly damaged, the increased blood volume may raise cardiac output via the Frank-Starling effect. In many instances this compensation is inadequate and fluid continues to accumulate in the interstitial space of most organs including the lung, where it may be lethal because it may interfere with oxygen transport.

REGULATION OF POTASSIUM BALANCE

Total body potassium in adult males is about 50 mEq/kg body weight. As explained in Chapter 1, 95% of the body's potassium is in the intracellular fluid; extracellular concentration of potassium is low, the normal range being 3.5–4.5 mEq/L.

Regulation of potassium excretion is as important to the well-being of the organism as is regulation of sodium excretion. Deficiency or excess of potassium may produce considerable morbidity or even death. Failure to excrete the 50–100 mEq of potassium normally ingested daily produces *hyperkalemia*, which can eventually cause paralysis of skeletal and cardiac muscle. Death may occur when serum K$^+$ rises to or above 8–9 mEq/L. Decreased total body potassium, secondary to negative potassium balance, may lead to *hypokalemia*, which also may produce skeletal and smooth muscle paralysis. (See Chapters 1 and 2, where the importance of K$^+$ in regulating the resting membrane potential and excitability is described.) In order to achieve potassium balance, the total amount of potassium lost by all routes must equal the amount ingested. Since only 5–10 mEq of potassium is lost daily in the stool and usually less than 5 mEq is lost in sweat, the urine is by far the most important route for potassium excretion.

The proximal tubule reabsorbs about 65% of the filtered potassium under all conditions of potassium balance. Probably, both active and passive processes are involved. Since the absorption rate is independent of the body stores of potassium, regulation of renal potassium excretion in relation to the body's needs is *not* a function of the proximal tubule. The regulation of potassium excretion occurs in the distal tubule and collecting duct.

In potassium-depleted animals, potassium reabsorption occurs in both of these segments. The amount of potassium in the final urine is inversely proportional to the total body potassium deficit. When the deficit is 100–200 mEq, urine potassium excretion is about 10–20 mEq per day. With deficits of 200–400 mEq, urinary potassium excretion drops to 5–10 mEq per day and, with greater deficits, less than 5 mEq per day is excreted. Thus, a greater deficit of potassium is required to achieve maximal potassium conservation than is required to achieve maximal sodium conservation. In the usual situation, 50–100 mEq of potassium must be excreted each day, and most of this amount enters the filtrate in the last half of the distal convoluted tubule.

The process of potassium secretion by the distal tubule is complex. As might be expected, however, potassium movement is a function of the permeability properties of the luminal and peritubular membranes of the cells making up the tubular epithelium and of the ion pumps present on each membrane.

The concentration of potassium in the distal tubular cells is at least 10 times that of plasma, and a chemical gradient exists for diffusion of this ion into either the filtrate or peritubular fluid. However, permeability of the lateral and basal membranes of these cells is high for potassium and chloride, but low for sodium. On the other hand, the luminal membrane permeabilities for sodium and potassium are almost equal. Therefore, the electric potential

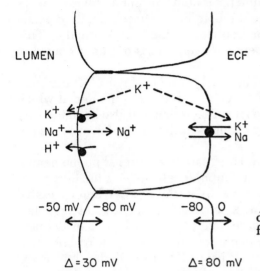

Fig 6–21.—Potassium secretion by distal tubular cell. *ECF* = extracellular fluid.

across the peritubular membrane is close to that predicted for a potassium diffusion potential. It has a value of about −70 to −80 mv. Therefore, potassium movement across the membrane is favored by a chemical gradient and opposed by a substantial electric gradient. High intracellular potassium concentrations are maintained by an active sodium-potassium exchange pump (Fig 6–21).

The luminal membrane potential is much less than the contraluminal membrane potential and has a value of 20–40 mv. This is due to the fact that the luminal membrane permeability for sodium is relatively high. Thus the lumen of the tubule is electrically negative compared to the peritubular space. It follows that the electric force opposing the movement of potassium out of the cell is considerably less on the luminal side than on the peritubular side and, as a result, net potassium movement secretion occurs in the direction of the filtrate rather than of the interstitial space.

The Nernst equation describes the distribution of an ion in fluid bathing a membrane or epithelium in the presence of an electric potential difference (PD) across the membrane. By applying this equation to the distal tubule epithelium, it is possible to estimate what the tubular fluid potassium (K+, TF) should be, given a PD of −50 mv and an extracellular fluid K+ concentration (K+, ECF) of 4 mEq/L.

$$PD = -61 \cdot \log \frac{K^+,TF}{K^+,ECF} \qquad (6)$$

$$-50 = -61 \cdot \log \frac{K^+,TF}{4}$$

$$6.6 = \frac{K^+,TF}{4}$$

$$K^+,TF = 26 \text{ mEq/L}$$

Under normal conditions, tubular fluid potassium concentrations are always less than this value. In fact, the ratio of tubular fluid/extracellular fluid potassium concentration is always less than predicted by the Nernst equation. This observation means that there must be an active process or pump removing potassium from the tubular fluid.

Net K^+ secretion will then be the difference between the rate of K^+ diffusion down the electrochemical gradient from cell to lumen and the rate at which K^+ is pumped from lumen to cell. The rate of secretion will also be influenced by changes in intracellular K^+ and by changes in the luminal membrane potential.

The regulation of K^+ secretion is in large part an intrarenal phenomenon. The potassium concentration of the distal tubular cells reflects the intracellular potassium concentration of the other cells in the body. When excessive potassium is ingested, the intracellular K^+ concentration rises and net potassium secretion occurs. Conversely, excessive potassium loss or inadequate intake cause intracellular potassium depletion; net potassium reabsorption occurs because the rate of potassium diffusion into the tubular fluid is exceeded by back absorption.

In addition to intrarenal control, potassium excretion is also influenced by aldosterone. Increased potassium intake initially elevates K^+ concentration, which directly stimulates the release of aldosterone from the adrenal cortex. Aldosterone then stimulates distal tubular potassium secretion, probably by increasing distal tubular intracellular potassium concentration by augmenting the activity of the peritubular membrane sodium-potassium pump.

Several other factors may influence potassium secretion. Both alkalosis and acidosis have an effect. In acidosis, hydrogen ions enter cells where they are buffered by the large impermeable anions. Potassium ions leave the cells in order to maintain electroneutrality. In spite of the increased plasma $[K^+]$ that results (about 0.6 mEq for each 0.1 pH unit change), there is a reduction in potassium excretion. Conversely, in alkalosis, intracellular buffering is reduced and potassium enters cells. Thus *alkalosis increases intracellular potassium concentrations,* including those of the distal tubule, and stimulates K^+ secretion. Plasma potassium is usually reduced in alkalosis. Conversely, *acidosis reduces intracellular potassium concentration,* reduces potassium secretion and may be associated with hyperkalemia.

The rate of potassium excretion is also related to the rate of delivery of filtrate to the secretory site. Since luminal fluid potassium concentration reaches its equilibrium value very rapidly (in less than 5 seconds), potassium excretion is dependent on the rate of flow of filtrate at the secretory site. Thus, an increase in flow rate will increase potassium excretion, and markedly decreased rates of flow will limit potassium secretion.

The collecting duct also may secrete potassium. Although most micropuncture studies have demonstrated that virtually all the potassium in the urine can be accounted for by distal tubular secretion, under some conditions potassium secretion may occur in the collecting tubule. Sodium depletion is the most important of these conditions. When potassium secretion does occur in the collecting tubule, it occurs by a sodium-potassium exchange mechanism.

CLINICAL DISTURBANCES OF POTASSIUM BALANCE

Potassium is plentiful in most of the food we eat. However, large amounts of potassium may be lost in diarrheal fluid, as the result of vomiting or as the result of use of diuretics. Patients with tumors of the adrenal cortex that secrete mineralocorticoid hormones such as aldosterone excrete large amounts of potassium when sodium intake is not restricted.

The chief symptom of potassium depletion is weakness and, ultimately, paralysis of muscles. The pathophysiologic basis for these symptoms is hyperpolarization of muscle cells, which is due to a marked increase in the ratio of intra- to extracellular potassium activities when extracellular potassium concentrations fall to very low levels. Smooth muscle of the intestine is also affected, and gastric atony and a small bowel ileus attend potassium depletion. The physiologic effects of potassium depletion on the electrophysiology of the heart are often seen in the ECG, which may show development of a U wave, reduction in amplitude of the T wave and S–T segment depression. Potassium depletion also decreases the ability of the kidney to concentrate urine. Bicarbonate resorption and ammonia secretion are stimulated by potassium deficiency (see below).

Hyperkalemia is unusual in clinical medicine because the kidneys are so efficient in excreting this ion. Even in chronic renal failure when the GFR has been reduced to 5 ml per minute or less, potassium excretion matches ordinary potassium intake. Only when renal failure causes urine volume to fall below 1 L per day is hyperkalemia likely to occur. In exceptional cases, however, when aldosterone secretion is reduced or absent, hyperkalemia may occur even when urine volumes are adequate.

Elevation in plasma potassium above 7 mEq reduces the membrane potential of excitable cells. Conduction of action potentials is impaired and muscle contraction cannot occur. Paralysis usually starts in the muscles of the lower extremities but, on occasion, the heart muscle may be affected first. The ECG demonstrates peaked T waves owing to very rapid repolarization. The QRS widens, the P–R interval lengthens and the P wave may disappear. Eventually the pattern may look like a sine wave and finally the heart stops.

DIURETICS

As described above, some diseases are characterized by a net gain of sodium and water because the regulatory system fails to allow the kidney to excrete the ingested load. When the retained fluid accumulates to the point where it interferes with the function of other organ systems, physicians administer drugs called diuretics, which inhibit tubular absorption of sodium and other electrolytes, thereby also causing increased water loss in the urine.

At least five types of diuretics are available.

1. CARBONIC ANHYDRASE INHIBITORS.—These agents act primarily on the proximal tubule, where they inhibit secretion of hydrogen ions into the tubule and decrease bicarbonate and sodium absorption.

2. THIAZIDES.—Most members of this group also have mild carbonic anhy-

drase-inhibiting activity, but their major effect is to inhibit absorption of sodium and other electrolytes in the distal convoluted tubule. The diuretic action of these agents is mild and, thus, they are used in conditions in which massive salt and water losses are not desired, e.g., hypertension. The thiazides produce significant losses of potassium either by inhibiting the luminal membrane potassium pump or by increasing the flow of filtrate past the secretory site.

3. ETHACRYNIC ACID.—This is an organic acid that blocks sodium chloride reabsorption in the thick ascending limb as well as in the distal tubule. It therefore inhibits the ability of the kidney to concentrate and dilute urine. Ethacrynic acid reacts with functionally important sulfhydryl groups of enzymes involved in the transport process. It produces a profound diuresis, limited only by the amount of filtrate that reaches the site of the drug action—the thick ascending limb of Henle's loop.

4. FUROSEMIDE.—The site of action of this widely used diuretic is the same as that of ethacrynic acid although the drug does not bind sulfhydryl groups. Furosemide is an analogue of sulfanilamide with minimal carbonic anhydrase activity. It is a powerful diuretic that has proved very safe for clinical use.

All of the above diuretics increase the flow of tubular filtrate past the potassium-secreting site. Furthermore, the concentration of sodium in the filtrate reaching the potassium secretory site is increased. Both factors increase the rate of potassium secretion and may lead to potassium depletion. This common complication of diuretic therapy can be avoided by the proper administration of potassium salts.

5. SPIRONOLACTONE AND TRIAMTERENE.—Two classes of diuretics inhibit sodium absorption without increasing potassium excretion. The spironolactones compete with aldosterone for binding sites. These agents may well reduce the renal transport pool for potassium by reducing potassium entry into cells of the distal tubule. The sodium diuresis produced is mild and they are not effective when little aldosterone is present. Triamterene and amiloride can act even in the absence of aldosterone. They increase sodium and water excretion, without increasing potassium excretion, by reducing the passage of sodium across the luminal border of cells of the distal tubule and collecting duct. The subsequent reduction in the transport of sodium is probably responsible for the decreased rate of potassium excretion. The combination of a thiazide diuretic and either triamterene or spironolactone is frequently used to obtain a good sodium diuresis and yet prevent potassium depletion.

OSMOTIC DIURESIS

Substances that are filtered by the glomerulus, but not absorbed, may retard absorption of water and salt by their osmotic effects when they are present in large amounts. Mannitol is such a substance. Let us suppose that mannitol has been rapidly administered intravenously in amounts sufficient to increase plasma osmolarity and, therefore, the osmolarity of the glomerular filtrate. As sodium and its anion are reabsorbed isosmotically in the proximal tubule, the concentration of mannitol increases because it is not absorbed. Since an osmotic gradient cannot be developed across the epithelium of the proximal tubule, the presence of mannitol retards filtrate absorption. As a result, the fraction of filtered fluid reaching the end of the proximal tubule and delivered to the more distal nephron is much greater than normal. Furthermore, the proximal tubule can only develop a concentration gradient for sodium of 35 mEq/L or less because it has a very high permeability for that ion. Thus the increased volume of filtrate reaching the end of the proximal tubule also contains an increased quantity of sodium (and chloride). The sodium load to the more distal nephron is also increased substantially. In the distal tubule, the situation is more complex. Water and sodium chloride absorption increase but not enough to offset the increased load. Thus, as the result of excretion of large amounts of a nonreabsorbable solute, both water and sodium chloride excretion increase. This process is called osmotic diuresis.

Under certain pathologic conditions, the plasma concentration of naturally occurring filtrable solutes may rise so high that the amounts delivered to the tubular cells exceed their capacity to absorb them and osmotic diuresis results. The best example of this is the hyperglycemia of diabetes mellitus in which plasma glucose levels are markedly increased and the entire filtered load cannot be absorbed. Then glucose behaves in a manner entirely analogous to mannitol.

THE KIDNEY AND HYDROGEN ION HOMEOSTASIS

The concentration of hydrogen ions in extracellular fluid is relatively small, about $10^{-7.4}$ M/L. Thus, the extracellular fluid may be considered slightly alkaline. It is crucial that this concentration be precisely regulated because the rates of many enzymatic reactions are critically altered by changes in hydrogen ion concentration. In physiologic and clinical work, it is convenient to discuss hydrogen ion concentration $[H^+]$ in terms of pH, where pH $= -\log [H^+]$.

Thus normal plasma has a pH of 7.4. Changes in pH to values outside the 7.0–7.8 range for more than brief periods are incompatible with life.

Intracellular metabolism of food forms two types of acid: volatile and nonvolatile. The *volatile* acid, a byproduct of oxidative metabolism, is carbonic acid, which is derived from hydration of carbon dioxide. Let us consider the reaction

$$CO_2 + H_2O \overset{CA}{\rightleftharpoons} H_2CO_3 \rightleftharpoons H^+ + HCO_3^- \tag{7}$$

All steps are reversible and the hydration and dehydration of CO_2 are catalyzed by the enzyme carbonic anhydrase (CA). The lung is responsible for elimination of the 10–20 M of CO_2 produced daily. The process, discussed in detail in Chapter 5, is so efficient that in normal individuals a significant acidosis from CO_2 retention does not develop even when CO_2 production is very large.

For the most part the *nonvolatile* acids produced in metabolic processes are phosphoric and sulfuric acids, which are derived mainly from protein breakdown. They are strong acids and, therefore, at body pH they dissociate into H^+ and SO_4^{2-}, HPO_4^{2-} and $H_2PO_2^{-}$. The liberated H^+ must be buffered to prevent marked changes in pH. Proteins are the primary intracellular buffer but, when the hydrogen ions diffuse into the extracellular fluid, they are buffered primarily by the bicarbonate-carbonic acid buffer system. Approximately 1 mEq/kg of body weight of nonvolatile acid is produced daily from the average American diet. However, since man is omnivorous and since his diet varies considerably from habitat to habitat, season to season, and according to tradition and individual preferences, the daily acid load may vary considerably. Ultimately, the kidneys must excrete an amount of hydrogen equal to the nonvolatile acid generated by metabolism in order to maintain the slightly alkaline pH of body fluids.

If all of the H^+ generated from nonvolatile acids entered body fluids and remained there, body pH would fall to levels inconsistent with life. To prevent this, three lines of defense operate: (1) large hydrogen ion loads are chemically buffered as they are generated, (2) the lung, by altering P_{CO_2}, increases the capacity of the major buffer system to maintain a nearly constant pH and (3) the kidneys excrete the acid and regenerate the chemical buffers.

SYSTEMIC REGULATION OF pH

CHEMICAL BUFFERS

The principal buffer in the extracellular fluid is bicarbonate. Its interaction with hydrogen ions is as follows:

$$HCO_3^- + H^+ \leftrightarrows H_2CO_3 \leftrightarrows CO_2 + H_2O \tag{8}$$

and applying the Henderson-Hasselbalch equation

$$pH = pK + \log \frac{base}{acid} \tag{9}$$

$$pH = 6.1 + \log \frac{HCO_3^-}{H_2CO_3 + \text{dissolved } CO_2} \tag{10}$$

Since $0.03 \times P_{CO_2} = H_2CO_3 + \text{dissolved } CO_2$, and normal $P_{CO_2} = 40$ mm Hg, the denominator is equal to $0.03 \times 40 = 1.2$. Normal $HCO_3^- = 24$ mM. Hence

$$pH = 6.1 + \log \frac{24}{1.2} \tag{11}$$

$$= 7.4$$

As an example of the buffering effect of this system, let us compare the addition of 12 mM/L of hydrochloric acid to 1 L of water and to 1 L of extracellular fluid. Let us assume that the extracellular fluid $[HCO_3^-]$ is 24 mM/L and the $[H_2CO_3]$ is 1.2 mM.

Water: 12 mM HCl contains 12 mM = 0.012 M hydrogen ion, therefore,

$$pH = -\log(0.012) = 1.92$$

Extracellular Fluid: 12 HCl + 24 $NaHCO_3$ → 12 NaCl + 12 $NaHCO_3$ + 12 H_2CO_3. There is already 1.2 mM H_2CO_3 in the fluid, therefore,

$$pH = 6.1 + \log\frac{12}{13.2} = 6.06$$

Thus bicarbonate substantially reduces the change in pH caused by the addition of hydrogen ions.

The body contains numerous other buffer systems (mainly proteins) that act in the same way as bicarbonate. However, the bicarbonate-CO_2 system is by far the most important physiologic buffer because its two components can be effectively regulated by the kidneys and lung.

RESPIRATORY CONTROL

Chemical buffering acts almost instantaneously, but chemical buffers alone cannot prevent pH from falling to lethal limits. As we have just noted, the addition of 12 mM HCl to plasma caused the following reaction:

$$12\ HCl + 24\ NaHCO_3 \rightarrow 12\ NaCl + 12\ NaHCO_3 + 12\ H_2CO_3$$

Carbonic acid is in equilibrium with water and CO_2 as follows:

$$12\ H_2CO_3 \rightleftharpoons 12\ H_2O + 12\ CO_2$$

If the system is closed, so that CO_2 cannot be lost,

$$pH = 6.1 + \log\frac{12}{13.2} = 6.06$$

In the body, however, respiration is stimulated by the increased CO_2 and by the fall in pH. This causes much CO_2 to be expired from the lung and the denominator in the Henderson-Hasselbalch equation may fall to about 0.8 mM/L. As a result, the pH increases:

$$pH = 6.1 + \log\frac{12}{0.8} = 7.28$$

Respiratory compensation is effective but almost never complete. Extracellular fluid pH returns *toward* normal but rarely makes it all the way.

The example shows that chemical buffers plus respiratory control could maintain pH *almost* within normal limits in the face of an acid load of 12 mM H^+ due to the buffering action of HCO_3^- and the loss of carbonic acid (CO_2) from the lung. But in the process, 12 mM $NaHCO_3$ buffer was lost (transformed to NaCl) and the further addition of acid would produce a more drastic pH change. It is the function of the kidney to replenish and conserve body stores of bicarbonate.

RENAL REGULATION OF pH

CONSERVATION OF BICARBONATE

About 5 M of bicarbonate is filtered through the glomeruli per day and virtually the entire amount (99.9%) is reabsorbed: 90% in the proximal tubule and 10% in the distal tubule and collecting duct.

The process of bicarbonate absorption begins with generation of a hydrogen ion inside the tubular cell (Fig 6–22). Two mechanisms have been proposed. First, cellular carbonic anhydrase may catalyze hydration of CO_2 to form carbonic acid (H_2CO_3); H_2CO_3 dissociates into a hydrogen ion, which is available for transport into the lumen; the anion HCO_3^- diffuses from the cell into the extracellular fluid. Second, water may be split to H^+ and OH^-; H^+ is transported into the lumen, whereas OH^- combines with CO_2 to form HCO_3^-, which diffuses into the extracellular fluid. The latter reaction is also catalyzed by carbonic anhydrase.

Hydrogen ion is secreted by an active process into the tubular fluid, where it reacts with bicarbonate to form carbonic acid. Carbonic acid is rapidly dehydrated to CO_2 in a reaction catalyzed by the carbonic anhydrase found in

Fig 6–22.—H^+ derived from H_2CO_3 or the splitting of water is secreted into the lumen of the tubule. Secreted H^+ is buffered by HCO_3^-, forming H_2CO_3, which is dehydrated in a reaction catalyzed by carbonic anhydrase *(Ca)* to CO_2 and H_2O. CO_2 diffuses into the body. Meanwhile, bicarbonate formed in the cell diffuses into plasma. Net effect is HCO_3^- absorption.

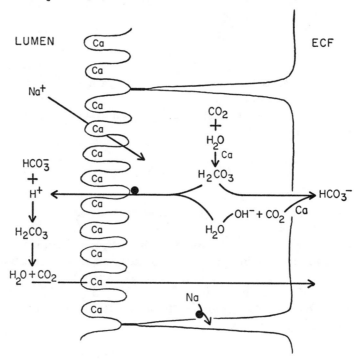

abundance in the microvilli of the early proximal tubule. The net result of these two steps is titration of HCO_3^- in the filtrate, with subsequent formation of CO_2 and H_2O. The CO_2 diffuses into the extracellular fluid and the water is reabsorbed from the lumen to extracellular fluid. Meanwhile in the cell, HCO_3^- formed either from dissociation of carbonic acid or from combination of OH^- with CO_2 diffuses into the interstitial space. Strictly speaking, bicarbonate ions are not directly reabsorbed. Rather, bicarbonate is eliminated from tubular fluid and regenerated in the cell water as a result of hydrogen ion secretion. The net effect of these processes is the same as if bicarbonate in the filtrate had been reabsorbed.

Although it was believed for many years that Na^+ reabsorption was closely linked to H^+ secretion in the proximal tubule, there is a growing belief based on several studies that this may not be the case. H^+ secretion may well occur as an independent process.

Bicarbonate absorption normally proceeds until its luminal concentration is reduced by about 15 mM/L to 10 mM/L. Since HCO_3^- reabsorption proceeds at a rate greater than that of Cl^-, the concentration of Cl^- increases reciprocally by the same degree, leaving the filtrate Cl^- concentration about 15 mM greater than that of interstitial fluid.

At least three major factors alter bicarbonate reabsorption by the proximal tubule: (1) an elevation in PCO_2 increases bicarbonate absorption, probably because intracellular pH is increased, (2) potassium depletion also increases intracellular hydrogen concentration and increases bicarbonate concentration and (3) conversely, a decrease in PCO_2 or potassium loading reduces bicarbonate absorption.

Bicarbonate reabsorption in the proximal tubule changes in the same direction as sodium absorption in response to a variety of stimuli. For example, extracellular fluid volume expansion not only reduces proximal sodium absorption but also bicarbonate absorption. Extracellular fluid volume depletion has the opposite effect.

REPLENISHMENT OF BICARBONATE

As discussed earlier, secretion of hydrogen ion is accompanied by intracellular generation of bicarbonate, which then diffuses into the extracellular fluid. In order to replenish extracelluar bicarbonate, which has been depleted by non-volatile acids, hydrogen ion secretion into the tubular lumen must exceed that required for recovery of filtered bicarbonate. In fact this happens, but the mechanism responsible for regulation of hydrogen ion secretion in this process is not yet understood.

The nephron is capable of maintaining a chemical gradient for H^+ between lumen and extracellular fluid along its entire length. In the proximal tubule, hydrogen ion concentrations are about 2.5 times greater than that of the extracellular fluid, and in the collecting duct a 1,000-fold concentration difference can be maintained. Thus, each liter of urine at pH 4.4, the lowest pH the kidney can achieve, contains only 0.04 mEq/L of hydrogen ion, a minute fraction of the 50 mEq per day that needs to be excreted. However, the urine does contain buffers, which take up large amounts of hydrogen ion and enhance the hydrogen ion excretory capacity of the kidney.

URINARY BUFFERS

The principal urinary buffers are *phosphate* and *ammonia*.

PHOSPHATE.—Dibasic phosphate can buffer hydrogen ions, as shown:

$$HPO_4^{2-} + H^+ \leftrightarrow H_2PO_4^- \tag{12}$$

This reaction has a pK of 6.8. Therefore, at the pH of glomerular filtrate (7.4), the ratio of the dibasic to monobasic form can be calculated from the Henderson-Hasselbalch equation:

$$7.4 = 6.8 + \log \frac{(HPO_4^{2-})}{(H_2PO_4^-)} \tag{13}$$

$$\therefore \log \frac{(HPO_4^{2-})}{(H_2PO_4^-)} = 0.6$$

and the

$$\frac{(HPO_4^{2-})}{(H_2PO_4^-)} = 4$$

Thus most of the phosphate in the glomerular filtrate exists as the dibasic

Fig 6–23.—Regeneration of HCO_3^- occurs when H^+ derived from H_2CO_3 or the splitting of water is secreted into the filtrate and buffered by monohydrogen phosphate (HPO_4^{2-}). The HCO_3^- left behind diffuses into the plasma. The net effect is secretion of hydrogen ion and generation of extracellular fluid *(ECF)* HCO_3^-.

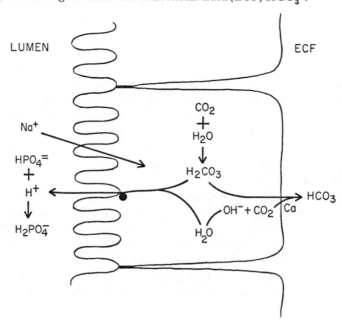

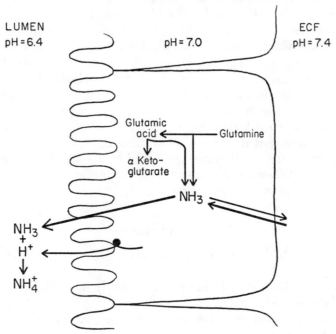

LUMEN
pH = 6.4

pH = 7.0

ECF
pH = 7.4

Glutamic
acid ← ── Glutamine

α Keto-
glutarate

NH₃

NH₃
+
H⁺
↓
NH₄⁺

Fig 6–24.—Ammonia secretion supplies a buffer for hydrogen ion secretion. In the lumen of the tubule, ammonium ion is formed very rapidly, trapping NH_3 in the filtrate.

form and is available for buffering. If the final urine pH is 5.0, then 98% of the remaining phosphate will be in the form of $H_2PO_4^-$ as a result of H⁺ secretion (Fig 6–23).

Even though 85% of filtered phosphate is reabsorbed, this buffer system still accounts for about 40% of the excretion of hydrogen ion. Other potential buffers that are filtered into the urine include creatinine, the hippurans and acetoacetate. However, these bases have low pKs and their physiologic significance is small. The amount of hydrogen ion excreted with a buffer can be determined by titrating the urine with NaOH to the plasma pH. The number of milliequivalents of hydrogen ion secreted and buffered is called *titratable acid* and is equal to the number of milliequivalents of NaOH required. The titratable acid is usually 20–40 mEq per day if enough buffer is present.

AMMONIA.—The remaining portion of the acid load is excreted as ammonium ion. Ammonia is formed in the cells of the nephron from the amino acid glutamine, which is transported into renal cells from the plasma by an active process. Glutamine is deaminated to produce glutamic acid and ammonia (NH_3), and glutamic acid is further deaminated to produce an additional molecule of NH_3 and alpha ketoglutarate (Fig 6–24). Other amino acids, especially glycine and alanine, may contribute an amine group to ammonia production through the process of transamination of alpha ketoglutarate to glutamic acid.

NH_3 reacts reversibly with H⁺ to produce ammonium

$$NH_3 + H^+ \leftrightarrow NH_4^+ \tag{14}$$

The pK of this reaction is 9.3 and, thus, in all body fluids more than 98% of NH_3 exists as NH_4^+. Free NH_3 is highly diffusible and readily passes through the membranes of the renal cells into the lumen of the tubule as well as into the interstitial fluid. Since tubular fluid is more acid than extracellular fluid, eq. 14 proceeds more rapidly to the right in the tubular fluid, and the NH_3 concentration is always lower than in the extracellular fluid. Thus, there is a chemical gradient in favor of a higher rate of diffusion of ammonia into the filtrate than in the opposite direction. High concentrations of NH_4^+ are achieved in the urine because the back diffusion of NH_4^+ is minimal since the cellular membrane is impermeable to this cation.

About 60% of the usual dietary acid load of 40–80 mEq per day is excreted as ammonium. When the dietary acid load is increased over a period of days, ammonia production increases dramatically.

SUMMARY

The response to the influx of a load of H^+ into the extracellular fluid is as follows:

1. The H^+ is immediately buffered, primarily by HCO_3^-; pH falls, but less than in the absence of buffer.

2. Ventilation increases, causing CO_2 to be "blown off," and pH returns toward 7.4.

3. The kidneys respond in a more leisurely fashion. Bicarbonate is generated from secretion of H^+ buffered within the tubule by phosphate and ammonia.

The response to an alkali load similarly involves the following:

1. Immediate chemical buffering, resulting in an increase in plasma bicarbonate.

2. Rapid respiratory response: hypoventilation with CO_2 retention.

3. Slow renal response, with excretion of bicarbonate and diminished secretion of ammonia.

DISORDERS OF ACID-BASE METABOLISM

Acidosis is any abnormal process that would by itself make arterial pH more acid than normal. *Alkalosis* is any abnormal process that by itself would make arterial pH more alkaline than normal.

Acidosis and alkalosis can be either respiratory or metabolic in origin. Whatever the primary cause, compensatory changes occur. We have already seen how metabolic acidosis is partly compensated by increased respiration. Similarly, respiratory acidosis, caused for example by chronic obstructive pulmonary disease, is compensated by renal retention of bicarbonate.

In the following examples, the physiologic response to an alteration in the acid-base status of a normal individual will be considered. In each case, the Henderson-Hasselbalch equation (eq. 2) is used to calculate the unknowns.

RESPIRATORY ALKALOSIS.—A young adult is carried from sea level by helicopter to a weather observation post at 12,000 feet. Because of the decrease in

P_{O_2} of inspired air, ventilation increases and P_{CO_2} is reduced to 30 mm Hg and thus the pH of arterial blood increases to 7.52. Because of the reduction in P_{CO_2}, the H^+ content of the renal tubular cells is reduced and less hydrogen ion is secreted. As a result, recovery of filtered bicarbonate is reduced and a fall in plasma bicarbonate ensues. If the final bicarbonate concentration is 20 mM/L, the arterial blood pH will be returned to a nearly normal value of 7.45. As with most other acid-base alterations, the compensatory physiologic response does not completely correct the original pH abnormality.

RESPIRATORY ACIDOSIS.—In a 55-year-old man, hiccups developed for an unknown reason. His physician prescribed a mixture of 5% CO_2, 20% O_2 and 75% N_2 for him to breathe. After 48 hours of treatment, the hiccups persisted and P_{CO_2} was found to be 55 mm Hg. Without renal compensation, a respiratory acidosis should have occurred, with an arterial pH of 7.26. However, the increase in P_{CO_2} raised intracellular H^+ concentrations and increased H^+ secretion by renal cells and total acid excretion by the kidneys. As a result, the plasma bicarbonate concentration was increased to 31.4 mM after renal adaptation, and pH of the arterial blood increased to 7.38.

METABOLIC ALKALOSIS.—A vegetarian, age 24, began ingesting twelve 0.6 gm tablets of $NaHCO_3$ per day as part of a food fad. If it is assumed that his diet alone was neutral with respect to hydrogen ion balance, the daily ingestion of 100 mM of $NaHCO_3$ would make his hydrogen ion balance negative by 100 mEq. In the absence of a renal response, this amount of base would be expected to raise plasma HCO_3^- concentration by about 5 mM/L in 2 days, and without a respiratory response the pH of arterial blood would increase to about 7.48. However, both the kidneys and the respiratory center would respond to this alkali load. A slight decrease in ventilation would occur sufficient to increase the P_{CO_2} to 44 mm Hg and return the pH to 7.44. In the presence of a normal arterial volume and in the absence of a marked deficit of total body potassium, reabsorption of filtered bicarbonate would be reduced and bicarbonate would appear in the urine. At equilibrium, the arterial pH, plasma bicarbonate and P_{CO_2} would be normal. The renal correction for the increased load of base is effective in this instance because the high dietary sodium assures a normal arterial volume and allows the rapid excretion of $NaHCO_3$.

METABOLIC ACIDOSIS.—A medical student ingested 10 gm NH_4Cl daily for 3 days, providing an additional H^+ load of 185 mEq per day. The response to this acid load is as follows: in the liver, NH_4^+ is metabolized to urea and H^+ is liberated only to be buffered by HCO_3^- in the extracellular fluid. The resulting fall in plasma pH stimulates the respiratory center, and an increase in ventilation reduces P_{CO_2}. This acid load might be expected to reduce the plasma HCO_3^- concentration to 14 mM and, if the respiratory response lowers the P_{CO_2} to 27 mm Hg, arterial pH after the respiratory response will be 7.37. In the kidney, H^+ secretion increases perhaps because of a fall in intracellular pH. As a result, virtually all filtered HCO_3^- would be recovered and, because urinary pH would fall to low levels, all available urinary buffers (HPO_4^{2-}, creatinine) would be fully titrated. Ammonia production would be stimulated to three to four times the normal, and most of the added acid load would be excreted as NH_4Cl. As a result, the student would achieve hydrogen ion balance with a normal pH, PCO_2 and HCO_3^- in several days.

GLUCOSE AND AMINO ACID REABSORPTION

The proximal tubule is capable of removing from the glomerular filtrate a variety of organic substances including glucose, amino acids and organic anions. When they are present in plasma in normal concentrations, glucose and amino acids are recovered from the filtrate virtually in toto in the first 2–3 mm of the tubule.

The mechanism of absorption of glucose and amino acids by the tubule is very similar to the mechanism of absorption of these substances in the proximal small bowel; there is co-transport of the organic solute with sodium across the luminal membrane of the epithelial cells and facilitated diffusion from the cell into the peritubular fluid down a chemical potential.

Figure 6–25 illustrates the current view of glucose transport in the early proximal tubule. A carrier for glucose and sodium is present in the luminal

Fig 6–25.—Glucose absorption begins when sodium and glucose form a complex with a carrier located in the luminal membrane. The carrier with either solute alone cannot translocate. Once in the cell, sodium and glucose dissociate and the empty carrier can translocate once more to the luminal side of the membrane. The energy for entry into the cell is provided by the electrochemical gradient for Na. Back diffusion into the lumen is limited because of the uphill electrochemical gradient. Glucose leaves the cell by facilitated diffusion along the lateral and basal membranes. *ECF* = extracellular fluid.

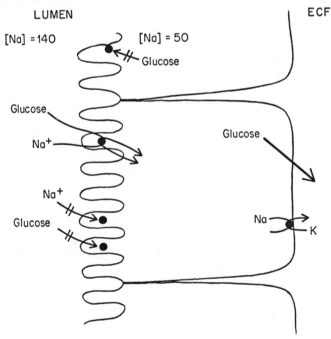

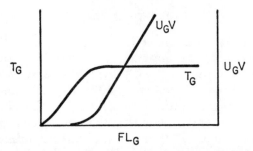

Fig 6–26.—As filtered glucose (FL_G) increases, reabsorbed glucose (T_G) increases until the reabsorptive mechanism is saturated. Then glucose appears in the urine in increasing amounts as FL_G continues to rise. U_GV = excreted glucose.

membrane with sites available for both glucose and sodium. The carrier can translocate through the membrane under two conditions: (1) when all of the sites for glucose and sodium are occupied and (2) when all the sites for glucose and sodium are empty. If glucose *or* sodium sites only are occupied, translocation does not occur. Since the sodium concentration of the tubular filtrate is about three times greater than that of the cell water, the rate of formation of glucose-sodium-carrier complexes will be much greater on the luminal side than on the cell side, and net movement of glucose and sodium via these carriers will be from tubular filtrate to cell fluid. The electrochemical gradient for sodium provides the energy for transmembrane movement. The sodium gradient is maintained by the active sodium pumps along the lateral and basal cell membranes.

Once the carrier has translocated, glucose and sodium dissociate and enter the cell water. Formation of the complex occurs at a slower rate because of the low intracellular sodium concentration. Glucose in the cell water diffuses down its chemical gradient into the extracellular fluid with the aid of a glucose carrier in the basal and lateral cell membranes. This carrier does not have sites for sodium.

Amino acid transport occurs by essentially the same process except that three carriers appear to be involved. One reabsorbs glutamate and asparate; the second, arginine, cystine, lysine and ornithine; and the third, the remaining amino acids required for protein metabolism.

The reabsorption mechanism for glucose has a limited capacity (Fig 6–26). If glucose is infused into a test subject, plasma glucose levels increase, the filtered load increases, but for a while only trace amounts of glucose appear in the urine. However, when the filtered load reaches a certain point, the reabsorptive mechanism becomes completely saturated and the amount of glucose reabsorbed remains constant. Glucose then appears in the urine and the amount increases rapidly as the filtered load increases further. The appearance of glucose in the urine indicates in most, but not all, patients that blood sugar levels are elevated. In a few individuals (about 1% of the population), glucose is found in the urine even at normal plasma levels because the glucose transport system of their kidneys has a decreased capacity.

UREA TRANSPORT

In contrast to glucose, urea movement into and out of the nephron is extraordinarily complicated. Micropuncture studies have shown that urea is reabsorbed passively in the proximal convoluted tubule, but not to the same degree as the rest of the filtrate. About 50% of filtered urea remains at the midpoint of the rat proximal tubule when 65% of the filtrate has been absorbed. In the pars recta and descending thin limb, urea may be secreted into the tubular fluid, and more is added in the thin ascending limb of Henle's loop. The movement of urea is probably passive, although evidence exists for an active secretory site in the pars recta. The amount of urea added between the end of the accessible portion of the proximal tubule and the beginning of the distal convoluted tubule is large—perhaps 60% of the original filtered load.

In the distal tubule and cortical collecting duct, very little urea is reabsorbed. However, the segment of the collecting duct that passes through the inner medulla is permeable to urea in the presence of ADH. As a result, urea clearance varies with urine flow or, more precisely, with the fraction of filtered water excreted. In antidiuresis, the excretion of urea is about 40% of the amount filtered, whereas in a diuretic state, 60% of filtered urea will be excreted.

ORGANIC ANION TRANSPORT

The cells of the proximal tubule can transport a variety of organic anions, including many carboxylic acids, phenols, sulfonic acids and heterocyclic compounds, from the peritubular fluid to the filtrate. The single transport system involved is extremely complex. The initial step occurs on the peritubular side of the cell where the organic anion is actively transported into the cell. This step is both sodium and potassium dependent and, therefore, some type of co-transport system involving these ions and the transported anion may be present. The organic anions are concentrated in the cell and then passively diffuse into the tubular fluid. The continuous movement of filtrate down the tubule maintains a high chemical gradient for the transported anion and provides for continuous entry of the ion into the urine from the tubular cell. As the tubular fluid moves downstream away from the secretory site, secreted substances may back diffuse into the cell and the interstitial fluid. Thus the apparent rate of secretion, as determined by the quantity of transported substance found in the urine minus that filtered, may be less than the true secretory rate.

The organic anion transport system is universally present in kidneys of virtually all species. Its function remains unclear since most products of normal metabolism are not secreted by this mechanism. Uric acid is an exception. In man and several species of apes, this end-product of purine metabolism appears to be reabsorbed in the early proximal tubule and secreted by the organic anion transport system in the more distal part of the proximal tubule.

The organic anion transport system also secretes ingested nonmetabolizable organic anions. A variety of useful drugs is secreted, including penicillin and aspirin. When drugs handled by the kidneys' transport system are prescribed, care must be taken to adjust the dose if renal function is impaired; otherwise the plasma concentration of the drug may rise to toxic levels.

BIBLIOGRAPHY

Brenner, B. M., Deen, W. M., and Robertson, C. R.: Glomerular Filtration, in Brenner, B. M., and Rector, F. C., Jr. (eds.): *The Kidney* (Philadelphia: W. B. Saunders Co., 1976).

Bricker, N. S., and Schultze, R. G.: Renal Function: General Concepts, in Maxwell, M. H., and Kleeman, C. R. (eds.): *Clinical Disorders of Fluid and Electrolyte Metabolism* (New York: McGraw-Hill Book Co., 1972).

Burg, M. B.: Mechanisms of Action of Diuretic Drugs, in Brenner, B. M., and Rector, F. C., Jr. (eds.): *The Kidney* (Philadelphia: W. B. Saunders Co., 1976).

Derks, J., Seely, J., and Levy, M.: Control of Extracellular Fluid Volume and the Pathophysiology of Edema Formation, in Brenner, B. M., and Rector, F. C., Jr. (eds.): *The Kidney* (Philadelphia: W. B. Saunders Co., 1976).

Fitzsimmons, J. T.: The physiologic basis of thirst, Kidney Int. 10:3, 1976.

Giebisch, G.: Some reflections on the mechanism of renal tubular potassium transport, Yale J. Biol. Med. 48:315, 1975.

Goldberg, M.: Water Control and the Dysnatremias, in Bricker, N. S., and Seldin, D. W. (eds.): *The Sea within Us* (New York: Plenum Publishing Corp., 1975).

Hierholzer, K., and Wiederholt, M.: Some aspects of distal tubular solute and water transport, Kidney Int. 9:198, 1976.

Klahr, S., and Slatopolsky, E.: Renal regulation of sodium excretion, Arch. Intern. Med. 131:780, 1973.

Kokko, J. P., and Tisher, C. C.: Water movement across nephron segments involved with the countercurrent multiplication system, Kidney Int. 10:64, 1976.

Leaf, A., and Cotran, R. S.: *Renal Pathophysiology* (New York: Oxford University Press, 1976).

Maffly, R. H.: The Body Fluids: Volume, Composition and Physical Chemistry, in Brenner, B. M., and Rector, F. C., Jr. (eds.): *The Kidney* (Philadelphia: W. B. Saunders Co., 1976).

Malnic, G., and Steinmetz, P. R.: Transport processes in urinary acidification, Kidney Int. 9:172, 1976.

Pitts, R. F.: Control of renal production of ammonia, Kidney Int. 1:297, 1972.

Pitts, R. F.: *Physiology of the Kidney and Body Fluids* (3d ed.; Chicago: Year Book Medical Publishers, Inc., 1974).

Robertson, G. L., Shelton, R. L., and Athar, S.: The osmoregulation of vasopressin, Kidney Int. 10:25, 1976.

Schultze, R. G.: Recent advances in the physiology and pathophysiology of potassium excretion, Arch. Intern. Med. 131:885, 1973.

Seldin, D. W., and Rector, F. C., Jr.: The generation and maintenance of metabolic alkalosis, Kidney Int. 1:306, 1972.

Shrier, R. W., and Berl, T.: Non-osmolar factors affecting renal water excretion, N. Engl. J. Med. 292:81;141, 1975.

Steele, T. H., and Rieselbach, R. E.: The Renal Handling of Urate and Other Organic Anions, in Brenner, B.M., and Rector, F. C., Jr. (eds.): *The Kidney* (Philadelphia: W. B. Saunders Co., 1976).

Stein, J. H.: The Renal Circulation, in Brenner, B. M., and Rector, F. C., Jr. (eds): *The Kidney* (Philadelphia: W. B. Saunders Co., 1976).

Suki, W. N.: Disposition and regulation of body potassium; an overview, Am. J. Med. Sci. 272:31, 1976.

Sullivan, L. P.: *Physiology of the Kidney* (Philadelphia: Lea & Febiger, 1974).

Tuber, C. C.: Anatomy of the Kidney, in Brenner, B. M., and Rector, F. C., Jr. (eds.): *The Kidney* (Philadelphia: W. B. Saunders Co., 1976).

Ullrich, K. J.: Renal tubular mechanisms of organic solute transport, Kidney Int. 9:134, 1976.

Vander, A. J.: *Renal Physiology* (New York: McGraw-Hill Book Co., 1975).

Windhager, E. E., and Giebisch, G.: Proximal sodium and fluid transport, Kidney Int. 9:198, 1976.

7

The Gastrointestinal System

EUGENE D. JACOBSON, M.D.

THE GASTROINTESTINAL SYSTEM is essential for life because it is the only normal avenue of nutrition. More Americans are affected by serious diseases of the gastrointestinal tract than by disorders of any other system except the cardiovascular. Thus three of the leading six causes of death from cancer are malignancies of the colon, pancreas and stomach. One tenth of the population suffers from peptic ulcers at some point in life, and millions of Americans have cirrhosis of the liver, gallstones, mesenteric vascular disease and ulcerative colitis. We would be hard pressed to find a person who has not experienced diarrhea, constipation, indigestion or vomiting at least once.

The major function of the gastrointestinal system is to absorb nutrients. Before this can be achieved, food must be moved from the mouth to the small intestine and then subjected to the action of organic and inorganic juices, which dissolve and digest it. Finally, unabsorbed materials must be moved into the large intestine and eventually expelled. There are many other important functions as well. The system must prevent absorption of toxic waste products and microorganisms. Gastric acid is bactericidal, and the small intestine sweeps itself clean of bacterial nutrients. The lining of the stomach and gut is highly resistant to damage by bacteria, digestive enzymes and hydrochloric acid. The stomach and intestine also serve as temporary storage sites, allowing time for dilution, digestion or desiccation of substances passing through them. The splanchnic organs are sites of reticuloendothelial elements and sources of antibodies. Gastrointestinal and associated tissues elaborate major hormones, including insulin, glucagon, gastrin, secretin and cholecystokinin. They also contain substances including histamine, serotonin, bradykinin and prostaglandins that have been implicated in disease. The gut contributes to fluid and electrolyte balance, and the liver is a major site of intermediary metabolism. Most drugs are absorbed from the gut and degraded in the liver. Some are excreted in the feces. Finally, the circulation of the gastrointestinal tract, liver and spleen receives about one quarter of the cardiac output, and the veins are a major blood reservoir.

Much of the gastrointestinal tract lies relatively dormant for much of the time. When a meal is ingested, the gastrointestinal tract awakens. Its juices flow, the stomach and small bowel move, hormones are released and digestion and absorption take place. Other parts of the body also are stimulated. The conscious brain is pleased. Body temperature increases. Cardiac output rises (perhaps the way to a man's heart is, after all, through his stomach). Thus normal gastrointestinal functions influence other systems and, when the gastrointestinal tract is diseased, the other systems suffer as well. In the state of nature one of the commonest motivations for activity is hunger.

The gastrointestinal system is structurally well suited to its functions. The basic design of the system is that of a long tube into which various glandular units secrete juices and around which muscular elements contract and relax to propel and churn the contents. The inner lining of most of the tube is convoluted and covered by finger-like villous projections on which are superimposed many microvilli. Together these structures increase the inner surface area of the tube two million-fold. This enormous surface increases the ability of the gut to rapidly absorb nutrients.

Secretion, motility and probably absorption in the gastrointestinal tract are regulated by the autonomic nervous system—particularly the parasympathetic division—and by hormones released, mainly from the mucosal lining of the stomach and gut, during digestion. These neurohumoral factors also influence the gastrointestinal circulation.

The sections that follow have been written to reflect a combination of that which is "tried and true" and that which is new—both selected for the relevance of the physiology to current clinical practice. The information is presented in two parts. The first deals with general cellular and tissue processes, including mucosal metabolism, membrane transport and the neurohumoral factors that regulate them. The gastrointestinal circulation is also discussed. The second part is devoted to organ functions; this is the meat of the chapter and describes the physiology of motility, secretion and absorption, organ by organ in the alimentary tract.

REGULATORY FACTORS: MUCOSAL METABOLISM AND MEMBRANE TRANSPORT

MUCOSAL METABOLISM

Each day hundreds of grams of nutrients and electrolytes move across the gastrointestinal mucosa accompanied by liters of water. The absorption and secretion of many of these substances require energy and this is derived from oxidation of glucose and fatty acids. Poisons that interfere nonspecifically with energy production stop active transport across the mucosa.

Mucosal metabolism is regulated by intracellular substances including enzymes such as ATPase and carbonic anhydrase, and nucleotides such as cAMP (adenosine 3',5'-cyclic monophosphate). By governing the generation of energy for active transport, these agents are rate limiting for secretion and active absorption. Some of them may also act directly upon membrane transport itself.

In the gastric mucosa there is an ATPase that may be involved in transport of sodium from the cell into the extracellular compartment during secretion. In the intestinal mucosa an ATPase activated by sodium and potassium ions may be involved in the transport of these ions. Added ATP activates this transport. Carbonic anhydrase assists gastric mucosal parietal cells to rid themselves of hydroxyl ions produced during active secretion of hydrogen ion.

cAMP also regulates mucosal metabolic reactions and transport. According to the "second messenger" theory of Sutherland, many blood-borne agents, including hormones, initiate changes in cellular metabolism by activating a membrane-bound enzyme, adenylate cyclase, which catalyzes the conversion of a small proportion of intracellular ATP into cAMP. Cyclic AMP triggers the complex enzyme systems responsible for oxidative phosphorylation, glycolysis and lipolysis—processes that provide the energy for active transport of some solutes across the gastrointestinal mucosa. In the stomach, accumulated cAMP may provoke secretion of chloride and hydrogen ions into the juice.

The energy requirements of the gastrointestinal mucosa are very high. The maximally secreting gastric mucosa has been estimated to generate more energy per gram than any other tissue in the body. The high energy production of the gastrointestinal mucosa is used not only in the processes of secretion and absorption but also in replication of mucosal cells, which are being continually sloughed into the lumen at a rapid rate. In the stomach, epithelial cells are formed in the neck of the gastric glands and migrate up the lining of the gastric pit to become the mucosal surface of the gastric lumen. During their migration they differentiate, enlarge and develop new enzyme systems. Their life is brief and their turnover time (the time required to replace the existing mucosal cell populace with new cells) is only 1–2 days.

In the small intestine, epithelial cells are produced in the bottom half of the crypts of Lieberkuhn and then migrate along the villous surface to the tip, from which they are sloughed. Their turnover time is 30 hours.

Few factors are known that affect cell turnover in the mucosa. Experiments in cats suggest that there is an oxygen countercurrent exchange mechanism in the villus. At the base of the villus, the inflow and outflow vessels are within 10 μ of one another, and some oxygen (which is lipid soluble) diffuses from arterial blood to enter the venule without passing to the capillaries at the villous tip. This reduces the oxygen tension at the tip of the villus, and this may be detrimental to prolonged cell life. If this mechanism existed in man, it would aggravate mucosal hypoxia in ischemic diseases of the gut, in which necrosis invariably begins at the tip of the villus.

Gastrin is another regulator of mucosal growth in the stomach and intestine. When gastrin is released, as after a meal, mitotic activity is increased.

In celiac disease, caused by a mucosal allergic response to a protein in wheat products, mature epithelial cells are destroyed and mitotic activity in the crypts is accelerated.

The luminal surface of intestinal epithelial cells is greatly infolded to form numerous microvilli, collectively termed the brush border, which contain digestive enzymes such as lactase; enzymes involved in energy production such as ATPases; and binding proteins, which serve as carriers for actively transported substances. The turnover time for the brush border is 18 hours and it is even less for the enzymes.

The high degree of cellular and metabolic activity of the mucosa may account for its vulnerability to damage by physical factors such as radiation; by ingested chemicals such as aspirin or alcohol; and even by the normal gut contents such as gastric juice in the upper gut or bacteria in the lower gut.

MEMBRANE TRANSPORT

The general physiology of cell membranes and their role in the exchange of substances between the cell and the fluid surrounding it are discussed in Chapter 1. Those aspects of special relevance to the gastrointestinal tract will be briefly noted here and are discussed more fully in the chapters dealing with specific organ functions.

FORMS OF TRANSPORT (Table 7–1)

The two major types are *active* and *passive* transport. Passive transport occurs without consumption of energy. All materials that cross membranes may be passively transported, including substances that usually are actively transported, if conditions are suitable. In the stomach and gut most items are exclusively passively transported. They include water, lipids, drugs, some electrolytes and vitamins and a few saccharides.

PASSIVE TRANSPORT. – The movement of passively transported materials is influenced by their concentration and electric gradients and by osmotic and hydrostatic forces (Chapter 1). Osmotic movement of water across the gastrointestinal tract membranes is important for two reasons. First, the volumes of fluid being transported are large, amounting to several gallons daily. Osmotic movement of water is a passive process requiring no work on the part of the body or its membranes. If the digestive tract were to move all of that water by an expenditure of metabolic energy, we would have to spend all of our time eating to provide the substrate for an enormous internal chore. Second, whenever water-soluble substances move through an enteric membrane – during absorption or secretion, by passive or active transport processes – water will follow the solute passively to restore osmotic balance across the

TABLE 7–1.–FORMS OF
GASTROINTESTINAL TRANSPORT

Passive Transport
  Water-soluble diffusion
    Concentration gradient
    Electric gradient
    Osmotic and hydrostatic pressures
  Lipid-soluble diffusion
    Lipid-lipid interactions
    Non-ionic diffusion
  Active Transport

---

membrane. One application of this phenomenon is found in the use of certain laxatives such as magnesium sulfate, which remain unabsorbed in the lumen of the gut and attract water from the tissue by osmosis. This increases the bulk of the stool and promotes bowel movement.

Hydrostatic pressure also moves water from capillaries in the gut into the extracellular spaces. This force is the difference between capillary pressure (about 15 mm Hg) and tissue pressure (less than 0 mm Hg). If the vascular pressure is greatly elevated, as during the generalized venous congestion observed in heart failure, the rate of water movement into the extracellular spaces becomes excessive and produces edema.

When water is flowing from one space to another across a membrane, it tends to pull solute particles along. This acceleration of passive movement of solute is termed *solvent drag.*

TRANSPORT OF LIPID-SOLUBLE SUBSTANCES.—Three major classes of important materials can penetrate the lipids of the intestinal cell membranes passively and are not restricted to passage through aqueous-filled pores. These are dietary lipids, weak electrolytes and certain vitamins.

1. Digested *dietary lipid* consists mainly of fatty acids, monoglycerides, free cholesterol and lysolecithin. Most of these substances, together with bile salts, form aggregates called *micelles* in the lumen of the upper gut. The micelles migrate to the intestinal membrane and release their digested lipid products into intestinal epithelial cells. Some lipids, including medium or short chain triglycerides, pass through the membrane without participating in micelle formation.

2. *Weak electrolytes* also move across a lipid phase, mostly as undissociated molecules. Their transit is not slowed as much by the lipid membrane as is the movement of strong electrolytes. Weak electrolytes in the form of weak acids such as aspirin or weak bases such as aminopyrine dissociate poorly in aqueous solution. The major impetus to their movement is a concentration gradient between the lumen of the gut (high concentration) and the blood (low concentration). Many drugs are compounded in pills as weak electrolytes to spare the patient the discomfort of an injection.

Another force accelerating transport of weakly electrolytic drugs is the hydrogen ion concentration gradient across the stomach. The concentration of acid in the lumen is one million times that in the blood. A weak acid such as aspirin dissociates poorly at a pH of 1.0 and dissociates well at a pH of 7.4 (Fig 7–1). Thus when a person swallows aspirin (and before the stomach can empty the drug into the duodenum), some of the aspirin penetrates the lipid lining of the lumen and enters the blood where the higher pH favors dissociation. This sets up a sink, since the dissociated aspirin cannot readily return the way it came. The ease with which aspirin enters the mucosal lining of the stomach also contributes to its ability to damage that lining. Another damaging agent that permeates the gastric mucosa easily is ethyl alcohol. The common combination of a hard night's imbibing, followed the next morning by ingestion of many aspirin tablets to relieve a hangover, exposes the individual to a serious risk of bleeding gastritis.

3. The lipid-soluble *vitamins A, D, E* and *K* are transported passively across the intestinal mucosa. Since they are lipid soluble, diseases leading to

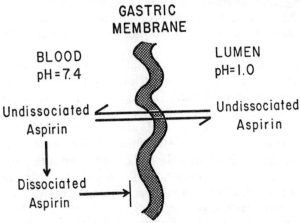

GASTRIC
MEMBRANE

BLOOD
pH = 7.4

LUMEN
pH = 1.0

Undissociated
Aspirin

Undissociated
Aspirin

Dissociated
Aspirin

Fig 7–1.—Diffusion of a weak electrolyte in response to pH gradient across the stomach. The weak acid aspirin is mainly undissociated in acid gastric juice and penetrates membrane in the undissociated state. At pH of blood, much more aspirin is in dissociated form and cannot recross membrane. Thus a high concentration gradient for undissociated aspirin is maintained. Hence net movement of swallowed aspirin is from juice to blood.

malabsorption of lipid also induce malabsorption of these vitamins as well as symptoms of avitaminoses (skin, eye, bone and blood abnormalities). Such diseases include those involving inadequate digestion of lipids (chronic pancreatitis), reduction in small intestinal surface area (small bowel resection), a badly damaged mucosal surface (celiac disease) or loss of bile salts (hepatitis).

ACTIVE TRANSPORT.—Most of the substances that are secreted into or are absorbed from the gastrointestinal tract move passively under the influence of electrochemical or osmotic gradients. The major contribution of active transport is absorption of substances when their luminal concentration is less than their plasma concentration. Intestinal active transport guarantees near total absorption of glucose, galactose, amino acids and sodium from the small intestine. Because of osmotic effects, much water and other solutes secondarily enter the body because of active transport. Complex active transport processes are responsible for uptake of vitamin B_{12} and iron, without which severe anemia may develop. In secretory organs, active transport mechanisms are responsible for secretion of specialized digestive juices, such as the highly acidic gastric juice or the alkaline pancreatic secretion. The characteristics of active transport and a model to explain the absorption of glucose by an intestinal epithelial cell have been described in Chapter 1.

REGULATORY FACTORS: HORMONAL AND NEURAL CONTROL

HORMONAL CONTROL

The first two proved hormones were secretin (1902) and gastrin (1905). These gastrointestinal hormones and a third, cholecystokinin, have recently been shown to be long chain polypeptides and have been synthesized. Shortly before, during, and for some time after a meal is eaten, these hormones are released into the circulation. They profoundly influence motility, secretion and absorption and are the most important direct regulators of gastrointestinal functions.

These hormones interact with one another and with the gastrointestinal parasympathetic nerves. Let us consider, for example, the following sequence of normal interactions at meal time: (1) vagal nerve stimulation releases acetylcholine near the antral mucosal cells inducing them to release gastrin into the blood, (2) vagal nerve terminals also release acetylcholine near parietal cells and near smooth muscle cells of the stomach wall, (3) gastrin reaches the same parietal and smooth muscle cells via the circulation, (4) acetylcholine and gastrin stimulate the parietal cells to secrete hydrochloric acid and the muscle cells to gradually empty that acid into the duodenum, (5) acid contacting the duodenal mucosa subsequently causes release of the hormone secretin into the circulation and (6) secretin reaches the parietal cells and inhibits secretion of acid. Complex neurohumoral synergisms and inhibitions are as characteristic of gastrointestinal endocrinology as of the endocrinology of other organ systems.

GASTRIN

Human gastrin is a polypeptide of molecular weight 2,114 and consists of 17 amino acids. The last four acids at the C-terminal end are tryptophan, methionine, aspartate and phenylalanine (Trp-Met-Asp-Phe-NH$_2$). This tetrapeptide possesses the biologic activities of the whole molecule. A pentapeptide, consisting of the tetrapeptide combined with a stabilizing fifth amino acid, has been synthesized for use in man.

Most gastrin originates in the antral mucosa (distal fifth of the stomach); in the middle of that mucosa are argentaffin-like cells (gastrin cells), which elaborate granules containing gastrin. When the antral mucosa is bathed by gastric acid at a pH below 2.0, the release of gastrin is inhibited. Since the resting stomach secretes little acid and most food has a pH above 2.0, secretion of gastrin is inhibited infrequently. Lesser sources of gastrin include other areas of the gastrointestinal mucosa and the pancreatic islets.

The major stimulus for secretion of gastrin into the circulation is the release of acetylcholine near the gastrin cells by terminals of the vagus nerves or by intrinsic cholinergic nerves in the mucosa in response to hunger and ingestion of food.

The two most important actions of gastrin are its stimulation of hydrochloric acid secretion by the stomach and its growth-promoting effect on the gastrointestinal mucosa. Thus, following surgical removal of the antral portion of the stomach, gastric secretion of acid declines (a desirable effect in hypersecreting duodenal ulcer patients), and the gastric mucosa atrophies (an undesirable effect that predisposes to gastritis).

Other effects of gastrin, which have been shown experimentally but may not occur at physiological concentrations, include its stimulation of lower esophageal sphincteric contraction; gastric and intestinal motor activity; pancreatic secretion of enzymes, water and bicarbonate; release of insulin; hepatic secretion of water and bicarbonate; contraction of the gallbladder; inhibition of intestinal absorption of sodium and glucose (and therefore of water and other electrolytes) and dilatation of gastrointestinal arterioles.

The subcellular mechanisms of these varied responses to gastrin are mostly unknown, since the biochemical reactions responsible for secretion, absorption and motor activity are incompletely understood. However, the ability of gastrin to act as a selective growth hormone for the inner lining of the stomach and gut involves stimulation of deoxyribonucleic acid (DNA), ribonucleic acid (RNA) and protein synthesis.

The concentration of circulating gastrin can be measured in man by radioimmunoassay and under basal conditions is 30–120 pg/ml. In patients with a gastrin-producing pancreatic tumor or hyperplastic islets (Zollinger-Ellison syndrome) serum concentrations of gastrin usually exceed 500 pg/ml. In some patients with pernicious anemia, the gastrin levels also are abnormally elevated. The latter finding is somewhat paradoxical, since pernicious anemia is accompanied by near or total loss of parietal cells and by inability to secrete acid. In the Zollinger-Ellison syndrome, on the other hand, there is a great increase in the parietal cell mass, and the stomach secretes excessive amounts of acid. The seeming paradox can be explained as follows. In pernicious anemia the lack of gastric acid results in loss of the mechanism whereby antral gastrin release is normally inhibited, i.e., lowering of the antral surface pH below 2.0. In the Zollinger-Ellison syndrome the circulating gastrin comes mainly from the abnormal tissue of the pancreas, not from the inhibited antral mucosa.

Gastric ulcer patients often have elevated serum gastrin concentrations. Unfortunately for diagnostic purposes, the serum gastrin of most patients with duodenal ulcers is usually within the normal range.

SECRETIN

Ernest Starling performed a classic experiment in physiology at the turn of the century. He showed that introduction of hydrochloric acid into a denervated loop of gut prompted pancreatic secretion of water and electrolytes. Since the loop and pancreas were connected only by the circulation, he concluded that a blood-borne substance from the intestinal mucosa stimulated pancreatic secretion. He then extracted the agent from the intestinal mucosa and injected it into the circulation, thereby again stimulating the pancreas to produce its juice. Starling had discovered the first hormone — secretin.

Secretin is a helical polypeptide of 27 amino acids. Chemical substitution of any of these amino acids inactivates the molecule.

Secretin is released from villus epithelial cells of the upper small intestinal mucosa when the pH of the surface drops below 4.5. Normally this occurs when the stomach empties its hydrochloric acid into the duodenum. Other natural stimuli for release of secretin include dietary fatty acids.

Secretin has many gastrointestinal actions. Its major effect is to stimulate pancreatic secretion of water and bicarbonate. It also evokes hepatic secretion of water and bicarbonate. In addition, secretin is a potent inhibitor of gastrin-stimulated secretion of acid by the stomach, of gastrin release from the antrum and of gastric motility. Thus, in a major way, secretin opposes the effect of gastrin at meal time. Gastrin stimulates gastric secretion of acid and the emptying of that acid into the duodenum, which subsequently causes release of secretin. The effect of secretin is to increase pancreatic and hepatic secretion of alkaline juice into the duodenum, to neutralize gastric acid, to inhibit gastric secretion of gastrin and hydrochloric acid and to delay gastric emptying of acid. In this manner the triggering of gastrin effects by eating is eventually turned off by secretin. This is negative feedback.

The antagonism of gastrin effects by secretin in many gastrointestinal organs exhibits characteristics of noncompetitive inhibition. The two hormones are chemically dissimilar, and they do not compete for the same receptor site on the cells they affect.

Another important effect of secretin is its stimulatory action on gastric secretion of pepsin. This is not easy to explain in terms of the large picture, since secretin inhibits gastric secretion of acid, which is necessary for activation of pepsin. Also, it has little direct effect on pancreatic secretion of enzymes.

Under experimental conditions secretin stimulates release of insulin and inhibits release of glucagon from the pancreatic islets. Secretin slows intestinal absorption of sodium and water, inhibits intestinal motility, potentiates the contraction of the gallbladder caused by cholecystokinin, relaxes the lower esophageal sphincter, dilates mesenteric arterioles and opposes the mucosal growth-promoting effects of gastrin. The physiological importance of these actions is not clear.

The biochemical pathways for these many actions are unknown. The role of elevated or depressed plasma levels of secretin in diseases is also unknown and awaits development of a good radioimmunoassay for the hormone. In the Zollinger-Ellison syndrome, very large volumes of gastric acid are dumped into the duodenum and appear to contact insensitive receptors for secretin release. The secretin levels in the blood are probably reduced here. Surprisingly, secretin may stimulate gastric secretion of acid and intestinal secretion of fluid in the Zollinger-Ellison syndrome. This strange endocrine disease probably involves abnormalities of many gastrointestinal hormones.

CHOLECYSTOKININ AND GASTRIC INHIBITORY PEPTIDE

The polypeptide chain of cholecystokinin contains 33 amino acids. At the C-terminal end of the molecule, we encounter the familiar tetrapeptide of gastrin, Trp-Met-Asp-Phe-NH$_2$. Therefore, the actions of cholecystokinin resemble those of gastrin, the differences being a matter of relative potency. For example, cholecystokinin is a feeble stimulant of acid secretion in the stomach, and gastrin is a far weaker stimulant of gallbladder contractions than is cholecystokinin.

Like secretin, cholecystokinin originates in the upper intestinal mucosa. Digested food products—especially fatty and amino acids—are the major substances to evoke release of cholecystokinin when they contact the mucosal surface. Acid gastric juice also prompts elaboration of cholecystokinin, but less effectively than food.

The primary gastrointestinal effects of cholecystokinin are strong stimulation of gallbladder contractions and pancreatic enzyme secretion. The ability of cholecystokinin to empty the gallbladder has been used as part of a diagnostic radiologic test (Graham-Cole) for gallstones obstructing the opening of the bladder. Another polypeptide, caerulein, derived from the skin of certain toads, is a chemical homologue of cholecystokinin and gastrin and is even more potent in provoking emptying of the gallbladder.

Cholecystokinin is the major stimulant of pancreatic secretion of enzymes, although both gastrin and acetylcholine also contribute. Thus ingestion of a meal signals the pancreas to secrete the main digestive enzymes at three times by three messengers: (1) before and during eating, nerve impulses traverse the vagi and acetylcholine is deposited near secreting cells, (2) when food is in the stomach, gastrin is released and (3) after gastric emptying and during digestion of fat and protein, cholecystokinin is elaborated. All in all, this is a neat arrangement whereby thinking about eating and the subsequent digestion of food regulate the appearance of digestive enzymes in the right place at the right time.

Cholecystokinin augments the effect of secretin in stimulating pancreatic secretion of water and bicarbonate. In part because the chemical structures of these two hormones are not similar, the augmentation is considered to operate via two different receptors in the pancreas and is termed "potentiation." Since the dumping of hydrochloric acid, fat and protein by the stomach evokes simultaneous release of both secretin and cholecystokinin, the latter hormone plays a significant role as regulator of the volume and alkalinity of the pancreatic juice. Since the digestive enzymes have neutral or slightly alkaline pH optima, cholecystokinin is the major messenger for digestion of fat and protein in more than one way.

Cholecystokinin and gastrin are homologous hormones, both structurally and in terms of their effects. However, when present in combination—as occurs at mealtime—they interact competitively. Thus cholecystokinin alone is a weak stimulant and gastrin alone, a powerful stimulant of gastric acid secretion; however, when the stomach is being driven to secrete acid by gastrin, addition of cholecystokinin inhibits the effect of gastrin. This antagonism is termed "competitive inhibition," partly because the active sites of the two hormones contain similar chemical structures. The explanation of this inhibition involves a single stimulatory receptor on the acid-producing parietal cell, which can be activated when molecules of either gastrin or cholecystokinin occupy the receptor. Addition of weakly stimulating molecules of cholecystokinin leads to occupation of some receptors by this hormone instead of the more powerful gastrin molecules. Competitive inhibition might be considered the Gresham's law of endocrinology.

Cholecystokinin has been reported to inhibit intestinal absorption of sodium and water and to stimulate pepsin secretion by the stomach, water and bicarbonate secretion by the liver, gastrointestinal motor activity and insulin release. These actions are of uncertain physiological significance.

Gastric inhibitory peptide is another polypeptide released by the intestinal mucosa when its surface is bathed by dietary fat. Its major effects are inhibition of gastric secretion of acid and gastric emptying.

OTHER HORMONES AND TISSUE SUBSTANCES

Glucagon is a gastrointestinal hormone, at least in terms of its organ of origin (pancreas), its food-related modes of release (hypoglycemia, ingestion of glucose, aminoacidemia), and its effects on gastrointestinal functions (stimulates hepatic flow of water and bicarbonate, stimulates pancreatic release of insulin, inhibits gastric secretion and motility, inhibits pancreatic secretion, dilates the intestinal circulation). However, these are relatively unimportant in comparison with the role of glucagon in regulating carbohydrate metabolism. Similarly, although insulin is produced and released by the pancreas, this vital hormone is far more a body-wide metabolic agent than a specific regulator of gastrointestinal functions.

Pituitary and adrenal hormones also influence gastrointestinal functions, although this is hardly their major effect. Growth hormone influences growth in the digestive tract, and its absence leads to atrophy of the gastrointestinal mucosa and shrinkage of the solid organs, with decreased gastric secretion and diminished motility. Absence of ACTH and adrenal cortical failure, e.g., Addison's disease, cause marked inhibition of the gastric acid secretory response to stimuli such as gastrin, acetylcholine and histamine. Conversely, chronic administration of either ACTH or adrenal cortical hormones increases gastric secretory responsiveness to these stimuli.

Certain tissue substances are somewhat like hormones. A hormone is manufactured in a specific organ and is released into the blood to act on its target cells. Tissue substances are produced in many organs and act locally. For example, mast cells, which are found in the mucosal lining of the stomach, small and large intestine (as well as in most organs in the body), contain at least three tissue substances: histamine, 5-hydroxytryptamine (serotonin) and heparin. Other tissue substances known to be present in the gastrointestinal mucosa include prostaglandins, bradykinin, kallidin and substance P. It is uncertain whether any of these agents is a significant normal regulator of cellular or organ functions in the gastrointestinal tract. They may assume medical importance in one of two situations: (1) they may be involved in a gastrointestinal disease process because of excessive production—when overproduced, the tissue substance gets into the circulation in its active form and produces toxic effects; and (2) some tissue substances have been synthesized commercially and are used as drugs because of particular effects on a gastrointestinal function.

Serotonin and bradykinin are produced in excessive quantities by metastasizing carcinoid tumors of the small intestine and may be partly responsible for the respiratory difficulties and flushing of the skin that occur in some patients with these tumors.

Histamine is present in the normal circulation in an inactive form. In severe allergic reactions (anaphylactic shock), histamine is released into the blood in its active form and contributes to many of the symptoms of these patients, e.g., respiratory difficulties, itching, skin wheals, hypotension. Histamine is also a powerful stimulant of acid secretion by the stomach. As a drug, histamine is widely used to test the ability of the stomach to secrete acid. This diagnostic procedure—the augmented histamine test—provides important information in the workup of patients suspected of having pernicious anemia, gastric and duodenal ulcers, gastric carcinoma and the Zollinger-Ellison syndrome.

Prostaglandins are naturally occurring fatty acids. However, when used in large doses therapeutically, they may cause diarrhea through a mechanism that appears to involve activation of adenyl cyclase and cAMP accumulation in the intestinal endothelial cells. As a result, there is excessive transport of chloride and accompanying water into the intestinal lumen. To date there are two diseases characterized by excessive production of prostaglandins and diarrhea: medullary carcinoma of the thyroid gland and a rare pancreatic tumor.

Prostaglandins are promising drugs for treatment of gastrointestinal diseases but their use is still experimental. Their effects on the gastrointestinal tract include (1) inhibition of gastric secretion, which might be useful in treating patients with duodenal and gastric ulcers; (2) relaxation of gastrointestinal sphincters, which might be useful in treating patients with contracted sphincters as in achalasia, congenital pyloric stenosis, pyloric obstruction secondary to an ulcer and spasm of the sphincter of Oddi; (3) constriction of the splanchnic arteries (prostaglandin $F_{2\alpha}$), which might be useful in treating massive gastrointestinal hemorrhage and (4) dilatation of the mesenteric arteries (prostaglandin E), which might be useful in treating intestinal ischemia. In the aforementioned diseases, current drug treatment is usually ineffective and surgical treatment is less than ideal.

NEURAL CONTROL

The sensation of hunger, the sensing of food and the processes associated with eating clearly involve the CNS and its sensory informers. The regulatory messages are relayed from the cerebral cortex and other higher centers to integrating centers in the vast unknown between the cortex and the medulla. Besides the mere fact that a man is hungry, his brain must integrate his emotional condition at the time he eats, his response to the kind of food available and his state of health. These factors influence gastrointestinal functions. Thus, for example, anger increases and emotional depression decreases the rate of acid secretion by the human stomach. The effect of acute, severe anxiety in inducing emptying of the bowel is sufficiently well known to be the butt of jokes.

Modulation by higher autonomic integrative centers involves inhibition of incoming stimuli. An inhibitory locus for both feeding and gastric secretion is located in the ventromedial nucleus of the hypothalamus. Electric stimulation of the amygdala and anterior and posterior hypothalamus also inhibits gastric secretion in primates. Furthermore, in experimental animals, sectioning through the pons to divorce the medulla from higher control increases the gastric secretory response to medullary impulses.

Medullary messages reach the gastrointestinal organs via autonomic efferent nerves. The parasympathetic preganglionic fibers to the salivary glands are conveyed by cranial nerves VII and IX to ganglia near the glands. The vagus nerves conduct the parasympathetic preganglionic fibers for the esophagus, stomach, pancreas, liver and biliary tree, small intestine and proximal half of the colon. Vagal efferent fibers terminate either adjacent to the effector cell (e.g., parietal cell, acinar cell, smooth muscle cell, gastrin cell) or adjacent to intrinsic plexuses of cholinergic nerves within the substance of each organ. An effector cell can respond to vagal impulses directly, to intrinsic nerves stimulated by the vagi or to intrinsic nerves stimulated by local, nonvagal factors. The terminal colon receives cholinergic input from presacral parasympathetic nerves and intrinsic plexuses.

In general, parasympathetic activation increases organized motor activity in the walls of the hollow gastrointestinal organs and relaxes the muscular valves between the organs; this results in greater mixing and accelerated forward movement of the contents. Parasympathetic activation stimulates secretion of saliva, gastric and pancreatic juices and bile; this results in the digestion of food into simpler molecules that can be absorbed by the small intestine. Unabsorbed food, some secreted substances, surface cells that have been sloughed and bacteria constitute the excreta; parasympathetic activation mobilizes the colon to rid itself of this waste. Vagal stimulation also evokes direct release of gastrin and indirectly (through stimulation of gastric secretion and emptying) causes the release of other intestinal hormones.

Surgical interruption of the vagal parasympathetic supply to the abdominal viscera used to be commonly employed to reduce gastric secretion in patients with peptic ulcers. Unfortunately, this procedure also has adverse effects on the digestive function of the gastrointestinal tract by interfering with normal motility and secretion of pancreatic and biliary juices. It is not difficult in medical practice to replace a natural disease with an iatrogenic disorder. Recently, a more limited operation (selective vagotomy) has been introduced to retain the gastric antisecretory effect and avoid the untoward effects of vagotomy.

Sympathetic postganglionic fibers usually hug the arteries coursing to each gastrointestinal organ. Sympathetic activation constricts the arteriolar smooth muscle and reduces blood flow to the organ. Gastrointestinal motility is generally decreased by sympathetic activity. Salivary secretion is increased.

In overwhelming stress states, such as severe heart failure or following extensive surgery, sympathetic overactivity may become chronic and lead to marked reduction in gastrointestinal mucosal blood flow. This sometimes results in cell necrosis, ulceration, hemorrhage, massive gangrene of the gut and death.

REGULATORY FACTORS: THE SPLANCHNIC CIRCULATION

The splanchnic circulation, i.e., the circulation to the alimentary canal, liver and spleen, is the largest systemic regional vascular bed and carries about one quarter of the cardiac output. As is the case for all regional circulations, blood flow through the splanchnic vessels is influenced by mechanical, neurohumoral and metabolic factors. These have been discussed elsewhere (section on regional circulations, Chapter 4) and the present description will consider only the blood flow changes induced by alterations in the function of the gastrointestinal tract and its associated glands.

BLOOD FLOW TO THE SALIVARY GLANDS

Salivary secretion is controlled almost exclusively by the parasympathetic and sympathetic nerves and their neurotransmitters acetylcholine and norepinephrine. During stimulation of salivary secretion, glandular blood flow increases; when secretion abates, blood flow declines. Acetylcholine is a direct vasodilator and increases blood flow. Catecholamines act on vascular alpha receptors to cause an initial constrictor response, which is rapidly replaced by a dilator beta response; the net effect of either sympathetic stimulation or circulating catecholamines therefore is increased blood flow. In addition to these responses, blood flow increases during secretion because of increased glandular metabolism. Oxygen is consumed and vasodilator substances (notably bradykinin) are produced. The relative importance of vasodilator nerves versus vasodilator metabolites has been disputed for more than a century and the issue has not yet been resolved.

BLOOD FLOW TO THE STOMACH

Gastric mucosal blood flow increases when gastric secretion is increased by secretagogues (histamine, gastrin, acetylcholine), and decreases when secretion is inhibited, e.g., by catecholamines, vasopressin, prostaglandin E_1, atropine or secretin. The action of these agents upon the circulation of the stomach is twofold: (1) by altering metabolism, they indirectly alter blood flow by increasing or decreasing the release of vasodilator metabolites and (2) some secretagogues (histamine, acetylcholine) directly dilate and some inhibitors (norepinephrine, vasopressin) directly constrict the gastric blood vessels. Thus histamine increases blood flow to the stomach, both because it increases secretion and mucosal metabolism and because it relaxes gastric arteriolar smooth muscle. One type of secretory inhibitor, prostaglandin E_1, decreases mucosal blood flow because it inhibits secretion and metabolism, whereas another inhibitor, vasopressin, decreases blood flow because of both its antisecretory and its direct vasoconstrictor properties. Thus the two major regulators of gastric mucosal blood flow are the rate of gastric secretion (and metabolism) and the constrictor or dilator properties of circulating agents. However, we cannot overlook the role of the autonomic nerves. The vagi stimulate gastric secretion (directly and via release of gastrin), and the splanchnic sympathetic nerves inhibit it. Gastric mucosal blood flow is increased during vagal stimulation mainly because of the increased mucosal metabolism and partly perhaps because of a direct vasodilator action of released acetylcholine. Correspondingly, sympathetic stimulation reduces mucosal blood flow partly because of reduced metabolism and partly because of a direct vasoconstrictor effect of released norepinephrine on mucosal arterioles. Hormones, like gastrin, secretin and adrenal corticosteroids, appear to influence blood flow primarily through their effects on secretion and mucosal metabolism.

The question of a relationship between mucosal circulation and peptic ulcer disease is intriguing. Mucosal ulcerations are provoked by the action of hydrochloric acid and pepsin on a mucosal lining whose resistance to damage has been lowered. An inadequate circulation contributes directly to lowering mucosal resistance. However, experimental findings indicate that mucosal circulatory insufficiency follows, rather than precedes, the event causing ulceration, e.g., topical damage with lipid-soluble agents such as aspirin. Furthermore, an acidic juice containing pepsin is necessary for ulcer formation. Chemicals causing ulcers include those that increase (histamine) and decrease (vasopressin) both blood flow and secretion. Finally, ligation of 90% of the blood supply to the stomach does not induce ulcers.

BLOOD FLOW TO THE PANCREAS

The pancreatic circulation, like that of the salivary glands and stomach, is linked loosely to secretion. By this we mean that increased secretion (and metabolism) are associated with increased blood flow to the pancreas, and decreased secretion is accompanied by decreased blood flow. In addition, the direct actions of powerful circulating vasodilators and neural transmitters must be remembered. Pancreatic secretion, metabolism and blood flow are enhanced by secretin, cholecystokinin and parasympathetic stimulation and are diminished by sympathetic stimulation and catecholamines.

Ingestion of a meal increases pancreatic blood flow. Activation of the vagi stimulates gastric acid secretion and emptying of acid and food into the duodenum. This causes mucosal release of the hormones secretin and cholecystokinin. Acetylcholine (released by the vagi and intrinsic pancreatic nerves), secretin and cholecystokinin stimulate pancreatic secretion and metabolism, causing local release of dilator metabolites. Acetylcholine and secretin are also direct vasodilators. The end result of the interaction of all these factors is increased pancreatic blood flow.

BLOOD FLOW TO THE INTESTINE

Intestinal blood flow comprises more than half the splanchnic flow. The process of seeing, smelling and tasting food increases cardiac output and intestinal blood flow. The subsequent eating and swallowing of the food augments and prolongs these effects which are integrated by both neural and humoral mechanisms including vagal stimulation and secretin release.

When food is presented to the gut, blood flow is increased. Presumably, metabolic activation during digestion and absorption of nutrients reduces tissue oxygen tension and releases local vasodilator metabolites. During exercise, blood flow to the gut is reduced, despite increased cardiac output, probably secondary to sympathetic neural vasoconstriction. The competition between leg muscles and intestines for blood may have an adverse effect on either digestion or exercise following a meal.

ORGAN FUNCTIONS: MOTILITY

Success of the gastrointestinal tract depends upon complex patterns of movement that are responsible for the transit of food and fluid from the mouth to the absorbing gut, for the mixing of digestive juices with that food and for the movement of unabsorbed materials to the anus. These movements are mainly regulated by autonomic nerves, especially parasympathetic motor nerves, which stimulate forward propulsive movements and relax the gastrointestinal sphincters. Sphincters are muscle thickenings that serve as valves and are located at the junctions between different parts of the alimentary canal; contracted sphincters impede forward and backward movement of visceral contents. Most of the parasympathetic preganglionic outflow to the gastrointestinal tract is delivered by the vagus nerves, which activate plexuses of intrinsic cholinergic nerves in its walls. Gastrointestinal motility also depends on the intrinsic excitability characteristics of its smooth muscle cells, on the sympathetic nervous system and on gastrointestinal hormones.

MOTILITY IN MOUTH AND PHARYNX

Chewing food involves complex integrated voluntary and reflex movements of the mandible, tongue and cheeks. These movements are integrated in the mesencephalon. It is widely held among the laity that thorough chewing of food is essential for digestion. Chewing increases salivary secretion of alpha amylase, but this digestive enzyme is not essential for the breakdown of starch. Bolting of food chunks is common among some animals and some people and, aside from the esthetic qualities of the practice, little evidence exists that it is harmful.

After being chewed, the food bolus is moved by the tongue upward and backward in the mouth toward the pharynx. This process usually initiates swallowing which is the major motor activity of the pharynx. The first step is the sealing of lips, nasopharynx (through elevation of the soft palate) and glottis. The swallowed food will then have only one avenue, namely, the esophagus. As the tongue finishes pushing food, the posterior pharyngeal muscles contract in a coordinated fashion around the bolus and impel it rapidly toward the esophagus, whose upper sphincter relaxes as the food arrives.

ESOPHAGEAL MOTILITY

In the adult, the esophagus is a hollow tube 25–35 cm long. Except after swallowing food or fluid, it is collapsed and the pressure inside its lumen is about the same as intrathoracic pressure (a few mm Hg above or below atmospheric, depending upon respiration). Sphincters are located at the upper and lower ends of the esophagus and are closed except when food is passing. At rest, the closed sphincters oppose either swallowing of air from above or reflux of gastric contents from below because the pressure at the sphincters exceed the pressures in the pharynx, body of the esophagus and stomach by several mm Hg.

The main function of the esophagus is to assist in rapid movement of swallowed substances from pharynx to stomach. Swallowed water reaches the stomach in a second or two, and solid food traverses the normal esophagus in 10 seconds, even in a patient lying on his side.

The upper esophageal sphincter consists of the cricopharyngeus muscle attached to the cricoid cartilage. This muscle is normally contracted and maintains a pressure of about 50 mm Hg within its lumen. When swallowing begins, there is a rapid elevation and then descent of the larynx. As the larynx moves up, the cricopharyngeus relaxes and is pulled up to open the upper esophageal sphincter; when the larynx descends, the muscle is activated and snaps back to close the sphincter. Thus there is a close coordination of swallowing and passage of food into the esophagus. Respiration is another influence on upper esophageal sphincteric activity. The cricopharyngeus contracts during inspiration and relaxes during expiration. These movements are controlled by neurons in the nucleus ambiguus in the medulla.

The active contribution of the esophagus to forward propulsion of a swallowed bolus of food consists of a series of coordinated contractions and relaxations. First, the upper esophageal sphincter relaxes as swallowed food approaches and closes behind the bolus when food has passed. Then the body of the esophagus moves the food down its length by a form of motility—peristalsis—found in most of the gastrointestinal tract. Peristalsis consists of a contraction of circular muscle at one point in the wall, followed by contraction of the adjacent distal (anad) longitudinal muscle. After this, the next distal batch of circular muscle contracts and the sequence is repeated successively down the tube. These movements are coordinated and give the appearance of waves moving smoothly along the length of the esophagus. Following onset of peristalsis, the lower esophageal sphincter relaxes and, after the bolus passes, the sphincter closes behind the moving food.

The mechanical movements of the esophagus are strongly influenced by vagal nerve activity but also exhibit independence from external nervous control. This independence is a function of the cholinergic neurons of the myenteric plexus, which is located between the longitudinal and circular muscle layers. These neurons comprise a continuous network of intrinsic neural regulators of esophageal motion. In addition, esophageal smooth muscle cells independently exhibit inherent rhythmic contractions whose frequency decreases down the length of the esophagus. Thus peristalsis is controlled by a balance between extrinsic and intrinsic cholinergic nerves and intrinsic myogenic automaticity. Integration of neural control of esophageal peristalsis occurs within the medulla, which receives sensory information from the walls of the pharynx and esophagus via afferent fibers in the glossopharyngeal and vagus nerves.

The lower esophageal sphincter is the distal 2 cm of the organ and is closed except during the approach and passage of a bolus. The sphincter is relaxed by vagal motor nerves and by fibers of the myenteric plexus. The latter intrinsic nervous system is activated either by the vagi or by more proximal intrinsic nerves, which respond to stretch of the esophageal wall as food is worked down toward the stomach. Another relaxant of the lower esophageal sphincter is the hormone secretin, which is released when acid gastric juice is emptied into the duodenum. Epinephrine, prostaglandins and cAMP also relax this sphincter. Another hormone, gastrin, acts oppositely to elevate resting pressure and to prompt contractions of the sphincter. In addition, the inherent

Fig 7–2.—Neurohumoral control of esophageal motility. Voluntary nervous system, parasympathetic nervous system, intrinsic nerves and gastrointestinal hormones influence movement of food from mouth to duodenum (see text).

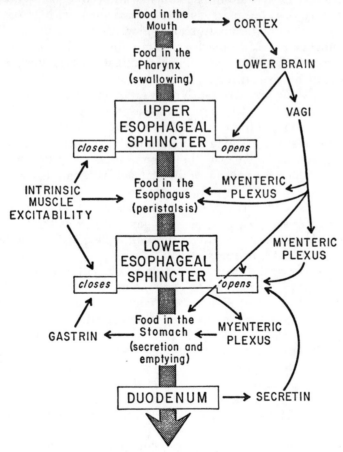

electric activity of the sphincteric smooth muscle cells contributes to tonic contraction.

Thus the mechanisms underlying the act of moving chewed food from mouth to stomach are interrelated. Control of swallowing and of esophageal motility involves the central nervous system, cranial nerves, intrinsic neural plexuses and gastrointestinal hormones (Fig 7–2).

Disturbances in this process are not uncommonly found in medical practice. If the lower esophageal sphincter fails to relax adequately as food approaches (achalasia), the patient will either regurgitate the food or, less commonly, may experience chest pain as the food is propelled past and stretches the spastic sphincter. Conversely, if the lower esophageal sphincter fails to close adequately after the food has entered the stomach, acid gastric juice will enter the lower esophagus and may cause an unpleasant sensation referred to as *"heartburn"*; if this reflux of gastric juice becomes chronic, the esophageal mucosa may become inflamed (esophagitis).

GASTRIC MOTILITY

The stomach is a way station that stores swallowed food and fluid for minutes to hours. It also churns its contents to mix them with gastric juice and dumps small aliquots of this mixture into the waiting duodenum.

The proximal half of the stomach exhibits an internal pressure at rest that is about the same value as that within the abdominal cavity. Ingestion of sizable volumes of food, fluid or air changes that pressure surprisingly little, since the muscular walls of the upper stomach relax following swallowing. This allows the stomach to accommodate large loads, a property of some advantage to the cave man, who ate infrequently but well, and to his descendents who eat more often and less well. The proximal stomach contracts at times, especially when the contents of the organ are being emptied into the gut. These contractions serve to reduce the volume of the stomach as it empties and to provide additional push to the moving gastric contents.

By contrast, the distal half of the stomach is much more motile than the proximal, and peristaltic waves seem to gather force as they approach the pylorus. Between contractions, pressures within the distal half of the stomach reflect intra-abdominal pressures.

Peristaltic waves of the stomach have been classed as types I, II and III, according to their pressure patterns. A saline-filled catheter with its open tip in the gastric lumen and its external end connected to a pressure transducer records two to four pressure changes per minute in adult subjects. These pressure changes last 2–20 seconds each. If the amplitude of the pressure change is less than 10 mm Hg, the contraction is defined as a type I wave; if the contraction exceeds 10 mm Hg, it is a type II wave. Type III waves occur when several areas of the stomach contract simultaneously, producing a slow tonic change, usually of more than 10 mm amplitude lasting more than a minute; type I or II waves may be superimposed on a type III contraction (Fig 7–3).

These pressure changes push food and fluid to and fro in the stomach. Much

MOTOR ACTIVITY

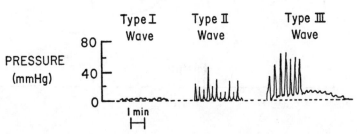

Fig 7–3.—Representative illustration of pressure waves of types I, II and III.

of the material that is slowly propelled toward the pylorus by a single wave undergoes retropulsion, i.e., the material is moved back into the proximal stomach to be sloshed about by another wave. The result of this sort of action is mixing of the solid food particles with the liquid and with the corrosive gastric juice. On each pass of a wave, a small fraction of the gastric content is emptied into the duodenum.

Gastric motility is controlled by several factors. The vagus nerves send efferent fibers, which synapse with either smooth muscle cells or intrinsic cholinergic nerves. The intrinsic nerves are arranged as plexuses between the muscle layers of the stomach walls and also synapse with the smooth muscle cells. In addition, these cells exhibit automaticity and are responsive to gastrointestinal hormones.

Stimulation of vagal efferents initiates relaxation of the upper half of the stomach. Curiously, stimulation of the sympathetic efferents has the same effect. Secretin and cholecystokinin induce relaxation whereas gastrin causes contractions in the upper stomach.

In the distal half of the stomach, electrophysiologic recordings from the muscle show a basic electric rhythm, which slowly depolarizes and repolarizes in a cyclic fashion. The intrinsic electric activity of the gastric longitudinal muscle determines the maximal rate at which the stomach can contract and also the direction and velocity of contractile waves. The slow electric rhythm does not itself initiate mechanical activity, but at a given point in the depolarization phase of a slow wave a number of spike potentials may be generated (Fig 7–4). Each mechanical contraction of the wall is preceded by spike potentials. Although the rate of the basic electric rhythm is the same from point to point along the stomach wall, these waves do not appear simultaneously at all points. The lag in appearance time is organized so that the basic electric rhythm seems to migrate from the midportion of the greater curvature toward the pylorus and toward the lesser curvature. This is also the path taken by peristaltic waves.

If the vagi are cut, the basic electric rhythm will persist, but the slow waves lose their coordination and the frequency of spike discharge diminishes. Correspondingly, mechanical contractions become less frequent and are weaker after vagotomy; conversely, vagal stimulation increases the force and frequen-

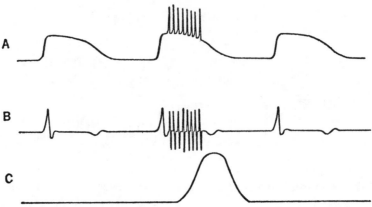

Fig 7–4.—Electric activity of the gut (schematic). **A,** slow waves recorded with intracellular electrode. Second wave shows bursts of action potentials during phase of depolarization. **B,** same as **A,** but recorded by extracellular electrode. **C,** muscle contraction. Note that onset of contraction is preceded by the appearance of action potentials. (Courtesy of Christensen, J. Reprinted by permission from the New England Journal of Medicine 285:85, 1971.)

cy of movement in the distal half of the stomach. Gastrin increases the frequency of the basic electric rhythm, frequency of spike bursts and frequency of mechanical contractions in the distal half of the stomach. On the other hand, the coupling of the basic electric rhythm to motor activity is not absolute; secretin, for example, inhibits mechanical contractions and slows the rate of firing of spike bursts but does not affect the frequency or organization of the basic electric rhythm.

Food entering the stomach stretches the walls and stimulates the intrinsic cholinergic nerve plexuses, which activate the electric and mechanical components of motility. Thus the processes related to hunger and its satisfaction influence lower gastric movements. Those movements are restricted in frequency and direction by the basic electric rhythm, which can, however, be altered by extrinsic and intrinsic nerves and the hormones released during feeding.

GASTRIC EMPTYING

Emptying of the stomach depends upon aforementioned factors that influence motor activity throughout the gastrointestinal tract: inherent smooth muscle excitability, intrinsic nerves, the vagi and hormones. In addition, the nature of the gastric contents affects emptying. Fat and protein in the ingested food, a very acidic juice and a hyperosmolar mix of juice and food delay emptying. The closer the contents approximate isosmotic saline the faster they depart from the stomach.

The mechanism by which the gastric contents regulate their own emptying involves a network of sensors in the *duodenal* mucosa. Stimulation of these chemo- and osmoreceptors by substances emptied into the duodenum slows gastric emptying.

The pyloric sphincter is a zone of elevated pressure between stomach and gut, which can contract further and hamper emptying. When gastric acid and lipid contact the duodenal mucosa, the hormones secretin, gastric inhibitory peptide and cholecystokinin are released and the pylorus contracts.

The rate of emptying also reflects the balance between motor activity in the distal stomach and in the proximal duodenum. Since the latter exceeds the former, gastric emptying is slow.

Thus the processes surrounding gastric emptying are marked by built-in negative feedback mechanisms (Fig 7–5). Vagal excitation increases emptying of gastric contents and causes release of gastrin; emptied gastric contents stimulate release of secretin, enterogastrone and cholecystokinin and activate mucosal receptors in the duodenum. The four hormones and the local nerves act to decrease distal stomach motility, increase pyloric tone and increase duodenal motility. The result is impedance of gastric emptying. It is hardly surprising, therefore, that Thanksgiving dinner should contribute its sense of plenty for hours via the full stomach. The major normal function of the stomach is storage.

When the delay in emptying becomes excessive, serious medical problems may develop. Two common disease states involving excessive delay are hypertrophic pyloric stenosis and pyloric obstruction. Hypertrophic pyloric stenosis occurs in infants and involves hypertrophy and spasm of the pyloric sphincter. The baby is hungry, vomits his milk and loses weight. Surgical correction of this congenital abnormality is the usual therapeutic measure. In adults with long-standing duodenal, pyloric or prepyloric ulcers, extensive

Fig 7–5.—Control of gastric emptying. Nervous and hormonal factors operate as negative feedback to decrease gastric emptying after emptying has been stimulated.

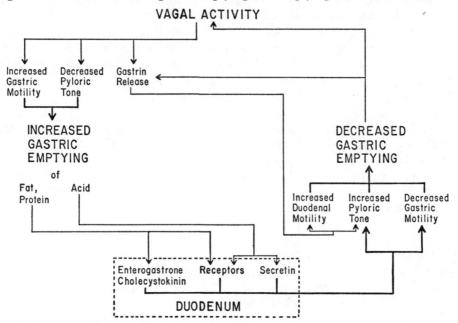

scarring, edema and spasm lead to obstruction of the pyloric channel. Medical management involves continuous aspiration of the gastric contents and intravenous alimentation in the hope that most of the obstruction is not caused by permanent scarring but by acute inflammation, swelling and edema, which will subside. This failing (as is often the case), surgical intervention will be required in the form of gastric resection and bypass procedures.

At the other end of the spectrum is excessive gastric emptying in the wrong direction, commonly and more simply termed "vomiting" or something still more descriptive. The symptom is usually of minor medical significance, but prolonged vomiting, especially in babies and in the elderly, can become life threatening owing to the effect of loss of gastric juice on body fluid, electrolyte and acid-base composition.

Vomiting is usually preceded by an aura, termed *nausea,* which prepares the individual for the ensuing ordeal. During this time there is a diffuse autonomic discharge. Sympathetic activation increases heart rate, sweating and salivation. Parasympathetic activation increases salivation, stimulates gastric motility and relaxes the upper and lower esophageal sphincters. The individual then takes a deep breath, closes the glottis and contracts the abdominal musculature, forcing the stomach against the descended diaphragm. This is coupled with considerable gastric motor activity, including even some antiperistaltic motions. The main thrust, however, comes from squeezing the stomach between diaphragm and belly wall, and the vomitus spurts up the esophagus, past the sphincters to the outside.

SMALL INTESTINAL MOTILITY

The small intestine is a 300 cm long hollow tube in the adult and constitutes far more than half of the internal distance from mouth to anus. Its surface area is in excess of two million square centimeters. Its functions, i.e., digestion and absorption of nutrients and rejection of waste, are essential to life. To assure these processes, the muscular walls of the gut must mix the nutrients (chyme) that it receives from the stomach with digestive juices, expose the chyme to the mucosal surfaces of the gut and propel the chyme down the length of the small intestine toward the colon.

Much of the time there is no chyme in the lumen, and the small intestine exhibits no contractile activity. During these resting periods, internal pressure of the gut is less than 10 mm Hg.

A single contraction usually involves a segment of the gut about an inch in length at a moment in time. Contractions reflect mostly the activity of the thicker circular muscle layer of the wall. They do not necessarily occlude the gut lumen, nor are they always concentric. As in the stomach, contractions can be classed empirically as type I, II or III waves (see Fig 7–3). In the fasting gut, the waves may show a repetitive pattern consisting of long (30-minute) periods of no wave activity, succeeded by long periods of periodic contractile activity and finally a long period of continuous bursts of contractions, which stop suddenly. Then the cycle is re-enacted.

These small intestinal contractions do not progress longitudinally along the gut. They may occur simultaneously at many points and lend the appearance of a string of sausage links to the barium-filled intestine at x-ray. These contractions are called *rhythmic segmentations* and are principally type I waves. Their purpose is to mix the chyme, thereby increasing the chances that undissolved material will be solubilized, that undigested nutrients will encounter digestive enzymes and that digested substances will be presented to the absorbing surfaces of the intestinal mucosa. Because of the sheer bulk of the intestinal chyme, the mixing process becomes a mechanical catalyst for digestion and absorption.

Similarly, the mucosal and submucosal smooth muscle initiates undulating movements of the inner lining of the gut and its finger-like villi. These motions not only stir up the broth but expose new mucosal surfaces for further absorption of nutrients.

In addition to the stirring motions, there are propulsive contractions—peristaltic waves—which move the chyme on to previously unexposed mucosal surfaces for continued absorption of liquid and solute. Peristaltic waves are also responsible for propelling the undigested, unabsorbed substances down to the colon. These propulsive waves are mainly type I and type III contractions.

The motor activity of the gut is regulated by factors enumerated previously in considering control of esophageal and gastric motility: smooth muscle excitability, intrinsic and extrinsic cholinergic neurons and hormones. In addition, there is an extensive sympathetic innervation to the small intestine and there are locally released chemicals that may modulate movements of the gut.

The intestinal smooth muscle has an unstable resting potential of about 60 mv, which is discharged with some regularity. This basic electric rhythm originates in the longitudinal muscle layer of the gut wall and spreads into the circular layer. It is observed whether or not the gut is contracting.

At any given site along the gut, the basic electric rhythm is constant, although its frequency may be increased by fever or slowed by hypothermia. The rate is faster in the upper half of the small intestine than in the lower, and total transection of the gut will decrease the rate below the cut. This indicates that the frequency of the basic electric rhythm in any given segment of gut is also influenced by the frequency of the more proximal intestine. A group of smooth muscle cells in a segment of gut generate their basic electric rhythm in unison, i.e., their intrinsic frequencies are the same. The adjacent anad group of muscle cells have a slightly slower intrinsic frequency, the next distal segment still slower, and so on down the gut. The intrinsic frequency of one group of muscles is also coupled to that of neighboring groups in such a way that the group with the faster frequency speeds up the rate of the slower. Thus the depolarization of this 10-foot muscular tube is coordinated in the manner of a long chain of coupled oscillators.

Like many other neuromuscular excitations, the basic electric rhythm has been related to cyclic changes in membrane permeability to sodium. Depolarization occurs when sodium moves passively down its electrochemical gradient into the muscle cell when its membrane permeability increases. Repolarization of the membrane is associated with return of sodium permeability to its resting state and to increased potassium permeability.

Spike potentials are another type of electric activity observed in the gut.

They appear irregularly (although always during the depolarization phase of the basic electric rhythm) and are lead to contraction of the muscle mass from which they originate. Contraction of intestinal smooth muscle does not occur without preceding spike potentials.

The interrelation between mechanical and electric activity in the gut appears to be this: spike potentials always precede a muscular contraction, and spike potentials occur during a specific phase of the basic electric rhythm. Thus the rate of mechanical contractions cannot exceed the rate of the basic electric rhythm. Similarly it follows that there is a gradient from duodenum to ileum, not only for the rate of electric depolarization but also for the maximal rate at which the intestinal muscle can move. The maximal rate and direction of intestinal motility are determined by the basic electric rhythm.

The initiation of spike potentials and mechanical activity depends upon neurohumoral factors. A pattern of reproducible motor responses to a stimulus is considered to be reflex when it is abolished by denervation. These reflexes involve intrinsic autonomic nerves in the wall of the gut and may also involve vagal afferent and efferent pathways. The *intestino-intestinal* reflex consists of a generalized inhibition of small intestinal contractions when one segment of gut is distended. In patients with paralytic ileus, part of the gut is distended and the rest is abnormally relaxed; medical treatment has included use of drugs to block the intestinal sympathetic receptors that appear to mediate the intestino-intestinal reflex. There is also a reflex that causes closure of the ileocecal junction when the cecum contracts; this reflex prevents colonic contents from being spurted back into the ileum.

The intrinsic cholinergic nerves release acetylcholine, which initiates contraction of the small bowel. The intrinsic adrenergic nerves release catecholamines, which relax the bowel wall.

The extrinsic cholinergic nerves are branches of the vagi. Their stimulation increases intestinal motor activity. Paradoxically, however, after surgical vagotomy, a frequent complaint of patients is diarrhea, which is often assumed but has not been proved to be due to excess intestinal motor activity. Vagotomy may actually cause diarrhea by abolishing the segmentative contractions of the small gut that normally impede forward propulsion of liquid contents in the lumen. Nevertheless, the diarrhea is usually stopped by use of an anticholinergic drug.

The extrinsic adrenergic nerves reach the gut traveling along branches of the superior mesenteric artery. Their stimulation inhibits intestinal motor function.

Among the circulating chemicals that depress small intestinal motility are epinephrine, norepinephrine and secretin. Substances that increase contractions of the small bowel include gastrin, cholecystokinin and angiotensin II. The physiological significance of these effects is not clear.

Locally released chemicals also affect intestinal motility. These include histamine, prostaglandins and serotonin. Their role in the normal control of contractions is uncertain. In disease states in which they are released in excessive quantities, their effect is usually to increase motor activity and contribute to diarrhea.

Abnormalities of small intestinal motility occur and may become medical problems. Paralytic ileus, with generalized intestinal muscle inhibition, can be life threatening; the gut becomes a still compartment in which large volumes of fluid and electrolytes may be sequestered.

Another small intestinal disorder involving abnormal motor function is Asiatic cholera. Again the abdomen is silent, save for periodic massive rushes of fluid along the gut to the outside world. The motor disorder in commonplace diarrheas is unclear. It is widely held that diarrhea is the result of excessive motility. In reality, the best-established mechanism in diarrhea involves excessive secretion of fluid and electrolytes into the lumen of a hypomotile gut.

BILIARY TRACT MOTILITY

Secretion of bile and its flow into the upper gut are essential for normal digestion and absorption of fats. The movement of bile also assumes medical importance in view of the commonly occurring diseases that interfere with this flow, namely gallstones, carcinomas of the biliary tract and pancreas, congenital anomalies of the tract and postsurgical strictures.

The mechanical factors that influence biliary flow include tension generated by the sphincter of Oddi, the contractions of the muscular walls of the bile ducts and the gallbladder and the force imparted by secreting liver cells. The secretory pressure may reach 40 cm of water at times. Since pressures in the gut lumen are usually low, the pressure gradient favors bile flow.

The gallbladder serves several functions. It absorbs sodium and water, thereby converting the large volume of watery bile it receives from the liver into a smaller volume of viscid, dark bile. The gallbladder stores bile during interdigestive periods, and its walls are able to relax so that increasing volumes can be accommodated with only small pressure changes. Following feeding, the gallbladder contracts rhythmically and slowly, forcing viscid bile into the upper gut. Fat in the duodenum is emulsified by the natural detergents in bile, and contact between that fat and the mucosa of the gut releases cholecystokinin which causes the gallbladder to contract and is the major regulator of biliary flow into the upper gut during digestion. Gastrin, released during feeding, is a weaker stimulant of gallbladder contractions.

The bile duct exhibits contractile activity that is not as well organized as an intestinal peristaltic wave. This activity causes bile to enter either the gallbladder or the gut, depending on where the resistance to flow is lower.

The terminal part of the common bile duct is the sphincter of Oddi, which regulates the entry of bile into the gut. The sphincter can oppose pressures up to 40 cm of water, thereby preventing bile from flowing into the gut. Contractions of the gallbladder can, however, overcome the sphincteric resistance to flow. Since the sphincter itself is surrounded by duodenal muscle, contractions of the duodenum also increase resistance to bile flow. Relaxation of the sphincter occurs when the gallbladder contracts. It is also controlled by cholecystokinin via an effect that is associated with increased intracellular cAMP. Thus in a rather satisfying conceptual way, dietary fat in the upper gut regulates its own rate of digestion and absorption by sending a messenger—chole-cystokinin—to initiate contraction of the biliary reservoir and relaxation of the valve that prevents bile from entering the gut between meals.

COLONIC MOTILITY

The colon has all the functions of the more proximal parts of the gastrointestinal tract. It is a conduit and also secretes, stores and absorbs.

The ileocecal junction is the sphincter between the small bowel and cecum. Most of the time the sphincter is closed to prevent cecal waste and bacteria from returning to the small bowel. When a peristaltic wave pushes material toward the sphincter, the valve relaxes and allows a spurt into the cecum. Then it closes. The action of the sphincter is governed by intrinsic mechanisms and is unaffected by extrinsic denervation.

Most of the colon, including the cecum, ascending, transverse and descending portions, appears saccular at x-ray. This seems to be due to incompleteness of the longitudinal muscle layer of the wall. Circular muscle tone produces periodic indentations, called "haustrations."

Pressure-sensing instruments and x-ray visualization of colonic motility reveal nonpropulsive phasic contractions at pressures of about 5 mm Hg, with occasional stronger contractions up to 50 mm Hg, persisting for as long as a minute, half the time at peak activation. Tonic contractions analogous to type III waves with superimposed phasic contractions also are seen. Rhythmic segmentation is the most commonly observed motor activity in the colon; this nonpropulsive motion serves to sift feces to and fro, allowing absorption of electrolytes and water. There usually is little forward propulsive movement in the colon, but three or four times daily, usually after meals, vigorous colonic contractions occur, causing anad propulsion of the colonic contents. They are termed "mass movement" and are usually engendered by the presence of food in the stomach. In this complex function, phasic colonic activity declines abruptly, and feces are dispatched along the colon fairly rapidly. Then the haustral contractions occur. Mass movements are responsible for transporting aliquots of the fecal content from the right side of the colon to the left.

Once feces fill the sigmoid colon, peristaltic activity moves the material into the empty rectum, where distention signals the need to defecate. Strong colonic contractions can have a similar effect even if the rectum is empty. The usual position of the body during defecation is sitting or squatting; this position provides mechanical advantages, namely, gluteal stretch applied to the external anal sphincter and abdominal compression by the thighs. Often individuals take a deep breath and attempt to exhale against the closed glottis (Valsalva maneuver); this procedure elevates both intrathoracic and intraabdominal pressures markedly, generating more squeeze upon the serosal surface of the rectum. The longitudinal rectal muscle contracts, thereby shortening the channel; the internal anal sphincter relaxes as peristaltic waves approach; the individual voluntarily relaxes his external anal sphincter and a successful "bowel movement" ensues.

Control of colonic mechanical functioning is related to the usual factors observed in other parts of the enteric canal, i.e., smooth muscle electric activity, intrinsic and extrinsic nerves and circulating chemicals.

There is a colonic basic electric rhythm of about six per minute, without any gradient from cecum to anus (which is probably the basis for the prolonged periods during which feces may remain in the colon). Spike potentials precede mechanical contractions and occur during the depolarization cycle of basic electric rhythm.

The intrinsic nerves are found in plexuses between the muscle layers and are mainly cholinergic, with some adrenergic representation as well. They receive signals, not only from other intrinsic nerves but also from mucosal sensors and from extrinsic nerves.

In Hirschsprung's disease there is a segment of colon devoid of intrinsic neurons. The segment becomes spastic and obstructs the colon, leading to accumulation of incredible quantities of feces. It appears that the intrinsic nerves of the colon act mainly to inhibit smooth muscle contractility.

Extrinsic cholinergic innervation is delivered by the vagi to the cecum, ascending and proximal transverse colon and by the pelvic nerves to the distal colon and rectum. These nerves contact either muscle cells or intrinsic neurons. When the extrinsic cholinergic nerves are stimulated, the predominant colonic response is increased mechanical activity. Administration of a cholinergic drug can induce a bowel movement.

Adrenergic nerves reach the colon traversing perivascular spaces; they originate in the superior and inferior mesenteric plexuses. Stimulation of these nerves relaxes the colon. Distention of one part of the organ (as by impacted feces, a tumor or much trapped gas) initiates a reflex that travels via sympathetic nerves to inhibit colonic contractions. Surgical sympathectomy in patients increases colonic motility.

Circulating substances also influence motions of the large bowel. Epinephrine depresses colonic motility. Gastrin increases contractility of the colon and relaxes the ileocecal sphincter and appears to be responsible for the mass movements described above. In the past, these were thought to be due to a gastrocolic reflex.

ORGAN FUNCTIONS: SECRETION

SALIVARY SECRETION

The salivary glands are not essential for life. Nevertheless, their secretion is important to the hygiene and comfort of the mouth and teeth, and it also contributes to normal digestion. People with congenital or acquired inability to secrete saliva suffer from the chronic problems of dental caries, inflammation of the buccal mucosa and a dry mouth.

The composition of saliva from different salivary glands varies. The largest of the salivary glands are the paired parotid glands, each located near the angle of the jaw and the ear. They secrete a fairly watery juice, whereas the smaller bilateral submaxillary and sublingual glands elaborate a more viscid saliva. Other, still smaller glands are located in the mucosa covering the palate, buccal areas, lips and tongue. Each salivary gland consists of a blind-end system of microscopic ducts, which branch out from grossly visible ducts. At the blind end of the ducts are the polygonal acinar cells; the ducts themselves are lined by columnar cells. Myoepithelial cells surround the secretory cells. Each microscopic duct with its acinar and duct cells operates as a unit termed the "salivon." The primary secretion of the salivon changes as it moves down the duct to join secretions from other salivons to become the final saliva of that gland.

The blood supplied to the salivary glands is distributed by branches of the external carotid artery. The direction of arterial blood flow within each salivary gland is opposite (countercurrent) to the direction of flowing saliva within the ducts of each salivon. The arterioles break up into capillaries around acini and in nonacinar areas as well. Blood from nonacinar areas passes through portal venules back to the acinar capillaries, from which a second set of venules then drains all the blood to the systemic venous circulation.

Both components of the autonomic nervous system reach the salivary glands. Parasympathetic preganglionic fibers are delivered by the facial and glossopharyngeal nerves to autonomic ganglia, from which postganglionic fibers pass to individual glands. The sympathetic postganglionic nerve fibers

Fig 7–6. – Autonomic nerve supply to major salivary glands.

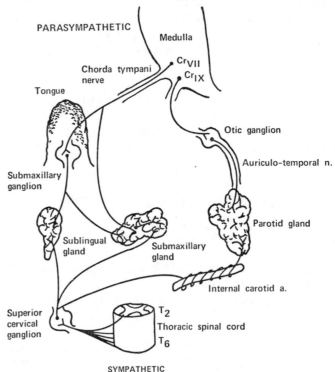

originate in the cervical ganglion, and reach the gland via the periarterial spaces (Fig 7-6). Parasympathetic and sympathetic mediators regulate all known salivary gland functions to an extraordinary degree. Their rather exclusive influence includes major effects, not only on secretion but also on blood flow, muscle activity, growth and metabolism.

FUNCTIONS OF SALIVA

Saliva serves two major functions: protective and digestive. The *digestive* contributions include dissolving and washing away food particles on the taste buds of the tongue to enable a person to taste the next morsel of food he eats. Saliva has slippery organic solutes that lubricate the food as it is chewed and assist its subsequent easy passage through the pharynx and esophagus to the stomach. Saliva contains an enzyme, alpha amylase, which breaks the 1,4 glycosidic bonds of chains of glucose molecules in polysaccharides such as starch. Perhaps 75% of the starch in a well-chewed crouton or piece of pasta al dente is digested to the disaccharide stage by salivary alpha amylase before the material reaches the duodenum, where remaining starch will be digested by pancreatic alpha amylase. Since starch is one of the most common components of the world's food, normal digestion may be assisted by salivating before and during eating.

Saliva has several *protective* actions. It dissolves and washes away retained food particles between teeth. These particles are nutrients for bacterial metabolism, and their fermentation leads to halitosis. Saliva also contains an antibacterial enzyme, lysozyme, which attacks the microbial cell wall. It is difficult to assess how effective this agent is. On the one hand, xerostomia (dry mouth) is associated with chronic infections of the teeth and buccal mucosa; on the other, the normally salivating mouth contains the second richest bacterial count of any bodily orifice.

In areas of the world where ingested water has elevated amounts of fluoride, salivary secretions contain increased concentrations of that anion, and the incidence of dental caries is reduced. Other absorbed halides, such as chloride and iodide, also appear in the saliva. In fact, iodide is actively secreted into the saliva and may there reach a concentration many times higher than in plasma.

Saliva protects the mouth by buffering noxious stimuli. For example, many people sip simmering solutions of chicken soup or coffee. The free flow of saliva dilutes and lowers the temperature of such hot fluids, thereby preventing a scalded tongue. Similarly, the salivary glands secrete copiously before vomiting and so protect the mouth by partially neutralizing the regurgitated acid.

Saliva also seems necessary for prolonged speech in the highest primate, as evidenced by the invariable glass of water on the podium of a public speaker. Finally, the moist mouth is essential for oral comfort, and the dry mouth is an early signal of dehydration, thirst or fear.

COMPOSITION OF SALIVA

The major constituents of saliva are water, electrolytes and some enzymes. The unique properties of this juice are (1) its large volume relative to the mass of the glands that secrete it, (2) its osmolality, (3) its high potassium concentration and (4) the specific organic materials it contains.

Compared with other secretory organs of the gastrointestinal tract, the salivary glands elaborate a remarkably large volume of juice per gram of tissue. For example, an *entire* pancreas may reach a maximal rate of secretion of 1 ml per minute, whereas at highest rates of secretion in some animals, a submaxillary gland can secrete 1 ml *per gram* of tissue per minute—a 50-fold higher rate. In man, the salivary glands secrete at rates several fold higher than other gastrointestinal organs per unit weight of tissue.

The osmolality of saliva is significantly lower than that of plasma at all but the highest rates of secretion, when the saliva becomes isotonic with plasma. As the secretory rate of the salivon increases, the osmolality of its saliva also increases.

The potassium concentration of saliva is two to 30 times that of plasma, depending upon the rate of secretion, the nature of the stimulus, plasma potassium concentration and plasma concentration of mineralocorticoids. Saliva has the highest potassium concentration of any digestive juice, and at maximal values its concentration approaches that of potassium within cells. These remarkable levels of salivary potassium imply the existence of an energy-dependent transport mechanism within the salivon.

When the salivary glands are secreting very briskly and the saliva is isotonic with plasma, salivary potassium and often bicarbonate concentrations exceed those of plasma; however, salivary sodium and chloride concentrations are much lower than in plasma. As the secretory rate tapers off, the concentrations of Na^+, K^+ and HCO_3^- decline significantly (Fig 7–7). Hence the osmolality of the saliva is low at low rates of secretion.

The peculiarities of the ionic composition of saliva depend upon the way the salivon produces its final juice. The acini elaborate an initial secretion whose potassium and sodium concentrations and osmolality equal those of plasma. As the initial fluid moves along the duct, an energy-dependent transport process reabsorbs sodium. Potassium exchanges in the reverse direction (potassium secretion) to some extent. Chloride moves from the juice into the tissue passively to accompany sodium and some bicarbonate exchanges for the chloride. The duct lining is relatively impermeable to water and, because more sodium leaves the duct than is replaced by potassium, the juice in the duct becomes hypotonic. The end result is a hypotonic saliva, high in potassium and bicarbonate and low in sodium and chloride.

The major clinical implication of this information relates to the non-eating patient with a neck or facial fistula draining saliva to the outside (as one recovering from resection of a malignancy). The parenteral fluid administered to this patient should contain supplemental potassium.

Previously it was noted that saliva contains alpha amylase and lysozyme. Another salivary enzyme, kallikrein, converts an extracellular protein into the

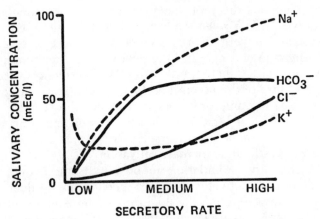

Fig 7–7.—Relationship of secretory rate to electrolyte composition of saliva.

nonapeptide bradykinin, which is a powerful vasodilator. Other organic materials normally present in saliva include blood group substances and some components of the blood clotting system.

The total protein concentration of saliva is about one-tenth that of plasma. The subcellular mechanisms involved in salivary enzyme secretion include ribosomal protein synthesis, migration and encapsulation of the protein and release of the resulting granules into the lumen of the duct. These processes will be discussed more fully when pancreatic enzyme secretion is considered.

Subcellular protein synthesis, acinar secretion and ductal modification of saliva are mainly under the control of neural transmitters: acetylcholine, released by parasympathetic nerves, and catecholamines, released by sympathetic terminals or brought to the glands by the blood. The secretory effects of these agents appear to be mediated by increased intracellular synthesis of cyclic nucleotides. Acetylcholine activates the enzyme, guanylate cyclase, which synthesizes cyclic guanine monophosphate (cGMP): catecholamines activate adenylate cyclase, which synthesizes cAMP. These cyclic nucleotides are involved in phosphorylation at the cell membrane, which binds calcium and increases the ability of the membrane to transport ions actively. When cAMP is injected into the blood supply of a salivary gland, the organ secretes.

Nervous Regulation of Salivary Glands

Secretion from the salivary glands is not primarily regulated by hormones, although mineralocorticoids from the adrenal cortex and ADH from the posterior pituitary will decrease sodium and increase potassium concentrations of the saliva. However, unlike other exocrine secretory organs of the gastrointestinal tract, initiation and maintenance of secretion by the salivary glands are almost exclusively dependent upon parasympathetic and sympathetic nerves. The parasympathetic system is the stronger stimulus for secretion; its fibers are distributed to all salivary glands and their transmitter, acetylcholine, either directly stimulates receptors on secretory cells or stimulates intrinsic cholinergic nerves, which release additional acetylcholine. Each acinar cell has five to ten axons converging upon its receptors. These receptors are bound to the cell membrane and may be identical with the membrane-bound enzyme, guanylate cyclase, that stimulates cGMP production within the cell. The electric response of the membrane to acetylcholine is rapid depolarization. Drugs such as atropine or scopolamine block the secretory receptor to acetylcholine and abolish secretion. Thus a common annoying side effect of these drugs is a dry mouth. Conversely, drugs that inhibit acetylcholinesterase, such as neostigmine, increase salivation.

Parasympathetic nerve stimulation not only starts and maintains secretion within the salivon, but activates transport mechanisms in duct cells, which change the initial secretion into the final saliva. These acute cellular responses to nervous stimulation are accompanied by an increase in glandular metabolism, with increased consumption of oxygen and glucose and increased production of active metabolites. A long-term metabolic effect of parasympathetic stimulation is growth of the gland.

Fig 7–8.—Effects of parasympathetic and sympathetic nervous stimulation on blood flow in salivary gland. Both increase secretion and metabolism and lead to augmented glandular blood flow.

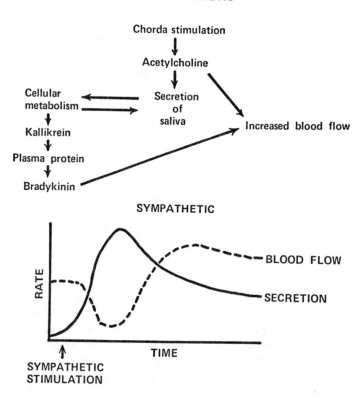

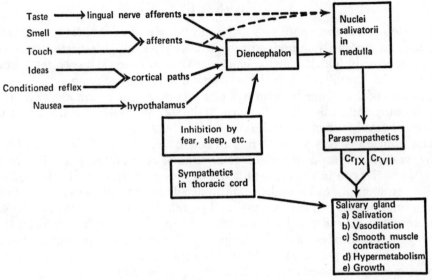

Fig 7–9.—Neural control of major functions of salivary glands. Cr_{VII} is the facial nerve, Cr_{IX} the glossopharyngeal nerve.

Some of the parasympathetic fibers cause arteriolar relaxation and increase blood flow to the salivary glands. The increased blood flow may also be maintained by dilator metabolites, such as bradykinin, which appear during active secretion. The direct and indirect effects of parasympathetic stimulation on the salivary circulation appear in Figure 7–8.

Parasympathetic nerves also evoke contractions of myoepithelial cells and these help to squeeze saliva from the salivon and macroscopic ducts into the mouth.

Sympathetic nerves pass to the salivary acini and, upon stimulation, evoke a secretory response. These nerves release catecholamines (norepinephrine, epinephrine), which affect two receptors on the secretory cell membrane, that is, the alpha and beta adrenergic receptors. Changes in salivary secretion seem to result mainly from activation of the beta adrenergic receptors.

Sympathetic stimulation causes a biphasic change in blood flow (see Fig 7–8). The earliest vascular response is a transient decrease in flow but this is followed by increased flow presumably due to stimulation of beta adrenergic receptors combined with effects of dilator metabolites.

Sympathetic stimulation and local release of catecholamines acting upon beta adrenergic receptors also change the composition of saliva, increase metabolism and growth of the gland and prompt the contraction of the myoepithelial cells around the salivon.

The dual autonomic regulation of these glands is unusual because both the parasympathetic and sympathetic systems stimulate secretory, metabolic, trophic, muscular and circulatory functions in similar directions. Their complementary effects are shown in Figure 7–9.

GASTRIC SECRETION

The stomach is important to our lives in three main ways: (1) it stores food and fluids temporarily, allowing us to eat large meals, (2) it secretes the intrinsic factor without which a fatal anemia may develop and (3) it secretes a digestive juice that converts ingested food into the semiliquid chyme.

Diseases of the stomach include peptic ulcers, gastric carcinoma and gastritis, bleeding and indigestion. Because of the frequency of these disorders, the subject of gastric secretion is of enormous medical interest.

The stomach is very variable in size and shape. It weighs less than 1% of body weight and has a mucosal surface area of about 800 sq cm. The major secreting cell type in the mucosa is the oxyntic or parietal cell. The parietal cell mass amounts to around one billion cells, which, however, constitute only a small fraction of the mucosal cell population. Together these cells can secrete up to 20 mEq of hydrochloric acid in a few hundred milliliters of juice per hour. In abnormal states, these cells are ultimately responsible for the occurrence of peptic ulcer disease, and therapeutic measures aim in one way or another to reduce or abolish parietal cell secretion.

The parietal cells are limited to the mucosa of the body and fundus of the stomach and are located deep within the oxyntic glands (pits), whose ducts open into the gastric lumen. The distal fifth of the stomach, or antrum, contains pyloric glands, which have no parietal cells.

Chief or peptic cells, which secrete pepsinogen, also are located in oxyntic glands. Epithelial cells and mucous cells occur throughout the inner lining of the stomach. The antral mucosa contains specialized endocrine cells, G cells, which produce gastrin.

The juice secreted by the stomach is actually the final mix of three separate secretions by different cells in its mucosal lining.

1. The parietal cells elaborate a solution containing about 150 mEq/L of hydrochloric acid and the intrinsic factor, a macromolecule that binds vitamin B_{12}. With maximal stimulation, the normal adult stomach can secrete several hundred milliliters of juice per hour, and in a usual day 1–2 L are formed. In a hypersecretory disease such as the Zollinger-Ellison syndrome, 8 L of highly acidic juice may be secreted.

2. The peptic cells secrete a small volume of juice with an electrolyte composition resembling that of extracellular fluid or plasma. The organic material in this juice is pepsinogen, the precursor of the proteolytic enzyme pepsin.

3. The third component of gastric juice, a complex organic gel composed of various macromolecules, is secreted in small volumes by mucous cells and is termed mucus.

The most direct way of collecting gastric juice or assessing mucosal function in patients is by nasogastric intubation. A pliable, lubricated plastic tube is passed through the nose and down the gullet, assisted by having the patient swallow water. The external end of the tube is then connected to a large syringe or gentle pump to suck the gastric juice. Aspirated material is usually analyzed for the concentration of hydrochloric acid, pH, pepsin content and volume collected per unit time. Modifications of the tube allow a wide range of procedures to be carried out in man, such as measurement of the potential difference across the mucosal membranes, evacuation of gastric contents in cases of attempted suicide, collection of epithelial cells for cytologic diagnosis of cancer, endoscopic visualization and biopsy of the mucosa and analysis of the electrolytes and macromolecules in the juice.

In the hospital, gastric secretion can be stimulated by histamine, gastrin or one of their analogues, and resting and stimulated gastric secretory output can be measured. Resting output of acid is normally 0–5 mEq per hour, and stimulated peak output is 5–40 mEq per hour for brief periods. In patients with pernicious anemia or gastric ulcer, these values are often reduced, whereas patients with the Zollinger-Ellison syndrome and many with duodenal ulcer exhibit elevated outputs of acid.

SECRETION OF WATER AND ELECTROLYTES

The events responsible for gastric secretion of water and electrolytes take place at the membranes of epithelial cells lining the mucosa. Although this lining consists of several dissimilar cell types, each of which contributes differently to the final product, it is convenient to picture secretory events as if there were only one cell type separating the extracellular and plasma compartments from the juice in the lumen of the stomach. The epithelial cell has two membranes: (1) the serosal membrane between the extra- and intracellular spaces and (2) the mucosal membrane between the intracellular space and the lumen.

Three factors operate at each membrane to regulate movement of water and ions: (1) the permeability of the membrane to the moving substances—membranes are relatively impermeable to sodium and potassium and are much more permeable to chloride; (2) the presence of active transport mechanisms or pumps for specific ions—the gastric mucosa actively transports hydrogen, chloride and sodium; and (3) the existence of electrochemical, osmotic and pressure gradients across the membranes.

It is possible to explain the movement of water and electrolytes involved in formation of gastric juice and the varying concentrations of electrolytes in the compartments of the mucosa in terms of the preceding three factors. They are depicted in Figures 7–10 to 7–15 and described in the text that follows.

HYDROGEN (FIG 7–10).—The most outstanding characteristic of gastric juice is its nearly incredible concentration of hydrogen ion. Few life forms can exist in an environment whose pH is less than 1.0; yet, in the vast majority of humans, the gastric mucosa shows no ill effects, although it is regularly bathed by a juice with a hydrogen ion concentration close to 150 mEq/L. The concentration gradient for hydrogen ion from juice to plasma is often greater than one million to one, which exceeds usual ionic chemical gradients across other living membranes by several orders of magnitude. Maintenance of this enormous gradient without damage to the gastric mucosa has prompted investigators over the last century to assign a special resistant property to the mucosa. This is usually thought to be a great impermeability of the mucosal membrane to hydrogen ion, despite the great permeability of other mem-

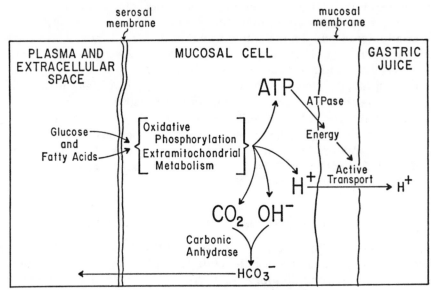

Fig 7–10.—Production and transport of hydrogen ion of gastric juice. Glucose and fatty acids are substrates for metabolic generation of hydrogen ion, which is actively transported into the juice.

branes to this smallest of ions. Since the mucosal membrane is also the site of a rather unusual hydrogen pump, it is equally plausible to ascribe resistance of the mucosa to autodigestion by acid to its ability to pump out the back-diffusing hydrogen ion.

Secretion of hydrogen ion into the gastric lumen requires two major intracellular steps: production and active transport of the ion, both of which consume energy and substrate. The specific substrates are glucose and fatty acids, which are metabolized by oxidative phosphorylation and the electron chain to yield four products involved in acid secretion: hydrogen ion itself, ATP, carbon dioxide and the hydroxyl radical. There is also some extramitochondrial oxidation of substrate by a dehydrogenase at the mucosal membrane, with release of protons (hydrogen ions) into the lumen.

A carrier in the mucosal membrane transports hydrogen ion into gastric juice against the concentration gradient. The carrier requires energy produced during the breakdown of ATP by ATPases. Within the cell, carbon dioxide combines with hydroxyl to form bicarbonate and with water to form unstable carbonic acid, which splits into bicarbonate and hydrogen ion. These reactions involving carbon dioxide are mediated by the intracellular enzyme, carbonic anhydrase. The bicarbonate then diffuses from the cell into the blood in about the same molar quantities as the hydrogen ion entering the juice. Therefore, venous blood leaving the stomach is more alkaline than is arterial blood entering the organ.

CHLORIDE (FIG 7–11).—Both serosal and mucosal membranes are permeable to chloride. Passive movement of chloride across these membranes is

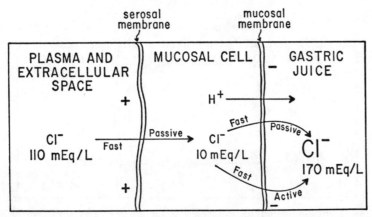

Fig 7–11.—Chloride transport across gastric mucosal cell. Serosal and mucosal membranes are relatively permeable to chloride, and the ion moves readily from extra- to intracellular spaces down concentration gradient. At the mucosal membrane, chloride accompanies hydrogen ion into juice. In addition, an active transport mechanism moves chloride into juice against its electrochemical gradient.

more rapid than passive movement of the cations and is nearly as rapid as that of actively transported hydrogen.

The concentration gradient for chloride across the serosal membrane favors its movement into the cell, whereas the electric gradient has an opposite effect. At the mucosal membrane, the concentration gradient is in the direction of chloride movement into the cell. The chloride content of gastric juice is significantly higher than that of plasma and, to achieve this, the gastric mucosa must possess a chloride pump which actively transports this anion into the juice. This pump appears to be loosely coupled with the proton pump so that both pumps tend to keep in step when acid secretion is stimulated. Active transport of chloride occurs in both stimulated secreting and resting stomachs (at a lower rate in the latter). The rate of movement of chloride ions across the mucosal membrane can be converted, by calculation, into the dimensions of an electric current, which turns out to be equal to the measured current across the membrane. Hence active transport of chloride is considered to be an electrogenic pump, i.e., it produces an electric potential difference across the stomach wall. In the resting stomach of human beings, this potential difference is about 40 mv, with the lumen negatively charged with respect to the serosal surface. When secretion commences, the magnitude of the potential difference decreases because positively charged hydrogen ions are actively and rapidly secreted into the gastric lumen along with chloride ions. Thus, as secretory rate increases, the difference in concentrations between chloride and hydrogen ions diminishes, although chloride concentration is always greater.

SODIUM (FIG 7–12).—Transport of sodium across gastric epithelial cells is influenced mainly by the relative impermeability of both mucosal and serosal membranes to the ion and by an active transport mechanism or sodium pump

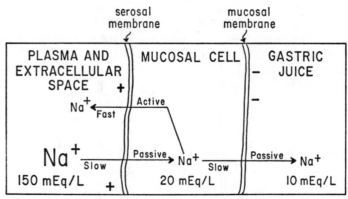

Fig 7–12.—Sodium transport across gastric mucosal cell. Both membranes are relatively impermeable to diffusion of sodium. A sodium pump at the serosal membrane actively transports ion from cell to extracellular space, which keeps intracellular concentration low.

in the serosal membrane. Poor permeation of the membranes by sodium means that passive movements of the ion into the cell and from cell to juice are slow processes, as opposed to the more rapid passive movement of chloride, for example.

In the serosal membrane a sodium-potassium pump pushes sodium rapidly from the intra- to extracellular space against both electric and concentration gradients. The pump couples extrusion of sodium with uptake of potassium. At the mucosal membrane there is an electrochemical gradient favoring movement of sodium from cell to juice: intracellular concentration of the ion is greater than its concentration in the juice during stimulation of brisk secretion (although not at lowest secretory rates), and the luminal surface of the mucosal membrane is negatively charged with respect to the inside of the cell. Thus sodium moves slowly from the cell into juice, partly to accompany secreted chloride and, to a lesser extent, to exchange for hydrogen ion, which is diffusing back from juice to cell. The net result of these membrane phenomena relating to sodium is the maintenance of lower concentrations of the ion in the juice and in the cell despite a high concentration of extracellular sodium.

POTASSIUM (FIG 7–13).—Like sodium, movement of potassium across either serosal or mucosal membranes is slow. At the serosal membrane, the concentration gradient for potassium is nearly 50 to one, with the greater concentration of the ion inside the cell. Maintenance of this gradient depends upon linkage of potassium transport to the sodium pump with potassium exchanging for sodium. Thus there is a net flux of potassium into the cell at the serosal membrane. At the mucosal membrane, there is a large electrochemical gradient favoring potassium movement into the juice; nevertheless, the impermeability of the mucosal membrane to potassium and the absence of active transport of the cation limit flux into the juice. Consequently, gastric juice contains a relatively low potassium concentration, although it is higher than that in extracellular fluid and plasma.

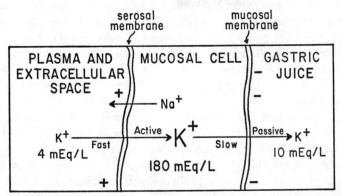

Fig 7–13.—Potassium transport across gastric mucosal cell. Both membranes are relatively impermeable to passive movement of potassium, but rapid transit of ion into the cell from extracellular space is coupled to the sodium pump. Consequently, intracellular potassium concentration is high.

WATER (FIG 7–14).—Water moves passively from the plasma and extracellular compartments across the serosal and mucosal membranes into the juice. Most of the membrane is composed of lipoproteins through which water is transported slowly. Water-filled channels penetrating the membranes are the main avenues for water movement. The serosal membrane behaves as though it were perforated by channels with an average diameter of 0.8 nm. The mucosal membrane behaves as though it contained more pores but with a diameter of only 0.4 nm.

The major force pushing water from the plasma space into the cell is a pressure gradient from the capillaries where pressure varies from 15–35 mm Hg to the cell where pressure is around zero. A model that fits fluid movement

Fig 7–14.—Water transport across gastric mucosal cell. Pressure gradient from vascular to intracellular compartment forces water movement across serosal membrane by bulk flow. Secretion of solutes into gastric juice attracts water osmotically across mucosal membrane by diffusion.

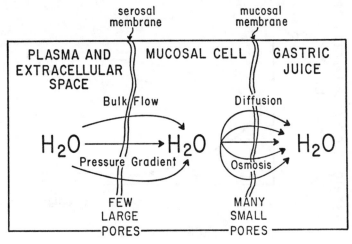

through the pores of the serosal membrane in response to a pressure gradient is the same model that has been applied to bulk flow of blood in arteries – Poiseuille's law (see section on hemodynamics, Chapter 4). The critical variables that determine bulk flow are the magnitude of the pressure gradient and the size of the pores. A gastric secretory stimulant such as histamine, which can apparently increase the radius of these pores, greatly increases bulk flow of water into the cell.

Osmosis is the major force attracting water from within the cell into the juice through the small pores of the mucosal membrane. As hydrochloric acid is secreted, water follows osmotically. In other words, water moves along its own concentration gradient to dilute the ions of the juice, which has a final osmolarity close to that of plasma. Diffusional flow of water across the mucosal membrane is described by Fick's law (Chapter 1). The limiting factors that regulate this flow are the number of pores and the thickness of the membrane.

INTERRELATION OF INORGANIC SECRETIONS. – The concentration of electrolytes in gastric juice is variable but can usually be related to the rate of secretion. The juice is a mixture of two components approximating one-tenth normal hydrochloric acid and extracellular fluid in varying proportions. At low secretory rates, the latter component predominates; at high secretory rates, the juice is mostly a dilute solution of hydrochloric acid. The relationship between volume rate of secretion and ionic concentrations in the juice is shown in Figure 7–15.

At low secretory rates, we find (1) a considerable concentration of sodium, although not as great as in plasma; (2) a potassium concentration that is small but above plasma values; (3) chloride concentration the same as that in plasma and (4) a small hydrogen ion concentration. As secretory rate increases to maximum, sodium concentration falls to low levels, potassium increases modestly, chloride increases from its value found in extracellular fluid to a concentration as high as 170 mEq/L, and the concentration of hydrogen ion increases rapidly to 150 mEq/L.

Fig 7–15.–Relationship of secretory rate to electrolyte composition of gastric juice.

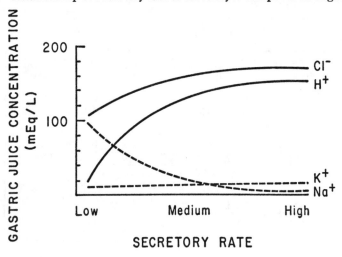

EFFECT OF MUCOSAL DAMAGE.—When the gastric mucosa is damaged by topical application of such substances as hyperosmolar aspirin, ethyl alcohol, bile salts and acetate, exogenous hydrogen ion placed in the lumen moves more rapidly into the tissue, not to return, and sodium appears in the lumen (under these conditions the two ions exchange about equally). This phenomenon is termed "breaking the barrier." After the barrier has been broken, hydrogen ion accumulates in the mucosa, intracellular pH is altered, enzymatic reactions are impaired and cellular energetics are compromised. An initial response is a slowing of all active transport. Impedance of the chloride and sodium pumps is reflected in a fall in the potential difference across the mucosa. Diminished activity of the hydrogen pump is part of a vicious cycle leading to further accumulation of this ion in the mucosa. Vulnerable mucosal cellular elements that can be disrupted by acid include mast cells containing histamine, serotonin and heparin, and lysosomes containing proteolytic cathepsins and hydrolases. The result is local ischemia, vascular stasis, hypoxia and tissue necrosis. Albumin and pepsin leak into the gastric juice and, if the stomach is secreting, bleeding occurs. The significance of this set of events relates to the common occurrence of diffuse and occasionally massive bleeding in people who regularly consume ethanol and aspirin. Furthermore, the mucosal lesions in topical damage may be forerunners of peptic ulcers, a disease whose cause and pathophysiology are unknown.

CONTROL OF ACID SECRETION (FIG 7–16)

Normal regulation of gastric acid secretion may be divided into stimulatory and inhibitory phases. There is a small basal secretion that varies greatly in volume from person to person. Acid secretion is turned on by hunger, by hostility or by the sight or thought of food. This "cephalic phase" of secretion involves nerve impulses relayed via the hypothalamus to the vagal nuclei in the medulla. Vagal efferents release acetylcholine at three receptor sites: (1) at the serosal membrane of the parietal cell, (2) at dendrites of intrinsic cholinergic nerves in the wall of the stomach and (3) at the G cells of the antral mucosa. The intrinsic nerves are stimulated to release their acetylcholine near parietal or G cells. The G cells release gastrin into the blood to circulate through the body and back to the parietal cells. Acetylcholine and gastrin directly stimulate the parietal cells to secrete acid.

The "gastric phase" of secretion is initiated when food is swallowed and stretches the wall of the stomach. The intrinsic nerves are stimulated to release acetylcholine near G cells and parietal cells. More acid is secreted. Stretching the stomach also causes reflex acid secretion. Both the afferent and efferent nerve fibers involved in the reflex travel in the vagus nerves.

The gastric mucosa contains large amounts of histamine. Injection of this substance greatly stimulates acid secretion, an effect that can be prevented by pretreatment with agents such as cimetidine that block histamine H2 receptors. H2 blockers also inhibit the acid secretory response to gastrin and other stimulants. It has been suggested that histamine is constantly being released in small amounts in the gastric mucosa and greatly magnifies the response of the oxyntic cells to the physiologic stimuli, acetylcholine and gastrin.

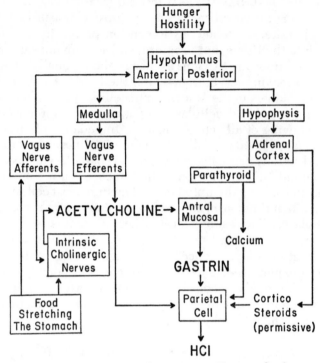

Fig 7–16.—Stimulation of gastric acid secretion. The two final common stimuli in the normal stomach are acetylcholine and gastrin. Calcium and corticosteroid hormones also influence rate of secretion.

Corticosteroid hormones and blood calcium concentrations also influence secretion. When calcium concentrations are increased, gastric secretion is stimulated. The corticosteroid hormones are permissive stimuli of acid secretion, i.e., their injection does not cause increased secretion; however, pretreatment with corticosteroids permits the stomach to secrete more copiously to a fixed amount of gastrin. This effect is of importance in dealing with patients who are on long-term corticosteroid treatment. Their stomachs secrete more acid in response to normal stimulation, and these patients are more prone to have duodenal ulcers.

TABLE 7–2.—INHIBITORY FACTORS IN GASTRIC
ACID SECRETION

Central Nervous System
  Decreased stimulation
  Emotional state—depression, anxiety
Gastrin interaction with calcium
Acid secretion
  Antral mucosal pH < 2.0
  Duodenal mucosal pH < 4.0
Long shots
  Sympathetic alpha and beta agonists
  Hormones and tissue substances—vasopressin, prostaglandins

The human stomach secretes little acid for most of each 24 hours. After hunger is satisfied by a meal, the stomach winds down. The normal inhibitors are listed in Table 7–2. These include CNS, hormonal and local factors. When the appetite has been sated by eating, hunger-driven impulses cease to trouble the stomach. The vagi stop firing. Certain emotions such as anxiety and depression also inhibit gastric acid secretion.

The hormone gastrin is part of a negative feedback mechanism controlling secretion of acid. Gastrin stimulates parietal cells directly, but it may also cause the thyroid to release calcitonin and thereby decrease blood calcium concentration and hence acid secretion. This indirect effect has been shown in rats but not yet in man.

Another negative feedback involves the acid itself. As the stomach secretes acid, the juice soon bathes the distal, nonacid-secreting portion of the mucosa. When the pH of the antral mucosal surface falls below 2.0, a local inhibitory signal prevents further secretion of gastrin by G cells. This deprives the stomach of another stimulus for its parietal cells. The acidic juice is eventually dumped by the stomach into the upper gut, along with whatever food is in the stomach. These materials contact the duodenal mucosa and cause release of three hormones, secretin, cholecystokinin and gastric inhibitory peptide, which inhibit secretion.

Sympathetic nerves are distributed to the stomach, and their stimulation inhibits secretion. Both alpha and beta adrenergic agonists also inhibit gastric secretion, as do vasopressin and certain prostaglandins. Their physiologic role is uncertain.

PEPSIN

The final gastric juice contains a proteolytic enzyme, pepsin, which is secreted by several gastric mucosal cell types (mainly the peptic cells) as a precursor, pepsinogen. Pepsinogen has a molecular weight of 42,500, pepsin a molecular weight of 35,000. Pepsinogen is synthesized within peptic cells, encased in granules and expelled from the cells by exocytosis, a process discussed more fully in connection with pancreatic acinar cell secretion of enzymes. Pepsinogen is produced and secreted in nongranular form by peptic and mucous cells of the oxyntic gland area, as well as by mucous cells of the pyloric gland area, and by duodenal mucosal Brunner's glands. Pepsinogens are of two unrelated immunologic types: group I pepsinogens are secreted by oxyntic gland mucosa and group II pepsinogens are secreted throughout the gastric mucosa and by Brunner's glands. Each group has isoenzyme subtypes.

Pepsinogen is converted to pepsin in the gastric juice when pH drops below 5.5; newly formed pepsin breaks peptide linkages in the pepsinogen molecule to produce more pepsin. At pH 5.5 to 3.5, pepsin has weak proteolytic activity, digesting mainly tertiary and secondary protein bonds; this alters the physical properties of proteins in solutions, e.g., their viscosity and optical rotation. At a pH under 3.5, pepsin becomes a powerful proteolytic substance and splits primary peptide linkages in proteins. About 15% of the digestion of dietary protein is attributable to pepsin.

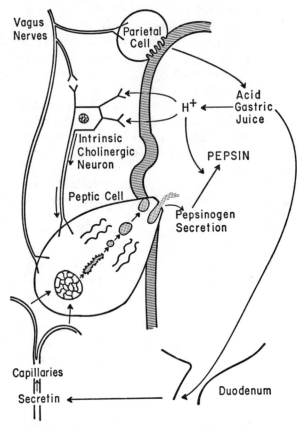

Fig 7–17.—Control of pepsinogen production and conversion into pepsin. Parasympathetic nervous activity and gastric acid are the major stimuli for pepsinogen secretion.

Pepsin has been implicated in the genesis of peptic ulcers. When the gastric mucosa is damaged by aspirin or ethanol, pepsin is released before superficial ulceration is evident. The accumulation of acid in the tissue during damage may activate pepsin to digest the mucosa from which the enzyme originates. There is an adage in gastroenterology that states: "No acid, no ulcer"; i.e., peptic ulcer formation does not occur in patients whose stomachs are achlorhydric. By the same token, peptic ulcers do not occur in the absence of pepsin.

Normal regulation of pepsinogen secretion and conversion to pepsin is closely related to gastric acid secretion (Fig 7–17). Vagal release of acetylcholine near peptic cells and near intrinsic nerves (which also release acetylcholine near peptic cells) constitutes the direct chemical stimulus for production of pepsinogen. In addition, vagal stimulation causes the parietal cells of the stomach to secrete hydrochloric acid. Some of the acid leaks back into the mucosa and stimulates dendrites of intrinsic neurons, which then release acetylcholine near peptic cells to activate secretion of pepsinogen. Cholinergic stimulation also causes the stomach to empty its acid into the duodenum, whose mucosa elaborates secretin. This hormone circulates through the blood to the peptic cells to evoke their secretion of pepsinogen. And finally, as noted above, conversion of pepsinogen to the active enzyme pepsin occurs only in a juice containing acid.

The preceding account has emphasized the great importance of the vagi and of the neurotransmitter acetylcholine in controlling gastric secretion. This is significant for the medical management of peptic ulcers since agents that block cholinergic receptors reduce gastric secretion. Moreover, some surgical approaches to the treatment of duodenal ulcers utilize vagotomy to reduce gastric secretion.

MUCUS

Mucus is a conglomerate secretion consisting of mixed macromolecules, including proteins, glycoproteins, polysaccharides and such biologically important substances as intrinsic factor, blood group substances and plasma proteins. Gastric mucus is secreted throughout the mucosa by epithelial and mucous cells. Physically, mucus is a gel whose stability in solution is dependent upon charged sulfate and carboxyl groups and the hydrogen bonds on the mucus macromolecules. Varying the pH of gastric juice can disturb these electrostatic forces sufficiently to precipitate mucus from solution. The role of mucus in the stomach seems to be protective, particularly against autodigestion of the mucosal membrane by pepsin. Abrading the lining of the stomach with a blunt object or exposing the lining to topical damaging agents causes secretion of mucus. Abnormally large amounts of mucus, consisting mainly of plasma proteins, are secreted in *Menetrier's disease;* this protein-losing gastroenteropathy can deplete the circulation of a significant proportion of its dissolved protein.

INTRINSIC FACTOR

Intrinsic factor is a mucoprotein with a molecular weight of 55,000 that is secreted by the parietal cells. The capacity of the stomach to secrete intrinsic factor approximately parallels its ability to secrete hydrochloric acid. Thus in patients with absence of intrinsic factor, achlorhydria is usually found and, in normal individuals, secretagogues stimulate secretion of both acid and intrinsic factor.

Intrinsic factor binds to ingested vitamin B_{12} in the lumen of the stomach, forming a nondialyzable complex, which remains intact as it travels from stomach to ileum. There are specific receptors on the ileal mucosa to which the complex attaches, permitting active transport of vitamin B_{12} into the circulation.

In pernicious anemia there is a lack of intrinsic factor coupled with an increase in circulating antibodies to parietal cells and to intrinsic factor. Many patients with pernicious anemia have greatly elevated plasma concentrations of gastrin.

PANCREATIC SECRETION

The pancreas is a slender, solid organ buried deep within the upper abdomen. It stretches from the second portion of the duodenum on the right to the spleen on the left and is close to the stomach, common bile duct and portal vein. Although its mass is equal to only one tenth of one per cent of body weight, the pancreas secretes more than a liter of digestive juice (the exocrine secretion) into the duodenum each day. The pancreas also secretes into the blood at least three hormones (the endocrine secretion), among which is insulin, essential for normal metabolism throughout the body. Thus the tiny pancreas is truly vital for the proper functioning of both the gut and the person. Pancreatic diseases may have profound effects on gastrointestinal secretion, digestion, absorption and motility, as well as on metabolic operations in all tissues. These diseases include such common dangerous entities as diabetes mellitus, acute and chronic pancreatitis and carcinoma of the pancreas.

Histologically the pancreas comprises a system of microscopic ducts that are terminal branches of gross main ducts. At the blind ends of the duct system are clusters of polygonal cells containing granules in their cytoplasm. These are called acinar cells. Their primary function is to secrete digestive enzymes. The small ducts are lined by columnar-shaped duct cells whose function is to modify the fluid and electrolyte composition of pancreatic juice.

Isolated clumps of cells, the islets of Langerhans are distributed throughout

Fig 7–18.—Anatomical relationship between pancreatic islet cells and acinar cells. Hormones from islets can reach acinar cells because of circulatory arrangement.

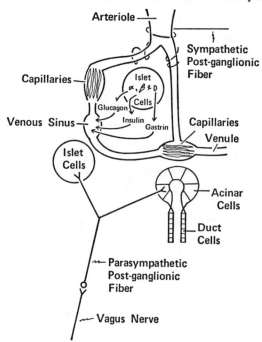

the pancreatic parenchyma. The islets contain three important endocrine cell types: (1) the alpha cells secrete glucagon; (2) the beta cells manufacture and secrete insulin; and (3) the delta cells secrete gastrin in amounts that probably do little normally in regulating gastric acid secretion, but may autoregulate pancreatic enzyme secretion.

In rare cases, delta cell hyperplasia occurs, either diffusely or as a discrete pancreatic tumor associated with other endocrine tumors. The results are excessive production of gastrin and onset of the Zollinger-Ellison syndrome. This disorder is life threatening, because of (1) massive hypersecretion by the stomach (up to 10 L of acid gastric juice daily) leading to upper intestinal ulcers, hemorrhage and perforation, and (2) unremitting diarrhea with fluid and potassium loss secondary to both the large volume of fluid dumped into the upper gut by the stomach and the direct interference by gastrin with intestinal absorption of sodium, glucose and water. If the source of excess gastrin is a single pancreatic tumor, surgical removal results in a cure. Unfortunately, the usual source is diffuse delta cell overgrowth throughout the pancreas, and total pancreatectomy is a difficult resection that leaves behind a poor candidate for long-term survival. Until recently therapy has centered upon a somewhat less dangerous approach, i.e., removal of the target organ, the stomach. In the last few years that histamine H_2-receptor antagonist, cimetidine, has been used successfully to control hypersecretion of acid.

The arterial supply of the pancreas comes from branches of the celiac and superior mesenteric arteries. Pancreatic blood drains to the portal vein. The arteries course countercurrent to the ducts within the substance of the organ and terminate in capillary beds around the acinar cells at the far end of the ducts. In addition, some arteriolar branches break up into capillaries around the islets of Langerhans. These capillaries converge into collecting vessels that pass to the capillary net around the acini. Blood then passes from acinar capillaries to the main venous system of the pancreas. This complex arrangement is depicted schematically in Figure 7–18. Since some blood courses first to the islets, which secrete gastrin into the blood, and then to the acinar cells, which secrete enzymes into the juice in response to stimulation by gastrin, the pancreas has the capacity to autoregulate part of its own exocrine secretion.

The efferent nerve supply to the pancreas includes both sympathetic and parasympathetic nerves. Sympathetic postganglionic fibers emanate from the celiac and superior mesenteric plexuses and accompany the arteries to the organ. Parasympathetic preganglionic fibers are distributed by branches of the vagi coursing down the antral-duodenal region. These fibers terminate either at acini, islets or intrinsic cholinergic neurons of the pancreas. In general, the sympathetic nerves inhibit and the parasympathetic nerves stimulate pancreatic exocrine secretion.

Afferent nerves run from the pancreas to the CNS and can carry the message of pain, for example, in response to an encroaching mass (pancreatic carcinoma) or inflammation (pancreatitis). At times, the pain may be extreme and is referred to the upper or entire abdomen or to the lumbar area.

MECHANISMS OF WATER AND ELECTROLYTE SECRETION

The human pancreas secretes over a liter of juice daily, consisting mostly of water. This movement of water across epithelial membranes occurs primarily in response to net transport of ions into the ducts and is another example of water being attracted passively into a gastrointestinal juice by osmotic forces. The final pancreatic juice, which flows from the main pancreatic duct at the ampulla of Vater, is isosmotic with plasma; however, juice secreted at its origins by acinar and duct cells is hypertonic and attracts water as it flows toward the gut through the 10 cm long human pancreatic duct system. Part of the impetus to flow is the diffusional force that develops because of the standing osmotic gradient between the original and the final juices. Other forces pushing the juice out of the gland are imparted by secreting cells and by smooth muscle lining the ducts.

Secretion of electrolytes by the pancreas involves metabolically dependent (active) transport processes. We could guess this from the fact that the bicarbonate concentration of briskly secreted pancreatic juice may be four times that of plasma and extracellular fluid. There is a potential difference across the pancreatic duct epithelium of several millivolts (lumen negative). Use of the Nernst equation indicates that the bicarbonate concentration of the pancreatic juice is too high to be explained by passive electrochemical forces. Evidence has been obtained that the duct epithelium contains two active transport mechanisms that modify the electrolyte composition of the juice. One is the familiar sodium pump, which helps to keep intracellular sodium concentration low and the potassium high. This pump maintains the concentration of sodium in the juice at about 150 mM over a wide range of secretory rates. In addition, the pump assists the movement of hydrogen ion into the blood.

The second active transport mechanism is an anion pump, which contributes to the elaboration of high bicarbonate concentrations in the juice. The bicarbonate is generated inside the acinar and duct cells by hydration of carbon dioxide and is catalyzed by carbonic anhydrase:

$$CO_2 + H_2O \xrightarrow[\text{anhydrase}]{\text{carbonic}} H_2CO_3 \longrightarrow HCO_3^- + H^+$$

The enzyme is not essential for the reaction, since the enzyme inhibitor acetazolamide does not stop pancreatic secretion of bicarbonate. The CO_2 in the reaction comes from cellular metabolism and from the blood perfusing the pancreas. The H^+ that results is transported into the blood. The reaction takes place near the cell membrane separating the cytoplasm from the lumen of the duct where there is also a HCO_3^--dependent ATPase. Activation of this enzyme by bicarbonate generates energy for active transport of the electrolyte into the juice in exchange for chloride (Fig 7–19).

The electrolyte composition of pancreatic juice that dribbles out of the main ducts is influenced by the rate of secretion (Fig 7–20). At very low rates of secretion, chloride and bicarbonate concentrations are about the same as in plasma. As the rate of secretion picks up, the bicarbonate concentration rises dramatically to peak values at highest rates of secretion; bicarbonate concen-

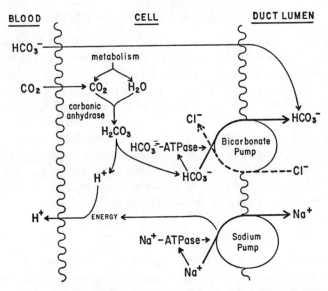

Fig 7–19.—Membrane mechanisms determining electrolyte composition of pancreatic exocrine secretion. Sodium pump drives sodium into juice and hydrogen ion into the blood. Bicarbonate pump drives bicarbonate into juice in exchange for chloride (see text).

tration in the briskly secreting pancreas exceeds 100 mEq/L. Simultaneously the chloride concentration declines reciprocally with the rising bicarbonate. Thus the sum of bicarbonate and chloride concentrations is fairly constant over the secretory range. Cation concentrations do not vary with the rate of secretion.

The major clinical implication of this information relates to postsurgical conditions in which the patient has a draining pancreatic fistula. His pancreatic secretion drains to the outside, and he cannot eat and drink normally because of the operation. Intravenous replacement of fluid and electrolytes must restore the large bicarbonate losses.

Fig 7–20.—Relationship of secretory rate to electrolyte composition of pancreatic juice.

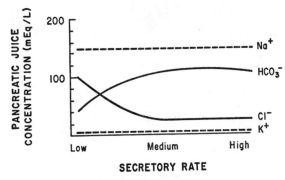

MECHANISMS OF ENZYME SECRETION

The acinar cells synthesize, store and secrete a collection of potent digestive enzymes. In patients with chronic pancreatitis and reduced ability to generate these enzymes, about half the ingested food goes undigested and is not absorbed. The enzymes of the pancreas participate in digestion of dietary proteins, carbohydrates and lipids. To some extent the nature of the usual diet influences the types of enzymes secreted by the pancreas.

The pancreas, like the stomach, secretes its proteolytic enzymes in precursor forms (called zymogens). The acinar cells release trypsinogen, chymotrypsinogen and procarboxypeptidase, which are converted into trypsin, chymotrypsin and carboxypeptidase. Also, as in the stomach, conversion from zymogen to active enzyme in the juice is often autocatalyzed, e.g.,

$$\text{trypsinogen} \xrightarrow{\text{trypsin}} \text{trypsin}$$

Trypsin and chymotrypsin break internal peptide linkages in proteins and polypeptides. Carboxypeptidase attacks the C-terminal end of polypeptide chains chipping off an amino acid with a free carboxyl group. Another important enzyme, alpha amylase, splits dietary starch into the di-

Fig 7–21.—Pancreatic acinar cell production of enzymatic protein (see text).

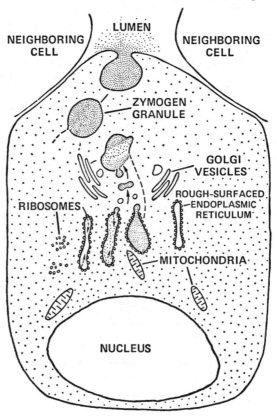

saccharide maltose and other small products by attacking the 1,4 glycosidic bonds that connect the glucose molecules of the starch chain. Pancreatic lipase attacks the ester linkages of fats, with production of fatty acids and glycerol. There are many other pancreatic digestive enzymes with complex actions. The processes of digestion will be more fully considered later in the chapter.

The acinar cell of the pancreas manufactures and releases digestive enzymes or zymogens in a well-defined manner (Fig 7–21). The steps involved occur over a period of a few hours and start with incorporation of amino acids into protein by ribosomes on the membrane of the rough-surfaced endoplasmic reticulum. This protein accumulates in the cysternae surrounded by the rough-surfaced reticulum. Next, some of the protein buds off from the cysternae and migrates to the region of the Golgi apparatus, which invests the naked protein with a membrane. More protein is attracted to the package to form a zymogen granule. The granules swollen with protein can be seen with a light microscope. When the pancreas is stimulated to secrete enzymes, the granules move to the apical portion of the acinar cell and attach to the membrane separating the cell from the duct lumen. Finally, rupture occurs at the site where part of the granule has fused to the membrane and the granular contents are dumped into the duct. The migratory steps of formed protein within the cytoplasm and final release of the enzymes into the juice depend upon energy derived from cellular oxidation of fatty acids rather than glycolysis.

In the whole body, the direct regulators of acinar cell production and release of protein are acetylcholine, cholecystokinin and gastrin. These secretagogues activate the membrane-bound enzymes guanylate cyclase (for acetylcholine) and adenylate cyclase (for cholecystokinin and gastrin). These activated enzymes, respectively, cause the accumulation of cyclic GMP and cyclic AMP and lead, via reactions that have not been fully characterized, to increased secretion of digestive enzymes.

CONTROL OF PANCREATIC SECRETION (FIG 7–22)

Between meals the pancreas is at rest. Nevertheless, it is susceptible to the same regulators that operate at meal times. Some nervous and hormonal stimulation of the pancreas seems continuous, since the resting pancreas secretes its juice at a respectable rate.

The vagi and the intrinsic cholinergic nerves release acetylcholine, whose major action is to increase secretion of enzymes by the acinar cells. The effect of acetylcholine is direct, appears to involve release rather than synthesis of enzymes and causes little change in the volume rate of pancreatic secretion. Unfortunately for easy understanding, the direct actions of acetylcholine are not the only pancreatic effects produced by vagal stimulation. Indirect effects of acetylcholine or vagal stimulation upon pancreatic exocrine secretion include (1) stimulation of gastrin release from the gastric antrum (and possibly the islets of Langerhans)—gastrin is a direct activator of pancreatic enzyme secretion; and (2) stimulation of gastric secretion of acid and gastric emptying of acid into the duodenum, which releases secretin and cholecystokinin

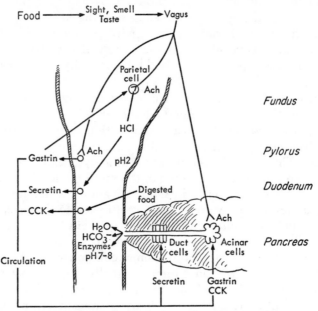

Fig 7–22.—Neurohumoral regulation of pancreatic exocrine secretion. Vagus nerves release acetylcholine (ACh), which stimulates (1) intrinsic pancreatic nerves to release ACh near acini, (2) antral G cells to release gastrin and (3) parietal cells to secrete acid gastric juice. The stomach empties its juice and food into the duodenum, which causes release of secretin and cholecystokinin (CCK). Secretin drives the duct cells to secrete an alkaline juice. Gastrin, CCK and ACh stimulate acinar cells to secrete enzymes.

from the duodenal mucosa—these two hormones cause pancreatic secretion of water, electrolytes and enzymes. It should be apparent that pancreatic neural and hormonal interaction becomes rather complex when the vagi are stimulated by the sight, smell and taste of a steaming bowl of Coquille de Saint-Jacques or when a vago-vagal reflex is evoked through stretching the stomach by eating too much of the delicacy.

The major sympathetic neurotransmitter to the pancreas is norepinephrine. This predominantly alpha adrenergic agonist inhibits pancreatic exocrine secretion of water, electrolytes and enzymes both through direct actions on secretory cells and via a reduction in pancreatic blood flow.

Cholecystokinin, gastrin and secretin all affect pancreatic secretion. The actions of cholecystokinin and gastrin are similar, as we would expect from their chemical similarity. They directly stimulate acinar cell secretion of enzymes. Cholecystokinin is the more potent agent and is released into the blood by the duodenal mucosa when its surface is bathed by solutions of lipid, protein or acid. Gastrin is released from the antral mucosa in response to vagal stimulation and food stretching the wall of the stomach when the pH at the antral surface exceeds 2.0. Thus events surrounding feeding are most conducive to release of cholecystokinin and gastrin and to pancreatic secretion of digestive enzymes.

Similarly, feeding ultimately causes the release of secretin when the acid stomach contents are emptied into the duodenum. Secretin stimulates primarily pancreatic secretion of water and electrolytes with little effect on secretion of enzymes. However, the pancreatic enzymes have pH optima in the range of 6.0 to 8.0; this range would be difficult to achieve in the face of gastric acid being emptied into the duodenum were it not for copious secretion of bicarbonate and water by the pancreas at meal times. The ability of secretin to drive pancreatic secretion is potentiated by cholecystokinin, gastrin and vagal excitation. All of these stimuli are present at meal times and amplify the effect of those small quantities of secretin released by gastric acid bathing the upper duodenal reaches.

BILIARY SECRETION

The exocrine secretion of the liver is called bile. Each day the liver secretes more than a liter of bile containing bile salts, mucin, hemoglobin breakdown products, phospholipids, cholesterol and electrolytes. Abnormalities in the production, composition or flow of bile into the upper small intestine may lead to jaundice, formation of gallstones or malabsorption of dietary lipids.

The bile salts are essential for normal emulsification of the oily and watery portions of the chyme in the upper gut and for the physical changes in luminal lipids (micellar formation) that precede absorption of fats. Bile salts are composed of bile acids conjugated to amino acids, such as glycine or taurine, in a complex with sodium. In human bile the common bile acids are cholic, deoxycholic and chenodeoxycholic acids. They originate during the metabolism of cholesterol and are conjugated in the liver to glycine or taurine before secretion into the bile.

The conjugation of two dissimilar molecules produces interesting new properties in the conjugate. Thus the pK of cholic acid is 6.0 but, when the bile acid is conjugated to taurine, only the sulfonic acid group of taurine is ionizable, and the pK of taurocholic acid is reduced to 2.0. At the neutral pH of jejunal chyme, taurocholate is fully ionized and cannot be absorbed across the lipid membranes of the jejunal mucosa. Bile salts are absorbed from the ileum after their digestive work is done.

The combination of bile acids and amino acids also produces a natural detergent, since bile acids are soluble in lipid and amino acids are soluble in water. Thus, as depicted in Figure 7–23, the conjugated molecules can effect emulsification of fat droplets in an aqueous solution. This is critical for the digestion of dietary fat, which is in the form of long chain triglycerides and which must be decomposed into monoglycerides and fatty acids before absorption will occur. The digestive enzyme lipase is water soluble and would work only slowly without emulsification of the fat.

The gastrointestinal travels of bile salts are complicated. After intrahepatic metabolism of cholesterol into bile acids, their conjugation to taurine and glycine and their complexing to sodium, the bile salts are secreted by the liver. The bile travels down the hepatic ducts into the common bile duct, which is usually closed by the sphincter of Oddi at the junction of the duct and the

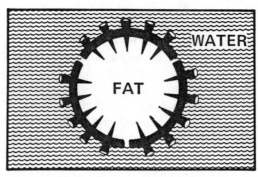

Fig 7–23.—Detergent properties of a conjugated bile acid. Cholic acid is soluble in oil and glycine is soluble in water. Glycocholate molecules surrounding a droplet of fat in water emulsify the mixture.

duodenal lumen. The bile is therefore diverted into a side branch, the cystic duct, and is stored in the gallbladder. The gallbladder actively transports sodium from the bile into the lining of the bladder and thereby extracts much of the electrolyte (chloride accompanies sodium) and fluid (water follows the ions) from the bile. The result is a great increase in concentration of bile salts in the bile which becomes darker and more viscid.

When dietary fat enters the duodenum, cholecystokinin, is released and circulates in the blood to cause contraction of the gallbladder and relaxation of the sphincter of Oddi. The bile is squeezed into the upper gut to participate in the emulsification of fat and in the formation of micelles (see p. 438). After digestion and absorption of the lipids, bile salts are swept down the gut to the distal ileum. The ileal mucosa possesses an active transport mechanism that is highly efficient in reabsorbing nearly all of the bile acid secreted into the upper intestine. Some bile salt is also absorbed passively in the upper gut. The bile acids, bound to plasma albumin, are carried in the portal circulation to the liver, which clears the bile acids from portal blood into the newly formed bile. This route—liver to biliary tree to duodenum to ileum to portal vein to liver—constitutes the *enterohepatic circulation.*

Disruption of the normal enterohepatic circulation of bile salts results in impaired digestion and absorption of dietary fat, in loss of calories and fat-soluble vitamins and in excess fat in the feces (steatorrhea). The causes of such disruptions are frequently iatrogenic. For example, following surgical resection of the distal two thirds of the stomach and anastomosis of the remainder of the stomach to the jejunum (Billroth II operation), the closed duodenum is a side arm of the new gastrojejunal canal. Abnormal bacterial overgrowth in this blind loop of duodenum results in deconjugation of bile salts by bacterial enzymes. The bile acids that are released behave like weak acids in the lumen of the gut and are absorbed by non-ionic diffusion, thereby lowering the concentration of bile acids in the upper gut, which diminishes formation of micelles and lowers the rate of absorption of fat. Furthermore, deconjugation of bile salts reduces their detergent properties and slows digestion of dietary lipid by pancreatic lipase. Another medically induced interruption of the enterohepatic circulation occurs with use of drugs such as neomycin or cholestyramine; both agents precipitate bile salts in the ileum and prevent their reabsorption. The result is rapid lowering of the pool of bile salts in the body and impaired digestion and absorption of dietary fat. Through use of isotopic labeling of the bile acid pool, it has been found that healthy people excrete less than one fifth of their bile acids daily; with malabsorption of bile acids, there is loss of three to five times this quantity. Even more severe degrees of bile acid malabsorption occur with resection of the distal ileum.

Hepatic secretion of bile is influenced by neurohumoral factors. At meal time when the flow of bile is needed for digestion and absorption of ingested fat, there is increased vagal activity and release of gastrin, cholecystokinin and secretin. These stimulate an increase in bile secretion, especially the inorganic components (water and electrolytes). Hepatic production of bile acids is stimulated mainly by increased delivery of bile acids from the ileum via the portal circuit, which occurs after a meal containing fat. Hepatic clearance of bile acids is about as efficient as ileal uptake. Consequently, the pool size of bile acids is small, there are multiple turnovers of the pool each day to meet digestive needs and the liver needs to synthesize only 1 mM of new bile acid each day to maintain a pool more than 10 times that large.

The inorganic moiety of bile is a juice resembling many extracellular fluids. When the ionic composition is plotted as a function of biliary secretory rate, as in Figure 7–24, the concentrations of chloride, potassium and bicarbonate are seen to be fairly constant and about equal to their plasma concentrations. The anomalous ion is sodium. Its concentration in bile is 50% greater than in plasma. On the face of it, the bile would appear to contain too much sodium for electrolytic and osmotic balance. It turns out that one third of the sodium is part of the organic complex with bile salts and does not behave like an electrolyte; the ionically effective concentration of sodium is the same as in plasma. However, in a patient recuperating from resection of the gallbladder with a draining T tube in the common duct, parenteral replacement must compensate for the high loss of sodium from the body through the T tube.

The bile is also the repository for other substances that may be involved in diseases. When hemoglobin is released from senile red cells and is metabolized, pigmented hemoglobin breakdown products are excreted in the bile.

Abnormalities in the handling of cholesterol, bile salts, calcium and bile

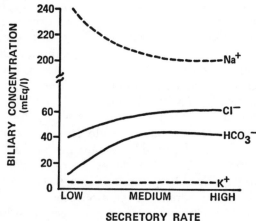

Fig 7–24.—Relationship of secretory rate to electrolyte composition of bile.

pigments may lead to the formation of gallstones; this may obstruct the common bile duct, causing local infection and spasm, pain, jaundice, interference with digestion and absorption of lipids and steatorrhea. Each year tens of thousands of Americans, including two of the last five presidents, have been forced to undergo surgery for gallstones. Beside the risk to life from the operation, chronic gallstones are associated with a higher incidence of gallbladder cancer, and frequent obstructions of the ducts may induce irreversible cirrhotic damage to the liver.

ORGAN FUNCTIONS: DIGESTION AND ABSORPTION

The overriding function of the gastrointestinal system is absorption of nutrients. This requires moving ingested food from mouth to anus and changing its physicochemical composition to permit absorption. Since much of our food is solid and semisolid, it is shredded and ground in the mouth, mixed with acidic fluid in the stomach and exposed to digestive enzymes, bile salts and alkaline fluid in the upper gut. The result is a fairly fast conversion of food into (1) an aqueous solution of hexoses, amino acids, salts and other water-soluble nutrients now isosmotic to plasma and (2) a lipid phase consisting of micelles and emulsified fat globules, both containing bile salts and digested lipids. In addition to the ingested materials, the gut must also handle secreted juices and sloughed cells. The quantities of ingested and secreted substances are sizable, amounting daily in adults to two gallons of water, 50 gm of salts and 500 gm of organic nutrients. It is remarkable that nearly all of this material is absorbed in the normal person within a few hours after presentation to the small intestine. Failure to eat leads to weight loss and wasting. Failure to digest or absorb food, as in malabsorption syndromes, leads equally to weight loss and wasting. Indeed, a recent innovation in the treatment of otherwise unmanageable and massive obesity is surgical removal or bypass of most of the small intestine.

The digestive processes within the gastrointestinal tract reduce complex organic food molecules to simpler molecules that can be absorbed rapidly. Enzymes secreted by the salivary glands, stomach, pancreas and intestinal mucosa accelerate digestion. The food is reduced physically to particles that dissolve or emulsify in water as a result of muscular action within the mouth and stomach, of dilution of food by secreted juices and of provision of a natural detergent for lipids (bile salts). From this view, the physical chemistry of the gastrointestinal tract utterly defeats the gourmet. Whether one dines on Beluga caviar and truffles or on hamburger and a bun, the small intestine will be presented with an essentially similar liquid chyme and the rectum with semisolid feces.

The efficiency of the small intestine is enhanced by the structure of its mucosal surface. Because of its convolutions, its villi and its microvilli, the surface area of the inner lining is amplified to two million square centimeters. During the several hours that chyme is propelled along the gut, muscular movements of the folds and the villi churn the liquid and expose it to the vast absorbing surface.

Despite these structural and biochemical advantages, the gut undergoes occasional dysfunction of its absorptive apparatus. The appearance of daily stools exceeding 200 gm of weight is termed diarrhea. Since nine tenths of the stool mass is water, diarrhea is malabsorption of water and dissolved electrolytes. Diarrhea may occur alone or in conjunction with excessive losses of fat (more than 10 gm daily) in the stools (steatorrhea) or with excess protein in the stools (azotorrhea).

WATER AND ELECTROLYTES

Most of the water and electrolytes presented to the intestine are derived from alimentary secretions. In an adult the daily volume of secretion includes 0.5 L of saliva, 2 L of gastric juice and 1 L each of pancreatic juice, bile and intestinal secretions. Eighty per cent of the sodium chloride and 50% of the potassium and calcium in the gut come from secretions. Any serious impairment of the absorptive capacity of the intestine or any great increase in secretion of fluid and salts causes diarrhea, dehydration and electrolyte imbalance. In babies, in the elderly and in debilitated patients, prolonged diarrhea can be life threatening. Severe untreated diarrhea, as in epidemic Asiatic cholera, rapidly kills the patient.

Absorption in the Small Intestine

Absorption of water and ions by the small intestine is the net result of two opposing transport processes taking place simultaneously at the mucosal membrane. Absorption occurs when the movement of materials from lumen into mucosa (insorption) exceeds the movement of materials from mucosa into the lumen (exsorption). Secretion is also a net effect and occurs when exsorption exceeds insorption.

Water and electrolytes are absorbed passively through the tight junctions and lateral spaces between cells and also transcellularly through the aqueous-filled pores of the intestinal epithelial cells. The absorption rate is faster in the more permeable jejunum than in the terminal small intestine.

The forces causing passive insorption of electrolytes include concentration and electric gradients, and the effect of water movement. Concentrations of sodium and chloride in the chyme may exceed their concentrations inside epithelial cells or in the extracellular fluid of the upper small intestine, especially after a meal of pickled herring, baked Virginia ham and potato chips. Similarly, the calcium concentration of jejunal chyme will be raised by drinking milk. The ions move through the pores from the region of higher concentrations in the lumen to lower concentrations within the tissue. Whatever the nature of the meal, passive movements of ions and water cause the chyme to become isotonic in the upper intestine.

The electric gradient across the mucosa of the human small intestine is small, being less than 10 mv, with the mucosal surface charged negatively.

In the upper small intestine, active insorption of glucose and amino acids leads to increased insorption of electrolytes. The mechanism is disputed but the insorption may be partly due to a solvent drug effect or to the sodium-cotransport mechanism discussed in Chapter 1. The insorption of the organic compound is associated with passive insorption of water by osmosis, and the solvent drags electrolytes along. This effect of glucose in causing absorption of water and electrolytes has been used in the treatment of Asiatic cholera, where insorption of glucose is normal despite a massive increase in exsorption of electrolytes and water. Infusion of a glucose solution into the upper gut maintains fluid and electrolyte balance in these patients despite their diarrheal state.

In the lower small intestine, concentration gradients for electrolytes are less likely to exist. By the time the chyme reaches the ileum, jejunal insorption and exsorption have reduced the volume, removed the glucose and amino acids and modified the osmolarity toward that of extracellular fluid. Furthermore, the ileal epithelium is less permeable than the jejunal epithelium. Thus there are no passive forces favoring ileal insorption of ions and water. Salvage of much of the electrolytes and water in ileal chyme occurs because of an *active* transport mechanism, the sodium pump located in the basolateral membranes of intestinal epithelium. The process is discussed in Chapter 1 (see also Fig 1–11). Sodium moves from the lumen into the ileal mucosa against both concentration and electric gradients. Chloride moves passively with sodium to maintain electric neutrality across the membrane, and water moves passively with the electrolytes to maintain osmotic balance.

The patterns of absorption and secretion of specific ions differ in the upper and lower small intestine (Fig 7–25). In the jejunum, most of the sodium is absorbed passively and the relatively small amount of actively transported sodium is coupled to secretion of hydrogen ion so that fluid remaining in the lumen becomes acid. Chloride, bicarbonate and potassium are also absorbed from the upper small intestine. In the ileum, a somewhat different pattern prevails: most sodium is actively transported, but acidification of the luminal contents does not occur because a chloride/bicarbonate exchange mechanism appears to be present as well as the Na/H exchange mechanism. The lumen

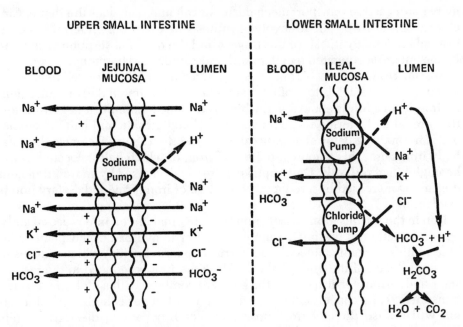

Fig 7–25.—Electrolyte transport mechanisms in upper and lower small intestines. In upper gut, most sodium absorption is passive. In lower gut, a sodium pump re-absorbs most of the sodium, and an anionic pump exchanges chloride absorption for secretion of bicarbonate.

contains hydrogen ion and bicarbonate, which combine to form water and carbon dioxide. Potassium is absorbed passively.

Hormones influence intestinal transport of electrolytes. Gastrin inhibits absorption of sodium, potassium and water. In the Zollinger-Ellison syndrome, in which plasma gastrin levels are elevated, the decreased ability of the gut to handle electrolytes and water contributes to the diarrhea found in this disorder. Antidiuretic hormone, secretin, cholecystokinin, glucagon and prostaglandins also interfere with normal absorption of fluid and electrolytes by the small intestine.

Intestinal absorption of water is a passive osmotic process. The water accompanies its solutes, which are being absorbed from the lumen into the mucosa, and the solution being transported across the mucosa is isotonic to plasma. Therefore, the rate of absorption of water depends upon the rate of absorption of solutes. Since more solute is absorbed in the upper than in the lower small intestine, water movement is also greater in the upper gut. Within a few hours after a usual meal of solid and liquid, most of the ingested water has been absorbed by the jejunum. Further details of water transport in the intestine are to be found in Chapter 1.

ABSORPTION OF CALCIUM.—Calcium is actively absorbed by the small intestine. The transport mechanism for calcium (1) is energy dependent, (2) absorbs calcium against electrochemical gradients, (3) exhibits a transport maximum when the luminal calcium concentration reaches 5 mM/L and (4) is competitively inhibited by magnesium and cobalt ions. The jejunum has a more potent active transport mechanism for calcium than does the ileum. Calcium absorption appears to involve synthesis of a binding protein by the intestinal mucosa. Synthesis of this protein and the rate of absorption of calcium by the gut are accelerated by vitamin D and parathyroid hormone.

ABSORPTION OF IRON. — Iron is essential for life, and prolonged deficiency of iron in the diet leads to profound anemia. The minimum daily requirement for iron in an adult is about 2 mg, which is also the amount lost in gastrointestinal, renal and uterine excretions and in sloughed skin and hair. The other major normal regulator of iron uptake, beside the dietary factor, is the total body iron store; this amounts to 4 gm of iron, of which half is located in the hemoglobin of red cells. Daily uptake of iron is decreased by an iron-deficient diet and is increased by a reduced body store of iron (provided dietary iron is available).

Iron in the ferrous form passes from the intestinal lumen into mucosal cells, where oxidation to the ferric state occurs. Ferric hydroxide phosphate combines with an organic molecule, apoferritin, to form ferritin, which diffuses to the serosal side of the epithelial cell. Here the iron is reduced to the ferrous state, and ferrous ferritin enters the blood to combine with a globulin, termed transferrin. If its concentration in the lumen of the gut is high, some ferrous ion will diffuse directly from lumen to blood, bypassing the ferritin step. Some dietary iron from ingested muscle and blood is absorbed attached to a porphyrin ring, which is part of the structure of hemoglobin and myoglobin.

Normally, absorption of iron is enhanced by dietary amino acids and sugars, which form soluble complexes with iron at neutral pH. The gastric juice also contains an iron-solubilizing factor, which may be the protein that binds vitamin B_{12}. Following total gastrectomy, iron absorption is greatly reduced by the loss of the gastric solubilizing factor, and iron deficiency anemia develops. Similarly, in pernicious anemia, in which gastric secretions are reduced and are abnormal, iron absorption is diminished.

The rate of absorption of iron from the duodenum is decreased by high rates of pancreatic secretion, since bicarbonate complexes with the ion to form macromolecular iron, which is poorly absorbed. Iron absorption is also decreased by events that depress development of red cells and manufacture of hemoglobin (erythropoiesis) in the bone marrow, such as radiation or starvation. Conversely, factors stimulating erythropoiesis in the marrow increase absorption of iron from the upper gut, e.g., hemorrhage or hemolysis. A more physiologic experience that alters intestinal absorption of iron is ascent to and descent from high altitudes. A trip to Lima, Peru, will increase erythropoiesis and uptake of iron from the gut; the trip from Lima back to New York will reduce erythropoiesis and absorption of iron.

ABSORPTION IN THE LARGE INTESTINE

The human large intestine is primarily a temporary storage bin for rejects from the small intestine. During their stay in the colon, waste materials are reduced in volume through colonic absorption of salt and water. The surface epithelium of the colonic mucosa is less than 1% of the surface area of the small intestinal lining and has a much lower aqueous permeability than does the epithelium of the ileum. Consequently, the absorptive capacity of the large intestine is not great, and less than a tenth of the total secreted and ingested water of the gastrointestinal tract is absorbed by the colon. Nevertheless, because of the long time that materials stay in the human colon, over 90% of the water (1 L per day), sodium (200 mEq per day), and chloride (100 mEq per day) presented to the large intestine is absorbed. Most of the sodium chloride and water absorption occurs in the proximal half of the colon where luminal salt concentrations are higher and where the feces are stored before evacuation.

The major transport mechanisms in the colon are depicted in Figure 7–26. These include a sodium pump, which causes that ion to be absorbed from the lumen against its electrochemical gradient (the electric potential difference across the colon is about 30 mv, with the lumen negative). Potassium moves passively into the lumen, driven by its electrochemical gradients. Water is absorbed osmotically along with the sodium. Bicarbonate is secreted actively into the colonic lumen and this movement is linked to active absorption of chloride.

Several drugs influence colonic water and electrolyte absorption. For example, mineralocorticoids increase potassium excretion and diuretics reduce absorption of water and sodium chloride. In diarrheal states, important losses of potassium and bicarbonate may occur. The potassium loss may average 10 mEq/L of excreted fluid. If the patient with diarrhea is not taking fluid by

Fig 7–26.—Electrolyte transport mechanisms in the colon. Sodium and chloride are actively absorbed. Bicarbonate exchanges for chloride. Potassium moves passively into the lumen.

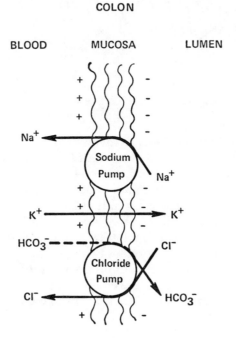

mouth, parenteral replacement will have to include supplemental potassium bicarbonate to avoid dangerous hypokalemic acidosis. Considerable colonic losses of salt and water with diarrhea may occur when excessive quantities of hydroxy-fatty acids are present in the colon, as in malabsorption syndromes marked by steatorrhea; here, colonic bacteria change oleic acid into 10 hydroxystearic acid, which is chemically similar to the active component of castor oil, i.e., ricinoleic acid.

CARBOHYDRATES

Carbohydrates constitute the core of common human diets. Cheap food is mostly carbohydrates. Even in the United States, with its affluence and culinary excesses, an average adult consuming 2,000 calories daily will eat 250 gm (1,000 calories) of carbohydrates including starch (potatoes, rice, spaghetti, pastry, bread), sucrose (table sugar), lactose (milk sugar) and fructose (fruit sugar). Ethnic diets in this country are often mostly carbohydrate. Each day the gut is confronted with 2 M of monosaccharides (glucose, galactose, fructose), easily the largest part of its solute load.

Most ingested carbohydrates require digestion to simpler molecules before their absorption will occur. This is accomplished by enzymes secreted in saliva, pancreatic and intestinal juices.

Starch is composed mainly of two subunits, amylopectin and amylose. The former consists of long chains of the monosaccharide, glucose, linked by oxygen atoms between the first carbon of one monosaccharide and the fourth carbon of the next, as

$$
\begin{array}{ccccc}
C^1 & C^1 & & C^1 & \\
 & | & & | & \\
 & C^2 & & C & \\
O & | & O & | & O \\
 & C^3 & & C & \\
 & | & & | & \\
 & C^4 & C^4 & C^4 & C^4 \\
 & | & & | & \\
 & C^5 & & C & \\
 & | & & | & \\
 & C^6 & & C & \\
\end{array}
$$

Amylopectin also contains branches involving linkages from the first carbon of one glucose to the sixth carbon of the next (one–six linkage). Amylose is a straight glucose chain of first to fourth carbon linkages.

The initial stage of digestion of starch is facilitated by salivary and pancreatic alpha amylase. This enzyme splits only the one–four carbon linkages, except the one–four linkages at the ends of glucose chains. The starch molecule is thereby split into a variety of smaller fragments: (1) the disaccharide maltose, the smallest fragment composed of two glucose molecules connected by the one–four carbon linkage; (2) maltotriose with three glucose molecules; and (3) alpha-limit dextrins, which are combinations of one–four and one–six linked glucose straight or branched chains. Thus alpha amylase only partially digests starch. Since carbohydrates are absorbed mostly as monosaccharides, additional digestion takes place in the gut. In patients with severe celiac disease whose microvilli are destroyed, further digestion of maltose does not occur. These patients excrete much of the ingested starch, but they absorb significant amounts of maltose.

The microvilli cumulatively are designated as the brush border. Their protein lining contains a class of enzymes called disaccharidases, one of which, maltase, breaks the one–four linkage between the glucose molecules of maltose, thereby releasing glucose. The disaccharidase invertase converts sucrose into glucose and fructose. Another enzyme, lactase, breaks the linkage between glucose and galactose. In more than one tenth of our population, there is a genetic defect manifest as a deficiency of brush border lactase. When a baby with an isolated lactase deficiency ingests milk, the lactose passes into the colon undigested. Colonic bacteria utilize this sugar to form lactic acid and carbon dioxide gas. The acid irritates the large bowel and provokes diarrhea. The gas distends the bowel and produces pain and flatulation. The undigested lactose is not absorbed and represents lost calories. The mother will complain that her baby is always hungry, drinks much milk, is losing weight, cries a great deal and has diarrhea.

Once digestion of carbohydrates has proceeded and monosaccharides are present in the lumen of the upper small intestine, absorption begins. The absorption of glucose is believed to be an active transport process because (1) glucose is transported against a concentration gradient; (2) its transport is slowed by metabolic inhibitors, diminished temperatures and chemically similar sugars; (3) the transport process has a maximum rate that is reached when glucose concentrations exceed 139 mM in the lumen of the gut and (4) glucose transport seems to depend on the presence of sodium ions in the intestinal lumen and on the active transport of sodium into the tissue. A model that has been proposed to explain the transport of glucose has been fully described in Chapter 1 and illustrated in Figure 1–8.

The other two common dietary monosaccharides are also transported by energy-dependent mechanisms. Galactose shares the same carrier as glucose. Fructose absorption is accelerated by metabolism within the mucosa, which injects this sugar into glycolytic pathways.

It should be noted at this point that much glucose absorption is passive. So long as the concentration of glucose in the lumen exceeds the concentration in the blood, the sugar will be absorbed passively down its concentration gradient. This situation will be found in the upper small intestine where concentrated nutrients are dumped by the stomach. Because of the greater permeability of the jejunal membrane, passive absorption along a concentration gradient is faster in the jejunum than in the ileum. Furthermore, in the upper gut, active transport of glucose often occurs at its maximal rate, since the concentration of the sugar is likely to exceed 139 mM. For these reasons more than three quarters of the ingested carbohydrate is absorbed in the upper half of the small intestine. The carriers of the ileum salvage the rest.

Gastrin, cholecystokinin and secretin inhibit absorption of both glucose and sodium. In patients with severe diabetes mellitus due to insulin deficiency, active absorption of glucose and sodium is increased and may be restored to normal rates by administration of insulin. In patients after adrenalectomy, both glucose and sodium absorption is depressed. In these cases, the hormones seem to affect sodium transport primarily and glucose only secondarily.

Intestinal absorptive capacity for carbohydrates is often assessed grossly in patients by measuring the rate of excretion of an ingested nonmetabolized sugar, D-xylose. This pentose is absorbed in the upper gut, cleared from the blood by the kidneys and excreted in the urine. In a patient with normal kidneys, a decrease in the amount of D-xylose appearing in the urine suggests serious malabsorptive disease in the small intestine, e.g., adult celiac disease.

In the presence of carbohydrate malabsorption, monosaccharides progress to the colon where they increase the osmotic attraction of water into the lumen. Colonic carbohydrate is also converted into organic acids, which irritate the large bowel, and into hydrogen gas and carbon dioxide. Consequently, the patient with malabsorption of carbohydrates suffers from diarrhea and flatulence, as well as weight loss because of loss of nutrients.

PROTEINS

Protein is expensive food. In primitive hunting societies and in the United States, many large four-footed herbivores have to be killed to load the table with a steady supply of protein. In less meat-oriented cultures, protein is ingested in fish, poultry, eggs, milk and grain products and vegetables.

Intestinal handling of protein is remarkably efficient. Within a few hours after ingesting a high protein meal, practically all the swallowed protein has been digested and absorbed. The feces contain a small amount of protein, but this is not of dietary origin; it comes from sloughed colonic mucosal cells, mucus and bacterial cells.

Since proteins are large molecules that will not cross the intestinal membrane rapidly, breakdown of ingested protein is an essential preliminary to absorption (except during the first few days of life when proteins, such as antibodies, are absorbed by pinocytosis from the colostrum in ingested mother's milk). This digestion of 100 gm of dietary protein and the same amount of endogenous protein from sloughed cells and secretions occurs mainly in the stomach and upper half of the small intestine. One fifth of the swallowed protein is digested by gastric pepsin and the remainder by pancreatic enzymes and intestinal hydrolases in the lumen of the upper gut and in the brush border. Pepsin, trypsin and chymotrypsin break primary nonterminal peptide linkages. Carboxypeptidases hydrolyze C-terminal peptide linkages. Enzymatic decomposition of proteins therefore yields amino acids and small polypeptides, which are absorbed swiftly by passive diffusion through aqueous pores and by active transport mechanisms located in the brush border membrane. The pores restrict passage of polypeptides with a molecular weight much above 200. Once inside the epithelial cells, proteidases hydrolyze the small polypeptides into amino acids.

Evidence favoring active transport for amino acids and dipeptides includes (1) transport against a concentration gradient, (2) specificity of the transport for the L form of the amino acid (expect D-methionine), (3) competitive inhibition of absorption of one amino acid by another actively transported amino acid and (4) dependence of amino acid active transport on sodium.

There are four separate carriers for active absorption of amino acids, each of which transports a group of amino acids. The four groups are neutral, basic, acidic (dicarboxylic) and imino (proline, hydroxyproline) amino acids. Members of a group competitively inhibit transport of other members of the group and may, in addition, inhibit active transport of some nonmembers. Galactose and glucose also competitively inhibit amino acid transport. Apparently, the kind of relationship between active transport of hexoses and of sodium applies to active transport of amino acids. The carrier for amino acids depends on the sodium pump and on the normal intra- to extracellular gradients for sodium and potassium. In transporting amino acids, the carrier consumes energy derived from hydrolysis of ATP.

Digestion of proteins and active transport of amino acids are both inhibited when intraluminal pH falls. Upper intestinal acidification is normally prevented by alkaline pancreatic secretion and by intestinal absorption of hydrogen ion.

Abnormalities of protein absorption occur in a number of clinical entities. In the Zollinger-Ellison syndrome, great volumes of gastric acid are dumped into the upper gut and interfere with intestinal handling of protein. An isolated defect in the transport of one amino acid, tryptophan, results in increased urinary excretion of indican, which is oxidized to a blue color in the diapers of babies with the blue diaper syndrome. In Hartnup's disease, intestinal absorption and renal reabsorption of all neutral amino acids are defective. The children have a rash, cerebellar ataxia and mental retardation.

VITAMINS

Vitamins are chemically dissimilar substances that are essential for health. Their absorption in the gut depends upon chemical and physical characteristics and intestinal capacities. The water-soluble vitamins are handled in the manner of water-soluble organic nutrients, i.e., they are either passively or actively transported. Lipid-soluble vitamins behave like other lipids in the gut, i.e., they are involved in micellar absorption.

Ascorbate, biotin, inositol, nicotinamide, para-aminobenzoate, pyridoxine and riboflavin are water-soluble molecules that diffuse through aqueous pores of the intestinal membrane in response to concentration gradients. Intestinal transport of these vitamins is passive.

Thiamine is a relatively large, water-soluble molecule that acts like a moderately strong base in aqueous solution. Hence it is poorly absorbed through the lipid membranes of the gut. With high concentrations of thiamine in the lumen, the vitamin is transported mainly by slow passive diffusion. At low concentrations, however, an active transport mechanism moves thiamine into the mucosal cells; thiamine absorption is slowed by chemically similar substances, by metabolic inhibitors and by lack of sodium. Despite the active transport mechanism, a maximum of only 5 mg can be absorbed each day from the normal adult gut. Nevertheless, it is not uncommon to find people who swallow many times this amount daily in the form of unnecessary vitamin pills and wastefully enriched breakfast cereals, as if to insure not only their own good health but also that of their fecal bacteria.

Dietary folate is a moderately strong acid and is too large a water-soluble molecule to penetrate the pores passively. In the gut, folate is hydrolized from its polyglutamic to its monoglutamic form, which is sufficiently small to pass through the pores. There also is an active transport mechanism for folate in the human gut. These digestive and absorptive processes are inefficient, and less than half the ingested folate is absorbed. In patients with a chronic malabsorption syndrome due to mucosal disease (e.g., adult celiac disease), an early indicator of the disorder is folate deficiency, with attending neurologic and hematologic abnormalities.

Vitamin B_{12} is absorbed by a process requiring binding of the vitamin to a glycoprotein (intrinsic factor) synthesized by gastric parietal cells. The vitamin B_{12}-intrinsic factor complex proceeds through the gut to the ileum, where it attaches to receptors on the brush border. Attachment requires calcium or magnesium and a normal pancreatic juice. The bound vitamin is then absorbed, but the precise mechanism remains unclear. Because of this rather tortuous absorptive process, the proportion of ingested vitamin B_{12} absorbed is minuscule. Furthermore, the process is vulnerable to disorder, and many diseases cause failure to absorb vitamin B_{12}. Resection of the stomach, atrophic gastritis, pancreatic insufficiency, malabsorption syndromes and ileal resection inhibit normal uptake of vitamin B_{12}. There is an intestinal fish tapeworm (*Diphyllobothrium latum*), which releases a chemical to break the bond between vitamin B_{12} and the intrinsic factor. With prolonged inhibition of vitamin B_{12} absorption, severe anemia ensues, with hematologic, cardiovascular, and neurologic signs and symptoms. In its classic form this condition is termed pernicious anemia and is caused by failure of the gastric mucosa to secrete the intrinsic factor, probably because of an autoimmune disorder that destroys parietal cells. If untreated, pernicious anemia kills its victim.

The lipid-soluble vitamins A, D, E and K are absorbed passively down their concentration gradients. Like fatty acids, monoglycerides and cholesterol, the lipid-soluble vitamin molecules are massed in micelles for transport across the mucosal membranes. Serious abnormalities in bile acid synthesis or in absorption of bile acids impair micellar formation and thus may also cause impaired absorption of fat-soluble vitamins. The first evidence of a chronic abnormality in bile acid synthesis or absorption may be a reduction in blood clotting factors owing to vitamin K deficiency. Symptoms of avitaminosis A, D, E and K may occur when there is severe generalized intestinal mucosal disease or after resection of the small intestine.

LIPIDS

Fats constitute about one fourth of the calories in an average American diet, and about 50 gm of lipid per day is ingested. Diets drastically lower in fat content are also low in calories, since a gram of fat contains twice the caloric value of a gram of either protein or carbohydrate; diets with much more than 50 gm of fat are somewhat unpalatable and may lead to obesity. The common sources of dietary fat are expensive foods such as meats, especially pork and lamb; milk products; eggs and oils. There is some evidence that excessive intake of saturated fats may contribute to the development of arteriosclerotic cardiovascular diseases in urban Americans.

An old wives' tale has it that fat is hard to digest. From the point of view of the gastrointestinal tract, this is correct. Digestion of carbohydrates is underway in the mouth during chewing and salivating, and a high protein meal may be mostly digested and absorbed from the intestine within an hour after it is eaten. Fat, on the other hand, may sit for hours in the stomach with little digestion and no absorption having occurred. Since fats are not soluble in the aqueous contents of the intestinal lumen, their digestion by water-soluble enzymes and their subsequent absorption depend upon chemical emulsification and formation of micelles, as well as the usual digestive and absorptive events involved with carbohydrates and proteins.

The physicochemical properties of water greatly influence digestion and absorption of lipids. Water molecules are strongly attracted to each other because of the asymmetric distribution of electrons within each molecule; the region of the oxygen atom has many electrons, whereas hydrogen atoms have few electrons. A water-soluble substance must have strongly polar hydrophilic groups to compete with the attraction of water molecules for one another.

Dietary lipids are polarized, i.e., their electrons are distributed asymmetrically over the lipid molecule. Consequently, lipid molecules orient in an aqueous solution in relation to the water molecules. Part of the lipid molecule is attracted to water and is soluble in it (the hydrophilic or polar part) and part of the molecule has no affinity for water (the lipophilic, hydrophobic or nonpolar part). Bile salts have the most marked combination of hydrophilic and lipophilic parts of their molecules and hence behave like detergents (see Fig 7–23). At a stable interface between the pure water and pure lipid portions of a mix, polar lipids form a film with hydrophilic parts oriented toward the water molecules and lipophilic parts excluded from interaction with water.

In our diet the fat is mainly in the form of triglycerides and cholesterol. A triglyceride is a molecule formed by the interaction of glycerol and long chain fatty acids via ester linkages:

$$
\begin{array}{c}
\text{H} \\
| \\
\text{H}-\text{C}-\text{O}-\text{(fatty acid)} \\
| \\
\text{H}-\text{C}-\text{O}-\text{(fatty acid)} \\
| \\
\text{H}-\text{C}-\text{O}-\text{(fatty acid)} \\
| \\
\text{H}
\end{array}
$$

The common fatty acids are long hydrocarbon chains containing 16 carbon atoms and no double bond (palmitate); 18 carbon atoms and no double bond (stearate); 18 carbon atoms with one double bond (oleate) and 18 carbon atoms with two double bonds (linoleate). Dietary triglycerides, long chain fatty acids and cholesterol molecules contain weak hydrophilic regions and tend to aggregate as an ion or as solid crystals in the aqueous solutions of the stomach and gut. As such, they are hard to digest and absorb.

Triglycerides are digested to form monoglycerides (glycerol backbone with only one fatty acid per molecule). Monoglyceride molecules and dietary phospholipids exist in the chyme as a bilayer, with the hydrophilic parts of each molecule on the outsides of the bilayer facing the water and the lipophilic parts facing one another inside the bilayer. Lipid bilayers alternate with layers of water to form a liquid crystal, which is the physical state of monoglycerides in the gut lumen. These liquid crystals of water, monoglycerides and phospholipids constitute a medium in which water-insoluble triglycerides, fatty acids and cholesterol will dissolve.

Bile salts and lysolecithin are the most asymmetric lipid molecules with respect to their polar and nonpolar parts. They dissolve well in both water and lipid phases. In watery solutions, bile salts behave like other polar substances so long as the bile salt concentration is low. Above a critical concentration, bile salts aggregate to form micelles, which consist of thousands of molecules oriented with hydrophilic parts facing the water surrounding the outside of the micelle. Single bile salt molecules in the aqueous phase enter the micelle and exchange for molecules leaving the micelle. Inside the micelle, the nonpolar sides of the bile salt (the sterol nucleus) molecules face one another. Water insoluble cholesterol, fatty acids and even small amounts of triglycerides can be dissolved in this nonpolar core.

Below the critical micellar concentration, bile acids are present in solution as single molecules, which coat the fat-water interface of the intestinal contents after a meal. This coat of bile acid molecules lowers interfacial tension and allows formation of tiny fat droplets, an emulsion. The emulsion is stable, because the bile acid-fat droplet has a negative charge and repels other droplets that do not coalesce into big fat blobs.

The effectiveness of bile acids as intestinal detergents depends upon their conjugation to glycine and taurine in the liver. Conjugated bile acids have a pK of 2.0 to 4.0 and dissociate well as anions in the upper intestinal chyme, where the pH is 6.0 to 8.0. Unconjugated bile acids, on the other hand, have a pK value of 6.0 and neither ionize well nor behave as effective detergents in the gut.

The oily mix of dietary triglycerides, cholesterol and phospholipids is dumped by the stomach into the duodenum. The oil blends with the watery intestinal contents and the bile, which contain micelles of conjugated bile acids, endogenous cholesterol and phospholipid. In order for this oil to dissolve in the micelles, digestion of the dietary triglycerides by pancreatic lipase, esterase and phospholipase is required.

Of these enzymes, the most important is lipase, which digests dietary triglyceride at the oil-water interface. Lipase breaks the ester linkages between glycerol and those fatty acids that are linked to the first and third carbon atoms of glycerol. Since the splitting off of acids is sequential, the reaction is

$$\text{triglyceride} \xrightarrow{\text{lipase}} \text{diglyceride} \xrightarrow{\text{lipase}} \text{monoglyceride}$$
$$+ \qquad\qquad\qquad +$$
$$\text{fatty acid} \qquad\qquad \text{fatty acid}$$

The terminal products of this digestive process are 2-monoglyceride and fatty acids. These products are much more soluble in bile acid micelles than are the original triglycerides. Under normal circumstances a small amount of glycerol is also liberated, especially from triglycerides containing medium chain (six to 12 carbon atoms) fatty acids. Glycerol can be absorbed by the intestinal mucosa either to be metabolized for energy, to be converted into a new monoglyceride in the epithelial cell or to pass into the blood and be carried away. The medium chain fatty acids are rapidly absorbed across the mucosa into the blood without having to participate in micellar formation (Fig 7–27). Medium chain triglycerides have been used therapeutically in patients with malabsorption of fat.

Pancreatic lipase is present in the upper gut complexed to another polypeptide, colipase, which protects lipase from the enzyme inhibitory action of bile acids. Colipase also alters the pH optima from 9.0 for lipase alone to 6.0 for lipase-colipase. Since the latter value is closer to the pH of the upper duodenum, colipase enhances the effectiveness of lipase at this site.

Pancreatic esterase requires bile acids for its activity. It splits ester linkages in water-soluble lipids and in micellar lipids. Its major dietary substrate is esterified cholesterol.

Pancreatic phospholipase A2 is secreted as an inactive precursor, which is

Fig 7–27.—Intestinal fat absorption. In lumen of gut, solubilization and digestion of most dietary lipids occur, with formation of micelles. After transport of lipids into the absorptive cell, re-esterification of lipids takes place, with formation of chylomicra.

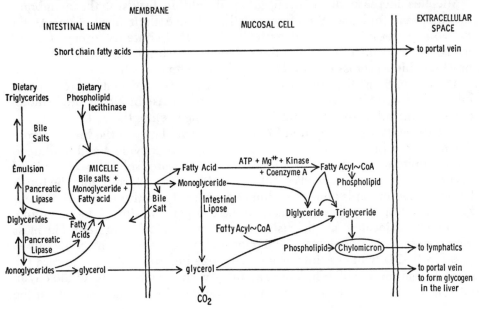

activated by trypsin. Phospholipase A2 digests about half of the ingested lecithin and other glycerophospholipids into lysolecithin and fatty acids.

Thus not long after ingestion of a meal containing fat, the lumen of the upper small intestine holds (1) an oil phase containing undigested triglycerides,
diglycerides, cholesterol esters, lecithin and other glycerophospholipids; (2)
an aqueous phase containing water-soluble molecules such as bile acids,
short chain fatty acids, glycerol and lysolecithin; and (3) micelles containing bile acids, long chain fatty acids, monoglycerides, de-esterified cholesterol and lysolecithin. The digestion by lipase, esterase and phospholipase
A2; the different solubilities of various classes of fats in lipid and aqueous
phases; and the powerful detergent properties of bile acids cause a continuous flow of fat from the oily phase into the micelles, and to some extent into
The micelles serve as the major vehicle delivering lipids to the absorptive
surface of the gut.

Absorption of lipid across the intestinal membrane is a passive process in
which the higher concentration of digested lipids in the lumen drives molecules by diffusion into the epithelial cells. Once inside the cell, the lipid is
exposed to intracellular processes that permit it to be carried away via the
lymphatics or the blood. Thus a concentration gradient is maintained as the
major driving force for transport of digested fats.

Although digested lipids are soluble in the lipid membranes of the intestinal brush border, there are two hindrances to their free diffusion into the cell:
(1) the lipid membrane itself and (2) the unstirred layer of fluid adhering to
the mucosal surface of the membrane. The unstirred layer has a higher concentration of lipid than does the inside of the cell and slows diffusion of fat
from the stirred main part of the luminal solution into the cell. Attractive
forces between the watery solvent and the dissolved lipid also slow entry of
lipid into the membrane.

Micelles increase the concentration of soluble lipids at the membrane
one-thousand-fold and therefore greatly exhance their absorption. At the
membrane the micelle disintegrates and its lipid contents (except for bile
acids) leave the micelle to enter the aqueous phase briefly as single molecules
and then diffuse across the lipid membrane to enter the cell.

At the membrane the micelle disintegrates and its lipid contents (except for
bile acids) leave the micelle to enter the aqueous phase briefly as single molecules and then diffuse across the lipid membrane to enter the cell.

As fatty acids are transported from the unstirred layer in the lumen across
the membrane of the brush border, complexing to coenzyme A and binding to
soluble proteins take place, which maintains the water solubility of the lipid
inside the cell. The absorbed complexed lipids are transferred from the brush
border to the endoplasmic reticulum, which contains the re-esterifying enzymes. New lipids are synthesized on the cytoplasmic side of the reticulum,
and then translocated into the cisternae along with beta lipoprotein. The mass
of new triglyceride, cholesterol and phospholipid grows to form a *chylomicron*
(see Fig 7–27), which achieves a certain size before it migrates toward the
serosal membrane which indents and fuses with the membrane of the chylomicron. Rupture occurs at the point of fusion, and the lipid contents of the

chylomicron are extruded into the extracellular space near a lymphatic capillary. The greater permeability of lymphatic capillary endothelium favors its uptake of lipid compared with vascular capillaries whose endothelium is less permeable. This is particularly evident after a fatty meal, when large amounts of lipid are extruded by enterocytes. Lipid appears to stimulate a large increase in lymphatic flow in the gut. Cholesterol and long chain fatty acids are carried from the gut exclusively by the lymphatics. Medium and short chain fatty acids and glycerol are transported from the gut only by the blood.

The chylomicron is the predominant vehicle for moving lipids from cells to lymphatics following a fatty meal. Some absorption of fat also occurs between meals, in which case the fat load presented to the gut is very small, and transport out of the enterocyte into the circulation utilizes another vehicle: the soluble, very low density lipoproteins, which also are synthesized at the endoplasmic reticulum and pass from enterocytes mainly to the capillaries of the portal circulation.

The great complexity of digestion and absorption of lipid makes this nutrient particularly susceptible to those diseases we lump under the heading *malabsorption syndromes.* Steatorrhea can occur with a large number of unrelated illnesses involving many mechanisms, e.g., lipase deficiency, bile acid abnormality, inadequate intestinal mucosal surface area and circulatory disturbance. With lipase deficiency, produced by chronic pancreatitis or ge-

Fig 7–28.—Malabsorption syndromes may develop from many different causes. Malabsorption of essential nutrients, especially fats, leads to widespread disease involving many systems of the body.

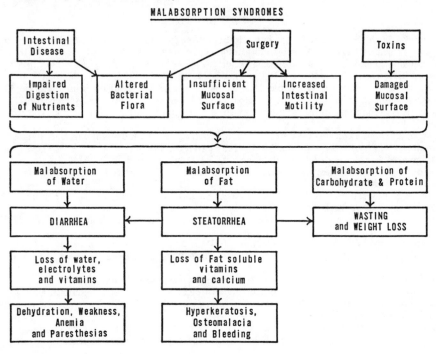

netic absence of the enzyme, triglycerides are not digested and the patient suffers weight loss and wasting. A bile acid abnormality may be produced by failure of the liver to manufacture bile acids (hepatitis); a block in the biliary tree preventing bile from reaching the gut (gallstones); failure of the ileum to reabsorb bile acids (ileitis); precipitation of bile acids in the lumen of the gut with neomycin; bacterial overgrowth and conversion of conjugated to unconjugated bile salts (blind loop syndrome) or inadequate release of cholecystokinin. When such abnormalities are present, micelles do not form and overall absorption of digested lipids is slowed; the patient undergoes weight loss, wasting and malabsorption of fat-soluble vitamins, which require micelles for their absorption. Unabsorbed fatty acids form soaps with calcium and potassium. The accumulated solutes draw water osmotically and lead to diarrhea and loss of water-soluble substances. The intestinal mucosal surface may be inadequate because of surgical resection or a bypass procedure, because of hypermotility following vagotomy or because of a destructive mucosal condition, as in celiac disease. In such cases all normally absorbed nutrients as well as fats are affected. Circulatory disturbances, which result in obstructed portal venous drainage, as in congestive cardiac failure, or in lymphatic blockade, as in Whipple's disease, may also impair fat absorption.

Whatever the underlying disease and by whatever mechanism it evolves, a severe malabsorption syndrome leads to steatorrhea, with loss of calories, lipid-soluble vitamins, carbohydrate, protein, water-soluble vitamins and electrolytes. The victim is wasted, malnourished and dehydrated; careful investigation will reveal abnormalities not only of the gastrointestinal tract but also of skin, bones, muscle, nerves and blood (Fig 7–28).

BIBLIOGRAPHY

Brooks, F. P.: Central Neural Control of Acid Secretion, in Code, C. F. (ed.): *Handbook of Physiology* (Washington, D.C.: American Physiological Society, 1967).

Christensen, J.: Motility, in Frohlich, E. D. (ed.): *Pathophysiology, Altered Regulatory Mechanisms in Disease* (Philadelphia: J. B. Lippincott Co., 1972).

Crane, R. K.: A Concept of Digestive-Absorptive Surface of the Small Intestine, in Code, C. F. (ed.): *Handbook of Physiology* (Washington, D. C.: American Physiological Society, 1967).

Daniel, E. E.: Symposium on colonic functioning: Electrophysiology of the colon, Gut 16:298, 1975.

Grayson, J.: The Gastrointestinal Circulation, in Jacobson, E. D., and Shanbour, L. L. (eds.): *Gastrointestinal Physiology* (Baltimore: University Park Press, 1974).

Grossman, M. I.: Candidate hormones of the gut, Gastroenterology 67:730, 1974.

Johnson, L. R.: Gastrointestinal Hormones, in Jacobson, E. D., and Shanbour, L. L. (eds.): *Gastrointestinal Physiology* (Baltimore: University Park Press, 1974).

Konturek, S. J.: Gastric Secretion, in Jacobson, E. E., and Shanbour, L. L. (eds.): *Gastrointestinal Physiology* (Baltimore: University Park Press, 1974).

Lanciault, G., and Jacobson, E. D.: The gastrointestinal circulation, Gastroenterology 71:851, 1976.

Levitan, R., and Wilson, D. E.: Absorption of Water Soluble Substances, in Jacobson, E. D., and Shanbour, L. L. (eds.): *Gastrointestinal Physiology* (Baltimore: University Park Press, 1974).

Makhlouf, G. M.: The neuroendocrine design of the gut, Gastroenterology 67:159, 1974.

Preshaw, R. M.: Pancreatic Exocrine Secretion, in Jacobson, E. D., and Shanbour, L. L. (eds.): *Gastrointestinal Physiology* (Baltimore: University Park Press, 1974).

Satir, B.: The final steps in secretion, Sci. Am. 233:28, 1975.

Schneyer, L. H., and Emmelin, N.: Salivary Secretion, in Jacobson, E. D., and Shanbour, L. L. (eds.): *Gastrointestinal Physiology* (Baltimore: University Park Press, 1974).

Schultz, S.: Principles of Electrophysiology and Their Application to Epithelial Tissues, in Jacobson, E. D., and Shanbour, L. L., (eds.): *Gastrointestinal Physiology* (Baltimore: University Park Press, 1974).

Sernka, T. J.: Mucosal Metabolism, in Jacobson, E. D., and Shanbour, L. L. (eds.): *Gastrointestinal Physiology* (Baltimore: University Park Press, 1974).

Simmonds, W. J.: Absorption of Lipids, in Jacobson, E. D., and Shanbour, L. L. (eds.): *Gastrointestinal Physiology* (Baltimore: University Park Press, 1974).

Walsh, J. H., and Grossman, M. I.: Gastrin, N. Engl. J. Med. 292:1324, 1975.

Weisbrodt, N. W.: Gastrointestinal Motility, in Jacobson, E. D., and Shanbour, L. L. (eds.): *Gastrointestinal Physiology* (Baltimore: University Park Press, 1974).

Weissmann, G., and Claiborne, R.: *Cell Membranes: Cell Biology and Pathology* (New York: H. P. Publishers, 1975).

Wood, J. D.: Neurophysiology of Auerbach's plexus and control of intestinal motility, Physiol. Rev. 55:307, 1975.

8

Metabolism and Body Temperature

WILFRIED F. H. M. MOMMAERTS, M.D., Ph.D.

METABOLISM

ALL VITAL PROCESSES involve myriad chemical reactions. Indeed each cell of the body can be considered to be a chemical factory subserved by the lung, gastrointestinal tract, cardiovascular system and kidneys, which exist mainly to deliver supplies and remove waste products from the factories. Much of the activity of the central nervous, endocrine and musculoskeletal systems is directed toward the acquisition and processing of food. The magnitude of the processes involved can be sensed from the fact that an average human being uses about 360 L of oxygen per day to "burn" several hundred grams of carbohydrates and fats and that about 3 million calories of heat are generated.

The chemical processes of the body collectively constitute *metabolism*. Thousands of chemical reactions are involved but they fall into two categories: *catabolic* and *anabolic*. Catabolic reactions involve the breaking down of larger, more complex molecules into smaller, simpler molecules. Energy is released in the process. The oxidative breakdown of glucose to carbon dioxide and water is a good example. Anabolic reactions involve the synthesis of complex molecules from simpler ones and require energy for their performance. The synthesis of proteins from amino acids exemplifies this.

In the cell, anabolic and catabolic reactions occur side by side. The structural components of the cell are constantly being broken down and replaced. Some of the energy released by catabolism is used to drive vital processes such as the sodium-potassium pump, muscle contraction and glandular secretion; some is used to drive the energy-requiring anabolic reactions; most is "lost" as heat.

The multitude of reactions involved in metabolism is tremendous and, for details of these, appropriate biochemical texts should be consulted. We shall consider here, however, the relation between the thermochemistry of such reactions and the general physiologic implications of energy.

Matter represents energy because of the potential energy of its interatomic and intermolecular forces. A given molecule A is associated with a given amount of potential energy E_A, which is dependent upon the types of atoms present and their structural interrelationships. If we convert molecule A into molecule B, e.g., in a biochemical reaction, the energies E_A and E_B will differ, and therefore energy $E_B - E_A$ will be liberated or absorbed when the reaction occurs. There is one specification to be applied. In the transformation (A) → (B), there will in general be a change in volume, meaning that work is done by or upon the atmosphere, and we cannot do anything about this as long as we let the events occur at constant pressure, which is usually their way. Accordingly, it is more practical to replace the change in energy E by that in the *enthalpy* H, which takes the volume alteration into account. Actually, the difference between the change in energy (ΔE) and of enthalpy (ΔH) in most reactions is generally small and is indeed negligible as long as there is no net formation or absorption of gases.

If a reaction runs its course without special arrangements or constraints, the enthalpy change will be manifested by production of heat, which can be measured using a calorimeter. Thus the heat production per mole of reaction under these conditions is the enthalpy change.

If molecule A is converted to B and heat is evolved, the reaction is termed *exothermic*. In this example, $H_B < H_A$ so that $H_B - H_A$ (ΔH) is negative. Since most spontaneous reactions are exothermic, we shall use the symbol $\overline{\Delta H}$ for $(-\Delta H)$ so as to have the convenience of positive figures. Reactions in which the system cools or takes heat from the surroundings are termed *endothermic*, i.e., ΔH is positive.

Employing special devices it is possible to use a chemical reaction to perform some type of work (mechanical, electric, osmotic, chemical). If the work done is w, then the amount of heat produced (g) will be less, such that $\overline{\Delta H} = w + g$. The production of directed work from chemical reactions requires a specific device or *transducer*, which confers some vectorial direction on the conversion. Examples are an electric element such as a battery, or the various molecular-transducing mechanisms that result in muscular contraction (see section on muscle, Chapter 2) or glandular secretion. Because of imperfections in transducers, only a fraction of the enthalpy change appears as work. But even if we had a perfect transducer, we would not in general be able to convert the entire ΔH into directed forms of energy. This is so for two fundamental reasons.

1. In the reaction A → B, there is a conversion of matter from one state and identity into another. This implies some difference in the order, mobility and randomness of the constituent atoms and molecular groups that is necessarily taking place when the reaction occurs. Thus a part of the heat liberated or absorbed is irrevocably connected with these probability aspects of the process, the change of randomness or *entropy S* associated with the reaction. For example, when water freezes to ice, it goes from a higher to a lower state of randomness (i.e., in ice the molecules of water have a more ordered structure with relation to each other) and the heat given off on that account can be nothing but heat because it derives from molecular rearrangements that cannot be

directed and thus cannot be converted to work. Only energies not so linked can in principle be converted into directed work, and this part will be called the *free energy F* (it should have been called the free enthalpy). Thus for any reaction

$$\overline{\Delta H} = \overline{\Delta F} + T\,\overline{\Delta S}$$

Since the entropy term (which is multiplied with the temperature [T] on account of the way the concepts became formulated) can be positive or negative, the possible work performance can be greater or less than the heat effect of the reaction would suggest. In the former instance, the system, when performing theoretically maximal work, would cool, i.e., draw heat from the surroundings, in order to perform work in excess of its enthalpy change. In the latter case, the system would necessarily deliver heat besides work, in addition to heat formed for whatever reasons.

2. Besides this fundamental aspect, there is a second reason. Theoretically, maximal work can only be obtained when the working process proceeds through a series of equilibrium states, i.e., when at any moment there is a perfect balance between the driving force and the force upon which work is to be done. Any deviation means that some of the available force is not transformed; it dissipates into heat. But the requirement of a perfect sequence of equilibrium conditions means that the process takes place at infinitely slow speed, therefore not at all. To proceed at finite speed, we must sacrifice some of the perfection in order to obtain an excess driving force, and this leads to dissipation of free energy into heat.

METABOLISM AND HEAT

With these insights, let us reconsider various aspects of metabolism.

1. *Anabolic* reactions require an energy input to proceed. They are coupled with energy-releasing driving reactions, and the overall event liberates heat. These energy aspects do not have a "functional purpose" by themselves. It is the specific substances resulting from the reaction that the body requires, but of course the thermal changes do enter into the overall caloric balance of the body. These patterns of synthetic reactions may be very complicated, e.g., formation of proteins on ribosomes, or they can be simple, e.g., synthesis of glutamine, which can be regarded as the simplest model of peptide synthesis. In either event there is heat production equal to the excess in enthalpy effect between the driving and driven reactions.

2. The typical *catabolic* reactions, such as anaerobic glycolysis and the oxidative metabolism of metabolic fuels, have the function of generating energy for performance of the mechanical, chemical, osmotic or electric activities of the cell. ATP plays a crucial role in all these processes. This substance is formed in large amounts, particularly in the mitochondria, during catabolic reactions. It acts as the energetic "cash" or "coinage" of the cells, since its function is to transfer energy and thus permit the coupling of catabolic reactions with cellular transduction processes or anabolic reactions:

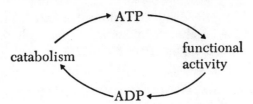

An example of the coupling of ATP with a transduction process is muscle contraction, already presented in some detail (section on muscle, Chapter 2).

3. Although we have indicated that metabolism gives rise to heat and work, we should also note that all the work performed within the body goes to heat as well. This is illustrated by the following examples: (a) All active cells accumulate certain ions, e.g., K^+, against a concentration gradient. This process is dependent upon metabolic reactions that generate ATP. But the accumulated ions leak out gradually (and also precipitously in small amounts during excitation). Thus, all the work performed to accumulate ions dissipates continually into heat. (b) Similarly, anabolic reactions involve heat and chemical work. But in the steady state, the synthesized substances break down as rapidly as they are formed, so that the total enthalpy change appears as heat. (c) The heart performs mechanical work where it propels blood against the opposing pressure of the aorta and pulmonary artery and this is external work from the vantage point of the heart itself. However, this work is frictionally dissipated when the blood courses through the vessels. Therefore, from the vantage point of the whole body, all of the heart's metabolism is converted into body heat, albeit again by a pathway that has a functional significance. (d) When muscles contract internally so as to maintain force or posture, metabolism is coupled with the presence of tension without net work. Sometimes actual mechanical work may be done by one muscle against other internal structures, and this will then again be dissipated into heat, not unlike the situation argued for the heart. This principle is highly developed in the function of shivering.

In summary, metabolism gives rise to heat and to various forms of work, but eventually all the work goes to heat as well. There is only one exception: when our skeletal muscles perform external work and this work is not returned to the body, as when bricks are piled onto a higher level, or when we walk up the stairs, then and only then is work abstracted from the body and only the dissipational and entropy-linked heat remains; the work may give rise to heat elsewhere in the universe, but that is another matter.

BASAL METABOLISM

All cells have a fundamental basal metabolism. They exist in a dynamic steady state, maintaining their improbable configuration against decay, by a continuous flow of chemical reactions. Life is a state of flux, in which substances are built to be decomposed, substances are accumulated and segregated only to leak back again toward the random distribution, which, if allowed to be attained, would signify death. This flow of metabolism, hence the flow of energy, constitutes the basal metabolism in the most elemental sense of the word. We may call it the Sisyphus aspect of life.

BASAL METABOLISM IN MAN

In a complex organism such as man, there is somewhat more to it, since it is not possible to restrict metabolism to those basic cellular functions, because certain maintenance activities must be continued. Thus "basal metabolism" or basal metabolic rate (BMR) is defined as the energy turnover under certain conditions specified by convention. These conditions are that the subject should be at complete rest and should not have exerted himself during the prior hour, since otherwise his restitution metabolism might still carry over. The subject should not be engaged in digestive and absorptive activity and should not be exposed to the specific dynamic effect of foods (see below); in practice, this means that no meals are to be taken following an early dinner the previous night. The test should be carried out at a moderate environmental temperature and, during its conduct, noise, discomfort and emotional impacts are to be avoided. The metabolism so measured includes the fundamental basal metabolism of all cells and some steady functions such as those of the kidney, heart and respiratory muscles. A certain amount of general skeletal muscle tone must also be maintained. Thus BMR, though composite in nature, is a well-defined and reproducible quantity. In sleep, the metabolic rate drops by about 10%, perhaps largely because of a reduction of muscle tone, whereas the rate always rises above the basal level during any muscular activity.

IN WHAT TERMS IS METABOLISM MEASURED AND EXPRESSED?

We can approach this from two standpoints. Since human metabolism in the aggregate is aerobic, we could determine oxygen consumption as a measure of metabolic rate, with perhaps carbon dioxide production for added characterization. Alternatively, since we are talking about an energy turnover, a suitable unit would be in calories per unit of time. In practice it is much more convenient to measure oxygen uptake and to derive heat production by calculation. We usually assume that the patient has used an average mix of foodstuffs and use a conversion factor of 4.82 kilocalories/L of oxygen. Even if a somewhat different mixture of foods is being taken, use of this factor does not introduce a great error.

SPECIFIC DYNAMIC EFFECT OF FOODS. — Among the factors that increase metabolism above the BMR are those occurring after ingestion of food. The effect is partly due to intestinal absorption reactions and to storage processes in the liver or elsewhere. After a carbohydrate or lipid meal, BMR rises only 3% or so, but after a protein meal BMR may increase 20–30% and remain elevated for several hours. This increment is called the specific dynamic action of the food in question. The high value for proteins is thought to be largely due to the oxidative deamination of amino acids derived from the ingested protein.

BODY SIZE.—Basal metabolism is related to body size. Large individuals turn over more energy than small ones. This has to be considered when comparing the BMR among individuals of the same species as well as in comparison of different species. The BMR correlates well with body weight (M) raised to the power 0.7, i.e., $M^{0.7}$, which is tantamount to correlating it with body surface area. This might be expected since body temperature has to be maintained against losses through the body surface. However, this explanation may be too facile, since the metabolic rate is independent of ambient temperature over a moderate range, and the relationship to surface area even applies to animals without temperature regulation. Some investigators have suggested that lean body mass might be a better normalizing factor. Nevertheless in clinical practice it has been found satisfactory to express BMR in terms of surface area obtained from nomograms that relate this variable to height and weight. The normal value is around 900 kilocalories (k-cal) per day per square meter. It declines with age and is generally 8% less in women than in men. Since the average human surface area is about 1.8 sq m, the average BMR is 1,500–2,000 k-cal per day, corresponding to an oxygen consumption of about 250 ml per minute.

METABOLISM AND ACTIVITY.—Clearly, caloric output is a function of activity. In a leisurely society a person might typically spend 8 hours sleeping (500 k-cal), 8 hours in sedentary and light activity (perhaps 800 k-cal) and 8 hours at work at whatever caloric output that entails. Table 8–1 gives examples of the caloric output in various activities, from which also the energy costs of different walks of life can be estimated. We often take 3,000 k-cal per day as the typical value for average activity not involving heavy physical labor.

Apart from various fluctuations from day to day, this metabolic rate must be balanced by the food intake if the person is in a steady state. In some populations the average available food falls far short of 3,000 k-cal, which means (unless the members of these populations have inherently smaller body sizes) that the steady state is maintained in a more emaciated condition and at a correspondingly lower level of bodily activity. In many of the technologically advanced nations, the reverse problem of overfeeding occurs.

METABOLIC TURNOVER

What is the meaning of the indicated metabolic rate in terms of the reaction through which a good fraction of energy metabolism funnels, i.e., the breakdown and restitution of ATP? This can be estimated along several routes, but let us take the most straightforward one. With average activities and a metabolic rate of 3,000 k-cal per day, about 25 M of oxygen is consumed per day. We know from biochemistry that oxidative phosphorylation yields about 6M ATP per mole of oxygen; thus our daily ATP metabolism amounts to 150 M per day. The 150 M or 100 kg per day contrasts with the 0.15 M ATP we may be estimated to contain; thus all our ATP turns over every one or two minutes.

TABLE 8-1.—CALORIC REQUIREMENTS FOR VARIOUS ACTIVITIES*

ACTIVITIES	K-CAL PER HR†	ACTIVITIES	K-CAL PER HR†
Domestic occupations		Physical exercise (cont.)	
Sewing	10–30	Running	800–1000
Writing	20	Cycling	
Sitting at rest	15	5 mph	250
Standing relaxed	20	10 mph	450
Dressing and undressing	30–40	14 mph	700
Ironing (with 5-lb iron)	60	Horseback riding	
Dish washing	60	Walking	150
Sweeping or dusting	80–130	Trotting	500
Polishing	150–200	Galloping	600
Industrial occupations		Dancing	200–400
Tailoring	50–100	Gymnastics	200–500
Shoemaking	80–100	Golfing	300
Bookbinding	75–100	Playing tennis	400–500
Locksmithery	150–200	Playing soccer	550
House painting	150–200	Canoeing	
Carpentry	150–200	2.5 mph	180
Joinery	200	4.0 mph	420
Cartwrighting	200	Sculling	
Smithery (light work)	250–300	50 strokes/min	420
Smithery (heavy work)	300–400	97 strokes/min	670
Riveting	300	Rowing (peak effort)	1,200
Coal mining (av. for shift)	200–400	Swimming	
Stone masonry	300–400	Breast and back stroke	300–650
Sawing wood	400–600	Crawl	700–900
Physical exercise		Playing squash	600–700
Walking		Climbing	700–900
2 mph	200	Skiing	600–700
3 mph	270	Skating (fast)	300–700
4 mph	350	Wrestling	900–1000

*Adapted from Mayer, J.: *Human Nutrition* (Springfield, Ill.: Charles C Thomas, 1972).
†Figures obtained for 150-lb subject. To be added to BMR plus 10% for specific dynamic action.

Since 1 k-cal is equivalent to 43 kg meter (kg-m), it follows that the daily energy turnover of the body under basal metabolism conditions approaches 10^5 kg-m, equivalent to lifting an American automobile to a height of more than 100 ft each day, or 100 automobiles to the top of Mt. Everest over a good lifetime. This is how much it costs, energetically, just to keep body and soul together. This too can be said to illustrate the meaning of the Sisyphus phenomenon.

PLACE OF MAN IN THE ENERGY BALANCE OF THE WORLD.—The sun enriches the earth annually with about 5×10^{20} k-cal of radiant energy. Of this, about 99.9% is absorbed directly by the atmosphere and the terrestrial surface, heating the periphery and being lost in the steady state by radiation. About 0.1% or 5×10^{17} k-cal per year is absorbed by the vegetation, which assimilates about 80×10^{12} kg of CO_2 and uses it to produce about 55×10^{12} kg of organic matter annually. Compared to these 5×10^{17} k-cal, the 10^{16} k-cal generated by the technologic use of fossil fuel is significant but not overwhelming.

The energy turnover in the bodies of the human population is of the order of 2×10^{15} k-cal, or nearly 0.5% of that of all living beings. Thus, of the total energy flux received from the sun, 99.9% goes to heat directly, 0.1% through the total biosphere, and 0.0004% through the human population. This part eventually goes to heat anyway, but along its pathways we live our lives and use the dissipated heat to maintain a constant body temperature.

SPECIAL FEATURES OF METABOLISM IN VARIOUS ORGANS AND TISSUES

Several of the body's organs are highly variable in their share of metabolism and it is not possible to give accurate figures for the distribution of metabolism unless the conditions of measurement are rigorously specified. Nevertheless, useful insights emerge if we compare the approximate distributions of metabolic activity and cardiac output in an average resting individual (Fig. 8–1).

In certain organs, specifically in liver, brain, heart and combined other viscera, there is a fair correspondence between metabolic and circulatory participations. By contrast, the kidney receives a much larger share of the circulation than is proportional to its metabolism. This share is explained by the fact that the total body requirement for its clearing action on the blood is disproportionate to the organ's own metabolism. The extreme in this regard is surely represented by the carotid bodies, which are perfused so richly that the Po_2 and Pco_2 that they measure are hardly altered during the passage. The organ that is most variable in a passive sense is the skin, where the blood flow rate is

Fig 8–1.—Approximate values for share of some major organs in circulation and metabolism of the body under resting conditions.

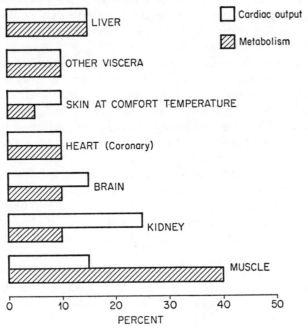

determined not by the organ's metabolic needs but by the requirements of temperature control (see section on cutaneous circulation, Chapter 4).

Several organs, notably the liver and kidney, are most versatile in their metabolic reactions. In the former these reactions are mostly in the realm of pure chemistry, whereas in the latter many are coupled with the special secretory and absorptive processes. Gastrointestinal epithelial cells also have a metabolism linked with transport functions and, since these are not continuous during a day, there is some measure of on-off control of metabolic rate, though not nearly of the magnitude found in muscle. Glandular cells, whether of internal or external secretion, are specialized in the synthesis or processing of the substances they elaborate and, in the case of holocrine glands, a large degree of cell renewal is needed in addition to the manufacture of one special product.

The brain deserves special mention because it requires a constant supply of glucose and is crucially dependent, even more than the heart, upon an uninterrupted supply of oxygen. Witness the grave depersonalization and vegetablization of patients whose brains have been oxygen-deprived for only a few minutes.

As a whole, the human body is an aerobic organism. At any given moment, anaerobic metabolism may be occurring in some part of the body, but at no time is lactate (end-product of anaerobic metabolism) excreted in the urine or otherwise in metabolically significant amounts. Thus we conclude that even when significant glycolysis takes place, it is compensated oxidatively elsewhere in the body, simultaneously or shortly afterward. For example, when glycolysis occurs in muscle, lactate escapes into the blood and is converted to glycogen in the liver via the Cori cycle (Fig 8–2). Some tissues exhibit aerobic glycolysis, e.g., retina, erythrocytes, thrombocytes and certain tumors. In these cases the supply of oxygen as well as nutrients may be plentiful, so that glycolysis is due to the proportion of oxidative and nonoxidative enzymes peculiar to these tissues rather than to oxygen lack.

Interesting is the case of the eye lens, which, being avascular, relies upon diffusion from the ocular humors for all of its nutrition. This is a highly specialized organ, which throughout life within its capsule continues the slow growth of the few nucleated cells at its periphery, resulting in a gradually increasing density of its core, the bulk of the lens consisting of tightly packed, optically quite homogeneous fibers. Glucose is its main caloric nutrient, largely used by way of glycolysis and also by some special pentosephosphate and sorbitol pathways. Maintained transparency of the lens is contingent upon normal metabolism. Apart from senility, cataract may result from metabolic disturbances such as galactosemia, and may occur in patients with congenital hemolytic anemia, characterized by diminished glucose-6-phosphate dehydrogenase in the erythrocyte and also, it seems, in the lens. Diabetes hastens the onset of cataract.

Less extremely, because of its lesser thickness, the cornea also occupies a somewhat special situation; it also is avascular except at the limbus. Its exchange of carbon dioxide and oxygen is largely through direct contact with the atmosphere; there actually is a net oxygen flux into the anterior chamber.

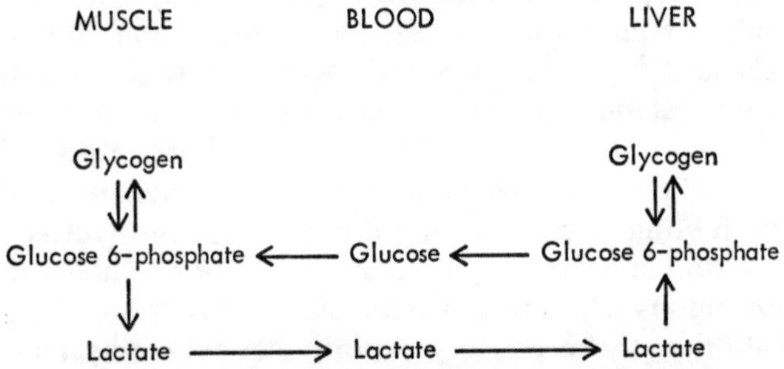

MUSCLE   BLOOD   LIVER

Glycogen   Glycogen
$\Downarrow\Uparrow$ $\Downarrow\Uparrow$
Glucose 6-phosphate $\longleftarrow$ Glucose $\longleftarrow$ Glucose 6-phosphate
$\downarrow$ $\uparrow$
Lactate $\longrightarrow$ Lactate $\longrightarrow$ Lactate

Fig 8–2.—Cori cycle showing how lactate is produced in functional anaerobiosis of active muscle, is reconverted to glycogen in liver and returned to muscle, where it is resynthesized to glycogen during recovery.

When the eyes are closed, there seems to be some maintenance of P_{O_2} from the palpebral conjunctiva but also from the aqueous humor. Among the functions requiring steady metabolic support is a transport of fluid out of the cornea by the endothelium (which also maintains Descemet's membrane). Without this transport the cornea would swell.

Connective tissues, cartilage, bone and teeth are generally distinguished by low cell density. For emphasis on the maintenance of the specific mechanical constituents of these structures, reference is made to biochemistry books.

Hibernating species, and some others as well, have a special tissue, brown fat, to which the function of chemical thermogenesis is ascribed. It is richer in mitochondria and higher in oxygen consumption than most other tissues, is stimulated by catecholamines and has a rich sympathetic innervation. In man, it occurs in the newborn infant and rapidly converts to ordinary adipose tissue.

SIGNIFICANCE OF MUSCLE IN BODY METABOLISM.—The only organs that can drastically alter their metabolic rates are the muscles. Their change from rest to activity can raise intrinsic metabolic rates by a thousand-fold, and this can be done on very short notice. At rest, their share in the body's metabolism and blood flow is likely to be less than 40% of the total, but in exercise it increases so much as to become entirely preponderant. For this reason, the level of muscular activity is the only really important variable of metabolism and of the circulatory and respiratory adaptations that this necessitates.

HORMONAL INFLUENCES ON METABOLISM.—Metabolism is increased by a number of hormones including thyroxine, catecholamines, androgens, estrogens and growth hormone. The most important hormone is thyroxine and clinical disorders of the thyroid gland resulting in hyper- or hyposecretion of thyroid hormones are not uncommon (Chapter 10).

HEAT AND BODY TEMPERATURE

We shall now examine the consequences of the fact that, coincident with metabolism, heat is produced. Elaborate mechanisms exist to dispose of this heat under many different conditions. The main variables are (1) the enormous increase in heat production induced by bodily exercise and (2) the wide range of environmental temperatures. Notwithstanding these great variations, mammals and birds can maintain their internal body temperatures with remarkable constancy and are therefore termed *homeotherms*.

We are familiar with the concept of temperature both intuitively and scientifically. Intuitive knowledge, based on sensory experiences, is subjective, as revealed by the feeling of chill during fever and by the well-known experiment of keeping one hand in very cold and the other in hot water, and then comparing impressions when both are transferred to a tepid bath. However, a general correlation of these sensory experiences with physical phenomena (e.g., the observation that bodies expand when in contact with heat) led to the development of thermometry and to the foundation of the concept of temperature in thermodynamics.

In the physiologic context, the main importance of temperature is that it determines the velocities of many phenomena, including biochemical reactions. Often, these increase two- or threefold in rate per 10 degrees C temperature rise (temperature coefficient, Q_{10}). No such generality applies to the effect of temperature on the position of chemical equilibria or on the concentrations of metabolic intermediates in steady or transient states. The general basis of the Q_{10} is that substances, before engaging in a reaction, must overcome an energy barrier by virtue of an activation energy and that they do this on account of their kinetic energy at a given temperature. The mean value of this kinetic energy increases in proportion to the absolute temperature in degrees K, but only a small fraction of the molecules displays the energy needed to overcome the barrier, and this fraction increases as a much steeper function of the temperature, hence the value of Q_{10}.

Unlike most simple reactions, biologic phenomena cannot increase indefinitely with temperature. At some point there will be denaturation of a key enzyme, or reversible or irreversible disarray of a responsible structural configuration. Individual biologic phenomena are often found to have optimum temperatures with respect to rate, but this is not necessarily the same as the optimal temperature for the life of the whole organism. Also, we should remember that at higher temperature the faster we live, the faster we die.

Regulation of body temperature can therefore be seen as an arrangement to maintain the body steadily at a temperature that represents a compromise — presumably an optimal one — of the thermal optima of all the body functions and their interactions. This is a late but important evolutionary development, foreshadowed by examples of social temperature control, as in bee communities or snake pits, reaching individual regulation of great perfection in homeothermic animals. We could imagine such arrangements to be possible at a wide variety of body temperatures, but it is remarkable that all mammals have settled for a very close range of temperature much like the human being, and birds for a similar span somewhat higher. The normal human body is said to maintain a temperature of about 37 C or 98 F, but this is a very incomplete statement. In fact, as we shall see, only the temperature of the central part of the body, the core, is regulated; the peripheral temperature fluctuates in the service of thermoregulation of the core.

At the basal rate, metabolism produces about 2,000 k-cal per day. If the body weighed 80 kg and had a specific heat of 0.83 k-cal per kg per degree, this basal heat production, if not removed, would lead to a temperature rise of the order of 30 C per day, thus reaching the boiling point in a couple of days if there were no self-limitation and death before that time. In intense muscular exercise, which may increase heat production up to 20-fold, the same rise would happen in hours instead of days. Shivering is an important adaptation to cold and can increase skeletal muscle heat production fivefold. A rare condition, malignant hyperthermia, sometimes develops during anesthesia and, by mechanisms unknown, unleashes a tremendously enhanced resting metabolism in the musculature. It would lead to thermal death if not noticed and checked.

Clearly, with such large heat productions inherent in life, mechanisms exist to dissipate it. Indeed, in temperate climates, the problem is to stay warm rather than cool.

MECHANISMS OF COOLING

The main mechanism that leads to cooling, and cooling only, is evaporation. This enables us to maintain body temperature even when the ambient temperature is higher, and heat would therefore flow toward rather than away from the body. The reverse would occur when water vapor condenses upon us when we enter unclothed from the cold into a moist greenhouse, an unimportant contingency. For each kilogram of water evaporated, 580 k-cal are removed; thus, the evaporation of a gallon could take care of a day's heat production. Evaporation takes place from the lung, the respiratory and oral passages and the skin. The skin is sufficiently permeable to water to allow for *insensible perspiration* (no visible sweat). Typically at comfortable external temperatures, about 7 k-cal per hour are lost in the respiratory tract and 10 through insensible perspiration, involving the evaporation of about 12 ml and 17 ml of water per hour, respectively. Together, this represents the evaporation of about 0.7 L per day, removing 400 k-cal or one fourth or one fifth of basal metabolism. The remainder of the heat is lost by conduction, convection and radiation.

HEAT EXCHANGE BY GRADIENT MECHANISMS

Whereas metabolism always produces heat and evaporation always removes it, the gradient mechanisms cause heat transfer either to or from the body, dependent on the temperature differences between the body surface and the environment.

The first gradient mechanism to be discussed is *radiation*. The wavelength of heat radiation depends on the temperature of the object. Heat radiation is located in the far infrared for the temperature range of the human body but reaches into the visible red when from a glowing stove, embers or an electric heater, and farther into the blue and beyond in the sun. Thermal radiation is an important factor in environmental physiology, contributing to the complexity of relating surrounding temperature to thermal balance and comfort. Indoors, we stand in radiative heat exchange with the walls of a room, which may include sources and sinks of heat. Thus we may lose heat in winter by radiation to a cold window but receive heat and comfort from the radiation emitted by a stove — factors that are important compared to the effect of the actual air temperature in the room.

The other gradient mechanisms, *conduction* and *convection,* can be treated together; both involve direct heat transfer to the material that is in direct contact with the body. Transfer to and from solids is by conduction only and is in proportion to the heat conduction rates in the materials. Convection contributes in liquids and gases and consists of mass movement whereby renewed warmer or colder material is brought in contact with the skin, this aspect being predominant over that of strictly diffusional conduction.

The relative contributions of the various heat exchange processes vary greatly with the circumstances. Some of the main features of thermoregulation can be seen from Figure 8–3, as follows:

For an unclothed motionless individual, 26–30 C would be comfortable ambient temperature. Within this range, metabolism is at its minimum, and its heat production is balanced by insensible evaporation (20–25%, as described) and by radiation and convection. Within this range, and also above it, regulation is by vasomotor adjustment, less or more blood being sent through the skin, which is the heat exchanger. At much lower temperature, e.g., at 10 C, heat loss by evaporation is little reduced, that by gradient transfer much increased; cutaneous vessels are maximally constricted and the skin is cold. These effects are somewhat met by increased metabolism, but this cannot suffice, and so the unclothed resting body cannot be in thermal balance at lower temperature. Therefore, a change in behavior is necessary. Either the body must shiver or engage in purposeful muscular activity, or we must put on clothes or blankets or huddle together, or combinations of these.

At higher surrounding temperature, metabolism in the peripheral tissues increases as their temperature rises to approach that of the core. Heat loss by gradient mechanisms diminishes and changes into heat uptake at about 35 C, which is now the temperature of the skin when the core is at 37 C. Evaporative heat loss then becomes the only remaining mechanism. Evaporation increases with increasing dilation of cutaneous vessels, more sharply with the appearance of sweating at about 31 C. When evaporation becomes insufficient or impossible because of an adverse combination of temperature and humidity, we exceed the limits of the environmental possibilities where human life can be sustained for anything but brief episodes. For example, in the sauna we can tolerate very high temperature, but only in dry air and for limited periods. On the other hand, life in some tropical rain forests may be marginally possible only when bodily effort is minimized.

Bodily exchange of materials in the form of food and excreta deserves mention, though the total contribution to the thermal balance is not large. Let us consider a pint of hot soup, e.g., 25 C above body temperature; this would add some 12 k-cal to the body, a minor factor in the balance though it may be a "heartwarming" experience on a cold day. Conversely, a pint of ice cream

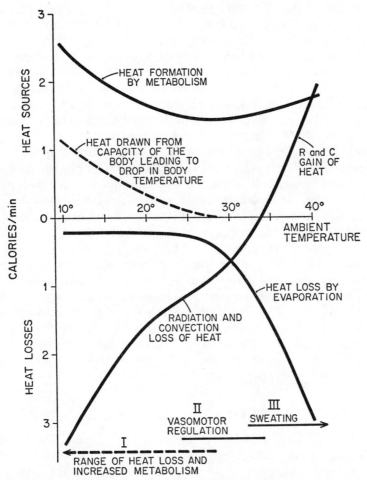

Fig 8–3.—Partition of heat exchange of resting unclothed human being at different environmental temperatures. Heat sources (plotted upward) are metabolism *(solid line)* and heat drawn from body stores. A positive value of latter means that heat is drawn from thermal capacity of the body, i.e., body cools off and is not in thermal equilibrium with environment; this part of curve is *dashed* to indicate that, unlike elsewhere, body temperature drops. Heat sinks or losses (plotted downward) are in form of evaporation and radiation *(R)* and convection *(C)*. Over range *I*, notwithstanding increased metabolism, heat is lost from body store (preventable by exercise or shivering, but these are not considered in the figure). Over range *II*, temperature control is by vasomotor regulation, and by evaporation without visible sweating. Over range *III*, cooling is by actual sweating.

would subtract more because, besides the 40 gradient difference, there would be the heat of melting of the ice; this would abstract perhaps 30 k-cal, still not a great deal in the total picture.

CORE TEMPERATURE.—As already implied, the temperature is not equal, and not equally maintained, in different parts of the body. The concept of core temperature alludes to a central and vital part of the body where the temperature is regulated accurately, contrasted to an outer shell in which the temperature fluctuates. Delineations are gradual and arbitrary, but we can fairly say that about two thirds of the body may be considered core; this is essentially the head and trunk, at least the deeper part. Non-core comprises the periphery of those parts, perhaps a half inch deep, and the extremities except for the centers of the more proximal parts of the limbs. Most typically, fingers, toes and ears are very much non-core, as are the male genitals for some not really understood reason. The function of temperature regulation is to keep the temperature of the core, which contains the vital organs, as constant as possible. This assures optimal function of such parts as the brain, spinal cord, heart, respiratory muscles, gastrointestinal system, liver, kidney, as well as for the growing fetus, which, if prematurely born, must be kept in a thermostatically controlled environment. It is particularly important to assure temperature constancy for various regulatory centers such as those that control respiration and hormone secretion. If these were at changing temperatures, their variable outputs could confound the regulation of vital functions.

Even within the core, temperatures may vary, depending on the specific location and on local metabolic activity in relation to the circulatory perfusion rate. Higher temperatures are encountered in the liver, where many chemical reactions occur. In exercise, thigh muscle temperature may reach 39 C despite increased circulation (and we may say that this makes more powerful exercise possible). Rectal temperature may vary by a degree or more, depending on the position of the thermometer with respect to veins carrying blood from the limbs when these are cold, and even arteries may be cooled in the pelvic region by heat exchange with the return blood.

IMPORTANCE OF CUTANEOUS CIRCULATION

The microcirculation of the skin and its control were described in Chapter 4 in the section on cutaneous circulation, which should be reviewed at this point. In the present context it is merely necessary to re-emphasize the key role that the skin plays in promoting heat loss, when circumstances threaten to increase core temperature, and in conserving heat when core temperature tends to fall.

Heat loss is markedly accelerated by reflex inhibition of the tonic sympathetic discharge to arteriovenous anastomoses. This greatly increases skin blood flow, raises skin surface temperature and therefore increases heat loss by the gradient mechanisms described above.

Heat conservation is promoted by increased sympathetic discharge to skin vessels, which reduces skin blood flow and temperature and thereby diminishes heat loss. Superficial fat is a very effective insulator when its circulation is reduced (whales, who live in a cold medium of high heat capacity, are insulated by a very thick layer of blubber). Thus the ability to reduce cutaneous blood flow enables us to convert the surface tissues from a good heat exchanger to a layer of "equivalent blubber" equal in its insulating properties to those of cork. This example emphasizes an overall principle of thermoregulation: by sacrificing the periphery, changing it from a participant into an inactive region of relatively low thermal conductivity, core temperature can be maintained with much less metabolism than would otherwise be needed. As with all adaptive mechanisms, this has its limits, and excessive exposure to cold leads to tissue damage in the form of frostbite.

INTEGRATION OF TEMPERATURE CONTROL

As we discussed in the section on cutaneous circulation in Chapter 4, the principal controlling centers lie in the hypothalamus where certain groups of neurons are very temperature sensitive. Neurons in the preoptic anterior hypothalamic area are particularly responsive to rises in temperature, whereas posterior hypothalamic neurons are especially sensitive to a fall in temperature. The interconnections of these neurons of the "heat regulating centers" and their mode of functioning are poorly understood. Evidence is accumulating that norepinephrine and serotonin are involved, although marked species variations occur. Injection of minute amounts of norepinephrine into the hypothalamus of cats causes a fall of core temperature, whereas similar administration of serotonin causes a rise. Evidence for a tonic action of norepinephrine was obtained by injection of the alpha receptor blocking agent phenoxybenzamine into the cerebral ventricles. However, norepinephrine and serotonin are not *essential* for maintaining a constant temperature, since adequate thermoregulation occurs under normal conditions even when these amines have been pharmacologically depleted. However, the amine-depleted animals stand up less well to very hot or very cold environments.

Whatever their fundamental mechanisms of action, these neurons achieve thermoregulation by appropriate control of sympathetic and somatic neural activity. Thus, a rise in core temperature leads to sympathetic inhibition and increased cutaneous blood flow. Sweating occurs if these changes are inadequate. Conversely, a reduction in core temperature leads to sympathetic activation and reduced skin flow. Shivering may develop if these measures fail to prevent a fall in temperature.

The hypothalamus not only responds to temperature changes of its neurons but also receives inputs from temperature receptors of the skin, although these appear to be of subsidiary importance in man.

The spinal cord is also temperature sensitive and may play a significant role in regulating heat balance, although its importance vis-à-vis the hypothalamus has not been established.

In summary, the following adaptations are at our disposal. For dissipation of heat there are the gradient mechanisms when the outside temperature is lower than that of the skin, and evaporation of sweat when it is not. For preservation of heat, there is the transformation of the dermal shell into an insulating layer. For generation of heat, there is voluntary work and shivering. These physiologic responses are aided or modified by clothing and by behavioral adjustments. The adaptations are engendered by changes in the signaling of skin, hypothalamic and spinal cord temperature sensors and are integrated by hypothalamic and spinal cord neurons. The sympathetic nervous system by its modulating effect on cutaneous circulation and on sweat gland activity plays a key role in the responses.

DISTURBANCES OF TEMPERATURE CONTROL

FEVER. — Elevated core temperature is such a common manifestation of illness that use of the thermometer to measure oral or rectal temperatures is the most widely used clinical investigative means to determine the presence of disease.

The causes of fever are many: infections, neoplasms, endocrine abnormalities, immunologic disturbances, tissue injury, poisons and drugs. Whatever the cause, metabolism increases and heat loss is reduced. Body temperature therefore rises, although the patient may actually feel cold, a feeling possibly induced by vasoconstriction in the skin. Indeed, particularly when temperature rises rapidly, shivering may occur and accelerate the temperature rise. Usually the temperature does not exceed 40 C, suggesting that some protective mechanism is activated at about this point. In patients in whom the temperature does rise to higher levels, brain dysfunction in the form of delirium and restlessness is evident.

The mechanism of fever is obscure but much interest has been aroused by the observation of Feldberg and others that injection of prostaglandin E into the cerebral ventricles produces fever in almost all species. It has also been found that administration of endotoxin and other fever-inducing compounds (pyrogens) increases the prostaglandin E_2 content of the CSF. Fever could be abated by use of aspirin, a known inhibitor of prostaglandin synthesis. Whether prostaglandins are involved in all types of fever in man remains to be established, as does the mechanism of action of prostaglandin on the neural control system.

HEAT STROKE. — This is due to failure of central control of heat regulation. It is particularly likely to occur in the elderly, in poorly conditioned individuals who exercise severely in extreme heat and in those taking atropine-like drugs that interfere with sweating. It usually happens after prolonged heat exposure and is characterized by rapidly rising core temperature, severe disturbances of brain function leading to coma, and absence of sweating. Cardiovascular collapse is common. Death occurs unless cooling can be rapidly effected.

HEAT EXHAUSTION. — This is due to loss of water and electrolytes secondary to heavy sweating. It is associated with prostration, increased heart rate, lowered blood pressure and laboratory signs of reduced extracellular and plasma volume. Body temperature is normal and the skin is cool and clammy (in marked contrast to the patient with heat stroke).

HEAT SYNCOPE. — This is a relatively benign condition in which heat exposure causes fainting due to inadequately compensated peripheral vasodilatation. It is particularly liable to occur after exercise because the dilated muscle vessels further reduce peripheral vascular resistance.

HYPOTHERMIA. — This is caused by exposure to cold. The very young and the very old are particularly susceptible, and underlying disease, which impairs the physiologic and behavioral mechanisms available to compensate cold, is frequently present. Thus it may occur in hypothyroidism, in cerebrovascular disease or in alcohol or barbiturate intoxication. Body temperature may fall below 32 C and this predisposes to ventricular fibrillation.

BIBLIOGRAPHY

Bligh, J.: *Temperature Regulation in Man and Other Vertebrates* (New York: American Elsevier Publishing Company, Inc., 1973).
Cabanac, M.: Temperature regulation, Ann. Rev. Physiol. 37:415, 1975.
Feldberg, W.: Body temperature and fever; changes in our views during the past decade, Proc. R. Soc. Lond. [Biol.] 191:199, 1975.

9

Neurophysiology

BRIAN W. PAYTON, M.D.

NEURAL ORGANIZATION AND FUNCTIONAL LOCALIZATION

THE CENTRAL NERVOUS SYSTEM includes both the brain and spinal cord, which are enclosed within the cranium and vertebral canal, respectively. Those parts of the nervous system that lie outside these bony structures, and which mainly comprise nerve trunks and their branches, are collectively referred to as the *peripheral nervous system*. A distinction also is made between the *somatic* nervous system and the *autonomic* nervous system. The former division includes both sensory and motor mechanisms innervating such tissues as skin and muscle throughout the body. Activity of the somatic nervous system is associated with sensation and perception, and with those movements of the body brought about by contraction of skeletal muscle. The autonomic nervous system, discussed more fully in Chapter 3, is primarily involved with visceral functions; its activity is not normally under "voluntary" control, nor is its activity perceived consciously in the same way as are somatic sensations.

The nervous system is made up of about 20 billion neurons plus other cells of non-neural origin. Many of the grosser aspects of function of the nervous system can be understood in terms of interactions between certain of these different, specialized neurons. Neurologic diseases or disorders are often associated with specific regions of damage within the central or peripheral nervous system. It is important to remember that, once damaged by injury or disease, neurons possess very little power of regeneration and repair. Functional losses therefore tend to be permanent, although some compensation may occur by the use of other nerve cells or mechanisms. Because of cell specialization, the type of functional loss depends on which cells are damaged. Although it is based on little evidence it is often said that we appear to lose a few thousand of these nerve cells every day. If this is so, only youth appears to think this becomes increasingly obvious with age! Whatever compensatory mechanisms may or may not exist, our relative lack of knowledge regarding regeneration and growth of neurons is a severe handicap in the treatment of many neurologic diseases.

A knowledge of the way nerve cells connect and interact, together with the sites within the nervous system where particular types of cells may be found, allows a very high degree of accuracy in localizing sites of injury and disease and in accounting for the associated signs and symptoms.

NEURAL ORGANIZATION

The cellular mechanisms whereby activity in a nerve cell can influence activity in another excitable cell, such as a nerve or a muscle cell, have been considered in a previous chapter, as has the ability of such cells to conduct an all-or-none signal, the action potential, from one location to another (Chapter 2). The nervous system does more than conduct coded information from point to point; it integrates various types of information and activity, not only for perceptual purposes but also to bring about appropriate motor activity or behavioral responses of the whole organism. Such responses will in turn modify the sensory information impinging on the CNS and again bring about further alterations in behavior. This complex interrelationship between the many different parts of the nervous system must always be borne in mind.

Nevertheless it is possible to consider certain subdivisions of function within the nervous system. This allows us to gain some understanding of the ways in which such interactions take place. Where our knowledge is more precise, we find there is an extremely high degree of order in the way connections or *synapses* are made between different nerve cells and the regions that are in-

Fig 9–1.— A, conventional symbolism for illustrating neuronal pathways and synaptic connections. **B,** connections of presynaptic terminals with postsynaptic cells are many and diffuse. Presynaptic terminals may derive from many separate neurons. Boutons may be located on surface of dendrites or cell body.

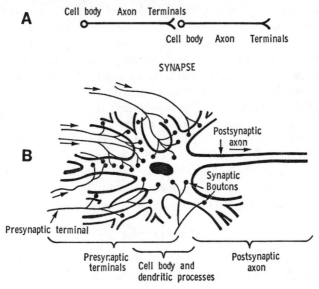

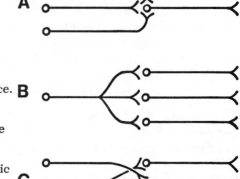

Fig 9–2.–Convergence and divergence.
A, convergence of three presynaptic
neurons on a postsynaptic cell. **B,**
presynaptic neuron diverging to synapse
with three separate postsynaptic cells.
C, combination of convergence and
divergence between pre- and postsynaptic
neurons.

terrelated. More surprisingly, the complex integration of activity that occurs
can often be related to some very basic and simple principles of interaction at
the synaptic level. Before taking some examples of more gross aspects of function within the CNS, it is worthwhile briefly reviewing the types of synaptic
connections and interactions that can be demonstrated at a cellular level.

CONVERGENCE AND DIVERGENCE

Schematically, connections between two neurons are often represented by
a diagram such as the one illustrated in Figure 9–1, A. This indicates that the
nerve axon proceeding from a cell body branches into fine terminals that synapse on the surface of another cell body, or its dendritic processes. The axon
of the second neuron finally terminates in a similar fashion. It is important to
realize that such a diagram implies that the presynaptic terminals branching
from a single axon make many synaptic contacts with the postsynaptic cell. It
should also be appreciated that almost invariably the postsynaptic cell is also
receiving synaptic input from many different presynaptic axons (see 9–1, B).
This phenomenon is known as *convergence* and is illustrated by Figure 9–2,
A. Most neurons do not send a single axon to synapse with just one other
nerve cell; rather, the axon usually branches and diverges to influence many
nerve cells (see 9–2, B), these need not necessarily be in close proximity within the CNS. Specific patterns of divergence and convergence of neurons occur
throughout all parts of the nervous system, but diagrams illustrating certain
pathways in the CNS do not necessarily indicate these additional connections, because they may not be directly relevant to the particular feature being
described or because they are not precisely known.

EXCITATION

The production of an action potential in a postsynaptic cell following release of a specific excitatory chemical transmitter has been carefully described previously, particularly in relation to transmission of activity between

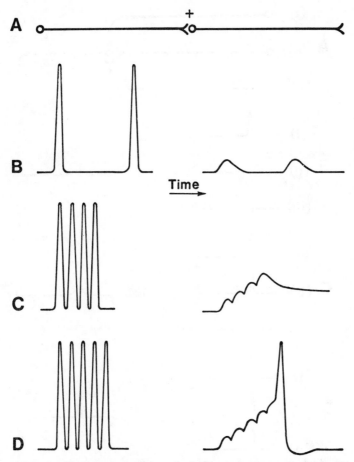

Fig 9–3. – Temporal summation. **A,** symbols for an excitatory synapse. **B,** two single presynaptic action potentials widely separated in time, leading to two separate excitatory postsynaptic potentials (EPSP). **C,** train of presynaptic impulses close together in time leading to greater depolarization of postsynaptic cell (temporal summation of EPSPs). **D,** further excitatory input leading to sufficient postsynaptic depolarization to generate postsynaptic action potential.

nerve and skeletal muscle cells. In this case we saw how a single presynaptic action potential in the neuron invariably gave rise to a postsynaptic action potential in the muscle. Although similar mechanisms are involved at excitatory synapses between neurons, a single action potential in a presynaptic neuron seldom gives rise to an action potential in the postsynaptic cell. A single action potential usually leads only to an *excitatory postsynaptic potential* (EPSP) in the postsynaptic cells. This is an inadequate level of depolarization to fire the postsynaptic cell (Fig 9–3, B). Even a short burst of presynaptic action potentials may still be an inadequate stimulus to fire the postsynaptic cell. The summation of the EPSPs will, however, bring the potential of that cell closer to its firing level, or facilitate it (see Fig 9–3, C) The postsynaptic response arising from an even greater number of presynaptic action potentials

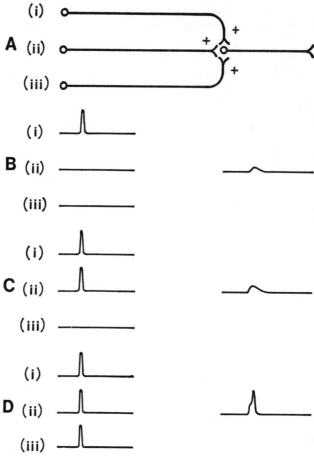

Fig 9–4.—Spatial summation. **A,** diagram for convergence of excitatory presynaptic inputs; **B,** single action potential in input *(i)* leads only to an EPSP; **C,** simultaneous activity in inputs *(i)* and *(ii)* leads to a larger EPSP (facilitation); **D,** simultaneous activity in all presynaptic inputs, leading to an EPSP adequate to initiate action potential in postsynaptic cell.

occurring in the same presynaptic cell may eventually depolarize the cell to a level that will cause initiation of an action potential (see Fig 9–3, D). The summation of EPSPs to a sequence of stimuli in time is referred to as temporal summation. Spatial summation refers to the summation of postsynaptic potentials brought about by multiple inputs from one or many different presynaptic cells and impinging on the postsynaptic cell simultaneously (Fig 9–4). To some extent the amplitude of the postsynaptic response to the input from a single cell will be proportional to the multiplicity of the boutons or synaptic connections from that source. Another neuron also receiving an input from the same cell may not be as densely innervated. Both temporal and spatial summation can act on the same postsynaptic cell at the same time, and in response to multiple inputs.

Inhibition

Inhibition, or the ability of presynaptic activity to stop or decrease activity in a postsynaptic cell is as important as excitation in neural integration. One of the cellular mechanisms for this involves liberation of a chemical transmitter from the terminals of certain presynaptic cells, and this hyperpolarizes the postsynaptic cell, producing an *inhibitory postsynaptic potential* (IPSP) (Fig 9–5, A and B). This hyperpolarization would reduce the depolarization induced by any excitatory inputs to the same cell. The activity in the postsynaptic cell is then dependent on both excitatory and inhibitory influences. Both temporal and spatial summation of inhibitory activity can occur (see Fig 9–5, C and D). This type of inhibition is referred to as *postsynaptic inhibition*. Another form of inhibition, *presynaptic inhibition* occurs when a reduction of neural activity in a cell is brought about by decreasing the amount of transmitter being released from the terminals of an excitatory input to that cell. In this

Fig 9–5.—Spatial and temporal summation of postsynaptic inhibition. **A,** diagram showing convergence of inhibitory pathways. **B,** single action potential in any one of three inputs will lead to hyperpolarization (IPSP) of postsynaptic cell. **C,** temporal summation of IPSPs following a train of action potentials in one of the inputs. **D,** spatial summation of IPSPs following activity in all inputs.

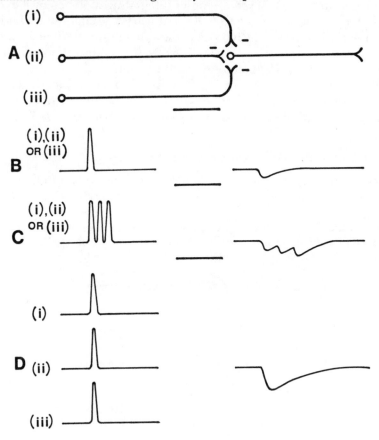

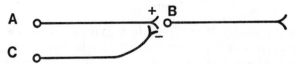

Fig 9–6.—Presynaptic inhibition. Cell *A* is an excitatory input to cell *B*; Cell *C* synapses with terminals of cell *A*; activity in cell *C* leads to reduction of excitatory transmitter released from cell *A*, if both *A* and *C* are active at the same time. This reduces or inhibits the influence of cell *A* on cell *B*.

case the terminals of an inhibitory input synapse with the terminals of the excitatory input (Fig 9–6). The transmitter released from the inhibitory input depolarizes the excitatory terminals and reduces the amount of excitatory transmitter liberated.

Regardless of the particular function of the nervous system with which it is associated, an individual nerve cell is either excitatory or inhibitory to the cells with which it connects, and it appears that any one cell releases only one transmitter substance. Features of neuronal organization such as convergence and divergence thus allow activity at any one site to influence many other sites, and excitation and inhibition can combine to modulate activity at these sites. Such mechanisms may be used for coding information, as we shall see when considering sensory systems, or for integrating motor control over other systems. Activity occurring at one site that causes excitation at another site can also be used to inhibit activity elsewhere if an inhibitory neuron is interposed in the second pathway (Fig 9–7, A). Neural connections can also exist such that activity in a single neuron could in itself lead to further activity in that cell (positive feedback; see Fig 9–7, B). Conversely, any such activity could be used to prevent further activity by interposing an inhibitory interneuron with appropriate connections (negative feedback; Fig 9–7, C).

TEMPORAL CONSIDERATIONS OF SYNAPTIC TRANSMISSION

The nature of synaptic transmission not only allows integration of different levels and sources of presynaptic activity but can introduce significant tempo-

Fig 9–7.—Simple examples of possible neural interactions. **A,** conversion of excitation to inhibition; **B,** positive feedback loop (this could be achieved without an interneuron); **C,** negative feedback loop.

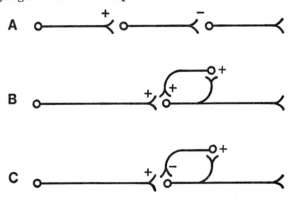

ral considerations in the integrative process. The distance between the pre-synaptic terminals and the region in the postsynaptic cell where the subsequent action potential originates may be as little as $5-20\ \mu$, and as much as a millisecond or more may elapse during this transmission. First there is the time delay between the presynaptic action potential and the start of the postsynaptic response (synaptic delay time), and then, in the case of excitation, there is the time taken for a postsynaptic response to reach a threshold-firing level (postsynaptic response time). In a comparable period (e.g., 1 msec.), an action potential in an axon will have been conducted as much as $300\ \mu$ to 100 mm, depending on its size and myelination. Interposition of chains of inter-neurons could therefore give rise to long delays and, in reverberatory circuits or feedback loops, long-lasting effects.

RELATIONSHIP OF NEURONAL CIRCUIT DIAGRAMS TO GROSS PHYSIOLOGIC FUNCTIONS

The simplistic neural diagrams outlining connecting pathways and synaptic sites that are frequently used in physiologic diagrams commonly indicate only a single neuron in each section of the pathway. The student should realize that in the physiologic situation we are almost invariably dealing with a population of neurons serving similar functions. Within each population each neuron has its own specific origin or distribution, although there may be some overlap in origin and distribution of individual cells. It is also important to bear in mind that, particularly within the CNS, various excitatory or inhibitory pathways usually modulate ongoing activity rather than initiate or completely extinguish it; i.e., excitatory inputs increase and inhibitory inputs decrease the firing frequency of postsynaptic neurons.

OTHER INTEGRATIVE MECHANISMS

So far we have discussed interactions among neurons, i.e., the excitable nerve cells capable of producing all-or-none electric activity. It must be remembered, however, that within the CNS are other cells, the neuroglia, which are even more numerous. Their function remains unclear. There is evidence that electric interactions can occur between neurons and glial cells and this could lead to modification of synaptic interactions and nerve conduction. It is generally assumed, though on little evidence, that glia serve a supportive and metabolic role in relationship to neurons, but this does not exclude a role in integrative processes.

The use of symbolic diagrams may lead us to think of the nervous system as a rigidly "wired" system. This might not necessarily be so; certainly time lapse cinematography of nerve cells in tissue culture gives us a much more dynamic impression of the nervous system. It is well known that the conduction velocity of action potentials is related to the diameter of the axon and the presence of myelin sheaths. Alterations in degrees of myelination or size of axons could affect synaptic interactions. Electrophysiologic observations tend to suggest that, over relatively short time scales (up to 24 hours), the patterns

TABLE 9-1.—SUBDIVISIONS OF CNS

REGION	SUBSTRUCTURE
Forebrain (prosencephalon)	Paired cerebral hemispheres Unpaired interbrain (diencephalon)
Midbrain (mesencephalon)	
Hindbrain (rhombencephalon)	Pons Cerebellum Medulla
Spinal cord	Spinal segments

of connection and integration seen in individual cells remain fairly stable, though it still remains possible that our techniques of investigation are inadequate to detect subtle alterations.

FUNCTIONAL LOCALIZATION

The general concept that individual nerve cells perform specific functions, and that cell bodies and the axons related to those functions tend to occupy specific regions of the nervous system, is very important to neurology. Much neurologic diagnosis, particularly of the localization of disorder or damage, depends on a good understanding of functional neuroanatomy. Because of this important relationship between function and structure and to avoid recourse to anatomical texts, the major anatomical divisions of the nervous system will be briefly described.

ANATOMICAL CONSIDERATIONS

The rostral end of the primitive nerve cord of the embryo develops to form three main regions: (1) most rostrally, the forebrain (prosencephalon), (2) the midbrain (mesencephalon) and (3) the hindbrain (rhombencephalon), which is continuous with the part of the nerve cord that becomes the spinal cord. All these structures are referred to collectively as the CNS. Further development

Fig 9–8.—Major subdivisions of CNS, as seen in sagittal section.

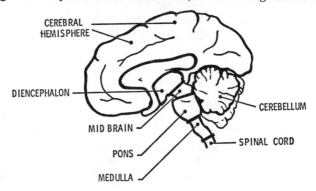

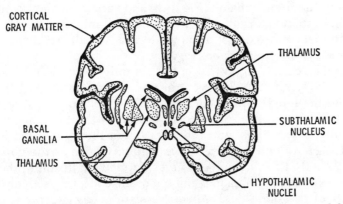

Fig 9–9. — Diagram of coronal section through cerebral hemispheres and diencephalon showing distribution of gray and white matter.

gives rise to the cerebral hemispheres (telencephalon) and diencephalon from the forebrain and to the pons, cerebellum and medulla from the hindbrain (Table 9–1). Because of the complex way the cerebral hemispheres and other substructures develop, recognition of these subdivisions in the mature brain is not always obvious (Fig 9–8). Sections through the CNS reveal areas of "gray" matter, which are the locations of the cell bodies of neurons and their associated synaptic connections. These areas of synaptic connections, particularly as seen in electron micrographs, are often referred to as the "neuropil." The gray matter has definite distribution patterns in different parts of the CNS. In the cerebral hemispheres and the cerebellum, it can be seen as a surface cortical layer. Elsewhere, large accumulations of neuron cell bodies form "ganglia" or nuclei such as the basal ganglia in the telencephalon, the

Fig 9–10. — Segment of spinal cord showing distribution of white and gray matter and emergence of spinal nerve roots.

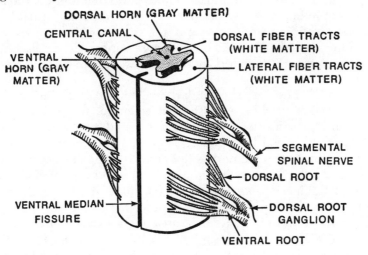

thalamus in the diencephalon (Fig 9–9) and various nuclear groups through-out the brain stem (a term that includes the midbrain, pons and medulla). The rest of the brain substance appears white and is made up of bundles of axons or fiber tracts connecting various regions.

In the spinal cord the gray matter lies centrally around the spinal canal and is bilaterally symmetric, has ventral and dorsal horns and is continuous throughout the length of the cord. The white matter, consisting of axons ascending and descending in the cord, surrounds the gray matter and so lies peripherally. The white matter fiber tracts run throughout the length of the cord and are frequently referred to as columns, such as the dorsal or lateral columns (Fig 9–10). The spinal nerves, formed by fusion of the dorsal and ventral roots carrying fibers into and out of the cord, respectively, are considered parts of the peripheral nervous system. They are predominantly the pathway for somatic motor and somatic sensory fibers controlling skeletal muscles and bringing sensory information from the various regions of the body.

FUNCTIONAL AND SOMATOTOPIC LOCALIZATION IN PERIPHERY AND SPINAL CORD

Although the nervous system develops from segmentally organized neural tissue, this segmentation is only obvious in the adult at the spinal cord level. Each cervical, thoracic, lumbar and sacral segment of the cord is related to a spinal nerve. The segmental distribution of the peripheral nerves derived from the spinal nerves becomes obscure following the sprouting of the limb buds and the formation of the brachial nerve plexus in the cervical and upper thoracic region, and lumbar and sacral spinal plexuses in the caudal parts of the body. Nevertheless, the segmental pattern of innervation continues to exist and is particularly obvious when we consider distribution of the sensory fibers in the spinal roots innervating the thorax and abdomen. As can be seen from Figure 9–11, sensory information from particular areas of the skin travels via afferent fibers entering the spinal cord at particular spinal levels. For example, sensory information arising in the region of the nipples will enter the cord at the level of T4 or T5, there being overlap of the adjacent sensory areas or *dermatomes*. Distribution of the nerve supply in these dermatomes is unilateral.

The segmental innervation of underlying muscle and other structures is less obvious but nevertheless exists. For example, the motor nerve axons supplying (and the sensory nerves originating in) the muscles on the anterior aspect of the thigh (quadriceps muscle) derive mainly from fibers in the second and third lumbar spinal nerves (L2 and L3). The corresponding dermatomes of L2 and L3 cover mainly the anterior and some medial and lateral aspects of the thigh and knee. Damage at the spinal nerve level, which could occur as the nerve passes between the appropriate vertebrae, might thus lead to sensory loss in the thigh and knee area and weakness in the quadriceps muscle group. The same features might also occur with damage to the spinal cord at the level of entry of the spinal roots. However, as we shall see when

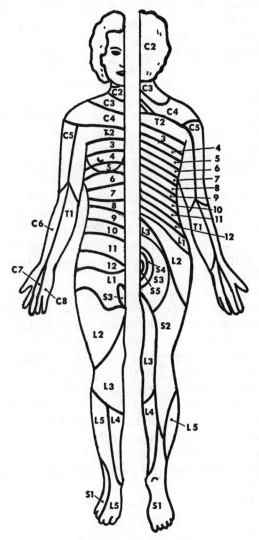

Fig 9–11.—Cutaneous distribution of spinal nerves, dermatomes.

considering spinal cord functions, additional sensory losses and motor weaknesses might then be expected.

Damage to specific peripheral nerve branches, as distinct from spinal nerves, produces different patterns of sensory or motor loss depending on the actual site of injury or pathologic lesion. The peripheral nerves carry axons, which may be related to many spinal segments. Also, as they branch to supply muscle or skin, axons carrying only sensory information may form separate bundles, or smaller peripheral nerves, that will be almost purely sensory in function, e.g., the cutaneous nerve branches. A knowledge of these aspects of neuroanatomy will enable certain types of functional loss to be related to damage or malfunction of particular peripheral nerve branches.

Our ability to localize sites of injury or malfunction in the periphery may seem obvious because different pathways and nerve bundles can usually be seen to be discrete. However, this general pattern of localization of nerve fibers, associated with particular regions of the body, continues throughout many parts of the CNS and is known as *somatotopic* or *topographic* localization. Many peripheral nerves are mixed in the sense that they carry both motor and sensory fibers, but separation of these fibers occurs in the spinal nerve roots. The sensory fibers forming the dorsal spinal roots enter the cord posterolaterally. The motor fibers emerge anterolaterally, forming the ventral root, which fuses with the dorsal root to form the spinal nerve (see Fig 9–10). Here, then, exist definite examples of functional and anatomical localization among nerve axons.

Correlation of pathologic studies with particular disorders of function, as well as anatomical and physiologic findings, has enabled us to outline fairly discrete pathways within the CNS, localizing not only particular motor or sensory areas and pathways, but also different types, or modalities, of sensation.

Figure 9–12 is a diagram of a transverse section of the spinal cord with its dorsal and ventral roots. Entering the dorsal root on the left side is the axon of a primary, or first order, sensory neuron conveying information regarding either touch, pressure or joint position sense. The cell body for this and all other first order sensory neurons is located in the dorsal root ganglion. After entry into the spinal cord, these fibers send branches into the posterior columns (dorsal funiculi), which ascend to higher levels. Other branches make synaptic connections with interneurons in the gray matter and are not included in the figure. Shown entering the right posterior root is a primary sensory neuron serving pain or temperature sense. At the level of entry, it synapses with an interneuron, which then synapses in the anterior horn of the gray matter with a motor nerve cell whose axon leaves via the ventral root.

Another branch of this temperature or pain fiber is shown synapsing with a second order neuron in the posterior part of the gray matter. This second or-

Fig 9–12.—Functional localization in spinal cord (see text).

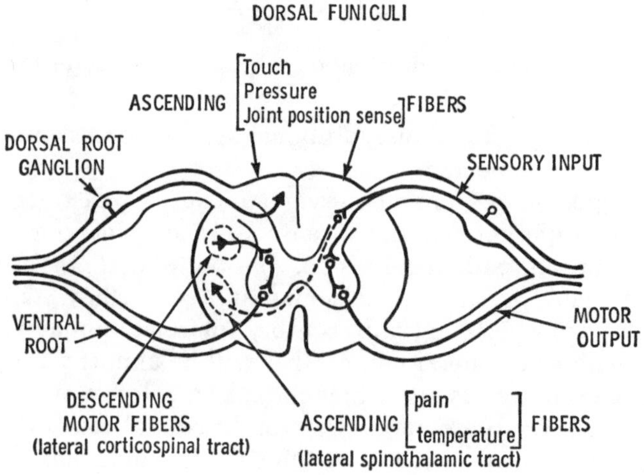

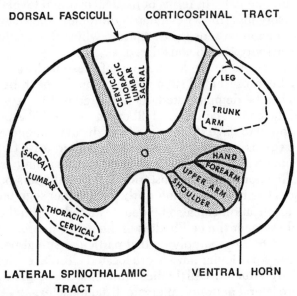

DORSAL FASCICULI   **CORTICOSPINAL TRACT**

LATERAL SPINOTHALAMIC   **VENTRAL HORN**
TRACT

Fig 9–13.—Topographic localization in spinal cord. Diagram of transverse section of cord at cervical level.

der axon is indicated in the figure by a dashed line crossing to the opposite side of the cord, where it too will then ascend, in this case in the anterolateral white matter (lateral spinothalamic tract) to higher levels. This branch is indicated by a dashed line, as it usually ascends one or two spinal segments before crossing to the opposite side. This separation of neurons of differing sensory modalities gives, at any spinal level, ascending fibers for some modalities of sensation located on the same side of the cord as their origin, whereas fibers of other modalities are carried on the opposite side of the cord. We shall see later that uncrossed fibers eventually cross over at a higher level in the CNS. Also shown on the left side of the cord is a descending fiber pathway (lateral corticospinal tract), most of whose axons leave to synapse with interneurons in the gray matter, which connect with motor nerve cells located in the ventral horn of the same side. These motor cells send axons out of the cord via the ventral spinal roots, which eventually synapse with skeletal muscle cells. These cells are called *lower motor neurons* and their axons form the *final common path* for motor activity.

The selected pathways illustrated in Figure 9–12 occur, of course, on both sides of the cord, and they continue to occupy specific locations in the cord as they ascend to, or descend from, higher levels of the CNS. Damage or lesions in the spinal cord therefore give rise to particular patterns of motor and/ or sensory dysfunction depending on their location and extent. If the lesion is unilateral, signs and symptoms can occur that relate to both sides of the body at and below the level of the lesion. Given the additional information that vibration sense travels in the dorsal funiculi and that other aspects of touch sense cross and travel in the lateral spinothalamic tract, the student should be able to work out the main consequences of a hemisection through the spinal cord on one side, in terms of particular spinal levels and laterality of motor and sensory loss (Brown-Séquard syndrome). The answer is given on page 495 in the section on sensory systems, in which the relationship of cord lesions to reflex activity is also considered.

So far in this description, the localization of these pathways in the spinal cord has mainly been presented from the functional aspect, but topographic localization also can be demonstrated. As indicated in Figure 9–13, if a high cervical level of the cord is considered, fibers in the medial aspect of the dorsal funiculi are found to have their origin from distal segments such as sacral and lower limb regions, whereas the more lateral fibers arise from upper limb and cervical segments. The fibers of the lateral spinothalamic tract also show topographic localization, the more posterior and lateral fibers being from distal segments and those lying more anteriorly from higher segments.

Topographic localization can also be demonstrated within the gray matter: (1) in any spinal segment the cell bodies of the motor neurons in the ventral horn are restricted to those supplying that particular spinal nerve; (2) motor neurons supplying a particular muscle are located close to each other; and (3) the more medially situated cell bodies innervate more proximal muscles, whereas the more lateral neurons innervate distal muscles.

LOCALIZATION AT HIGHER LEVELS

Similar examples of functional and topographic localization can be demonstrated at brain stem and cortical levels. Some examples such as the respiratory and cardiovascular "centers" in the brain stem have been presented elsewhere in relation to those particular systems. The cortex of the cerebral hemispheres provides a classic example of localization. It has long been known that particular regions of the cortex are intimately involved with certain functions of the brain. A notable example is the somatosensory cortex lying posterior to the central sulcus, which divides the frontal from the parietal lobe (Fig 9–14). Damage to this area leads to marked loss of sensory perception. Ascending sensory fibers follow definite pathways to this region, and, as we shall see later, definite, meaningful patterns of integration occur here and in other synaptic regions of the pathway as it approaches the cortex.

Another important sensory area of the cortex is located in the posterior part of the occipital lobe and is extremely important for visual perception. (It should be noted that the term perception implies conscious appreciation of sensation: the distinction is made because, as we shall see, much sensory information is not necessarily associated with perception.) Another example of a sensory projection area is the auditory cortex of the temporal lobe.

Motor areas have also been described. One of these, the somatomotor cortex, lies anterior to the central sulcus and occupies the posterior part of the frontal lobe adjacent to and parallel with the somatosensory cortex (see Fig 9–14). Another motor area, Broca's area, is intimately associated with speech and lies anterior to the most inferior part of the somatomotor cortex in the frontal lobe (other areas associated with the more perceptual aspects of speech lie in the parietal and temporal lobe cortex close to the auditory cor-

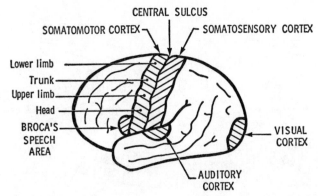

CENTRAL SULCUS
SOMATOMOTOR CORTEX — ⌐ SOMATOSENSORY CORTEX

Lower limb
Trunk
Upper limb
Head
BROCA'S
SPEECH
AREA

VISUAL
CORTEX

AUDITORY
CORTEX

Fig 9–14.—Functional and somatotopic localization in cerebral cortex (lateral view of left cerebral hemisphere).

tex). Cortical areas associated with speech have been used as examples here because, unlike the other instances of projection areas presented, they are predominantly unilateral, being in the left hemisphere in right-handed individuals. The presence of such areas with definite functional significance does not exclude the presence of widespread distribution of sensory and motor connections throughout the cortex and diffuse cortical connections associated with ascending and descending pathways.

In the case of somatomotor and somatosensory areas, topographic localization of cells intimately connected with particular regions of the body also occurs (see Fig 9–14). Specific visual cortical areas also correspond to particular parts of the fields of vision. Images of objects to one side of the visual axes are formed on the nasal half of the retina of one eye and the temporal half of the other retina, but nerve fibers originating from each of these regions project only to one hemisphere (see Fig 9–33). Each particular retinal area also projects to a restricted area of cerebral cortex.

REFLEXES AND SPINAL MECHANISMS

A *reflex* is an involuntary response elicited by a peripheral stimulus that is transmitted to the CNS by an afferent nerve and then to the effector organ by an efferent nerve. The reflexes involving particular neuronal connections within the spinal cord illustrate many of the integrative processes that have been outlined. From a clinical standpoint, knowledge of the presence, absence, exaggeration or diminution of spinal reflex responses is useful in determining the integrity of the various individual neural pathways involved. Detection of alterations in spinal reflexes is therefore an important part of a neurologic examination.

STRETCH REFLEXES

The muscle stretch reflex is one of the simplest of spinal reflexes to test and describe. It involves the involuntary shortening of muscle in response to a

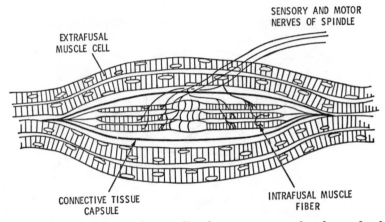

SENSORY AND MOTOR
NERVES OF SPINDLE

EXTRAFUSAL
MUSCLE CELL

CONNECTIVE TISSUE
CAPSULE

INTRAFUSAL MUSCLE
FIBER

Fig 9–15.—Diagram of muscle spindle showing encapsulated intrafusal muscle fibers. This connective tissue capsule is continuous with connective tissue lying between extrafusal fibers.

sudden increase in its length, usually brought about by tapping of its tendon. The stretch reflex should not be confused with the "tendon" reflexes associated with a decrease in muscle activity following an increase of tension in the tendons themselves. The stretch reflex is a response to a change in length of a muscle and it tends to maintain that muscle at a constant length. Stretch reflexes occur in all muscles but are particularly obvious in those involved in maintaining posture, where such a mechanism can easily be seen to be purposeful.

The basic neuronal circuitry of the stretch reflex is very simple. Lying within muscles and alongside the ordinary muscle cells are specialized receptive structures called muscle spindles. The number of spindles in a particular muscle is correlated with its function. Muscles involved in movements that require fine control have a higher density of spindles than those involved in coarser movements. Each spindle is made up of a fibrous sheath enclosing 2–12 fine specialized muscle cells called intrafusal fibers; this distinguishes them from ordinary skeletal muscle cells or extrafusal fibers (Fig 9–15).

The fine detail of these fibers will not be considered here, but there are two specialized sensory nerve endings associated with them (Fig 9–16). One of these is the annulospiral (primary) ending of a stretch-sensitive neuron,

Fig 9–16.— Sensory innervation of single intrafusal fiber.

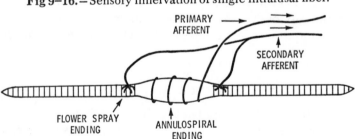

PRIMARY
AFFERENT

SECONDARY
AFFERENT

FLOWER SPRAY
ENDING

ANNULOSPIRAL
ENDING

whereas the other, the flower spray (secondary) ending, is of a different stretch-sensitive neuron. The first of these is so called because the sensory ending winds around the central portion of an intrafusal fiber. The flower spray endings lie on either side of the annulospiral ending (see Fig 9-16). Because the sheath surrounding the intrafusal fibers is continuous with the connective tissue between the extrafusal muscle cells, stretching the whole muscle stretches the spindle, and the increase in length is transmitted mechanically to the intrafusal fibers and their sensory nerve endings. The mechanical distortion leads to a conductance change in the membrane of the nerve terminals such that they depolarize and give rise to a generator potential. If this is large enough, it will initiate action potentials. There are slight differences in the way these primary and secondary endings respond. The primary endings are sensitive to changes in length and to the velocity of that change, whereas the secondary endings respond particularly to changes in length only. For the purposes of this description we shall consider only the role of the activity induced in the primary endings. The secondary endings have been pointed out because they serve as examples of how, even within a sensory modality such as "muscle stretch," specialization of individual cells can be selective to particular aspects of that modality and, of course, all of these will be involved in complex patterns of integration.

Activity induced in the primary endings is conducted along the peripheral nerve axons and enters the spinal cord via the dorsal roots. Within the cord, branches of these axons course through the gray matter and synapse directly with motor neurons, cell bodies located in the ventral horn of the gray matter (Fig 9-17). When activated, these cells initiate contraction of extrafusal muscle cells in the muscle from which the sensory information was derived. This, then, is the basic pathway for the stretch reflex and, because it involves only one interneuronal synapse, it is described as a monosynaptic reflex arc. As the muscle contracts, the spindle returns to its original length, the afferent sensory discharge decreases and this in turn reduces the excitation of the motor neurons.

When simplified schematic diagrams such as Figure 9-17, are viewed, it must be borne in mind that, in the normal physiologic state, we are invariably dealing with a population of sensory axons from many muscle spindles. Each individual sensory fiber provides inputs to some ventral horn cells by diver-

Fig 9–17.—Monosynaptic pathway for stretch reflex.

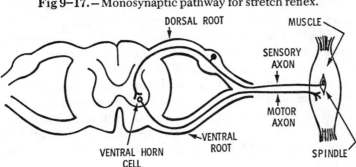

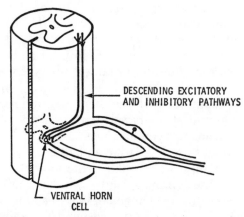

DESCENDING EXCITATORY
AND INHIBITORY PATHWAYS

VENTRAL HORN
CELL

Fig 9–18.— Descending pathways converging on ventral horn cell.

gence of its terminals within the gray matter. Within the pool of ventral horn cells receiving this input, each individual neuron also receives inputs from many sensory neurons. Thus there is both divergence and convergence. Finally, we must remember that we are not dealing with a single action potential traveling into the CNS via a single sensory axon, but trains of action potentials in many axons. Because of this divergence and convergence of similar inputs, both temporal and spatial summation takes place at the synapses.

We pointed out, when introducing examples of localization of function within the spinal cord, that the same ventral horn cells also receive inputs from descending motor fibers via the lateral corticospinal tract and that these play an important role in initiating motor movement (i.e., inducing activity in the final common path). There are many other inputs to these motor cells; in particular, both excitatory and inhibitory pathways descending from regions of the brain stem will influence the level of excitation of these cells and modify their response to other inputs such as that arriving from a stretch receptor (Fig 9–18). If descending excitatory pathways were very active, a brisk or exaggerated response might be expected in response to stretch. The same effect might be achieved by reduction of any descending inhibitory influences (dysinhibition).

ROLE OF THE GAMMA EFFERENT SYSTEM.— Before considering examples of how alterations of stretch reflex activity may be accounted for, we must consider some other features of muscle spindle function. It will be recalled that the stretch-sensitive nerve endings are intimately associated with the intrafusal muscle fibers. These fibers are also contractile and receive their own motor innervation by a set of neurons known as the gamma efferent system (Fig 9–19). This term is used because of the small diameter of their axons and to distinguish them from other motor neurons referred to previously, termed alpha motor neurons. The cell bodies of the gamma neurons also lie in the ventral horn of the gray matter. Although the source of input to the gamma motor neurons has not been completely determined, it appears likely that it may be the reticular formation of the brain stem and the extrapyramidal system (see later).

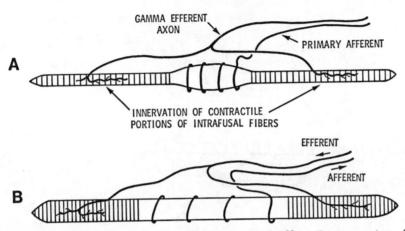

Fig 9–19.—**A,** motor innervation of intrafusal muscle fiber. **B,** contraction of intrafusal fiber causes no change in total length but stretches annulospiral ending and leads to sensory discharge.

Although the intrafusal fibers are contractile, the central region, in which the annulospiral stretch-sensitive nerve endings are located, is not. Because of the small size of the intrafusal fibers, they generate force insufficient to cause significant shortening of the total length of the spindle, particularly as the capsule is intimately connected with the connective tissue of the surrounding mass of skeletal muscle. Consequently, when the gamma efferent system is activated, contraction of the ends of these fibers stretches the central portion associated with the stretch-sensitive annulospiral ending of the primary afferent. If it is assumed that the spindle length stays constant, activation of the gamma motor system will give rise to a sensory discharge in the primary afferents, which in turn will reflexively bring about contraction of the extrafusal muscle cells and shortening of the whole muscle. This would decrease the length of the spindle and reduce the stretch of the annulospiral terminals. This, of course, leads to a decrease in activity of the afferent fibers and so reduces the excitatory drive on the motor neurons supplying extrafusal muscle fibers. The end result of gamma motor neuron stimulation would then be a shortening, owing to reflex contraction involving alpha motor neuron stimulation, and a return of the spindle afferent activity to its pre-existing level (Fig. 9–20). Initiation of movement by means of increasing gamma efferent activity has been proposed, but probably a more meaningful way of viewing this activity is as follows.

The sensitivity of the sensory endings to varying degrees of stretch is limited to a particular range; during muscle contraction the length of the whole spindle (or tension in the spindle) could vary greatly depending on the degree of shortening of the whole muscle. By altering the drive to the gamma efferent system under different degrees of stretch and length of the whole muscle, the sensitivity of the sensory endings in the spindle could be set to their optimal level for detecting either increases or decreases in length during changes in posture or movement.

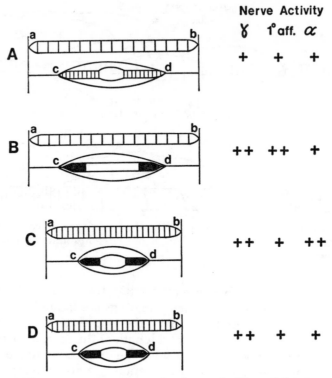

Nerve Activity

Fig 9–20.—Reflex contraction of extrafusal muscle fibers following activation of gamma efferent system. Distance *a-b* represents total length of whole muscle; *c-d* represents length of spindle. **A,** at this particular resting length of *a-b*, we shall designate a level of activity of one plus to gamma efferent neuron (γ), primary afferent stretch-sensitive neuron (1° aff.) and activity in motor neuron (α) to extrafusal muscle.

B, shortening of contractile elements of intrafusal fiber occurs in response to increase in gamma motor neuron activity. Initially, lengths *a-b* and *c-d* remain constant, but stretch-sensitive endings in central portion of intrafusal fiber are activated as spindle length *c-d* remains constant; therefore, an increase in primary afferent activity occurs. At first, no change will occur in alpha motor neuron activity.

C, increased primary afferent discharge in B leads to increased alpha motor activity, which will decrease lengths *a-b* and *c-d*. Shortening of spindle length *c-d* reduces afferent discharge.

D, reduced activity in primary afferent neuron decreases excitation of alpha motor neuron. Note that activity of afferent and efferent components of stretch reflex have returned to initial level but overall length of muscle has been reset to a shorter distance by continued gamma motor activity.

ALTERATIONS IN STRETCH REFLEX ACTIVITY

Testing a series of stretch reflexes is an integral part of a complete neurologic examination. Any alteration in responses can be used in conjunction with other findings to locate and determine any underlying pathology. The stretch reflexes commonly tested include those of muscles acting on the jaw, elbow, wrist, knee and ankle. Each represents innervation by different levels of the neuraxis. The knee jerk, for example, involves the second and third lumbar spinal levels and is the contractile response of the quadriceps muscle following a stretch stimulus to the tendon that attaches it to the tibia.

As with all clinical investigation, the entire picture has to be taken into account. Nevertheless, it is worthwhile considering some of the disorders of function that might account for abnormal reflex activity. Since there is marked individual variation in the degree of such reflexes, absence or exaggeration is not necessarily significant. Because stretch reflexes involve sensory neurons, motor neurons and muscle cells, disorders involving any of these may affect the response. For example, primary muscle diseases often lead to wasting and muscular weakness; diminution of responses would therefore be expected. Damage to the primary motor neuron can occur in specific instances, e.g., poliomyelitis, in which case muscle wasting and diminished stretch reflexes would also be expected, because these neurons form the final common path for those muscle cells. For the same reason, voluntary movement would also be expected to be weaker. In disorders of peripheral nerves, other types of sensory loss as well as diminished stretch reflexes would be likely to be apparent, since these disorders do not show much selectivity among axons subserving different functions.

Although the information from muscle spindle activity is conducted to higher regions of the CNS, it does not reach consciousness. Other proprioceptive (internal sensory) mechanisms exist for detecting movement and position sense. Localized damage at the appropriate spinal segmental level would affect not only reflex responses but sensory perception and voluntary movement in those segmental areas, and, as we shall see, reflex responses at lower levels.

Because of the descending pathways that impinge upon ventral horn cells, damage or malfunction within the CNS at a segmental level above that involving the reflex can markedly affect the response. Damage to such descending pathways often leads to marked increase in the reflex response owing to a reduction of inhibitory influences. During sleep most stretch reflexes are absent because of a decreased level of excitation within the cord. Often when reflexes cannot be elicited, it is possible to demonstrate their presence by asking the subject to grasp his hands together and try to pull them apart. Such a procedure (Jendrassik's maneuver) greatly facilitates all motor neuron responses in the cord. This phenomenon is termed reinforcement.

Alteration of reflex activity by disorders higher in the CNS is invariably associated with other disordered functions. The total picture depends on the exact location and pathologic nature of the lesion.

Disorder of areas related to control of the gamma efferent system alters reflex responses and also muscle tone, which is very dependent on normal stretch reflex pathways as well as on the gamma efferent system.

RECIPROCAL INHIBITION

If contraction of muscle is to produce movement, the muscles that cause the opposite movement (antagonists) must relax. For example, contraction of the

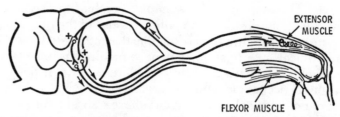

Fig 9–21.—Reciprocal inhibition. Diagram to show how antagonist muscles are inhibited during muscle contraction. Inhibition of flexors during an extensor response is brought about by further divergence of the afferent neuron to an inhibitory inter-neuron, which synapses with ventral horn cell of a flexor muscle.

quadriceps muscle during a knee jerk is accompanied by relaxation of the hamstring muscles. To achieve this, the primary afferent neurons from stretch receptors in a muscle synapse not only with the motor neurons supplying that muscle, but also indirectly with the motor neurons of the antagonist muscle. The pathway is indirect because, if these motor neurons are to be inhibited, an interneuron must be interposed to bring about a change from excitation to inhibition (Fig 9–21). In a reflex involving contraction of extensor muscles, inhibition of the flexors occurs and it should again be remembered that the particular alpha motor neurons receiving the inhibitory input are the same involved in a monosynaptic stretch reflex for that muscle. Likewise, the exten-sor motor neurons receive a similar inhibitory input that is activated during

Fig 9–22.—Examples of input-output relationships at spinal level. **A,** sensory input making direct synaptic contacts with motor neurons at a restricted segmental level but ascending and descending connections to interneurons. **B,** sensory input making synaptic connections with motor neurons at many segmental levels but via an interneuron. (Modified from Ramón Y. Cajal.)

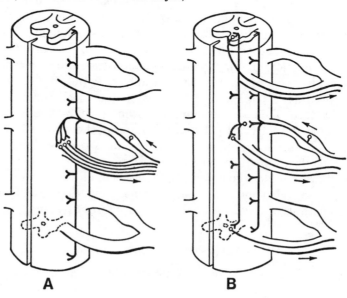

stretch reflex of the flexor muscles. These interneurons also diverge to make synaptic connections with pools of neurons, as outlined previously.

Using specific staining techniques that selected individual neurons and reconstructing what could be seen in many serial sections of the spinal cord, Ramón Y. Cajal (1852–1934) was able to shape common patterns of connections for certain motor, sensory and interneurons, although their specific functions were not necessarily known. Figure 9–22 is a modification of one of Cajal's diagrams showing how in some instances (see Fig 9–22, A) direct connections between sensory inputs and motor outputs may be restricted to a limited segmental level, although the primary input diverges to provide ascending and descending pathways innervating interneurons at many segmental levels. In other cases (see Fig 9–22, B), widespread connections to motor outputs at many segmental levels may be made via interneurons.

Some idea of the importance of interneurons in the CNS can be gained by looking at the numbers in the spinal segment of the cord. For each spinal segment there are approximately 375,000 nerve cell bodies in the gray matter, the number of sensory neurons entering the cord at each segment is around 12,000 and only about 6,000 motor neurons leave via the ventral roots. Most of the neurons therefore are interneurons.

WITHDRAWAL OR FLEXION REFLEXES

Withdrawal or flexion reflexes further exemplify integration of motor responses to sensory stimuli. We are all aware that painful or noxious stimuli, such as treading on a pin or touching a hot object, leads to a sudden withdrawal of the appropriate limb. The pathways involved in these responses are not simple monosynaptic pathways such as that for the stretch reflex; at least one interneuron is interposed between sensory and motor neurons (Fig 9–23, A). During excitation of the flexor muscles, inhibition of the extensor muscles (reciprocal innervation) occurs because of the participation of an inhibitory interneuron in the reflex (see Fig 9–23, B). It can easily be appreciated that if we were considering flexion and withdrawal of the lower limb in response to stepping on a pin, many flexor muscles both at hip and knee would be involved. Also, if the body is to retain its balance, other muscle groups must be activated to take the redistribution of weight, particularly the contralateral extensor muscles. This can be achieved by divergence of the sensory input to an additional excitatory interneuron that crosses to the opposite side of the cord and synapses with appropriate ventral horn cells (see Fig 9–23, C). In association with this crossed extensor response, the contralateral flexor muscles will be inhibited (see Fig 9–23, D). The pathways illustrated in Figure 9–23 are the simplest required to integrate and coordinate such activity. As indicated previously, at each of these synaptic sites pools of neurons participate, and temporal and spatial summation of activity in these and other pathways determine the final behavioral response.

The ability of the spinal cord to integrate various sensory inputs and translate them into purposeful motor responses depends on specific patterns of connection between many different elements, each serving special functions.

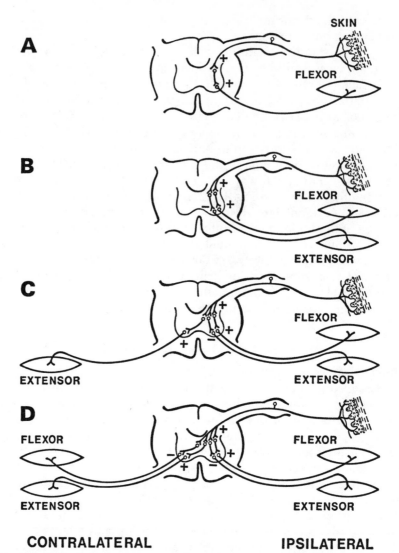

CONTRALATERAL   **IPSILATERAL**

Fig 9–23.—Pathways for flexor and crossed extensor reflexes. **A,** excitatory inter-neuron between sensory input and motor output to ipsilateral flexor muscle. **B,** recip-rocal innervation of ipsilateral extensor via inhibitory interneuron. **C,** excitation of contralateral extensor by excitatory interneuron. **D,** inhibition of contralateral flexor by inhibitory interneuron.

Activity in each element may be utilized for other integrative purposes and at other sites. Much of our understanding of the principles of integration at the spinal cord level comes from the work of Sir Charles Sherrington (1857–1952). It is a tribute to his genius that these principles were discovered before our knowledge of the cellular mechanisms of synaptic integration. Almost all of Sherrington's work involved well-designed experiments, with careful ob-servation of mechanical responses to applied stimuli. By appreciating how simple integrative processes can be achieved by such patterns of connection, it is not difficult to envisage how even more complex behaviors might occur, even at the spinal level.

SENSORY SYSTEMS

Consideration of the sensory pathways within the CNS is important not only for clinical diagnosis, but for insight into how the specificity of neurons and their synaptic connections allows coding of information. This coding can be explained by the type of synaptic integrative mechanisms previously presented.

Our current understanding of central sensory mechanisms is much better than that for motor control. This is partly because experimentally it is much easier to apply controlled sensory stimuli, similar to those that occur physiologically, than controlled "physiologic" movements. This is particularly true for the visual system where stimuli such as light intensity, color, size, shape and position relative to the body can be controlled to a fine degree. It is useful for the student of medicine to realize that this particular special sense is not only one of the most important perceptual senses, but it plays a significant role in many forms of behavioral response. Neurologic examination of the visual system reveals information regarding the integrity of much of the nervous system as well as of other physiologic systems. Our knowledge of the cellular integrative mechanisms in the visual pathways is primarily due to the beautiful work of David Hubel and Torsten Wiesel. Their findings show many similarities to the information processing and coding characteristic of the somatic sensory systems. Before considering the integration of neural activity associated with visual perception, certain aspects of the physiology of peripheral receptors and their pathways will be described.

SENSORY MECHANISMS IN THE PERIPHERAL NERVOUS SYSTEM

Peripheral sensory receptors frequently show marked structural specialization and are often associated with other complex anatomical structures. These may play an important role in the transduction process, i.e., conversion of physical stimuli such as touch, pressure and heat into neural responses, the action potentials. We saw this in relation to the stretch receptor and muscle spindle, but the term somatosensory system is generally used for the sensory systems (excluding the special senses) that are eventually perceived consciously. The eye and ear provide classic examples of this anatomical specialization with respect to visual and auditory receptors. Another example is the Pacinian corpuscle, a small structure that envelops the terminal of a particular type of sensory neuron and, because of its microanatomy and physical properties (Fig 9–24) makes this sensory nerve very sensitive to rapid, small fluctuations in pressure (vibration). Other pressure-sensitive sensory endings also play a role in "vibration" sense. They, too, may show definite structural specialization, e.g., Merkel's disks, Krause's end-buds, but it is not always possible to describe how such specializations relate to function.

CLASSIFICATION AND SPECIFICITY

Whether they reach consciousness or not, peripheral receptors are often classified as *exteroceptive* if they relate to sensation elicited on the surface of the body in response to external stimuli; *proprioceptive* if they relate to sensory information such as joint position sense, muscle contraction and tension; *visceral* if arising from internal organs and *deep sensory* if related to structures lying deep to the skin. This last category would include pressure receptors in tissues and/or in relationship to bone, as well as certain pain receptors (Table 9–2).

Sensory receptors may also be classified according to their response to a particular type of stimulus. It is a general rule that each particular primary sensory neuron will respond only to its specific modality of sensation, e.g., a mechanoreceptor will not be excited by thermal stimuli. Obviously, if certain types of stimuli are great enough, damage to that individual nerve axon may lead to an initial burst of activity; similarly, electric stimuli could initiate activity. However, within normal physiologic ranges, activity will not be induced except by a specific mode of stimulus. Should activity arise in any particular nerve fibers from nonphysiologic stimuli, and should it reach consciousness, it would be perceived as the sensation to which the nerve ordinarily responds. The site of origin to which the sensation is referred will depend on the distribution of the nerve axon or axons affected. It is not uncommon for people who have had limb amputations to complain of sensation, possibly

TABLE 9–2.—CLASSIFICATION OF PERIPHERAL
RECEPTORS°

According to Site	
Exteroceptive	Superficial receptors in skin (touch, pressure, pain, warmth, cold)
Proprioceptive	Receptors within muscles and joints; responsive to stretch, tension or movement
Viceral	Receptors within viscera of other organs; usually responsive to stretch or distention
"Deep"	Deep pressure receptors within tissues or in association with bone
According to Function	
Mechanoreceptors	Responsive to touch, pressure, stretch or mechanical deformation
Thermoreceptors	Responsive to temperature changes: cold receptors—increased neural activity on cooling; warm receptors—increased neural activity on warming
Chemoreceptors	Oxygen and carbon dioxide receptors in aortic and carotid bodies
Nociceptors	Pain receptors

°There is overlap between the two types of classification, and there are examples of some of these and other receptors within the CNS and in relation to the special senses (e.g., light receptors).

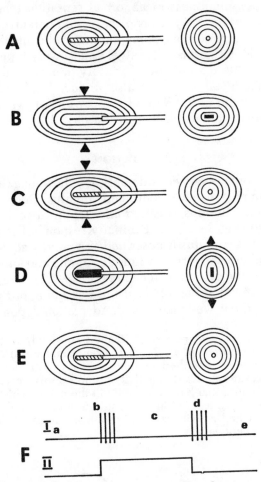

Fig 9–24.—Diagram of firing patterns of a rapidly adapting sensory receptor, the Pacinian corpuscle. Trace *F, I* represents electric activity in nerve; trace *F, II* indicates timing of applied stimulus. Diagrams **A, B, C, D** and **E** show longitudinal and cross-sectional views of capsule during different stages of compression and removal of stimulus. The corresponding firing patterns can be seen in **F,** *Ia, b, c, d* and *e,* respectively. *a,* in absence of deformation of corpuscle and nerve ending, no activity is seen in nerve. *b,* deformation following applied mechanical stimulus in one plane deforms capsule and terminal and initiates activity in nerve. *c,* if stimulus is maintained, pressures within capsule are able to redistribute themselves such that, although external shape of capsule remains distorted, terminal reverts to its original shape and no activity occurs in axon. *d,* removal of stimulus allows redistribution of forces (elastic and hydraulic) within capsule, again causing distortion of terminal; this again gives rise to activity in nerve. *e,* pressures redistribute to allow return of terminal to its original shape, and no neural activity is elicited.

pain, in the nonexistent limb. This is probably partly explained by abnormal activity in the peripheral cut ends of the nerves originally supplying the limb. A related phenomenon is encountered by most of us at one time or another: "pins and needles" is usually the result of temporary occlusion of blood supply to a peripheral nerve caused by local pressure. Initially, there may be marked sensory loss, numbness or abnormal sensation (paresthesia) in the area of distribution; as recovery takes place, action potentials may be initiated, giving rise to bizarre sensory perception from the affected area. It is bizarre because we do not normally receive, as a sensory input, that particular qualitative and quantitative mixture of sensory information. Diseases affecting peripheral nerve metabolism may also produce localized disturbances of sensation. These may be in the form of sensory loss or, when the damage or degeneration is actually taking place, paresthesia.

ADAPTATION

Peripheral sense organs vary considerably as to how long they will generate action potentials in response to a continued stimulus. Decreased responsiveness to a continuous stimulus is called *adaptation*. A receptor, such as the Pacinian corpuscle (Fig 9–24), that initially responds but then fails to respond to a continuous stimulus is described as a rapidly adapting or "phasic" receptor. On the other hand, primary muscle spindle and joint position receptors are slowly adapting or "tonic" receptors (Fig 9–25). They show a slightly higher frequency of action potentials at the beginning of the response, but thereafter the response is well maintained. Adaptation depends on specializa-

Fig 9–25.—Diagram of firing pattern (**A**, **B** and **C**) of slowly adapting receptor, the primary afferent of stretch receptors. Trace **D** *I* represents electric activity in nerve; trace **D** *II* indicates timing of applied stimulus. *a*, at any particular length of the ending, there may be background level of activity in the sensory nerve. *b*, following applied stretch, there is an initial burst of activity in the nerve, which then declines to new but higher background level of firing. *c*, on return to original length, there is an initial period when neural activity is even less than original at that length; then it returns to same background level.

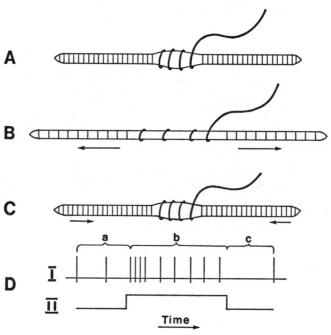

tion of sensory terminals and their associated structures, as well as on features of accommodation of excitable cell membranes. Phasic receptors play a major role in signaling *changes* in a particular mode of sensation, whereas tonic receptors signal information regarding steady states. Adaptation should not be confused with our ability to suppress perception of sensory information until attention is drawn to it. Central sensory mechanisms are responsible for this.

Receptive Fields

The sensory area supplied by an individual neuron is called the receptive field for that neuron. The relationship is analogous to that between a dermatome and the spinal dorsal root. For touch receptors in the skin there is considerable overlap of receptive fields (Fig 9–26). Moreover, the size of the fields varies considerably in different locations on the body surface, e.g., they may be small (millimeters) in the finger tips or centimeters in diameter in areas over the back. This is what we might expect if we are to have fine tactile discrimination limited to certain areas. The overlapping of receptive fields would at first sight appear to cause a loss of discrimination. Overlap is a common feature of most sensory receptive fields but, as we shall see, synaptic connections are such that discrimination is maintained or even enhanced.

One neural mechanism used to achieve this is termed lateral or afferent inhibition. Figure 9–27 shows a simplified scheme of lateral inhibitory connections at the first synaptic level of a main sensory pathway. All three overlapping receptive fields are supplied by an individual primary sensory neuron. Each of these is shown making straight-through excitatory connections with a second order sensory neuron. In addition, each first order neuron excites an inhibitory interneuron, which is shown producing presynaptic inhibition of the neurons associated with the adjacent receptive fields. In Figure 9–27, for the sake of clarity, they are shown as presynaptic inhibitory pathways; they could be making postsynaptic inhibitory connections with the second order neurons. Let us consider the following possible situations. First, discrete stimulation of the central portion of receptive field 2, will enhance the activity of second order sensory neuron 2. At the same time it will, by means of the inhibitory interneurons, diminish activity likely to occur in second order neurons 1 and 3; there is then marked "contrast" in activity between second order

Fig 9–26.—Diagram illustrating overlapping of cutaneous receptive fields of three first order sensory neurons (see text for description).

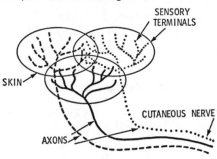

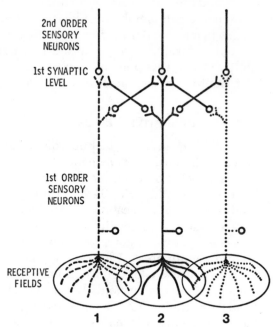

2nd ORDER
SENSORY
NEURONS

1st SYNAPTIC
LEVEL

1st ORDER
SENSORY
NEURONS

RECEPTIVE
FIELDS

1   2   3

Fig 9–27.—Diagram of lateral inhibition in sensory pathways (see text for description).

neuron 2 compared with 1 and 3. If a larger area of the receptive field is involved, this "contrast" becomes less and localization is less discrete. Lateral inhibitory connections do not exclude convergence or divergence of neurons at this or other levels. The presence of such interneurons allows a further integrative possibility. Descending, or other, inputs (not shown in Fig 9–27) to the interneurons could further modulate the influence of one order of neurons on another.

Lateral inhibition will also be encountered in our consideration of integration at the different stages in the visual pathways.

ASCENDING SENSORY PATHWAYS

We have already given some indication of the functional and somatotopic localization of sensory pathways within the CNS (see Figs 9–12 to 9–14). There are two major pathways within the spinal cord for somatosensory information to ascend to the cerebral cortex: (1) the anterolateral or spinothalamic system, and (2) the lemniscal or dorsal column system. A knowledge of their function and location is extremely important in neurologic diagnosis.

ANTEROLATERAL SYSTEM

Primary neurons carrying information regarding temperature, pain or gross aspects of touch enter the spinal cord along with other sensory neurons via the dorsal roots. Some of these fibers give off collateral branches that synapse in the dorsal horn of the gray matter and eventually play a role in reflex path-

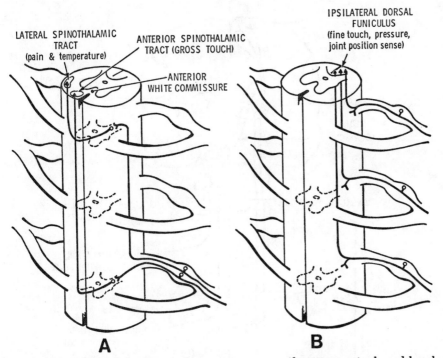

Fig 9–28.—Major subdivisions of somatosensory pathways at spinal cord level. **A**, anterolateral system. Anterior spinothalamic tract carrying gross touch fibers and lateral spinothalamic tract carrying pain and temperature fibers. First order neurons may ascend a few segments before synapsing, and second order neurons cross to contralateral side. **B**, dorsal column or lemniscal system. Branches of first order neurons ascend in ipsilateral dorsal funiculus.

ways. In addition, many branches of fibers carrying pain and temperature sensation ascend for a number of segments before synapsing with second order sensory neurons. All these second order neurons then send axons to the contralateral side of the cord, crossing corresponding fibers in transit from the other side in the anterior white commissure. Having crossed, these axons then ascend in the lateral spinothalamic tract (Fig 9–28, A). Fibers carrying information regarding gross aspects of touch may synapse with interneurons at the level of entry, but they also synapse at many levels (above and below) with second order sensory neurons. These also cross via the anterior white commissure and finally ascend in the anterior (ventral) spinothalamic tract (see Fig 9–28, A). Within the anterolateral columns of white matter are other ascending sensory systems carrying information on similar modalities, which project to different regions of the brain stem and do not reach consciousness. They will not be discussed here.

LEMNISCAL SYSTEM

The primary sensory fibers carrying information regarding fine touch, pressure and joint position sense also enter the cord via the posterior root. Branches of these primary sensory neurons turn rostrally and enter the dorsal funicu-

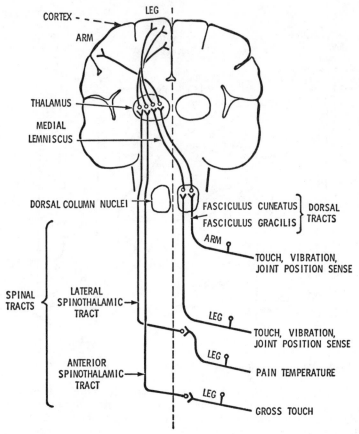

Fig 9–29.—Diagram summarizing somatosensory pathways and indicating relay sites for three orders of sensory neurons. Some aspects of functional and somatotopic localization are also shown.

lus. These axons are still those of first order sensory neurons and have not yet synapsed. They ascend on the same side of the cord as their origin (see Fig 9–28, B) and synapse with second order sensory neurons in the dorsal column nuclei (n. gracilis and n. cuneatus) at the junction of spinal cord and medulla. These second order sensory neurons send axons through the brain stem as a medial ribbon of fibers (lemniscus=ribbon), which cross over in the medulla (sensory decussation) to terminate on third order sensory neurons in the thalamus, a large nucleus in the diencephalon. During their passage through the midbrain, the fibers of the medial lemniscus meet the fibers of the anterior and lateral spinothalamic tracts. Thus all this ascending sensory information eventually terminates on the contralateral side of the brain. From the thalamus, thalamocortical fibers radiate out to the somatosensory cortex of the parietal lobe (Fig 9–29).

In summary, these sensory pathways involve three ascending orders of neurons and result in contralateral representation of the body in the cortex. These pathways also demonstrate definite functional localization and, in the lemniscal system in particular, very definite topographic localization.

CLINICAL CONSIDERATIONS OF SENSORY PATHWAYS IN SPINAL CORD

Pain, temperature, vibration and joint position sense are useful modalities for clinical diagnosis because they ascend in discrete pathways. Fibers carrying touch sensation ascend in both the dorsal columns (lemniscal system) and the spinothalamic tracts (anterolateral system). Although the fine discriminatory aspects of touch are mainly mediated by the lemniscal system, disorders of either pathway alone may make it difficult to detect disorders of tactile sensation because of cord lesions. For example, the main features of a lesion causing hemisection of the cord (Brown-Séquard syndrome) would be (1) loss of joint position and vibration sense on the same side as the lesion and below it; (2) loss of pain and temperature sense on the opposite side of the body but, because these fibers ascend a few spinal segments before crossing, the levels of the different modality losses for each side might be different; (3) alteration of the fine discriminatory aspects of touch sensation on the same side; (4) loss of voluntary movement on the same side and (5) depending on the type and duration of the damage, possible exaggeration of reflexes on the same side below the level of the lesion, owing to loss of descending inhibitory pathways to the alpha motor neurons.

One form of neurosyphilis, tabes dorsalis, particularly involves the central axonal processes of dorsal ganglion cells. Those axons in the dorsal funiculus are extremely long compared with those that synapse in the dorsal horn and, perhaps for this reason, are more markedly affected. Therefore, the degeneration that occurs selectively affects sensations such as joint position sense and vibration sense. The latter is detectable clinically and the former will lead to particular disturbances of movement and locomotion owing to loss of proprioceptive information.

Two other conditions involving the spinal cord are worth mentioning at this stage. One of them, *syringomyelia,* is of unknown etiology but frequently involves the cervical and upper thoracic regions of the cord. The nature of the lesions is related to abnormal development of glial tissue and cavitation in regions surrounding the central canal. As the sensory fibers serving pain and temperature cross in the anterior white commissure, loss of these sensory modalities may occur in the hands and fingers. Subsequent compression of ventral horn cells leads to weakness in the upper limbs.

The other condition is produced by *vitamin B$_{12}$ deficiency.* This vitamin is essential for proper maturation of bone marrow cells and for nutrition of the nervous system. Consequently, B$_{12}$ deficiency commonly causes anemia and neurologic disorders. In the cord, long fiber tracts are particularly affected and degeneration of descending corticospinal and ascending posterior column tracts may lead to both motor weakness and sensory losses. However, peripheral nerves are also involved, and clear-cut signs and symptoms related to specific localization of malfunction are not necessarily apparent. This should remind the student that diagnostic features dependent on very exact localization of the pathologic lesions are not necessarily as apparent as the previous descriptions might lead us to believe. After all, lesions are not in themselves necessarily very localized.

INTEGRATIVE ACTIVITY AT SYNAPTIC LEVELS

So far we have described how activity in a single neuron indicates a particular aspect of a simple modality of sensation. The intensity of that sensation is related to the amount of its activity, i.e., it is frequency coded. Convergence of activity of these lower order neurons onto higher order neurons at successive synaptic levels allows such information to be coded in additional ways. This is best understood by a consideration of what happens at successive relay levels in the visual pathways.

RETINAL INTEGRATION

It will not be necessary here to consider in detail the anatomy and physiology of the eye itself other than to emphasize certain important features. The eye focuses an inverted and reversed image of the exterior world on a layer of light-sensitive neurons on the inside and back of the eye. These photoreceptors form the outermost of the three nerve cell layers that go toward making the retina. It is in this receptor layer that light energy is transduced to neural activity, which is transmitted via a synaptic layer to a second order of neurons called bipolar cells; these cells in turn synapse with a third layer of neurons,

Fig 9–30.—Diagram of three basic neural layers of retina. Light passing through neural layers initiates activity in photoreceptor layer, which synapses with bipolar cells; these in turn synapse with ganglion cells. Axons of ganglion cells form optic nerve and enter the brain.

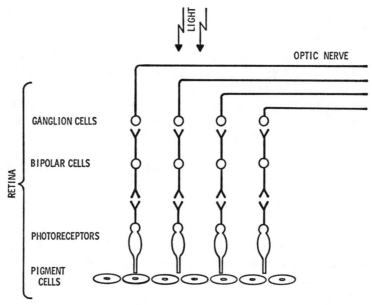

called the ganglion cells, from which arise the axons that form the optic nerve and conduct information into the brain (Fig 9–30). We should note that in the case of vision the possibility for integrative activity at two synaptic levels exists before sensory activity is conducted through other central pathways to the cortex. These central pathways are analogous to those of the somatic sensory system. The axons in the optic nerve synapse with a further order of neurons in the lateral geniculate body, which sends axons to synapse with cells in the visual cortex.

Additional neurons also occur in the retina: the horizontal and amacrine cells. They play an important role in forming lateral connections between the three main neuronal layers, and any critical description of ganglion cell activity must take their role into account. However, it is possible to illustrate some general principles of connecting pathways and the activity within ganglion cells, without reference to these; such an approach will be taken here.

Our understanding of the way in which patterns of light and shade falling on the retina can be analyzed at successive synaptic relay stations is predominantly due to the studies of Hubel and Wiesel—one of the most significant advances in our conception of integrative neural mechanisms since the work of Sherrington. Surprisingly perhaps, it involves the same general concepts of cellular interaction and connections. The work of Stephen Kuffler on the patterns of neural activity in the retinal ganglion cells first showed that analysis of visual information starts in the retina. This account also will start by considering the type of neural response that can be observed at the ganglion cell level (i.e., two synapses from the receptor).

Each ganglion cell has a receptive field, an area of retina within which light stimuli will evoke alterations in its firing pattern. Because of the optical properties of the eye, particular points on the retina correspond, of course, to a particular position in space outside the body from which the light stimulus emanates. This projection on the retina is called the visual field. Each ganglion cell receptive field then relates first to a specific area of retina and indirectly to a particular part of the visual field.

The receptive fields of ganglion cells are predominantly of two types. They are mapped by discretely recording the electric activity of a single cell by means of a fine microelectrode and at the same time shining small pencils of light on the retina.

The first type of ganglion cell field is one in which the cell responds by a marked increase in frequency of action potential discharge when the light beam is restricted to a circumscribed disk falling on the retina in the vicinity of that ganglion cell. In the periphery of the retina, this area may be as large as 1 mm, which would include thousands of receptor cells. In the central, or macular, region of the retina, the receptive field size is much smaller, but the main feature is that almost all ganglion receptive fields relate to a considerable number of receptors and intermediary bipolar cells. Ganglion cells that are excited by discrete disks of light show certain other features as well (Fig 9–31, A). If the light falls on the area surrounding this "on" region, the response of the ganglion cell is markedly diminished. Indeed if the light only falls into the concentric surround region, the ganglion cell responds with an

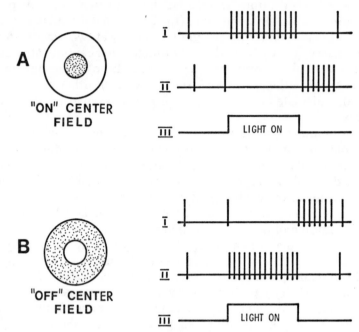

Fig 9–31.—Receptive fields of ganglion cells. In both **A** and **B**, trace *I* indicates firing pattern of ganglion cell when light falls on central part of field; trace *II*, pattern when light falls in surround area; trace *III*, duration of light stimulus. Receptive field **A** has a central "on" response, whereas light falling in surround area decreases firing frequency, but there is marked increase in firing frequency when light goes off in this area. Responses in receptive field **B** are reversed.

even lower frequency of firing than in the absence of any light. The whole receptive field can be described as having an "on" center and an "off" surround. Equal illumination of the entire area evokes only a very weak "on" response.

The other type of ganglion cell receptive field is the reverse of this. It has an "off" center and an "on" surround (see Fig 9–31, B), and general illumination of the whole area produces only a weak "off" response.

Before considering what "meaning" we may associate with the activity of a single ganglion cell, can we explain how such responses may come about in terms of what we know regarding the cell structures within the retina, the connections we can detect histologically and our knowledge of basic mechanisms of actions between neurons? Knowing the diameter of individual receptor cells and taking the central part of the receptive field as being 1 mm, we could calculate that 50,000–100,000 receptors would be activated by such a disk of light falling on the retina, the whole receptive field enclosing almost a million "individual" receptors. Even if this calculation were to be an overestimate of one or two orders of magnitude, it would be impossible to draw an exact representation of the pattern of connections. We do know from histologic sections that many receptors synapse with a single bipolar cell and many bipolar cells make contact with a single ganglion cell. Schematically then, we can draw a diagram showing how appropriate connections can account for the "on" response of the central part of the receptive field (Fig 9–32) and for the inhibitory surround.

Receptors in the "on" center connect to the ganglion cell by converging on excitatory bipolar cells, which in turn converge and excite the ganglion cell. Although we cannot directly show this for the individual ganglion cell under consideration, such pathways can be demonstrated histologically and physiologically and therefore supply us with a plausible explanation for the phenomena. Our "off" center cells can be explained by a similar pattern of connections, but with reversal of the excitatory and inhibitory natures of the pathways.

When viewing such a diagram, we must also realize that the same receptor cells and bipolar cells are involved in the receptive fields of other ganglion cells.

We can therefore provide a rational explanation to account for this type of response, but how does this fit with the way we "see" things at a perceptual level? We do not, after all, see the world as made up of light and dark doughnuts. Is there, in fact, anything that we perceive that can be related to these findings? There are two features of such a pattern of connections and their activity that relate to some general features of vision.

First let us consider what the response of a ganglion cell would be if the whole receptive field were illuminated. Whether it was an "on" center cell or an "off" center cell, very little activity would be conducted to the brain, as the effect of illuminating the surround would cancel out most of the response due to central illumination. Such a light stimulus causes a slight response, which is determined by the field center, i.e., in an "on" center field, there is a slight

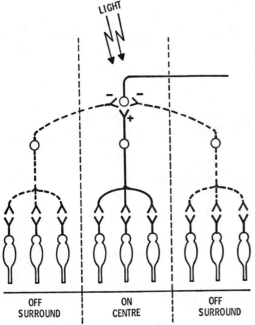

Fig 9–32. — Diagram showing connections that would account for an "on" center, "off" surround field of retinal ganglion cell.

increase in the level of activity in that ganglion cell, but general illumination of the whole retina produces minimal input of activity to the brain. As we shall see, owing to the connections in subsequent synapses, the neuronal activity following such a stimulus becomes less and less until at the cortical level hardly any cells show an increase in activity to such a stimulus.

This phenomenon, in fact, relates to more general aspects of visual perception, the second feature of the pattern of connections referred to earlier. We all know that it is not diffuse general visual stimuli that "excite" us but, rather, patterns of light and shade—in other words, the contrast between light and dark areas. The receptive fields of ganglion cells respond best when there is a contrast of illumination between the different parts, but still these connections do not appear to extract "form." In order to get some insight into how this may come about, we must follow the activity in these ganglion cells to higher levels in the CNS.

LATERAL GENICULATE INTEGRATION

The next synaptic relay station for the ganglion cell axons that make up the optic nerve is the lateral geniculate body or nucleus (LGN), a part of the thalamic complex. Optic nerve fibers from both eyes terminate here. The particular optic nerve fibers that reach the LGN of any one side are those arising from the temporal half of the ipsilateral retina and the nasal half of the contralateral

Fig 9–33.—Diagram of basic visual pathways from retinal ganglion cells to visual cortex. Note that nerve fibers of ganglion cells in temporal half of each eye do not cross to contralateral side, whereas fibers originating from nasal halves of retina cross in optic chiasma. Because the image falling on retina is reversed, visual stimuli arising to either side of visual axis *(VA)* of each eye will initiate activity in contralateral visual cortex.

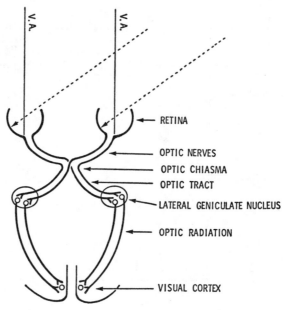

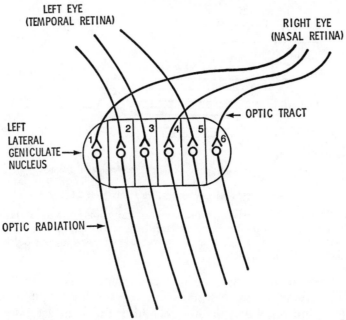

Fig 9–34.—Simplified diagram to emphasize layered cell structure of lateral geniculate nucleus and separation of inputs from contra- and ipsilateral sides of origin.

retina (Fig 9–33). Thus the right visual fields are served by the left LGN and vice versa. A knowledge of the gross aspects of the visual pathways and the relationship of each part of the pathway to the visual fields is very important for localizing damage. Differing visual field defects will arise depending on whether the pathway is interrupted at the retina, optic nerve, optic chiasma, optic tract or optic radiation levels. The student might find it interesting to predict the visual field losses to be expected following damage at these various levels and to verify them by reference to a standard neurologic text.

The LGN receives information regarding events occurring in the visual field on the opposite side of the visual axes of both eyes. The cells in this nucleus are arranged in layers, and almost all cells in each layer have connections only from one eye and not from the other (Fig 9–34). A few cells do receive an input from both retinas but this is not the general rule.

In general, the response patterns and receptive fields of lateral geniculate cells are similar to those described for ganglion cells, and, as might be expected from other features of the nervous system, we find neighboring lateral geniculate cells topographically arranged so that their receptive fields are located close to other neurons with receptive fields from similar parts of the retina. However, they definitely differ in one respect, as has been alluded to: they show even less response when the whole receptive field is illuminated; in other words, they are even more sensitive to contrast. Again, can we provide an explanation of how this is brought about? If we look at Figure 9–35 we could explain it by assuming that ganglion cells having similar but overlap-

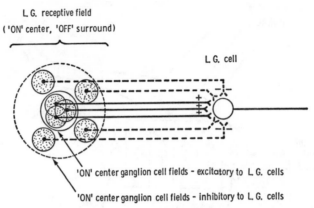

L. G. receptive field
('ON' center, 'OFF' surround)

L. G. cell

'ON' center ganglion cell fields - excitatory to L. G. cells

'ON' center ganglion cell fields - inhibitory to L. G. cells

Fig 9–35.—Diagram of possible connections of ganglion cells to lateral geniculate *(LG)* cells to account for enhanced lateral inhibition. In diagram only "on" centers for a few ganglion cell fields are shown; the mosaic of overlapping fields would be more marked than indicated.

ping receptive fields converge to excite geniculate cells. In addition, other ganglion cells having the same type of "on" center, but which overlap with the "off" surround regions, converge to inhibit the same lateral geniculate cell. The converse explanation would apply for an "off" center geniculate cell.

So far we have been considering the cat visual system, which is not particularly sensitive to color. The monkey appears to have well-developed color perception, as does man, and a brief consideration of the types of responses obtainable on monkey lateral geniculate cells will demonstrate an additional feature of integrative activity related to contrast. In these geniculate cells, we can find examples in which the "on" or "off" center of the receptive field responds best to light of one wavelength, e.g., red, whereas the complementary surround producing the opposite response is most sensitive to green light. Such a finding for lateral geniculate cells suggests that the connections are established not only according to spatial relationships but also by selectively receiving inputs from ganglion cells with red or green photoreceptors in their receptive fields. The reader will appreciate that once again it becomes increasingly difficult to show by means of a diagram how this could be brought about, but similar patterns of connections involving convergence of different inputs can be conceptualized.

To illustrate the relationship between an increase in visual contrast and the use of color-sensitive visual fields, anyone interested in photography will appreciate that, in black and white photography, particular attention has to be paid to lighting and shadows, whereas in color photography, as long as the lighting is of adequate brightness, contrast by illumination levels is not as important. We can appreciate the difference in clarity when watching a football game, or a surgical operation, between color and black and white TV.

Although a number of types of geniculate cells and their differing receptive fields have been described, the foregoing examples will suffice for our purposes. However, none of them appears to respond in a way that can be directly associated with form. Can we continue with this type of analysis, investigating the types of responses to be obtained at the next synaptic stage, and will we be able to use the same concepts of convergence of excitatory and inhibitory pathways on subsequent postsynaptic cells to explain our findings?

CORTICAL INTEGRATION

The next synaptic site for the lateral geniculate cells is area 17, or the striate area of the visual cortex. The geniculate fibers arrive at this site via optic radiation, or the geniculocalcarine tract. Hubel and Wiesel considered such an analysis of activity in these cortical cells and are still following this line of enquiry.

Their results show that cortical cells exhibit a wide variety of types of receptive fields, quite different from those exhibited by retinal ganglion or geniculate cells. As a general feature, hardly any cortical cells respond to general diffuse illumination of the retinas. The stimuli that many cells seemed to "prefer" were moving lights, often with a particular direction of movement.

One finding is that a small spot of light shone on a particular area of the retina will produce a very small "on" response in one type of cortical cell. If the same spot of light is shone on an adjacent area of retina, it too can produce an "on" response in that cortical cell, but only when there is a particular spatial relationship between the two areas. Moving the illumination to another adjacent area can again produce an "on" response, but only when all such adjacent areas lie in a straight line. The axis, or orientation, of this line in the visual field will be referred to shortly. The total "on" receptive field for that cell thus forms a straight line. Parallel to that "on" line are adjacent "off" areas (Fig 9–36, A). We could create such a receptive field by a series of very specific connections of ganglion cells with doughnut-shaped fields, to geniculate cells with similar receptive fields and so to cortical cells by convergence of

Fig 9–36.—**A,** receptive field of a simple cortical cell. **B,** such a field could be accounted for by convergence of lateral geniculate cells, whose receptive fields were linearly adjacent, with respect to the retina, to a single cortical cell.

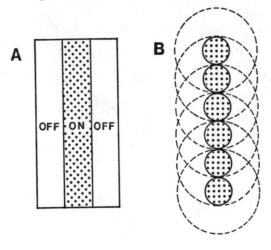

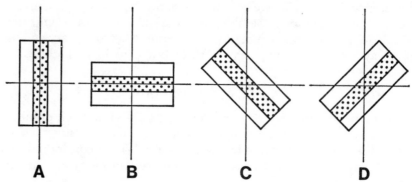

Fig 9–37.—Series of receptive fields of four individual simple cortical cells, each with "on" bar center and "off" surrounds but with separate axis orientation.

lower level fields related to particular parts of the retina (see Fig 9–36, B).

The type of cortical cell we have just described responds maximally to a line, or bar, of light falling on adjacent regions of the retina. We are now starting to find cells whose "preferred" stimuli are related to form, and this correlates well with the observation that destruction of this part of the visual cortex leads to loss of form vision. An even more interesting feature is that these cells not only respond most effectively to bars of light, but the orientation of the bars is equally important. Thus some cells respond only when the bar is vertical, others when it is horizontal and still others at other particular orientations (Fig 9–37). Again, "on" bar or "off" bar cells can be found.

Although what is observed depends on many cells, we have throughout been trying to assess the significance of activity within a single cell. It should be remembered also that, although convergence of activity from many different cells has continuously been referred to, the same lower order cells are being used for connections to other higher cells with differing receptive fields.

It has been suggested above (see Fig 9–36, B) that the firing pattern of the

Fig 9–38.—Alternative diagram of converging lateral geniculate (*LG*) cells to account for receptive field of a particular cortical cell.

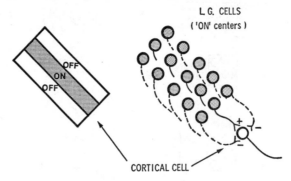

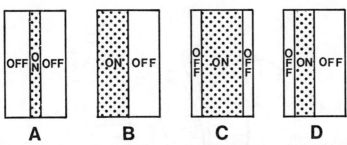

Fig 9–39.—Additional examples of receptive fields of simple cortical cells. Note variations in relative sizes of "on" and "off" regions and asymmetry in **B** and **D**.

simple cortical cells responding to bars of light could be accounted for by convergence of a single row of lateral geniculate cells with similar central and surround responses. Alternatively, or in addition, it could be accounted for by populations of lateral geniculate cells, related to the surround region, having an inhibitory input to the cortical cell (Fig 9–38). This pattern of convergence appears more likely because cortical cell fields are slightly larger than lateral geniculate cell fields, and it also would account for other "simple" cortical cell fields (see Fig 9–39). These fields show asymmetry of "off" and "on" regions as well as linear and axis orientation features. In Figure 9–39 only one orientation is shown, but examples of such cells with all orientations can be found.

Another feature of simple cortical cells is that those with asymmetric fields show preference to movement of the bar or edge of light across the field in a particular direction at right angles to the axis. Let us consider the example of the receptive field shown in Figure 9–39, B. This cell is responding best to an edge between a dark area to the right and a light area to the left of the axis. This cell actually shows an even higher frequency of action potentials when the edge of light moves from the "off" to the "on" region. The reason for this can be seen by referring to Figure 9–31, A. In the retinal ganglion cell, there was a burst of activity as the light was switched off in the "off" surround region. This response would summate with the increased response of the cell when light falls on the "on" center region. Such frequency variations are retained throughout the ascending relay stations. The cell depicted in Figure 9–39, B, would "prefer" an edge moving from the "off" region to the "on" region, i.e., right to left. An edge of light moving in the opposite direction does not fall first on an "off" region, so the response will not be as marked.

Some "simple" cortical cells respond to activity originating in either eye so long as the stimulus is falling on corresponding parts of the retina. This was indicated in Figure 9–33, in which lateral geniculate cells are shown converging on a single cortical cell. One other feature related to cortical cells—and topographic location—is that cells showing a particular axis orientation and corresponding to a specific area of the retina (and so, of course, to the visual field) are arranged in a columnar fashion in the cortical gray matter (Fig 9–40). Cells in adjacent columns show different axis orientation but receive input from a neighboring region of the retinal field.

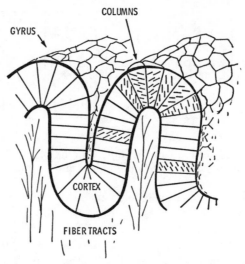

Fig 9–40.–Columnar organization of visual cortical cells. Physiologically, cells with same axis orientation and closely related topography of their receptive fields exist in columnar fashion. This is not visible histologically, but this area of cortex is rich in connections in the direction of the columns.

Within these columns are cortical cells whose preferred stimulus cannot be accounted for by convergence of known types of lateral geniculate cells. They have, among their features, the same axis orientation as other cells in the column and also receptive fields similar in size and topographic location. However, they do not have "off" surround regions; rather, they respond to edges between dark and light, which, so long as they are of the correct left or right relationship with respect to axis orientation for that cell, will continue to excite the cell regardless of where the edge falls in the total field. The response of such cells can be accounted for by convergence of inputs from neighboring simple cells in the same column. Such cells are called "complex" cells (Fig 9–41, A).

Investigation of other areas of the visual cortex reveal cells whose responses relate to further abstractions of form or pattern vision, e.g., cells whose

Fig 9–41.–Examples of receptive field of **A,** a complex cell, and **B,** a hypercomplex cell.

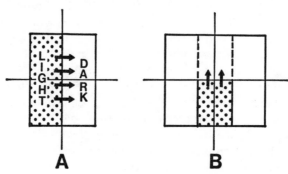

preferred stimuli are tongues of light or dark, or corners (Fig 9–41, B). Such responses are interpretable by further convergence of complex cells onto even higher orders of cells ("hypercomplex" cells).

IMPORTANCE OF CENTRAL INTEGRATIVE MECHANISMS

In previous chapters, examples of the relevance of particular neurophysiologic and neuroanatomical features to clinical situations have been frequently indicated. The student may be expecting similar material at this point. Unfortunately, it is not easy to directly relate this knowledge in a similar way. Why then has a substantial proportion of the available space been devoted to a consideration of this system and such integrative mechanisms? To answer this question, let us consider the informational significance of a burst of action potentials in one particular cortical cell. This could be indicated by the following: "There is a bar or edge of light at a particular point in space with reference to the body. This bar has a particular orientation to the vertical and horizontal axis and is moving at right angles to that axis." To communicate that information verbally takes around 13 seconds; to read it, about 5; the nervous system signals it—as well as all the parameters and coordinates—in less than 100 msecond, and it uses many cells but only five or six levels of synaptic relay stations. Helmholtz, the great German physician, physicist and physiologist, pointed out that if, with the 26 letters of the alphabet, we can create a world, we should not be surprised what we can do with 20 billion nerve cells.

It is not, however, just for intellectual satisfaction that such findings are significant. They also provide us with additional insight into brain mechanisms and allow further opportunity for questions, which, although indirect, relate to areas in which advancement in knowledge could have tremendous practical significance. For example, when we have a fairly accurate neural wiring diagram, together with its functional properties, questions such as: How determined is such a system? Can it be changed? What factors might influence any changes? can be asked and, hopefully, answers at very basic levels can be obtained.

PAIN PATHWAYS AND PERCEPTION

Although specific pathways in the cord and connections to the thalamus play an important role in our appreciation of pain, there are many features that do not fit with a simple concept of one set of specific sensory neurons related only to noxious stimuli. In the previous section on tactile receptors, we saw that descending pathways might modify synaptic transmission at lower levels and so modify ascending sensory information. There is also every reason to believe that further modification occurs at higher levels in the CNS. It is well known that the sensation of pain and the behavioral response elicited by certain noxious stimuli vary with the context in which the stimuli are presented. Severe wounds in battle and sports injuries are common examples. Major surgery performed under hypnosis and following acupuncture is also well authenticated. Intense pathologic localized pain, as, for example, in the "phantom limb" syndrome, is not necessarily alleviated by local anesthesia or interruption of the pathways known to be involved.

The psychologic aspects of pain perception make physiologic findings difficult to interpret. In 1965 a new theory of pain perception was introduced by Melzach and Wall. These workers postulated a "gating" mechanism whereby the different specific, primary sensory inputs entering the spinal cord, and known to be involved with noxious and other stimuli, are modified by interneurons present at the cord level. The gating is also modified by descending central pathways. Consequently, various factors will then determine whether the ascending "pain" pathways are activated. This "gate control" theory goes a long way to elucidate some of the previously unexplained features of pain perception and is mentioned here because it uses the same relatively simple concepts associated with integration, i.e., interneurons, excitation, pre- and postsynaptic inhibition and feedback mechanisms, that have been emphasized throughout this chapter.

MOTOR SYSTEMS

In the section on integrated spinal reflex mechanisms, some of the simpler aspects of sensory information and its influence on motor systems were considered (e.g., knee jerk, withdrawal and extensor reflexes). It was possible to describe the principles of this activity in terms of relatively simple circuits made up of differing types of neurons. Final control of muscle tone and the contraction, or relaxation, of muscles was seen to be mediated via the final common path, the alpha motor neuron. Activity of these neurons could be influenced by many discrete pathways. Analysis of more general aspects of purposeful, or "voluntary," movement, and its relationship to the position of the whole body in space (posture), requires that many different parts of the CNS be considered. This includes incoming sensory information, which will vary according to the different stages of any particular movement. Some of this information will modify outgoing motor activity at spinal levels and some of it will be analyzed at higher levels and will influence the final common path via pathways descending through the cord.

GENERAL ASPECTS

Figure 9–42 illustrates very simply the interconnections between the major parts of the CNS that are most closely involved in integrating motor activity. The directions of the arrows do not necessarily indicate whether the influence on the cells within the structure named is excitatory or inhibitory. Within each region various patterns of synaptic integration may occur, and the output will subsequently modify activity in other regions.

Unlike the sensory system, where we can easily consider that sensory activity is initiated in peripheral receptors, it is much more difficult to pinpoint a region for initiation of motor activity because it is so intimately related to incoming sensory information. Nevertheless it is convenient and conventional to consider the primary motor cortex as an area that is intimately involved in

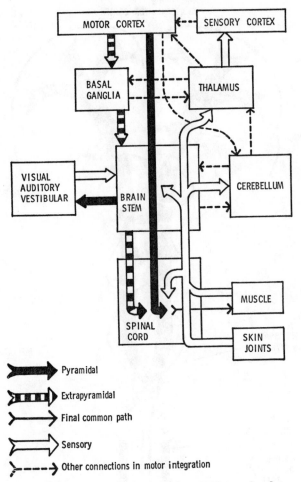

Fig 9–42.—Major interconnections within the CNS involved in control of motor activity.

such a function. Reasons for this are (1) that damage to this region causes definite impairment of voluntary movement and (2) that artificial stimulation leads to coordinated, if not purposeful, motor responses. It should be noted in Figure 9–42 that there is a major direct pathway descending from the motor cortex through the brain stem to the spinal cord and influencing the final common path. This corticospinal path is bilateral and is also termed the pyramidal tract (Fig 9–43) because it forms pyramidal protuberances as it courses through the medullary portion of the hindbrain. Most of these descending fibers, often referred to as *upper motor neurons,* terminate on interneurons at their respective segmental levels, but some connect directly with alpha motor neurons.

Also descending from the motor cortex are pathways that relay first in the basal ganglia, a group of nuclei in the telencephalon. Among the more important nuclei in this group are the caudate nucleus and the putamen. Many neu-

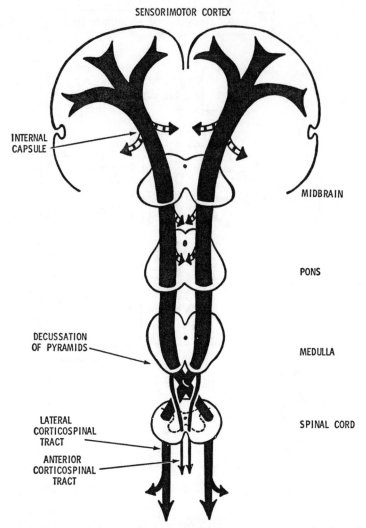

SENSORIMOTOR CORTEX

INTERNAL
CAPSULE

MIDBRAIN

PONS

DECUSSATION
OF PYRAMIDS

MEDULLA

SPINAL CORD

LATERAL
CORTICOSPINAL
TRACT

ANTERIOR
CORTICOSPINAL
TRACT

Fig 9–43.—Diagram of descending pyramidal system. Corticobulbar fibers (not shown) relay and exit at brain stem levels. Most corticospinal fibers cross at medullary level and continue in lateral columns of the cord. Note collaterals (*banded arrows*) to extrapyramidal system at basal ganglia and brain stem levels.

rons in the basal ganglia send axons that synapse with diffuse nuclei in the brain stem, the reticular formation. Also within the brain stem lie more well-defined nuclei such as the substantia nigra, red nuclei and vestibular nuclei. From the brain stem a large number of discrete pathways, e.g., rubrospinal, tectospinal and vestibulospinal, descend to influence directly or indirectly the alpha, or lower, motor neurons in the spinal cord. These pathways, or this system, are referred to as extrapyramidal (Fig 9–44). As will be seen later, pyramidal and extrapyramidal disorders produce specific physical signs that allow localization of pathologic symptoms.

The integrity of function of the motor cortex very much depends on the many interconnections between it and the sensory cortex. Consequently many descriptions of the nervous system refer to these areas collectively as the *sensorimotor* cortex. The motor cortex also receives input directly from parts of the thalamus and from the cerebellum.

The cerebellum receives input from ascending somatic sensory pathways and also from special senses such as the visual, auditory and vestibular systems in particular (see Fig 9–46). It also receives information from the motor cortex and is thus able to compare the activity of both motor and sensory systems. By means of its efferent fibers, it can modify motor activity either at brain stem levels or via the sensorimotor cortex.

Fig 9–44.—Composite diagram of major distribution of extrapyramidal system. (See text for description). Basal ganglia: *CN* = caudate nucleus; *P* = putamen; *GP* = globus pallidus; *SN* = substantia nigra. *T* = thalamus. The breaks in pathways at brain stem levels represent many nuclei, e.g., reticular formation, vestibular nuclei. The location of such nuclei at various brain stem levels in this diagram does not indicate precise positioning.

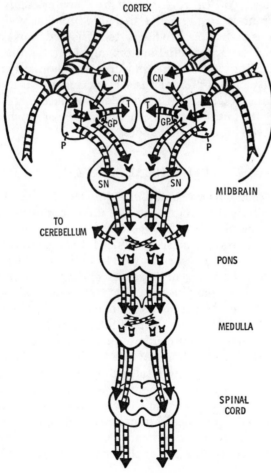

The vestibular mechanisms of the inner ear are sensory systems for detecting rotational movements of the head, the head's position in space and also alterations in velocity. Such information is important for coordinated movement. Sensory pathways from the vestibular apparatus therefore make extensive connections within the brain stem and with the cerebellum. Likewise, visual and auditory information is integrated into the motor control systems.

In summary, there are five main subsystems intimately involved in control of movement: (1) pyramidal system, (2) extrapyramidal system, (3) cerebellum, (4) vestibular and other cranial nerve systems and (5) general somatosensory mechanisms, but particularly proprioception from muscles and joints.

The student should realize that these subdivisions are somewhat arbitrary. The complexity of the interconnections is far greater than Figure 9–42 suggests and, in many instances, there is overlap of function and connection between such subdivisions. For example, collateral branches from the corticospinal (pyramidal) pathways make connections with the basal ganglia and with the nuclei in the brain stem, performing functions more readily associated with the extrapyramidal system. Likewise, the basal ganglia, which receive extensive input from cortical regions and are conventionally associated with the term extrapyramidal, also receive thalamic input, as well as sending output to the thalamus and so back to the cortex. The thalamus is not usually considered part of the extrapyramidal system any more than is the cerebellum, which has definite links with brain stem structures normally referred to as extrapyramidal. The main purpose of this section is to give the student a general scheme of functional relationships. More detailed accounts are to be found in more advanced texts, but they will be better appreciated when the more general framework given here is understood.

PYRAMIDAL SYSTEM

The simplified diagrams presented in Figures 9–42 and 9–43 indicate that the pyramidal tract provides a more or less direct pathway connecting motor cortex to the final common path. The cortical nerve cells involved are located not only in the precentral gyrus (see Fig 9–14) but also in cells in regions anterior to this, the "premotor" areas, as well as in the somatosensory cortex. As mentioned, collateral branches of these descending nerve fibers also provide significant inputs to basal ganglia and to brain stem nuclei.

As these fibers descend from the cortex to course through the diencephalon and reach the midbrain, they form part of the *internal capsule*. This structure lies between the thalamus and caudate nucleus medially and the lenticular nucleus laterally (Fig 9–45). Somatotopic localization of these fibers can be demonstrated at this level and persists as they descend through the midbrain and brain stem. At the level of the medulla, most of the fibers decussate to the opposite side to form the corticospinal tracts in the lateral columns of the spinal cord where somatotopic localization can again be demonstrated (see Fig 9–13). A few descending fibers remain uncrossed and descend in the anterior white matter of the cord. They finally cross at the level of their segmental

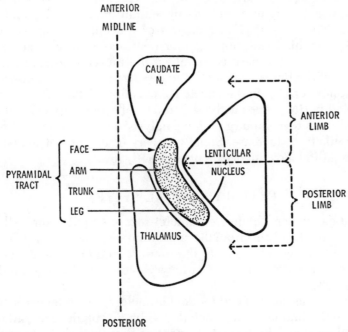

Fig 9–45.—Horizontal section through base of brain to show location of pyramidal tract in the internal capsule.

termination. Part of the pyramidal tract contains motor fibers related to the cranial nerves and is referred to as the corticobulbar tract.

Evidence that the motor cortex and pyramidal tracts play a major role in control of motor activity comes from many sources. Electric stimulation of these areas produces contraction and movement. The nature of the response depends on the degree of localization of the stimulus source, intensity of stimulation, actual site stimulated and state of cortical activity at that time. In general, stimulation results in coordinated muscle movements, not just contraction of a single muscle or group of muscle fibers. The responses involve reciprocal inhibition of antagonist muscles. Surgical removal of the motor cortex, or interruption of the pyramidal tracts, does not necessarily give rise to complete paralysis of the regions supplied. Movements can be initiated by other pathways; however, the ability to perform finely controlled movements is lost.

CLINICAL CONSIDERATIONS

Damage to the pyramidal system is not uncommon and the results parallel those in experimental procedures. The term "stroke" usually refers to some form of cerebrovascular "accident." This may be occlusion of a cerebral blood vessel, leading to anoxia and nerve cell damage in the region of supply, or hemorrhage from a blood vessel, which likewise will cause local tissue damage. The internal capsular region receives its blood supply from small branches of the middle cerebral artery; these are a common site for hemorrhage or thrombosis, which may interrupt the pyramidal tract as it descends through the capsule. This may lead to paralysis of the opposite side of the body (hemiplegia). At this level, upper motor neurons to the cranial nerves may also be involved, so the paresis can affect the muscles of head and neck as well as those of the limbs. During the acute stage a condition known as shock exists and reflex activity on the affected side is diminished, as is muscle tone. After a varied period, reflex activity (stretch reflexes) reappears and may be exaggerated despite voluntary motor paralysis. Muscle tone of the affected area is also likely to increase. Some recovery of movement may occur, but usually fine skilled movements are permanently lost.

EXTRAPYRAMIDAL SYSTEM

The basal ganglia and brain stem nuclei receive inputs from collaterals of the pyramidal tract and also from the sensorimotor cortex (see Fig 9–44). The output of the basal ganglia goes to the brain stem and to the thalamus, thus modifying motor activity not only at lower levels but also, via the thalamus, at cortical levels.

In addition to the caudate and lenticular nuclei in the telencephalon, other nuclei within the brain stem are included as basal ganglia or as part of the extrapyramidal system. The more important of these are the substantia nigra, reticular system and red nucleus.

A number of descending spinal pathways that can modify pyramidal and reflex activity at the spinal level arise from some of the brain stem nuclei. Activity in these descending pathways can be further modified at the brain stem level by inputs from the cerebellum, the special senses and ascending sensory pathways. The descending extrapyramidal fibers end on interneurons, the same ones that receive pyramidal input and are involved in spinal reflex activity.

The extrapyramidal system is involved in two major feedback loops: (1) from basal ganglia to thalamus, to cortex, to basal ganglia, and (2) from basal ganglia, to brain stem nuclei, to cerebellum, to thalamus, to cortex, to basal ganglia (see Fig 9–42).

CLINICAL CONSIDERATIONS

Disease of the extrapyramidal systems mainly causes a disorder of muscle tone and unwanted or purposeless movement. Several syndromes have been described. One is characterized by tremor and by some degree of increased muscle tone, "rigidity." The tremor is more severe at rest but may improve during voluntary movements, which, however, are not well controlled. There is a reduction and a slowing of voluntary movement. This condition, *paralysis agitans* or *parkinsonism,* may have different causes, but it relates directly or indirectly to decreased function of the cells in the substantia nigra. These cells normally contain a high concentration of dopamine, presumably their neurotransmitter. Interestingly, drugs that decrease brain dopamine levels may produce parkinsonism, and administration of L-dopa, a precursor of dopamine, often decreases the signs and symptoms. Atropine, which inhibits transmission at some cholinergic synapses, also improves certain features of parkinsonism, although it is not clear which of the possible pathways and feedback loops between the basal ganglia and the thalamus this may involve. Surgical destruction of the globus pallidus, and/or the appropriate thalamic nuclei to which it projects, also has counteracted what appears to be a decrease in output from the substantia nigra and has produced clinical improvement. The features of parkinsonism and its treatment illustrate how a better understanding of the interconnections within the nervous system can lead to rational therapy that will at least control the symptoms of neurologic disease.

Lesions at other sites within the extrapyramidal nuclei can give rise to very bizarre spontaneous movements over which the patient has no control (chorea, athetosis or hemiballismus). The exact etiology of such conditions is not clear and may be due to a variety of causes.

CEREBELLAR SYSTEM

As was pointed out above, the cerebellum has intimate relationships with the extrapyramidal system and motor cortex. It also receives considerable somatosensory information. The stretch-sensitive neurons in muscle pass directly to the cerebellum, but the information they convey is not transmitted to the cortex and so does not reach consciousness.

Sensory information carried by fibers from muscle tendons does reach consciousness and also diverges to the cerebellum along with axons conveying information regarding joint position.

Input to the cerebellum issues, then, from three primary sources (Fig 9–46). The first input comes directly from the cortex. The second comes from the brain stem and this includes collaterals and relays from the pyramidal and extrapyramidal systems, and also some direct input from vestibular pathways and special senses. As pointed out earlier, the function of the vestibular apparatus, which includes the labyrynth and semicircular canals, is to detect static, rotational and acceleratory movements of the head. Obviously, such information is very important in much motor activity. Information concerning the orientation of the head with respect to the rest of the body at any given moment must also be integrated. Consequently, proprioceptors in neck muscles and joint receptors in the neck have considerable input to brain stem nuclei and to the cerebellum. The third primary source of input comprises the ascending spinal pathways and is mostly direct. The brain stem contribution to the cerebellum may be modified by such ascending pathways as the spino-olivo-cerebellar.

The cerebellar afferent pathways diverge to supply inputs to both the cerebellar cortex and a group of subcortical nuclei within the cerebellum. The output from these nuclei forms feedback pathways both to the midbrain nuclei and to the cerebral cortex via the thalamus. This feedback system modifies descending motor activity in the cord and motor outflow of the cranial nerves. It should be noted that within the cerebellum there is also an internal

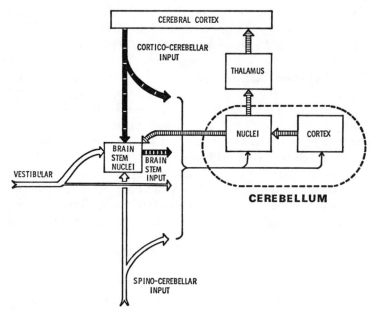

Fig 9–46.—Basic connections of cerebellum (see text for description).

feedback pathway involving the cerebellar cortex, which influences outflow from the cerebellar nuclei. These intracerebellar circuits are among the best understood in the entire CNS, largely as the result of studies by Sir John Eccles and co-workers. Lateral inhibitory mechanisms play an important role. It is perhaps worth noting that there are basically four cell types in the cerebellum and about seven basic patterns of connection (there are, of course, millions of cells), and yet the cerebellum is extremely important in bringing about coordinated, purposeful movements.

The general functions of the cerebellum are directly related to both control of posture and coordination of movement. It appears to do this by surveying both sensory input and motor output, appropriately modifying motor activity as necessary.

Some idea of the role of the cerebellum in performing such functions can be obtained by considering some particular examples.

The human being does not expend a tremendous amount of muscular energy in maintaining an erect stationary posture (we may have difficulty in telling that to the Marines!). The center of gravity in this position is such that it is mainly maintained by the skeleton and associated ligaments. This position is not, however, completely stable, and slight shifts of the center of gravity continuously occur and cause initiation of stretch reflex activity in postural muscles; this will correct any minor deviations. It will be remembered that the stretch reflex is intimately associated with gamma efferent control of the intrafusal fibers. The cerebellum exerts very definite controlling influences over this pathway and, as has been pointed out, the central connections of the afferents from muscle spindles terminate in the cerebellum.

Some of the features relating to control of movement can be seen by considering what is involved when we reach out to touch a discrete object in front of us level with our shoulder. Assuming the object to be less than an arm's length in front and that the arm is initially at the side of the body, an elevation of the upper arm at the shoulder must take place and the elbow must be flexed. The degree of movement at the shoulder and elbow determines the distance of the finger tip from the body. Most of us would perform such a movement quite smoothly, with perfect coordination of the muscle groups involved. Let us consider next just the flexion at the elbow: we do not overflex and then compensate by extending. The movement is likely to gradually increase in velocity and then slow as the finger reaches its target. To do this, extensors of the elbow have to come into play to cause braking of the movement, and yet initially the extensors of the elbow are inhibited to allow flexion (reciprocal innervation). Cerebellar influences on the outgoing patterns of motor activity appear to bring this coordination about.

It should be noted that the situation is very different if we consider the movements of a pianist's fingers during the execution of a fast and intricate piece of music. No time exists for such feedback pathways to be effective. In this case the movements involve a considerable degree of learning, and it is thought that the programmed pattern of activity is laid down at a cerebral cortical level.

Postural and movement control functions have common features if we consider the cerebellum as some form of error detector, error being deviation from what is required. During movement, error may not occur so long as changes in velocity and position are being continuously monitored and motor output appropriately modified.

CLINICAL CONSIDERATIONS

From what has been said regarding function, some features of cerebellar dysfunction might be predicted. In fact, our main reason for describing function in the above way results from what is observed in cerebellar disease.

Abnormal cerebellar function is usually characterized by (1) abnormal muscle tone, hypotonia; (2) instability, patients overcompensating for disturbances of posture; (3) poorly coordinated movements; (4) often tremor as a movement is performed (intention tremor).

These motor disturbances affect corticobulbar as well as corticospinal pathways, so that disturbances of eye movements and speech also occur.

VESTIBULAR AND OTHER CRANIAL NERVE SYSTEMS

The foregoing brief outlines of the pyramidal, extrapyramidal and cerebellar subsystems have referred to a role in motor activity for structures innervated by the cranial nerves. The relationship of the vestibular and cerebellar systems has been particularly emphasized.

The extraocular muscles that control direction of gaze by abducting, adducting, elevating, depressing and rotating the eyes within the orbit are supplied by motor divisions of the third, fourth and sixth cranial nerves. Motor and sensory nuclei of these nerves are located in the brain stem and have intimate connections with the corticobulbar projection of the pyramidal tract, with other brain stem nuclei and with the cerebellum.

The role of the visual system in motor coordination is somewhat indirect; blind people, although restricted in some aspects of motor activity, do not normally show disorders of coordination of movement provided other sensory systems are intact. We are all familiar with blind pianists, and there are blind golfers and baseball players.

Vision provides information regarding the position of external objects relative to the body. This is interpreted from the position of the objects in the visual field, direction of gaze (i.e., the direction of the visual axes in relation to the orbits) and position of the head, as indicated by the proprioceptors of the extraocular and neck muscles. The visual axes of the eyes are parallel when regarding distant objects but converge when fixating on near objects; when changing the direction of gaze, both eyes move in a coordinated manner. These movements require integration of sensory input and motor output and involve the cerebrum, cerebellum and brain stem.

CLINICAL CONSIDERATIONS

The pathways involved in coordinating vision with eye and other movements follow the general pattern that has been described. Pyramidal (in this case, the corticobulbar tract), extrapyramidal (particularly brain stem nuclei) and cerebellar connections are involved. Manifestations of disorders of eye movement, as with other motor disturbances, relate to the location of the diseased site.

Each of the three cranial motor nuclei that form the final common path to the extraocular muscles occupies a specific location within the brain stem (Fig 9–47). Each muscle is responsible for particular eye movements; consequent-

Fig 9–47.—Diagram showing cranial nerve innervation of extraocular muscles of right eye and different levels of localization of their motor nuclei in brain stem. (Corticonuclear fibers to IV nerve nucleus are predominantly uncrossed; crossing takes place in nucleofugal fibers.)

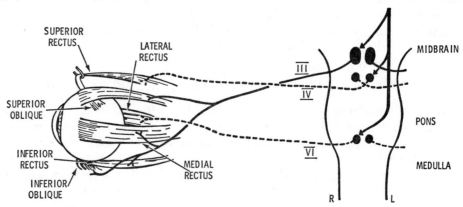

ly, inability to perform these movements can be related to particular cranial nerve nuclei. This often helps to localize the lesion.

Cerebellar disease causes poor coordination of all movements, and the eyes are commonly involved. They are unable to fixate steadily on an object when the gaze is directed away from the midline. Instead they oscillate rapidly, the movement in one direction being slightly more rapid than in the other. This condition is termed nystagmus and is a form of intention tremor. It can be lateral, vertical or rotary in direction. Various types of nystagmus are described, some not necessarily of cerebellar origin, but clinical differentiation between the different types is possible.

GENERAL SOMATOSENSORY MECHANISMS

The role of ascending spinal pathways in different aspects of motor activity has been pointed out. Some of these pathways, such as the spinocerebellar tracts, are almost exclusively used for motor coordination. Other ascending systems that pass via the thalamus to sensory cortex play a role in sensory perception (see section on sensory systems). Some of that same sensory information is also used, via collaterals, at brain stem levels for motor integrative purposes.

CLINICAL CONSIDERATIONS

The loss of somatosensory information may cause disorders of movement, as has been pointed out in relation to posterior column damage by neurosyphilis (tabes dorsalis). The proprioceptive loss gives rise to characteristic gait disturbances and, when visual clues are absent, e.g., when the eyes are closed, the posture can be very unstable.

HIGHER FUNCTIONS OF THE NERVOUS SYSTEM

It has been shown that many functions of the CNS can be understood in terms of particular patterns of activity in specific neurons, the pathways by which such activity may influence other cells, the sites of such pathways and the locations where integrative activity can occur. The models that we can build in this way do more than provide an explanation for the observed physiologic facts; they also give us insight into more general aspects of function. The question that now arises is: How far will the same principles help our understanding of the "higher functions" of the nervous system?

Most of us will understand the term higher functions as referring to concepts such as consciousness, emotion, learning, memory and intellect. Most people would admit that these are among the most interesting features of behavior and that the CNS is the organ mainly responsible for them. The lack of information in this area in many physiologic texts commonly used by medical students does not necessarily mean that those general concepts of CNS function, outlined previously, are not relevant to these functions. Rather, it indicates that our present stage of knowledge makes it difficult to build useful models.

The aim of this section is to indicate that general principles such as localization of function and specific connecting pathways relate to some of the concepts of "higher" nervous activity. Investigation of these functions presents many problems. Not least is the difficulty in defining the terms used. Terms such as learning and memory are abstract. They are inferred from observations of a subject performing a task, and so are very dependent upon such variables as motivational state and arousal levels of the subject. Lesions in particular parts of the CNS that may seem to disrupt learning or memory may actually be affecting the ability to perform some set of tasks. Learning may be demonstrable when a different type of performance is required. It is often difficult to determine whether an animal (or human being) cannot perform because it is unable to learn new information or will not perform because of inappropriate motivation, inadequate arousal or inability to make the necessary motor responses.

For ethical reasons it is often impossible to undertake appropriate experiments in human subjects. However, detailed clinical findings correlated with surgical or autopsy results have been a rich source of knowledge.

SPEECH

The development of speech and language as a form of communication is one of the features that differentiates human beings from other animals and there is a good deal of evidence concerning the localization of certain aspects of these functions (see Fig 9–14). Disturbances of speech and language therefore provide important clinical clues as to localized pathologic lesions. The verbal and visual aspects of speech and language are intimately connected anatomically and functionally.

Speech and writing obviously depend in part on pure motor functions as well as on auditory and visual recognition of symbols. In addition to the motor and special sensory cortical areas, discrete parts of the cerebral cortex are involved with perceptual, interpretative and motor expression of such symbolism. Disorders of these cortical areas lead to several types of speech and language dysfunction that fall into the general category of *aphasia*. This term refers to a weakness or loss of the faculty of transmission of language in any of its forms, excluding a pure motor loss involving the muscles involved in speech or a primary auditory or visual sensory loss.

In most individuals the cortical areas essential for normal language function are located in the left cerebral hemisphere. This is often referred to as the dominant hemisphere, since it also contains the motor centers that control the right side of the body, the preferred side for skilled motor activities for most individuals.

Three important cortical areas are involved in language functions (Fig 9–48): (1) part of the inferior convolutions of the frontal lobe—often referred to as Broca's expressive speech, or motor aphasia, center; (2) posterior part of the superior temporal gyrus, adjacent to the auditory cortex—often referred to

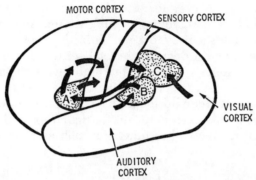

Fig 9–48.—Location of speech and language centers and their interrelationships with sensory and motor areas. *A,* Broca's motor aphasia center; *B,* Wernicke's receptive aphasia area; *C,* inferior parietal lobule.

as Wernicke's receptive aphasia area; and (3) inferior lobule of the parietal lobe. As suggested by their alternative names, damage to Broca's area causes deficits in motor production of speech, whereas damage to Wernicke's area results in disturbances of auditory aspects of language recognition. The third, parietal, area serves as an interpretive area for visual and other sensory recognition associated with language.

These areas are often referred to as association areas and are analogous to other cortical areas where interpretive and integrative activity associated with higher functions occurs.

Definite pathways can be shown linking the anterior and posterior language centers.

CONSCIOUSNESS

There are many observable attributes of consciousness, other than the continuum from deep sleep to alert wakefulness, that must be taken into account in any presentation of this subject. However, it is worthwhile noting that one of the main physiologic correlates of consciousness is an increased level of activity and excitability of the cerebral cortex. This level of activity is very dependent on diffuse nuclei located in the midbrain and pons, which are part of the reticular system that we considered earlier in relation to the extrapyramidal system. The reticular system receives general and special sensory input and has afferent and efferent connections with the basal ganglia. The more rostral part of the reticular system is termed the reticular activating system because, via thalamocortical pathways, it has a marked influence on the excitability of wide areas of cortex. Surgical section above the brain stem produces changes in behavior and in electric activity of the brain very similar to those found in sleep. Clinically, lesions involving this level of the brain stem can also induce drowsiness and lethargy. Lesions of other areas of the brain stem can produce insomnia.

It is as if the reticular activating system measures the overall quantity of ascending sensory information, as well as descending motor activity, and then feeds excitatory influences back to cortical areas to allow it to continue with such activity and to be receptive to incoming sensory information. Although such a mechanism relates mainly to a generalized sensory input, there is evidence of some selectivity. A decrease in total sensory input, be it tactile, visual or auditory, is conducive to sleep, but high levels of sensory input, e.g., noise, often can be ignored. It is interesting to note, however, that, although a mother may be able to sleep during high levels of background noise, the cry of her baby may easily awaken her. This particular example also shows that consciousness also involves other functions such as learning and memory.

EMOTION

This aspect typifies the semantic difficulties we often encounter in studying more abstract concepts of brain function. Behavioral responses that we associate with emotion are common in a wide variety of species, some of which we would hardly describe as showing much evidence of well-developed "higher" functions of the CNS. It is in this field that we must be most cautious in assuming that various "emotional" behaviors in animals correlate with subjective aspects of emotion experienced by human beings.

Behavioral manifestations of emotion frequently involve both somatic and visceral efferent activity. For example, during anger, muscle tone and heart rate are increased and violent muscular activity may occur.

The hypothalamus, the region of the diencephalon below the thalamus, has long been known to be associated with visceral or autonomic control. It is not essential for visceral reflex activity but influences it in much the same way that cortical and brain stem centers modify spinal reflex activity. A dozen or so particular nuclei can be identified in this region. Their patterns of connection are complex and poorly understood.

Responses obtained by stimulating discrete regions of the hypothalamus do not always give rise to such clear-cut behavioral responses as occur on stimulation of many other regions of the brain. Despite these difficulties, experimental and clinical evidence clearly indicates that the hypothalamus is intimately involved in the behavioral expression of emotions, control of autonomic activity, temperature regulation and control of food and water intake.

The difficulty in interpreting experimental findings in relation to hypothalamic function does not negate the general principles of function already discussed. After all, as previously indicated, a simple monosynaptic reflex, such as the muscle stretch reflex, will depend on many other influences affecting that particular synaptic site as well as on other pathways affecting the input, e.g., the gamma efferent influence on muscle stretch receptors. Any system that involves such structures as multisynaptic pathways, complex afferent and efferent pathways and feedback loops will be much more difficult to investigate. When, as in the hypothalamus, the nuclei are numerous and very closely associated, the pathways are not always discrete and often unknown and the area is relatively inaccessible, then it is not surprising that more meaningful interpretation of data is difficult to achieve.

LEARNING AND MEMORY

Current concepts of the physiology of higher functions of the brain owe a great deal to Setchenow and Pavlov. Their discovery that certain complex behavioral responses can be explained in physiologic terms and are amenable to physiologic, as distinct from psychologic, investigation may not now appear to be particularly revolutionary. However, we still have little idea of the cellular mechanisms that account for such responses. Scientific disciplines, be they physics or physiology, are deterministic and, although we may find no difficulty in accepting a concept explaining how a dog may salivate at the sound of a bell, many of us will find it difficult to accept a similar or related concept to explain why people may fall on their knees and pray at the ringing of a bell.

Previous examples of how particular behavioral responses are associated with increased activity of neurons in discrete parts of the brain did not exclude the possibility of widespread, but perhaps less obvious, neural activity in many other regions. The more complex the function the less discrete such activity appears and the more difficult it becomes to interpret its significance. As also was pointed out, there are many facets of these higher functions. In the case of memory, we have to consider immediate memory, short-term memory, long-term memory, registration and recall, to name but a few facets. We would not be surprised, therefore, that many different areas were involved.

The role of a particular part of the brain in the function of memory can be illustrated by studies on temporal lobe function. Penfield, a neurosurgeon, examined the responses of conscious patients to localized electric stimulation of the temporal lobes. One of the many interesting findings was that the stimulation frequently evoked memories unconnected with the experimental situation. These memories sometimes related to experiences of many years earlier and were often very detailed. However, the responses were unpredictable, and the response to a second stimulus of apparently the same strength, location and duration could evoke a memory quite different from the preceding one. These findings are extremely difficult to interpret; they do not necessarily imply that memories are stored in the temporal lobe.

Another intriguing finding in relationship to memory and learning emerged from the observations and experiments of Sperry following section of the corpus callosum in human beings and animals. The corpus callosum is an extensive fiber tract that connects the cerebral hemispheres. Occasionally this structure may be sectioned as a therapeutic procedure to prevent the spread of abnormal excitatory activity from one hemisphere to the other. Some very surprising changes have followed. In many respects the patient's behavior is normal, but it is possible to demonstrate that in some respects the halves of the brain work independently. Something learned by one-half will be completely unknown by the other, which is then unable to act when presented with the same information. This is best understood if we consider a particular example.

First, we must remember that the sensory and motor areas of the left half of the body are located in the right hemisphere and vice versa. If a normal subject is handed an object such as a plastic number, he can, without looking, identify that number by tactile information alone and write down with either hand what the number was. If the same experiment is carried out after the corpus callosum is divided, and the object is placed in the left hand, the subject is unable to indicate verbally what the object is. It should be remembered that the centers for speech lie in the left hemisphere in right-handed individuals. Although the object is recognized by the left hand, the subject is unable to identify the object verbally, as there is no communication between the left and right hemispheres. However, the subject can indicate recognition of the object by drawing it or choosing a like object, as long as this is done by the left hand. Cross communication still exists for visual information, provided it falls in the appropriate visual field. Likewise, verbal communication can inform the nonverbal, or nondominant, hemisphere what the dominant hemisphere "knows." Because of these other possibilities for cross communication, most of the abnormalities of behavior will be unnoticeable unless tested for in a very controlled way. Experiments along these lines indicate that "memories" or "learning" occurring in one hemisphere can be transferred to the other via the corpus callosum.

Physiologic models are inevitably influenced by comparison with known physical systems, current ideas of brain function often reflect analogous concepts in contemporary physics. Descartes' concept of nervous function was related to mechanical and hydraulic models of the nervous system because these were the best understood physical systems of his time. Our present ideas relate much more to our understanding of electric circuits, not because nervous activity involves utilization of electric energy but because of the way signals are conducted from point to point and can be modified in particular ways. This concept of the nervous system often leads us to think of it as a rigidly "wired" system. This is true to some extent but not to the extent that electronic circuitry is. Mountcastle puts this well when he says, "So far as the public behavior of man is concerned, there is nothing that could not in principle be duplicated by a mechanical device, nor is there anything inherent in the structure or function of brains that is outside the limits of the physical world; nothing which, in principle, could not be duplicated by an automaton, provided it were made of organic molecules and not electronic devices, for brains are a triumph of the organization of the former, not the latter."

Our present concept of the nervous system includes plasticity—an ability of some of the pathways or connections to change or for new connections to develop. This is not a particularly new idea; it is implicit in the theory of conditioned reflexes. In the light of recent knowledge of synaptic functions, one of the interesting questions is whether such new "connections" are the result of facilitation in pre-existing pathways or whether they involve the actual development of new anatomical connections. Both mechanisms may of course be involved. So far it has been very difficult to show any such changes, but it is very difficult to account for many physiologic findings without assuming that such changes can take place. Synapses, of course, are the site where such changes could most easily occur. Our ability to test these hypotheses requires that we know precisely the various connections for the system under investigation.

It has also been proposed that memory may be stored in some biochemical way, but early behavioral experiments supporting this view have been hard to duplicate. The discovery that genetic "memory" was stored in the form of DNA led to a renewed interest in the hypothesis. Experimental studies in this field have been subject to serious criticism, from the point of view of both experimental design and interpretation. For example, alteration in RNA content need not necessarily imply production of a new molecule coding the memory but may only imply new growth. New protein synthesis would fit a hypothesis supporting synaptic modification equally well.

NEURAL TRANSMITTERS IN THE CNS

Cellular and membrane aspects of conduction of the nerve impulse and synaptic transmission have been considered earlier in this chapter. In the case of the conduction of action potentials along the nerve axon, the cellular and membrane mechanisms involved are basically similar, regardless of the type of axon or its function. By comparison, synaptic transmission mechanisms show marked differences at different synaptic sites. The brain uses several different chemical transmitters. Each is specific for particular synapses, and each has its own specific release, storage, synthesis and degradation mechanisms. In addition, a transmitter may be excitatory or inhibitory to the postsynaptic cell.

Drugs whose primary action in the nervous system is on conduction mechanisms have a restricted use because they nonselectively depress all types of activity. Such drugs are useful as local anesthetic agents because they can be applied either topically or by injection at selected peripheral sites. Some slight selectivity of action occurs because small diameter axons are more readily affected than are larger ones. Consequently, conduction in pain fibers is blocked sooner than in other sensory fibers. In practice, this selectivity is probably insignificant. The use of these compounds on CNS structures is almost exclusively reserved for blocking sensory conduction at spinal cord levels so as to produce larger areas of anesthesia than is practicable by peripheral injection.

At any one synaptic site, there is likely to be only one particular excitatory or inhibitory chemical transmitter, although these may differ at different sites. Consequently pharmacologic agents affecting synaptic transmission mechanisms present a greater potential for selectively affecting particular functional pathways or mechanisms.

It has long been known that certain compounds appear to selectively affect central nervous system functions. A classic example is morphine, which has been used as a sedative to induce sleep and to abolish pain for several thousand years. We are still unable to completely describe this drug's effect in terms of specific sites or mechanisms of action.

Even though general anesthetic agents have been known for over a hundred years and have revolutionized the practice of medicine, much still has to be learned about their mechanisms. Likewise, we are unable to account for the actions of other central depressant drugs, such as the barbiturates, in a satisfactory manner. Most of our knowledge suggests that their action is in the nature of a general depression of all aspects of neural activity. There is some evidence that systems involving multi- or polysynaptic pathways are particularly liable to be interrupted or affected. In the case of the useful clinical actions of general anesthetic agents, the reticular activating system appears to be an important "target" area.

In the peripheral nervous system, including both somatic and autonomic divisions, only two particular transmitters, acetylcholine and norepinephrine, appear to be released from neurons during synaptic transmission. Nevertheless the transmitter-receptor interactions show marked differences at different sites and have led to pharmacologic classification of muscarinic and nicotinic actions for acetylcholine and alpha and beta actions for norepinephrine. This classification is based in part on the fact that some drugs block the activity of a particular transmitter at one site but not at another. From a therapeutic viewpoint, this has allowed greater specificity of action of drugs affecting different physiologic processes.

Identification of transmitter substances at synapses within the CNS is much more difficult than in the peripheral nervous system. The criteria that normally need to be fulfilled include demonstration of the following: (1) that the particular substance occurs in the nerve cells under study, (2) that the substance is released from those cells on stimulation or during activity, (3) that presynaptic stimulation produces postsynaptic responses similar to application of the compound to the postsynaptic cell and (4) that pharmacologic agents affect the response to both presynaptic stimulation and to suspected transmitter application in similar ways.

Because of technical difficulties, it has only been possible to identify a transmitter substance meeting all these criteria at a few central synapses. The best examples are the synapses between axon collaterals of the alpha motor neurons and Renshaw cells in the spinal cord. Acetylcholine is almost certainly the transmitter at these synapses. Another example is the synapses between the Purkinje cells in the cerebellar cortex and the cells in the cerebellar subcortical nuclei. These synapses are inhibitory, and gamma aminobutyric acid (GABA) appears to be the inhibitory transmitter.

Despite the difficulties involved in positive identification, there is convincing evidence from many different types of study that norepinephrine, 5-hydroxytryptamine and dopamine, as well as acetylcholine and GABA, are probably transmitter substances within the CNS.

CLINICAL CONSIDERATIONS

The use of L-dopa, a precursor of dopamine, in treatment of parkinsonism and its probable substitution for a diminished dopaminergic pathway from the substantia nigra to the corpus striatum was mentioned in the section on extrapyramidal systems (section on motor systems).

The 1950s saw the introduction of many new compounds that appeared to have definite selective actions on the CNS and were of immense practical use in treatment of psychiatric disorders. The most important of these were the phenothiazine derivatives, the monoamine oxidase (MAO) inhibitors and the tricyclic antidepressants. The obvious effectiveness of these compounds in relieving or alleviating certain mental disorders has not only stimulated the search for further compounds but also has provided impetus for seeking biochemical causes for psychopathology.

The phenothiazine compounds were initially synthesized during a search for new antihistamine compounds but were found to have extremely useful effects in calming, or "tranquilizing," severely disturbed psychotic patients. The mechanism of action remains unexplained, although it is known that they interact with several neurotransmitter substances.

The tricyclic antidepressant drugs such as imipramine and amitriptyline are another group of phenothiazine analogues synthesized for possible sedative, antihistaminic or antiparkinsonian activity. These drugs proved most useful in treatment of endogenous depressions.

Another compound with tranquilizing activity similar to that of some of the phenothiazine compounds is reserpine. This compound has a marked effect in releasing biologic amines from storage sites in nervous tissue, resulting in depletion of those stores. The sedative action of reserpine appears to be related to this depletion of catecholamine stores, but whether in the whole brain or at particular anatomical locations is not precisely known. It can be reversed by administration of catecholamines.

The MAO inhibitors inhibit one of the enzymes directly involved in the breakdown of catecholamines, and they would therefore be expected to potentiate synaptic activity at synapses liberating catecholamines. They too have clinical uses in treatment of depressive illnesses, but again it has not been possible to completely explain their beneficial effect by such a mechanism.

The use of drugs to modify "higher" nervous activity is probably the largest remaining area in medical practice that is almost purely empirical. This is inevitable in the light of our current lack of understanding of the neurophysiology of the higher nervous system. This unsatisfactory situation is compounded by the fact that these drugs are among the most widely and readily prescribed in the Western world.

INVESTIGATION OF FUNCTION AND DYSFUNCTION

For obvious ethical reasons, our ability to investigate the function of the human brain is restricted mainly to behavioral studies or to correlation of loss or alteration of function with pathologic findings. Consequently, most of our knowledge of the function of the nervous system comes from experiments performed on other species. The greater complexity and versatility of the human nervous system are major features that distinguish man from the other animals; this makes extrapolation of physiologic findings from other species to man difficult. Fortunately the evidence so far suggests that, although there are obvious species differences, the general mechanisms and principles of function of nervous systems appear to be common. Although animal studies may not always provide the right answers about details of function in the human nervous system, they continue to provide otherwise unobtainable clues. When findings generalize from one species to others, it becomes reasonable to believe that they may apply to human beings as well.

Although it is beyond the scope of this text to provide a comprehensive survey of neurophysiologic research studies, the student should be aware of the more important approaches being taken to further our understanding of the nervous system.

CURRENT TRENDS IN NEUROPHYSIOLOGIC RESEARCH

The importance of topography has been emphasized throughout this text; it is essential for neurologic diagnosis. Current concepts of function increasingly take into account the diffuse distribution patterns of functional pathways and extensive interconnections among various regions of the brain. Much remains to be done and a large area of neurophysiology is still devoted to "mapping" different pathways and identifying the sites where their activity is integrated.

The techniques that may be used are numerous. Neuronal activity can be induced by artificial localized electric stimulation, and simultaneous observation of behavioral responses can give clues to the function of the region stimulated. Because of the multiplicity of interconnections involved in complex responses, interpretation of the results can be very difficult. Functional pathways may also be "mapped" by combining electric recording techniques with electric stimulation. By accurately measuring the timing between the applied stimulus and any evoked electric response, and knowing the distance between the stimulus and recording electrodes, it is often possible to determine whether such pathways are direct or indirect, i.e., polysynaptic.

Many experimental variables may influence results obtained in this way. For example, as the strength of the artificial stimulus increases, more and more nerve fibers in the vicinity of the stimulating electrode are activated; this may in turn decrease our ability to localize a specific pathway, as a response in one region may be counteracted by inhibiting influences. Anesthetic agents necessary to carry out many of these types of experiment can markedly affect experimental findings by modifying neural activity and have frequently led to fallacious interpretation of results.

Fine microelectrodes can be used to record activity in single neurons following both artificial and physiologic stimulation. These studies are usually referred to as single unit analyses. Extra- or intracellular recording techniques may be used. These are particularly useful for following integrative activity in neurons. Many of these types of experimental procedures also involve subsequent histologic identification of the actual recording and/or stimulus site. Sometimes the actual cell may be identified. There is, of course, the problem here of developing an understanding of a functional system based on what is happening in a single or a few neurons out of a population of thousands of neurons involved in that function. However, as indicated earlier in the visual system in particular, such analyses can be very revealing.

Similar microelectrodes can be used to apply pharmacologic agents to individual neurons within the CNS. Simultaneous monitoring of the electric activity of that cell can provide useful information related to the synaptic transmission mechanisms involved. However, postsynaptic receptors may be activated by many chemicals applied in this way, making it impossible to identify the natural transmitter by this technique alone.

Histologic techniques are also available for tracing neural pathways. It has long been known that if nerve fibers are cut, those parts of the axons that have been separated from their cell bodies undergo degenerative changes that subsequently can be identified by special staining techniques. The cell bodies themselves may also show histologic changes. In these types of investigation, localized lesions are made and then, after a period long enough to allow for degeneration, the animal is killed. The nervous tissue is then cut into thin sections for microscopy, which are stained for visualization of any degenerated fibers, and the pathways can be reconstructed from serial sections.

More recently, techniques have been developed for specifically staining neurons containing catecholamines. The pathways of such neurons can therefore also be followed by serial sections.

In some instances it is possible to inject individual cells with markers, such as fluorescent dyes or other stains. These markers diffuse into the axon and dendritic processes of the cell. Unfortunately, this technique does not allow long pathways to be followed, but it can be extremely useful in determining other aspects of the cell's geometry. The same micropipet used for injecting such material can be used for stimulating or recording. This often enables functional identification to be correlated with structural features.

Certain organisms, particularly invertebrates, have anatomical characteristics that make them useful for neurophysiologic investigation. Not only is the total number of nerve cells in these species fewer but the cell bodies or axons are frequently large compared with mammalian neurons. This facilitates the use of intracellular recording techniques to study integrative mechanisms. It also is frequently possible to visually identify particular neurons from animal to animal. The functions and connections of such neurons can then be studied in great detail. An individual large cell is often suitable for biochemical analyses, useful for studying transmitter synthesis and biochemical aspects of nerve cell differentiation.

Examples of research in these species, which have played extremely important roles in leading to our present understanding of neural mechanisms, include (1) studies on the giant axon of the squid, which have been of fundamental importance in uncovering excitability and conduction properties of nerve fibers; (2) studies on the lobster and crayfish nervous systems, which have revealed the chemical nature and membrane properties of inhibitory synaptic mechanisms; and (3) studies on the leech nervous system, which have challenged previous concepts of glial cell function and uncovered neuron-glia interactions. These "simpler" nervous systems, which can be "mapped" in great detail, provide useful models for studying correlates of learning and conditioning. The Aplysia, a type of sea slug, has also been used extensively for such studies.

The regenerative properties of many nonmammalian nervous systems, including some vertebrates, provide useful opportunities to study regeneration, growth and other "plastic" capabilities of nerve cells.

The development of tissue culture techniques, which allows growth of isolated aggregates of nerve cells under highly controlled conditions, makes it possible to investigate some of the basic features of growth and synapse formation. Such information will probably be essential for understanding cellular correlates of "plasticity" and learning.

SPECIAL CLINICAL INVESTIGATIVE PROCEDURES

As with clinical examination of other physiologic systems, the history and physical examination are of prime importance in detecting abnormalities of function and in localizing any causative pathologic conditions in the nervous system. It must be realized that, although most of the procedures referred to below assist in the diagnosis of localized lesions, not all dysfunctions produce localizing signs.

ELECTROENCEPHALOGRAPHY

The underlying electric activity in the brain can be recorded from superficial electrodes placed on the scalp. The electric changes recorded in this way are reflections of a wide spectrum of synaptic activity in many different neurons. Unlike data from electromyograms, these alterations in electric activity are difficult, if not impossible, to explain in terms of specific cellular correlates.

A series of particular patterns of electric activity, as seen in the EEG, can be described for a variety of conditions, e.g., at rest with the eyes open or closed, or during sleep. Deviations from these patterns can be described for a variety of clinically abnormal conditions (Fig 9–49). By utilizing many electrodes over much of the scalp (and, of course, the underlying brain), regional deviations in patterns of activity can also help to localize underlying pathologic conditions.

RADIOLOGIC METHODS

X-ray studies are of great value in revealing anatomical abnormalities and pathologic distortions of the brain and spinal cord. "Plain" x-rays are usually not very helpful because neural tissues, blood vessels and CSF all have similar x-ray absorption coefficients. However, increases in x-ray contrast can be readily achieved by injecting air (pneumoencephalography) or radiopaque contrast media into the CSF around the spinal cord (myelography) or into blood vessels supplying the brain (cerebral angiography).

The necessity for these measures, which are relatively unpleasant and carry

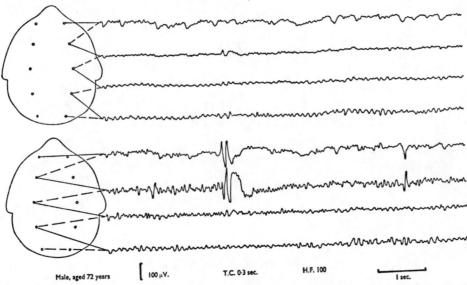

Male, aged 72 years [100 µV.   T.C. 0·3 sec.   H.F. 100   I sec.

Fig 9–49.—EEG in patient with meningioma in left frontal region. Position of pairs of recording electrodes for each trace is indicated at left of figure. Differences in waveforms between the two sides can be seen. Recordings from left frontal region show marked abnormal activity. (From Pyrse-Phillips, W.: *Epilepsy* [(Bristol: John Wright & Sons Ltd., 1969), p. 53]. Used by permission.)

a small risk, has been diminished by the recent advent of the EMI scanner. This device uses computer-assisted techniques that allow detection and localization of contrast between tissues with only slightly different x-ray absorption coefficients.

BIBLIOGRAPHY

Adrian, E. D.: *The Physical Background of Perception* (London: Oxford University Press, 1967).
Bennett, M. V. L. (ed.): *Synaptic Transmission and Neuronal Integration,* Society of General Physiologists Series, Vol. 28 (New York: Raven Press, 1974).
Curtis, B. A., Jacobson, S., and Marcus, E. M.: *An Introduction to the Neurosciences* (Philadelphia: W. B. Saunders Co., 1972).
Guyton, A. C.: *Textbook of Medical Physiology* (Philadelphia: W. B. Saunders Co., 1976).
Holmes, G., in Matthews, B. (ed.): *An Introduction to Clinical Neurology* (Baltimore: Williams & Wilkins Co., 1968).
Keele, C. A., and Neil, E. (eds.): *Samson Wright's Applied Physiology* (12th ed.; London: Oxford University Press, 1971).
Kuffler, S. W., and Nicholls, J. G.: *From Neuron to Brain* (Sunderland, Mass.: Sinauer Associates Inc., 1976).
Magoun, H. W. (ed.): *Handbook of Physiology, Section 1: Neurophysiology* (Washington, D.C.: American Physiological Society, 1959–60), 3 vols.
Millon, T. (ed.): *Medical Behavioral Science* (Philadelphia: W. B. Saunders Co., 1975).
Mountcastle, V. B. (ed.): *Medical Physiology,* Vol. I (13th ed.; St. Louis: C. V. Mosby Co., 1974).
Ruch, T. C., and Patton, H. D. (eds.): *Physiology and Biophysics* (20th ed.; Philadelphia: W. B. Saunders Co., 1973).
Schmitt, F. O., and Worden, F. G. (eds.): *The Neurosciences: Third Study Program* (Cambridge, Mass.: M. I. T. Press, 1974).

10

Endocrine Physiology

CLARK T. SAWIN, M.D.

GENERAL CONCEPTS

As ORGANISMS evolved to larger and more complex forms, the individual cells in each organism became physically distant from most other cells. Complex organisms such as animals cannot survive and reproduce without a high degree of integration of all cells in the body. In large part the nervous and endocrine systems provide the needed integration. Although these two systems are discussed in separate chapters, they are functionally so closely intertwined that neuroendocrinology has emerged as a discrete field of study.

Classically, the endocrine system was considered to be composed of the various glands that secrete substances "internally" (hence endocrine) into the blood rather than externally, or into the gut. These glands, e.g., thyroid or adrenal glands, act through their secretions, the *hormones* (from the Greek, ὁρμάω, I excite). We now recognize that the source of some hormones is not a discrete gland but rather a collection of cells located in an organ with other functions. Since the functional component of the endocrine system is the hormone and not the gland, it is better to consider the endocrine system as composed of hormones and the cells that secrete them than to restrict the definition to recognizable endocrine glands and their secretions.

What Is a Hormone?

A hormone is a chemical messenger that is *secreted* by a group of cells into the blood and then acts on other cells to *regulate* (but not initiate) their functions. The result is improved *homeostasis,* or maintenance of a constant internal environment. More broadly, this can be viewed as preservation of the organism both as an individual and as a species. The homeostasis results from a more appropriate and better coordinated response of the various tissues of the body to any change in internal or external environment. Clearly, if hormonal stimulation is to produce effective homeostasis, there must be appropriate mechanisms for turning hormonal secretion on and off. There are many spe-

532

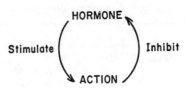

Fig 10–1.—General concept of endocrine negative feedback control in a simple closed-loop system.

cific ways in which this is done, but the important overall concept is that of a *negative feedback control system,* i.e., the action of a hormone directly or indirectly regulates the secretion of that same hormone. (Fig 10–1).

What Kinds of Hormones Are There?

There are two main groups (Table 10–1). One group is "sticky" or lipid-like and poorly soluble in water; these hormones circulate in the blood largely bound to a protein that "solubilizes" them. Members of this group include the steroid hormones (and vitamin D) and thyroid hormones. The second group is soluble in water and includes the peptide and glycoprotein hormones and several hormones, including the catecholamines, which are derived from amino acids.

How Are Hormones Made?

The steroid hormones are all derived from cholesterol by various enzymatic steps (Fig 10–2). The thyroid hormones are derived from the amino acid tyrosine, as are the catecholamines. Peptide hormones, such as insulin, are often cleavage products of a larger peptide prohormone (Fig 10–3). Glycoprotein hormones, on the other hand, are composed of two subunits, α and β chains, each of which is a peptide chain combined with a variable number of carbohydrate residues, such as sialic acid.

How Are Hormones Secreted?

In some cases, e.g., hydrocortisone, the hormone is secreted about as fast as it is made and appears to diffuse out of the cell, although this may be too simple a concept. In other cases the hormone is stored in the form of granules, which are released when a proper stimulus arrives at the endocrine cell. Often, as in the case of insulin, a microtubular-microfilamentous system aids and regulates secretion of hormone granules (Fig 10–4).

How Do Hormones Act?

Hormonal actions may be recognized and studied at many levels, such as that of the cell, the organ or the entire animal. To be relevant to the understanding of endocrine physiology or of clinical disease, hormonal actions must

TABLE 10–1.—PRINCIPAL HORMONES AND ORIGIN*

HORMONE	MAIN SITE OF ORIGIN†
BOUND TO PROTEIN IN BLOOD ["STICKY"]	
Cortisol (hydrocortisone, compound F)	Adrenal cortex (fasciculata)
Aldosterone	Adrenal cortex (glomerulosa)
Estradiol (E_2)	Women: ovarian follicle, corpus luteum, feto-placental unit
	Men: testis and peripheral conversion from testosterone†
Estrone (E_1)	Women: same as estradiol plus peripheral conversion from E_2
	Men: testis and peripheral conversion from androstenedione
Estriol (E_3)	Peripheral conversion (liver) from E_1 and E_2, feto-placental unit
Progesterone	Corpus luteum, placenta
Testosterone	Testis (Leydig cell)
Dihydrotestosterone (DHT)	Peripheral conversion from testosterone
Thyroxine (T_4)	Thyroid follicle
Triiodothyronine (T_3)	Thyroid follicle and peripheral conversion (liver) from T_4
Vitamin D	Skin, diet
1,25-Dihydroxyvitamin D (1,25[OH]$_2$D)	Kidney
NOT KNOWN TO BIND TO PROTEIN IN BLOOD *(largely peptides and glycoproteins)*	
Hypothalamic-releasing and -inhibiting factors (hormones)	
Thyrotropin-releasing hormone (TRH)	Hypothalamus, brain
Gonadotropin-releasing hormone (GnRH) (also called luteinizing hormone releasing hormone or factor: LRH or LRF)	Hypothalamus
Growth hormone-releasing factor (GHRF)‡	Hypothalamus
Somatostatin (or growth hormone release-inhibiting factor)	Hypothalamus, ?brain‖, ?gut, ?pancreatic islet, etc.
Corticotropin-releasing factor (CRF)§	Hypothalamus
Prolactin-releasing factor (PRF)§	Hypothalamus
Prolactin-release inhibiting factor (PIF)§	Hypothalamus
Melanocyte-stimulating hormone-releasing factor (MRF)‡	? Hypothalamus
Melanocyte-stimulating hormone release inhibiting factor (MIF)‡	?Hypothalamus
Neurotensin	?Hypothalamus
Substance P	?Hypothalamus
Endorphins	?Brain, anterior pituitary
Thyrotropin (thyroid-stimulating hormone or TSH)	Anterior pituitary
Follicle-stimulating hormone (FSH)	Anterior pituitary
Luteinizing hormone (LH)	Anterior pituitary
Growth hormone (GH) or somatotropin	Anterior pituitary
Somatomedin§	Liver
Corticotropin (ACTH)	Anterior pituitary

(continued)

TABLE 10–1. (cont.)—PRINCIPAL HORMONES AND ORIGIN°

HORMONE	MAIN SITE OF ORIGIN†
Prolactin (PRL)	Anterior pituitary, ?placenta
Melanocyte-stimulating hormone (MSH)	?Anterior pituitary
Lipotropin	Anterior pituitary, ?brain
Melatonin	Pineal
Vasopressin (or ADH)	Neurohypophysis
Oxytocin	Neurohypophysis
Calcitonin (or thyrocalcitonin)	Thyroid parafollicular cell
Parathyroid hormone (PTH)	Parathyroid gland
Thymin (or thymosin or thymopoietin)§	Thymus
Epinephrine	Adrenal medulla, ?brain
Norepinephrine	Sympathetic nervous system, adrenal medulla, ?brain
Insulin	Pancreatic beta cell
Glucagon	Pancreatic alpha cell
Renin§	Kidney
Angiotensin I	Blood
Angiotensin II	Lung, ?other tissues
Erythrogenin§	Kidney
Erythropoietin§	Blood
Relaxin§	Ovary
Inhibin§	Testis, ?ovary
Chorionic gonadotropin (CG)	Placenta
Placental lactogen (PL) (also called choriosomatomammotropin or choriomammotropin)	Placenta

°Does not include gut hormones or local hormones such as bradykinins.
†Peripheral conversion = made in tissue not usually considered endocrine, although precursor is secreted by recognized gland.
‡Structure proposed but not proved.
§Structure not known.
‖? = uncertainty as to actual secretion by indicated tissue.

be understandable at the level of the whole organism. In a broad sense the action of a hormone is a result of its circulating levels in the blood and the responsiveness of the various tissues. Since hormones do not act on all tissues of the body, despite their ubiquitous presence in the blood, tissues that do respond must have a specific mechanism for doing so. We now know that tissues responding to a hormone have receptor sites specific for that hormone.

The receptor site for peptide and glycoprotein hormones is on the responding cell's plasma membrane. Binding of the hormone to the receptor often causes activation of adenyl cyclase, also in the plasma membrane, and generation of cAMP within the cell (Fig 10–5). Cyclic AMP is sometimes referred to as an intracellular "second messenger," the hormone itself being the first. The rise in cAMP then activates a cAMP-dependent protein kinase, which presumably brings about the various actions of the hormone. Although probable and reasonable, this sequence has been proved in only a few instances. Some peptide hormones, such as growth hormone, prolactin and insulin, do not seem to activate adenyl cyclase, and insulin may, if anything, decrease cAMP levels in responding tissues. The second messenger for these hor-

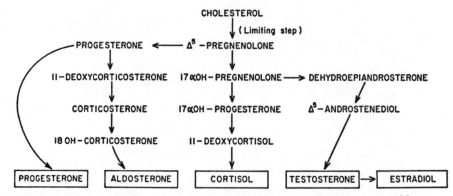

Fig 10–2.—Biosynthetic pathways of principal biologically active steroid hormones; actual hormones produced vary in different steroid-producing tissues.

mones is not clear, although some workers have speculated that it might be cyclic GMP (guanosine monophosphate).

The receptor site for the group of lipid-like hormones is inside the cell rather than on the surface (Fig 10–6). The steroid and thyroid hormones bind to specific cytosol proteins and are transported to the cell's nucleus; most of the actions of these hormones seem to be mediated by subsequent nuclear events, although the actual linkage system is not yet well defined.

How Is Hormonal Action Regulated?

The simplest negative feedback control is one in which a hormone's effect shuts off the hormone's secretion, e.g., parathyroid hormone (PTH) causes a rise in the level of serum calcium, which in turn directly suppresses PTH secretion. Thus hormonal action modulates and is modulated by the rate of hormonal secretion. Most hormonal control systems are more complex than

Fig 10–3.—Complete proinsulin molecule is cleaved to form connecting peptide (or C-peptide) and insulin, which has an A chain and B chain connected by disulfide bridges.

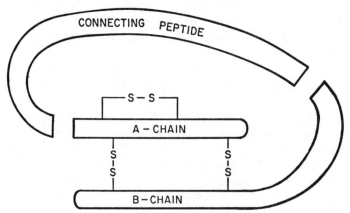

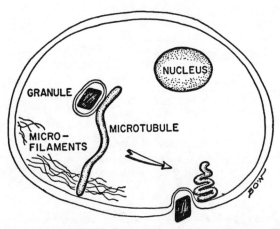

Fig 10–4.—Granular hormones move to cell surface and leave cell. One speculation is that microtubules may help by contraction (or by pointing the way) and that somehow microfilaments may inhibit secretion process.

this closed-loop system and include provision for some open-loop components. For instance, ACTH stimulates hydrocortisone secretion from the adrenal cortex (Fig 10–7). Hydrocortisone can act directly on the anterior pituitary to decrease ACTH secretion but also acts on the hypothalamus to do the same thing indirectly. Furthermore, under environmental stress the brain can directly stimulate the hypothalamus and anterior pituitary to secrete large amounts of ACTH and override the suppressive effect of a high level of hydro-

Fig 10–5.—Hormone acting by stimulating adenyl cyclase in cell membrane and causing formation of cAMP. Cyclic AMP binds to the regulatory subunit *(R)* of an inactive cAMP-dependent protein kinase *(RC)* and causes release and activation of the catalytic subunit *(C)*. The active protein kinase *(C)* catalyzes phosphorylation (usually of a protein), which presumably mediates actions of hormone.

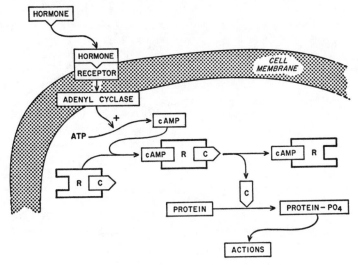

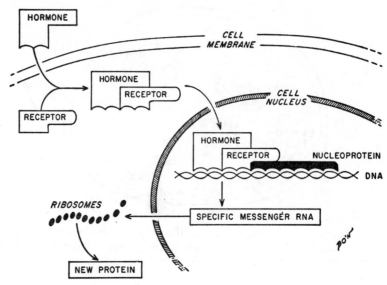

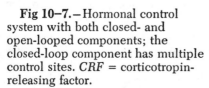

Fig 10–6.—Hormone acting after binding to intracellular receptors: one possible mechanism. Note change in receptor induced by hormone and its subsequent binding to specific sites in the nucleus. Other possibilities exist and the exact events are not clear.

Fig 10–7.—Hormonal control system with both closed- and open-looped components; the closed-loop component has multiple control sites. *CRF* = corticotropin-releasing factor.

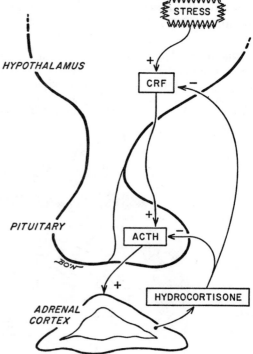

cortisone—all of which is an appropriate response of the organism as a whole but is not a simple closed-loop control system.

Hormonal action can also be regulated after the endocrine cell has secreted. For example, the cell could secrete a precursor that is converted elsewhere to the active hormone: e.g., one thyroid hormone, thyroxine, is converted to the more potent hormone, triiodothyronine, by the liver. The conversion site is thus a locus for potential control. Another type of control occurs when hormone on a responsive tissue regulates the further action of that hormone. Insulin appears to do just this by decreasing its own receptor sites. Or one hormone may modify the action of another hormone acting on the same tissue. Estrogen, for example, enhances the action of progesterone on the uterus. Or the hormone might be metabolized by the responsive tissue to a more or a less active form, thus increasing or decreasing its action. In some tissues, testosterone is converted to a more active form, 5 α-dihydrotestosterone. In other tissues progesterone can be changed to 20 α-hydroxyprogesterone, which is inactive. Again, these are sites for regulation of hormone action. Study of the regulation of these peripheral control mechanisms is in its infancy and little is known about them, but they are clearly important in the fine-tuning of hormonal actions.

How Do We Test for Endocrine Abnormalities or Disease?

Since there may or may not be any anatomical abnormality (and there often is not), we look for excessive or deficient hormone action. Excessive action is usually, perhaps always, due to an abnormally high level of circulating hormone. Deficient action may be due to a diminished level of hormone in the blood or to an impaired tissue response, as in an unusual form of dwarfism in which the tissues fail to respond to growth hormone. An abnormal hormone level in the blood is the usually recognized cause of endocrine disease. We might think the diagnosis simple; measurement of the serum hormone level would do. However, the range of values of any hormone in the general population is often wide, with some normal values being two or more times as much as others. A "normal" value for one patient may be too little for another. To individualize the measurement, the best approach is to test not only for an abnormal hormone level but also for a disordered hormonal control system. If the hormone level is *grossly* high or low, the control system has clearly been unable to compensate properly and a simple hormone measurement is enough. But the control system may still be partially defective when the hormone level itself is within the "normal" range. In this case specific testing of the control system will reveal the abnormality.

Suspicion of a hormonal deficiency calls for a *stimulation test*. Sometimes the patient may already have done this as an *endogenous* stimulation test. For example, a patient suspected of having adrenal insufficiency may have a low normal plasma level of hydrocortisone. A simultaneously high level of plasma ACTH indicates that the body senses a hydrocortisone deficiency and is trying to correct it by secreting more ACTH. However, the extra ACTH is unable to stimulate more hydrocortisone secretion (because of adrenal disease) and so there is a deficient response. Often, the stimulation test is done by the physician; this is the more usual *exogenous* stimulation test. In the instance just noted, it is difficult to measure endogenous plasma ACTH; the exogenous stimulation test consists of injecting ACTH and seeing whether or not the plasma level of hydrocortisone rises.

Suspicion of excessive hormonal secretion calls for a *suppression test*. As before, the patient may have already done an *endogenous* suppression test. The finding of a high value for PTH in a patient with mild hypercalcemia is not a normal finding; we would expect a low normal or low PTH value. A high value indicates excessive PTH secretion or hyperparathyroidism. *Exogenous* suppression tests done by the physician also are useful. For example, giving thyroid hormone should suppress thyroid function; if it does not, something is wrong with the thyroid's control system and the patient usually has hyperthyroidism.

The remainder of this section discusses most of the glands and hormones usually considered in the endocrine system. It should not be considered inclusive of all endocrinology. There are other hormones and factors that are hormone-like, but these will not be covered in any detail. The gastrointestinal tract secretes many hormones, which are thoroughly discussed in Chapter 7, and only briefly alluded to here. Some circulating substances such as bradykinin and histamine usually are not considered hormones, since most of their actions occur locally before they get into the blood stream. Other circulating factors are so ill defined that they are known only by their actions. For example, serum contains several factors that stimulate cell growth but little more is known about most of them; these factors presumably qualify as hormones. Though it is conceptually difficult, we should remember that all these hormones and substances circulate in the blood at the same time and act on one or another tissue constantly. Thus is the organism's integrity maintained.

HYPOTHALAMUS, ANTERIOR PITUITARY AND POSTERIOR PITUITARY GLANDS

The anterior pituitary gland was once called the "master gland" because it secretes numerous hormones, many of which control the secretion of other hormones (Table 10–2). In reality, the anterior pituitary is itself under the control of the hypothalamus to which it is closely linked by the pituitary portal system. Various hypothalamic-releasing and -inhibiting factors are carried by the blood flowing in this system to the anterior pituitary gland where they stimulate or inhibit the secretion of its hormones (Fig 10–8). Three of these hypothalamic factors are small peptides: thyrotropin-releasing hormone (TRH) is a tripeptide; gonadotropin-releasing hormone (GnRH) is a decapeptide (10 amino acids); and somatostatin, or growth hormone-inhibiting hormone, is a tetradecapeptide (14 amino acids). Presumably the other factors also are small peptides.

The control systems for secretion of the anterior pituitary hormones are complex and not fully understood. The controlling hypothalamic hormones

TABLE 10-2.—HORMONES OF HUMAN ANTERIOR
PITUITARY GLAND*

HORMONE	CELL OF ORIGIN	PRINCIPAL ACTION
Growth hormone (GH)	α	Growth, protein synthesis
Prolactin (PRL)	η	Lactation
Thyrotropin (TSH)	β_2, ?γ	Thyroid hormone secretion
Follicle-stimulating hormone (FSH)	δ†	Growth of ovarian follicle, spermatogenesis
Luteinizing hormone (LH)	δ†	Ovulation, corpus luteum formation, ovarian and testicular hormone secretion
Corticotropin (ACTH)	β_1†	Glucocorticoid secretion
Melanocyte-stimulating hormone (β-MSH)	?β_1†	?Darkening of melanocytes
Lipotropin	?β_1	?Mobilize FFA

*After Romeis and Ezrin (see Halmi, N. S., 1974); α and η cells contain acidophilic granules; ? = function in normal man not proved.
†In man FSH and LH may be secreted by the same cell as may ACTH and MSH/lipotropin.

are made and secreted by peptidergic hypothalamic neurons. These neurons in turn are affected by other hypothalamic and CNS neurons and respond to one or another of the neurotransmitters, norepinephrine, dopamine and serotinin, and perhaps others. For example, the secretion of growth hormone is thought to be partly regulated by dopaminergic neurons, which in turn control secretion of somatostatin or growth hormone-releasing and -inhibiting hormones by peptidergic neurons. Feedback control of anterior pituitary hormones is thus possible at many levels (two or more hypothalamic levels as well as at the anterior pituitary itself) and, under some circumstances, there may be positive or stimulatory feedback rather than the usual inhibitory negative feedback. For example, moderate levels of estradiol, the principal estrogen secreted by the ovary, can enhance luteinizing hormone (LH) secretion, whereas at high levels estradiol is inhibitory. The relationship of the hypothalamus and anterior pituitary is a striking example of the close linkage between the nervous and endocrine systems.

The hypothalamic hormones are not limited to the hypothalamus and are found in other areas of the brain such as the pineal gland and the organum vasculosum lamina terminalis (OVLT); TRH actually occurs throughout most of the brain, albeit in low concentrations. The functions of these hormones outside the hypothalamus are unknown; they may be neurotransmitters that came to be hormones as well by evolutionary accident. Other small biologically active peptides that resemble hormones, such as neurotensin (a 13 amino acid peptide) and substance P (an 11 amino acid peptide) also are found in the hypothalamus. Their function too is unknown, but it is an intriguing fact that neurotensin and substance P as well as somatostatin have recently been found in the gastrointestinal tract, resembling the corollary discovery of the gastrointestinal hormones, gastrin and vasoactive intestinal peptide (VIP), in the hypothalamus. The neural and hormonal relationships of the hypothala-

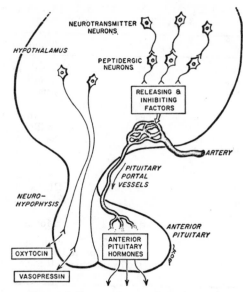

Fig 10–8. — Diagram of hypothalamic connections to pituitary gland. Hypothalamic-releasing and -inhibiting factors are secreted by peptidergic neurons in response to neurotransmitters; these factors then flow through pituitary portal vessels and regulate release of anterior pituitary hormones. Neurohypophysial neurons are a variety of peptidergic neuron and directly secrete oxytocin and vasopressin.

mus and gastrointestinal tract clearly need further exploration as much as does the role of these various peptides in the function of the brain.

The specific anterior pituitary hormones are discussed later in this section, with the exception of melanocyte-stimulating hormone (MSH). This hormone can cause darkening of the skin in high concentrations but probably what has been measured in blood as MSH is in fact lipotropin. Lipotropin is a larger 91 amino acid peptide, which reacts like MSH in the MSH assay; it is probably a precursor or prohormone of MSH (if MSH exists as such at all), since the entire β-MSH molecule is contained within lipotropin. Lipotropin itself was isolated from the pituitary and named because some investigators thought it was involved in release of free fatty acids (FFA) from adipose tissue. An exciting recent discovery is that several fragments of lipotropin other than MSH also occur in the brain and pituitary and bind to receptors in brain tissue that also bind morphine. These lipotropin fragments, now named *endorphins* because they are endogenous analogues of morphine, range in size from 5 to 30 amino acids; they probably modulate perception of pain, they may be involved in the secretion of growth hormone and prolactin, and changes in their production may even play a role in mental illness.

The linkage between the *posterior pituitary* and hypothalamus has long been recognized as another area of close neuroendocrine relationship. The posterior pituitary is in fact a neuronal extension of the hypothalamus and is also called the neurohypophysis. Its hormones, vasopressin (also called antidiuretic hormone or ADH) and oxytocin, are largely synthesized in the hypothalamus (see Fig 10–8). Once synthesized, these hormones are bound to a larger peptide, neurophysin, and transported to the posterior pituitary where they are stored until released by appropriate stimuli. Oxytocin is discussed later in the section on hormones and reproduction, and vasopressin is considered in the renal physiology section (Chapter 6).

THE ADRENAL GLAND

In 1849 and 1855, Thomas Addison described patients having "general languor and debility . . . and a peculiar change of colour in the skin . . ." At autopsy these patients had diseased adrenal glands. Although Addison knew nothing of hormones, he made the association between adrenal disease and clinical illness and realized that the adrenal glands were essential for life.

The adrenal gland is not a single gland but is embryologically and functionally two separate glands, the outer adrenal cortex and the inner adrenal medulla.

ADRENAL CORTEX

Histologically, this portion of the gland has several layers: the outer zona glomerulosa (actually a broken, scattered layer in man), the middle zona fasciculata and the inner zona reticularis.

The two principal hormones of the adrenal cortex are the steroid hormones, *hydrocortisone* and *aldosterone* (Fig 10–9; see also Fig 10–2). Hydrocortisone is made largely in the zone fasciculata and is a glucocorticoid, i.e., an adrenocortical steroid with major effects on glucose metabolism. Aldosterone is made in the zona glomerulosa and is a mineralocorticoid, i.e., an adrenocortical steroid with major effects on sodium and potassium balance.

GLUCOCORTICOIDS

In man, the principal glucocorticoid is hydrocortisone, also called *cortisol* or compound F (after an older nomenclature devised before the steroid struc-

Fig 10–9.—Structures of hydrocortisone (cortisol) and aldosterone. Note numbering of carbon atoms in the 4-ringed steroid nucleus, presence of 3-hydroxyl groups in hydrocortisone and hemiacetal group in aldosterone.

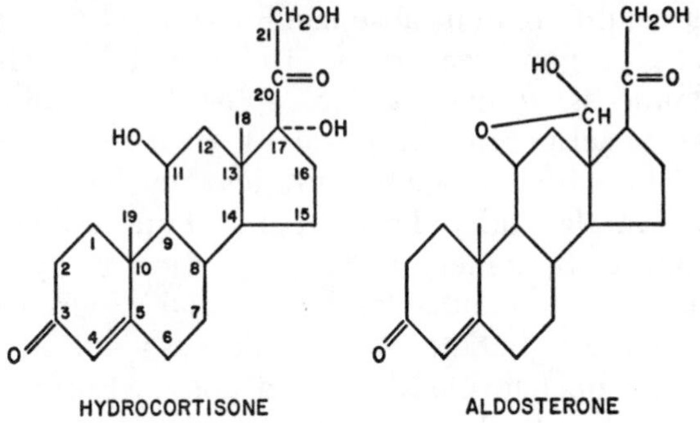

HYDROCORTISONE   ALDOSTERONE

tures were known). This is not true for all species; in the rat, for example, the principal glucocorticoid is corticosterone. Although glucocorticoids are so named because of their effects on glucose metabolism, these are by no means their only or even most important effects but may simply be the ones best studied.

SECRETION AND METABOLISM.—When cortisol is secreted into the blood (about 20 mg per day), it is about 90% bound to plasma proteins, mostly to a specific corticosteroid-binding globulin (CBG) or *transcortin*. Normal mean levels are about 15 μg/dl in the morning and 5 μg/dl in the evening. If cortisol does not act on a responsive tissue, it is converted to inactive metabolites mainly by the liver. These metabolites are largely reduced and conjugated products such as tetrahydrocortisol glucuronide.

ACTIONS.—Cortisol seems to have many actions on different body tissues. Many of its effects are not well defined and are still being actively investigated. Sometimes large pharmacologic doses have actions that are difficult to relate to the effects of physiologic levels of cortisol; the pharmacologic effects may reflect truly physiologic actions that are simply difficult to detect. It is important to note that cortisone, which has a molecular structure identical to cortisol except that it has an 11-keto group instead of cortisol's 11 β-hydroxyl group, is inactive and is effective in vivo only because the body converts it to cortisol.

When cortisol in the blood reaches a responsive tissue, it binds to a specific cytosol-binding protein (see Fig 10–6). This protein has been found in many tissues including muscle, kidney, liver, thymus, adipose tissue, blood vessels, lung, retina, testis and various parts of the brain. However, a reasonable connection between binding of cortisol to the cytosol protein and a specific biologic action has been defined for only a few tissues (e.g., liver and thymus).

The cortisol-cytosol binding protein complex then goes to the cell's nucleus, and the cortisol or the complex itself is transferred to a nuclear binding site. The precise nuclear events are unclear, but they result in an initial increase in RNA synthesis, particularly mRNA, and in most cells an increase in various specific proteins (e.g., enzymes), which then bring about the actions of cortisol in that particular cell. In most cases, however, the specific mRNA and protein have not been identified, leaving open the possibility of cortisol acting through a different mechanism in some tissues.

The *"permissive actions"* of cortisol are a case in point. Cortisol is said to act permissively when a physiologic amount of it allows another biologic process to occur that would not in its absence. For example, epinephrine is a potent stimulator of glycogen breakdown, or glycogenolysis, in the liver via a cAMP-mediated process; without cortisol, epinephrine no longer works. Just what cortisol does to act permissively is not known and "permissive" covers much ignorance. There is some relation to adenyl cyclase and cAMP. Cortisol does *not* directly stimulate adenyl cyclase, but it may support cAMP-dependent cellular responses by somehow promoting the activity of adenyl cyclase or steps subsequent to generation of cAMP. These permissive actions are clearly of great importance but there are few data as yet to connect them to the cytosol and nuclear mechanisms described above. Furthermore, there are fragmentary reports of cortisol acting without nuclear mediation, e.g., a rapid direct stimulation of adenyl cyclase and blockade of kinin generation in leukocytes.

Mediation by the cell nucleus is the best current model of cortisol's actions, but at present it is not well enough developed to explain directly its known effects on tissues or intact organisms. The presumption is that these tissue effects (discussed next) *are* mediated by nuclear effects of cortisol, but in most cases the presumption is unproved.

The overall effect of cortisol on *carbohydrate metabolism* can be viewed as helping to supply glucose to critical tissues when this is needed. During fasting, glucose is in short supply, and yet it is essential for normal brain function and must therefore be derived from the body's carbohydrate stores. Most comes from the liver. Without cortisol, hepatic glucose production is low and there may be serious hypoglycemia; diminished brain function may result. Cortisol affects several processes that ensure glucose supply to the brain. These actions can be called *permissive* because the cortisol level does not rise during fasting and yet it is needed for the glucose-maintaining processes to function well. These actions include the following:

1. Enhanced breakdown of glycogen (glycogenolysis) in liver and muscle. The liver then releases glucose into the blood more effectively in response to glucagon or catecholamines, and the muscle has an energy source.

2. Enhanced synthesis of glucose from precursors (gluconeogenesis) in liver. Most of the glucose formed is derived from amino acids and lactate. Cortisol affects several steps in this process: (a) it increases release of amino acids from muscle by decreasing amino acid uptake and incorporation into protein and perhaps by increasing protein breakdown (proteolysis) into amino acids; (b) it increases hepatic uptake of amino acids and lactate; and (c) it increases conversion by the liver of amino acids and lactate to glucose by enhancing the activity (and sometimes synthesis) of enzymes such as phosphoenolpyruvate carboxykinase that speed both conversion of amino acids to glucose precursors and conversion of glucose precursors to glucose. These enzyme activities may also be increased by cortisol's removal of an inhibitory effect of insulin on the enzymes.

3. Decreased uptake of glucose by tissues that do not require it. This action of cortisol has been worked out best for the thymocyte in which decreased glucose uptake is due to nuclear events leading to new protein synthesis; the protein then somehow causes inhibition of glucose uptake and subsequently cell lysis. In other tissues that are more important in terms of total glucose uptake such as muscle and adipose tissue, it is more difficult to show this effect but it probably does occur in adipose tissue. Part of this effect may not be a direct action of cortisol but, rather, inhibition of insulin's effect on stimulating glucose uptake.

4. Enhanced triglyceride breakdown (lipolysis) to FFA in adipose tissue. The FFA then not only provide an alternate energy source for tissues (e.g., muscle) capable of using them but also enhance gluconeogenesis by the liver.

The result of these effects is that stored glucose is released, new glucose is made and the glucose supplied goes to the brain, which needs it, while it is shunted away from other tissues that can burn FFA or ketoacids instead.

Other effects of cortisol seem unrelated to its actions on glucose metabolism and yet may be more important for survival. The tissues acted on are diverse and include, for example, the kidney (enhances efficiency of sodium and water retention and excretion), the lung (aids in maturation of fetal lung) and the uterus (may play an important role in intiating birth). Perhaps the most important of these other actions involve the brain, the cardiovascular system and the mechanisms involved in inflammatory and immunologic responses.

In the brain, cortisol's actions are almost completely undefined. About all we know is that many parts of the brain have cortisol-binding sites and that in the absence of cortisol a patient may become depressed, apathetic or even psychotic.

Without cortisol, blood pressure falls and the heart does not pump as well, eventually shrinking in size. Cortisol reverses these effects and so helps to maintain blood pressure, enhances the action of other vasoconstrictors and improves cardiac contractility. Except that these tissues have cortisol-binding sites, the mechanisms are unclear.

The most common clinical use of cortisol or its synthetic analogues is to decrease inflammation or to block immunologic responses in patients who do not have adrenal disease. These effects may not be physiologic, although it is tempting to regard them as exaggerations of subtle physiologic actions that we are not bright enough to detect without pharmacologic magnification. At the high doses used, glucocorticoids can decrease formation of antibody-producing cells, inhibit formation of lymphocytes involved in cell-mediated immunity and cause destruction of existing lymphocytes. The inflammatory response is blunted at several levels; cortisol can cause inhibition of leukocyte and macrophage migration, enhanced vasoconstriction and decreased kinin formation. Glucocorticoids may also decrease inflammation by stabilizing lysosomes, thus preventing release of lysosomal proteases that cause tissue breakdown.

Future investigation will clarify these nonglucose actions of cortisol and allow us to put them in proper perspective.

CONTROL OF CORTISOL SECRETION. — The secretion of cortisol is controlled by the plasma level of ACTH, which is secreted by the anterior pituitary gland. The rate of ACTH secretion is controlled by the level of cortisol and by corticotropin releasing-factor (CRF) from the hypothalamus.

Cortisol secretion responds to changes in plasma ACTH in a few minutes. Since ACTH is secreted in short bursts, cortisol secretion follows suit so that with frequently timed blood samples, we can show peaks of cortisol secretion throughout the day.

HOW DOES ACTH ACT? — The initial event is binding of ACTH to a specific receptor in the plasma membrane of the adrenal cell (Fig 10–10). There follows a sequence of activation of adenyl cyclase, increased cAMP synthesis, activation of protein kinase and ribosomal protein synthesis and enhancement by a specific protein of the conversion of cholesterol to Δ^5-pregnenolone. Increased cortisol synthesis follows apace. This general scheme has good evi-

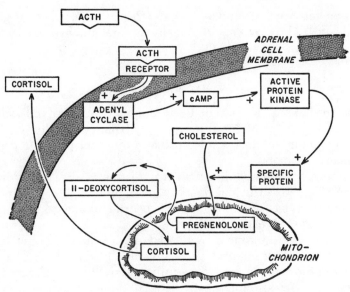

Fig 10–10. — Simplified scheme of stimulation of cortisol synthesis and secretion by ACTH. Note importance of mitochondrion.

dence to support it but several questions remain unanswered. What is the specific protein? Is the enhancement of Δ^5-pregnenolone synthesis due to biochemical stimulation or to stimulation of mitochondrial transport? Is cAMP the only intracellular messenger? Does ACTH enhance secretion in addition to synthesis?

HOW IS ACTH SECRETION REGULATED? — Without hypothalamic CRF, no ACTH would be secreted. It might seem then that the only problem would be the regulation of CRF secretion. However, the action of CRF on ACTH secretion can be modulated by the amount of cortisol present in the pituitary ACTH-secreting cells; a higher cortisol level blunts the effect of CRF. Since CRF is not yet reliably measurable, it is difficult to say what regulates its secretion. However, ACTH secretion is clearly stimulated by many kinds of stressful stimuli such as trauma, surgery and medical school examinations; these stresses act via the CNS and probably stimulate CRF secretion (see Fig 10–7). Cortisol probably acts on the hypothalamus to blunt CRF secretion as well as on the pituitary to blunt its action.

ACTH secretion is also related to the time of day. Plasma levels of ACTH and cortisol are higher in the morning than in the evening. This *diurnal variation* of ACTH and cortisol secretion is mainly determined by the *sleep-activity cycle* (the light-dark cycle of day and night seems to be a minor factor). Thus, if we decide to sleep during the day and stay up during the night, the diurnal cycle of cortisol will shift by one-half day after about one week. Since flying halfway around the world is essentially the same thing, the "jet lag" or "jet fatigue" that lasts a few days may be related to the several days needed for the shift in the diurnal cycle of cortisol.

Thus the plasma ACTH level is determined by the cortisol level, by the current sleep-activity cycle and by various stresses imposed at any given time. The actual cortisol level is a result of many internal and external factors and is in fact largely neurally controlled.

MINERALOCORTICOIDS

The "primordial sea" that circulates through all tissues of the body consists largely of salt and water. Maintaining a proper balance of sodium is critical for survival, and the mineralocorticoids are an important part of a complex sodium-regulating system.

The main mineralocorticoid in man is *aldosterone*, discovered in 1953 by Simpson and Tait. Its principal effect is to act on the kidney to increase sodium resorption. It has some intrinsic glucocorticoid activity but, in the concentrations found in vivo, this is slight.

About 125 μg of aldosterone is secreted daily. It is mostly bound to albumin and not to a specific binding-protein. It is metabolized by the liver along pathways similar to those for cortisol; a usefully measurable metabolite is tetrahydroaldosterone (THaldosterone).

ACTIONS.—The main site of action of aldosterone is on the kidney where it increases sodium resorption in the distal tubule and probably in the proximal collecting duct. It also causes loss of potassium and hydrogen ions into the urine by an "exchange mechanism" in which a resorbed sodium ion is exchanged for an excreted potassium or hydrogen ion. However, although this description is generally correct in that aldosterone's action includes excretion of potassium and hydrogen ions, it cannot be explained by a simple exchange mechanism; there are too many experimental situations in which sodium resorption and potassium excretion can be dissociated. Aldosterone also acts on the salivary gland, sweat glands, colon and possibly the small intestine to increase sodium resorption. Thus its overall action is to keep sodium in the body.

There may well be actions on still other tissues such as blood vessels or the heart because aldosterone is thought to play some role in maintenance of normal blood pressure. At the extreme, for example, excessive aldosterone can cause hypertension. However, it has not yet been possible to prove a direct cardiovascular effect of aldosterone separate from its effect on sodium metabolism.

HOW DOES ALDOSTERONE ACT?—The initial events are the same as those in cortisol's action: binding to a specific cytosol-binding protein, transport to the cell's nucleus, binding to nuclear material (DNA, nucleoprotein, or both) and a subsequent increase in RNA synthesis, especially in messenger RNA (mRNA) (Fig 10–11). The result is synthesis of a specific protein, *aldosterone-induced protein* (AIP), which then induces sodium resorption. Whether or not this mechanism applies to potassium is not clear; it may well not, since actinomycin D blocks the effect on sodium resorption but not that on potassium excretion.

HOW DOES AIP INCREASE SODIUM RESORPTION?—There are several possible mechanisms and some controversy remains over which one is correct.

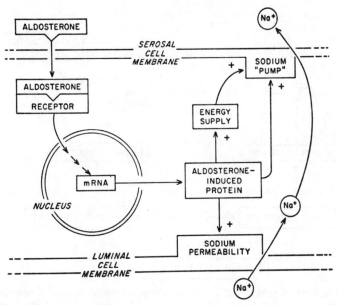

Fig 10–11.—Possible mechanisms of action of aldosterone on sodium transport across renal tubular cell.

Possibilities include the following: (1) AIP either is itself an enzyme or it activates an enzyme that enhances the permeability of the renal tubule's luminal surface to sodium (this postulated enzyme is thus a *permease*); (2) the AIP could somehow enhance the cell's energy supply by acting, for example, on glucose metabolism—since sodium transport accounts for most of the kidney's metabolic activity, enhancing the energy supply could indirectly increase sodium transport; and (3) the AIP could directly affect the sodium transport mechanism by which sodium is pumped out of the tubular cell's serosal surface into the blood. Since aldosterone can rapidly activate the renal tubular sodium-potassium-dependent ATPase activity, a component of the sodium "pump," the last possibility (3) may be more likely but the issue is by no means settled.

CONTROL OF ALDOSTERONE SECRETION.—The principal direct stimulator of aldosterone secretion in normal man is angiotensin II. This is an octapeptide generated from the decapeptide, angiotensin I, by the action of converting enzyme, particularly in the lung. Some evidence suggests that angiotensin III, which has one less amino acid than angiotensin II, is the actual stimulator of the zona glomerulosa.

However, angiotensin is only a part of a larger and more complex control system (Fig 10–12). Ultimately, circulating blood volume determines aldosterone secretion. Blood volume may be decreased by hemorrhage or by dehydration and loss of salt and water. In the kidney the decreased volume causes increased secretion of *renin*, an enzyme that catalyzes the conversion of circulating renin substrate, a peptide made in the liver, to angiotensin I. The stimuli for renin secretion may include decreased delivery of sodium (or chloride) to the macula densa; decreased distention of the renal arterioles and

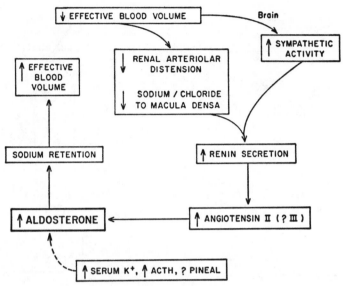

Fig 10–12. – Normal control of aldosterone secretion. Return of decreased effective circulating blood volume to normal turns off the aldosterone-stimulating system.

increased sympathetic nerve activity may also be stimuli for renin secretion (see Chapter 6, section on regulation of extracellular volume and sodium balance).

The increased renin secretion then causes increased generation of angiotensin I, which leads to increased angiotensin II and aldosterone secretion. The regulatory cycle is completed when the extra aldosterone causes sodium retention. Since increased body sodium causes a parallel increase in body water, the volume deficit is then corrected and the whole aldosterone-stimulating system is shut off.

Although complex, there is still more to the regulation of sodium balance than this. Because sodium and its accompanying water are so important to the body's integrity, teleologically the defense of body sodium almost demands redundant controls.

First, aldosterone secretion is stimulated not only by angiotensin but also by ACTH, as well as by a rise in serum potassium and perhaps by some hormonal factor found in or near the pineal gland. These factors are probably minor when the renin-angiotensin system is working normally. If, however, the kidneys are removed, aldosterone may be secreted at a normal rate; the higher serum potassium found in the anephric state seems responsible. In fact, one hypothesis is that the level of potassium inside the zona glomerulosa cell is the final regulator of aldosterone secretion and that any stimulus to aldosterone secretion does so by raising the intracellular adrenal potassium concentration.

Second, aldosterone is not the only controller of sodium balance. Most of the renal resorption of sodium occurs in the proximal tubule, probably without the need for aldosterone. Furthermore, after large doses of aldosterone in man have caused retention of 400–500 mEq of sodium, sodium retention ceases and any extra sodium is simply excreted in the urine even if aldosterone injections are continued. This "escape phenomenon" remains unexplained but it is obviously important for homeostasis. Some investigators have attributed it to an as yet unknown hormone, also called "third factor" (glomerular filtration of sodium and aldosterone are the first and second factors affecting renal sodium resorption), whereas others believe the phenomenon is simply caused by shifts in intrarenal blood flow.

Aldosterone is then a major, but not the only, regulator of sodium balance. Its secretion increases when blood volume falls, during dietary salt or water restriction, and when the body is in the upright position (due to pooling of blood in the legs). The renin-angiotensin and sympathetic nervous systems, acting in concert with changes in potassium, control its secretion.

ADRENOCORTICAL DISEASE

When the control systems cannot properly regulate adrenal cortical hormone secretion, clinical disease results. There may be a deficiency or an excess of cortisol, aldosterone or both, and the defect may be in the adrenal gland itself or in the control systems regulating adrenal hormone secretion.

ADRENAL INSUFFICIENCY.—In Addison's disease, mentioned earlier, the adrenal gland is destroyed by an autoimmune process or tuberculosis and the patient is deficient in both cortisol and aldosterone. This is *primary* adrenal insufficiency. Plasma ACTH and renin levels are high (an endogenous stimulation test) and the adrenals do not respond to ACTH or to angiotensin with a rise in cortisol or aldosterone (exogenous stimulation tests).

Sometimes there is aldosterone deficiency alone; the patient has weakness, lightheadedness, a high serum potassium and a somewhat low serum sodium. An enzymatic block in aldosterone synthesis may be responsible and there is then a high renin level. More often, however, renin deficiency is the cause of isolated aldosterone deficiency.

Similarly, there may be an isolated cortisol deficiency; as a primary adrenal disorder, this is more common in children and is probably caused by a defect in the enzymatic pathway leading to cortisol. Much more commonly, however, isolated cortisol deficiency is due to ACTH deficiency (*secondary* adrenal insufficiency). Occasionally this occurs because of pituitary or hypothalamic disease but more often because a physician has given and then withdrawn pharmacologic glucocorticoid therapy in a patient without endocrine disease. Glucocorticoid therapy suppresses ACTH secretion and, on withdrawal of corticoids, it may take months for it to return to normal. During this time the patient is at least partially adrenally insufficient.

Once any of these deficiencies has been defined, proper therapy consists of supplying what is missing, preferably by mouth since it is easier for the patient. The glucocorticoid used may be cortisol, cortisone or a synthetic such as prednisone. The mineralocorticoid cannot be aldosterone since it is not effective when given by mouth; the synthetic 9α-fluorohydrocortisone usually is employed.

EXCESSIVE SECRETION OF ADRENAL HORMONES. — This may arise from primary adrenal disease; for example, tumors of the adrenal glands may secrete too much aldosterone (primary aldosteronism or Conn's syndrome) or too much cortisol (a form of Cushing's syndrome). Or excessive secretion may result from too much stimulation; there may be too much renin secretion, resulting in high aldosterone levels (secondary aldosteronism) or too much ACTH secretion with high cortisol levels (Cushing's disease).

Primary aldosteronism presents with hypokalemia, alkalosis and hypertension; plasma aldosterone is high and plasma renin low (an endogenous suppression test, since a low renin is normally associated with a low aldosterone). Administration of a mineralocorticoid such as 9 α-fluorohydrocortisone causes little or no suppression of aldosterone (exogenous suppression test). Treatment is removal of the adrenal tumor. Sometimes the disease is not due to a tumor but to bilateral adrenal hyperplasia; the cause is unknown and treatment is difficult. Agents such as spironolactone that block aldosterone's actions may be helpful in this case or in patients with a tumor but in whom surgery might be risky. The hypokalemic alkalosis then disappears and the blood pressure usually (but not always) returns to normal.

In secondary aldosteronism there also may be hypertension and hypokalemic alkalosis if the excessive renin secretion is due to abnormal renal function, as in malignant hypertension or renal artery stenosis. Or the high renin and aldosterone levels may be appropriate responses to another disease associated with relatively poor renal perfusion such as nephrotic syndrome or hepatic cirrhosis with ascites; there may be hypokalemia but not hypertension. In secondary aldosteronism treatment is directed at the primary disease except that use of spironolactone may palliate some of the symptoms.

Excessive ACTH secretion owing to a pituitary tumor or to excessive CRF secretion by the hypothalamus causes classic Cushing's disease, first described by Cushing in 1912. We see obesity of the face and body, plethora (redness of the face) and frequently a high blood glucose. These findings are due to too much cortisol and are no different from those in patients with cortisol-secreting adrenal tumors or who are taking large doses of glucocorticoids. The term *Cushing's syndrome* applies to any patient with cortisol excess; *Cushing's disease* is used when the cause is excessive ACTH secretion from the pituitary. A variant is the Cushing's syndrome caused by ectopic ACTH secretion from a nonendocrine tumor such as lung carcinoma; these patients have mainly hypokalemic alkalosis since they do not have time to become obese and the high levels of cortisol have a moderate mineralocorticoid effect.

In all patients with Cushing's syndrome, a high plasma cortisol may be diagnostic. The plasma cortisol is, however, frequently equivocal and an exogenous suppression test is needed. The drug usually given is dexamethasone, a synthetic glucocorticoid that suppresses cortisol secretion in normal subjects and does not interfere with cortisol assays. If the plasma cortisol is not suppressed normally, the patient has Cushing's syndrome. Further tests are then needed to define whether the origin is adrenal, pituitary or pulmonary, and treatment is generally directed at the source of the disease, e.g., adrenalectomy for adrenal tumor, pituitary radiation for excessive pituitary ACTH secretion or use of cyproheptadine (a blocker of serotonin's action) for excessive CRF secretion.

ADRENAL MEDULLA

Sometimes viewed as an endocrine gland, the adrenal medulla is actually an integral component of the sympathetic nervous system. It is not essential to life provided the rest of the sympathetic nervous system is functioning well, since patients who have had total adrenalectomy and are treated only with glucocorticoids and mineralocorticoids have no difficulty. Yet it is the body's source of epinephrine, the first hormone to be isolated (1898) and once thought to be *the* adrenal hormone.

Epinephrine and *norepinephrine* are the two *catecholamine hormones;* epinephrine is made only in the adrenal medulla, whereas norepinephrine is made both in the adrenal medulla and in sympathetic nerve endings throughout the body (Fig 10–13). Both are derived from tyrosine. The small vesicular chromaffin granules in the medulla and nerve endings not only have a critical enzyme, dopamine β-hydroxylase, needed for synthesis of the catecholamine hormones, but can also take up and store these hormones until a nerve impulse results in their release. Both hormones are rapidly metabolized by an O-methyltransferase to the *metanephrines,* which are then converted to *vanillylmandelic acid (VMA)* by a monamine oxidase, so that only 2–5% of secreted catecholamines appear in the urine unmetabolized.

The *general actions* of the catecholamines, including the concept of alpha and beta adrenergic responses and their relationship to cAMP, are discussed in Chapter 3.

Fig 10–13.—Synthesis and storage of epinephrine and norepinephrine in adrenal medulla and sympathetic nerve. Epinephrine is made only in adrenal medulla but may be taken up and stored in vesicles of either tissue.

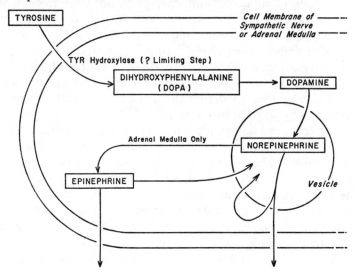

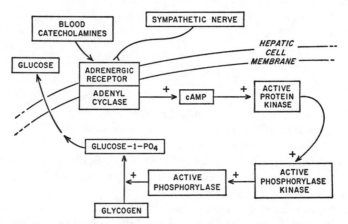

Fig 10–14.—Catecholamine-induced glycogenolysis in liver. Each active enzyme has been converted from an inactive form.

Here we shall mention only the more important metabolic actions. These include stimulation of both lipolysis and glycogenolysis and of oxygen consumption. In times of stress, such as external attacks on the organism or during heavy exercise, metabolic fuels such as glucose and FFA are needed by all tissues, and maintenance of body temperature is of obvious importance in these settings, at least in warm-blooded animals. Rapid stimulation of glycogenolysis and lipolysis will supply the needed glucose and FFA within minutes. In man, these effects are largely beta adrenergic and are probably initiated by binding of the catecholamines to a specific beta adrenergic receptor on the plasma membrane of hepatic and adipose cells.

Stimulation of glycogenolysis by epinephrine was the first described instance of a cAMP-mediated hormonal response. Sutherland later received the Nobel prize for this work (Fig 10–14). Although there is little question that epinephrine can stimulate glycogenolysis, it may not do so in vivo. More likely, norepinephrine released by hepatic sympathetic nerves is more important than circulating epinephrine from the adrenal medulla. Nevertheless, other effects of epinephrine can help maintain the blood glucose, principally its alpha adrenergic action on the pancreatic beta cell by which it inhibits insulin secretion, thus allowing blood glucose to rise. Catecholamine-stimulated lipolysis is also a cAMP-mediated process and, although sympathetic nervous stimulation may account for much of the FFA released under stress, circulating epinephrine and norepinephrine may also be important. Exactly how catecholamines induce thermogenesis, i.e., raise the body temperature, is not clear. The effect may be largely metabolic and secondary to oxidation of rapidly released glucose and FFA, particularly the latter, but it may also be due to a separate, perhaps hypothalamic, effect of the catecholamines.

There are only a few recognizable *disorders* involving catecholamine hormones. A *pheochromocytoma*, a tumor of the adrenal medulla or nearby sympathetic nervous tissue, may secrete large amounts of catecholamines and be a rare cause of hypertension. Measurements of urinary catecholamines, metanephrines or VMA usually make the diagnosis, and treatment is adrenalectomy. Sometimes there is a defect of the entire sympathetic nervous system, so that insufficient catecholamines are made, which results in hypotension on standing (orthostatic hypotension). The patient complains of lightheadedness and may actually faint. The condition is difficult to treat but catecholaminelike drugs may help. Finally, there is a poorly defined illness that may be due to increased sensitivity of beta adrenergic receptors. The patient has tachycardia and hypertension and appears nervous; a beta adrenergic blocking agent, propranolol, may be useful.

THYROID GLAND

Goiter, or an enlarged thyroid gland, has been a recognized disease for several thousand years. It was not until the late 19th century, however, that physicians recognized thyroid deficiency in man and treated it successfully with extracts of the thyroid gland. This treatment corrected the lethargy, dull mental state, puffy face, intolerance for cold and hypothermia in hypothyroid patients and it also was noted to increase oxygen consumption. Clearly, whatever was in the thyroid extract had potent effects on many tissues, including the CNS, and had something to do with maintaining body temperature (thermogenesis) and oxygen consumption.

In 1914 Kendall isolated *thyroxine*, and its structure was defined by Harington in 1926. Later, after realizing that there probably was another thyroid hormone, Gross and Pitt-Rivers isolated *3, 5, 3'-triiodothyronine* in 1952.

Synthesis and Secretion

The two thyroid hormones, thyroxine and triiodothyronine, are conveniently abbreviated to T_4 and T_3, respectively, since one contains four iodine atoms, the other three (Fig 10–15). They are synthesized by the iodination and coupling of tyrosyl residues in a large thyroid glycoprotein, thyroglobulin (TGB), which has a molecular weight of 600,000–1,200,000. Iodine absorbed

Fig 10–15.—Structure of thyroxine and triiodothyronine. The actual structure is three-dimensional so that the iodine at 3′ is not equivalent to the one at 5′; and 3′ iodine confers much biologic potency, the 5′ iodine none.

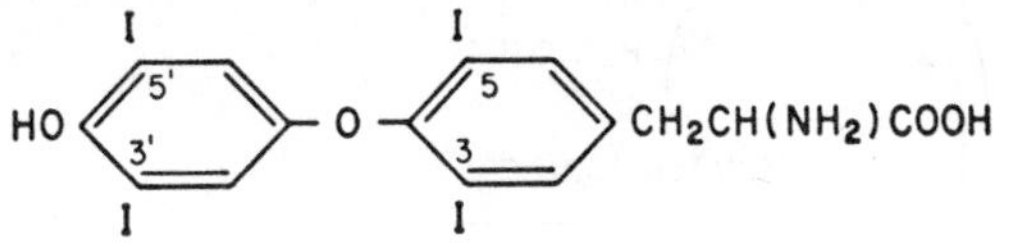

THYROXINE (T₄)

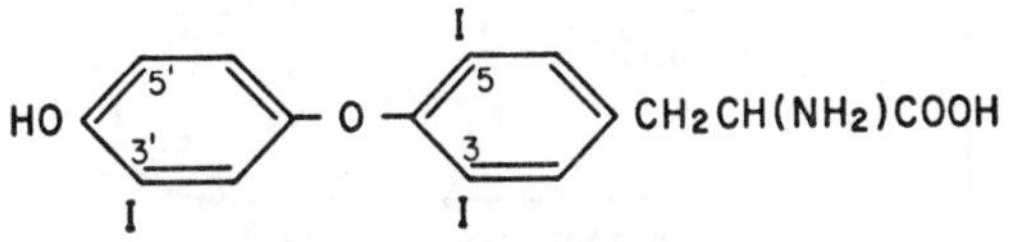

TRI-IODOTHYRONINE (T₃)

from the diet circulates in the blood as the iodide anion (I⁻) and is actively concentrated by the thyroid cell by a sodium-dependent iodide pump (Fig 10–16). Once in the cell, the iodide is oxidized to some as yet unknown form and is then attached to a tyrosyl residue in the TGB molecule, forming a mono-iodotyrosine residue (MIT). This whole process is catalyzed by a thyroid peroxidase and is called the organification of iodide. Another iodide goes through the same process and is attached to the same tyrosyl residue, forming 3, 5-diiodotyrosine (DIT). By another poorly defined process, one DIT then attaches to either another DIT or MIT residue via an ether linkage to form T_4 or T_3. This "coupling" reaction is probably enzymatic and may involve the peroxidase just mentioned. Most of the iodination occurs at the border of thy-roid follicular cell and the central colloid store, which is surrounded by the follicular cells. The T_4 and T_3 formed are an integral part of the TGB molecule and, as such, are stored until needed, mostly in the follicular colloid.

For secretion of T_4 and T_3 to occur, TGB must be broken down, since the intact molecule is too large to pass from the thyroid follicle into the blood (Fig 10–17). Microvilli on the follicular cell-colloid border aid in resorption of small droplets of colloid into the cell. There a protease, probably lysosomal, breaks down the TGB molecule so that T_4 and T_3 can be released to the blood stream. Much MIT and DIT is also released but most does not enter the blood because MIT and DIT are deiodinated in the thyroid cell and the iodine is immediately reused for organification.

Once the thyroid hormones are secreted into the blood, they are, like the steroid hormones, almost completely bound to plasma proteins, mainly to a specific thyroid hormone-binding globulin, or TBG (not to be confused with the TGB, or thyroglobulin, noted above). With a normal total serum T_4 level of 7 μg/dl, the amount *not* bound to protein (the free T_4), is only about 2 ng/dl (or 0.03%). T_3 is less avidly bound to protein and, with a normal total serum T_3 of

Fig 10–16.—Biosynthesis of thyroid hormones. Iodide pump normally maintains an iodide gradient of 25 to 1 (inside the cell versus outside). *MIT* = monoiodotyrosine; *DIT* = diiodotyrosine.

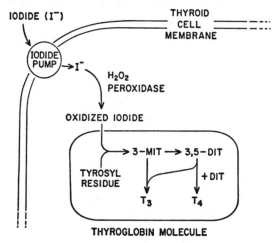

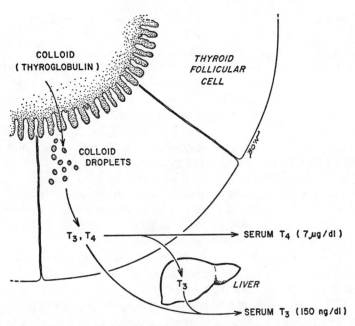

Fig 10–17.—Secretion and peripheral conversion of thyroid hormones. TSH stimulates colloid droplet formation and hormone release. Most of serum T_3 is made in the liver.

150 ng/dl, about 0.5 ng/dl (0.2%) is free T_3. It should be noted that most of the circulating T_3 is *not* secreted by the thyroid gland; about two thirds of it comes from deiodination of T_4 to T_3 by other tissues such as the liver. T_4 can also be metabolized to 3, 3', 5'-triiodothyronine, which is similar to T_3 except that the iodine atom removed is on the benzene ring nearer the amino group rather than on the more distal one. This compound is therefore called *reverse-T_3*. It is metabolically inactive and has at most only 5% of the potency of T_3. Whether T_4 is metabolized by the liver to T_3 or to reverse-T_3 is clearly important in the biologic action of thyroid hormones.

ACTIONS

The principal observable actions of the thyroid hormones in the whole animal are stimulation of (1) protein synthesis, growth and development; (2) oxygen consumption; and (3) central nervous system development and function. In general we can say that whenever a process is affected by thyroid hormone the process moves more quickly. At present there is a good deal of controversy over the active form of thyroid hormone and over the biochemical mechanisms responsible for the observed actions. It is useful to note that in many ways thyroid hormones resemble steroid hormones: they bind to proteins in serum, they bind to cytosol and nuclear receptors in target cells, their actions generally have a lag time of perhaps an hour or two, rather than the few minutes seen with most of the peptide hormones, and their actions can often be considered permissive.

The initial step in thyroid hormone action is binding of the hormone to specific cellular receptors. Receptors exist in both cytosol and nucleus and are found in at least liver, kidney, heart, lymphocyte, brain and anterior pituitary cells. The function of the cytosol receptor is not clear, but it is reasonably certain that binding of thyroid hormones to the nuclear receptor is necessary for most (though not all) of their actions. T_3 is two to five times more potent than T_4 and some investigators have suggested that T_3 is the *only* active thyroid hormone and that T_4 acts simply as a precursor for T_3. However, the nuclear-binding protein binds both T_3 and T_4. Although T_3 is bound more strongly, the T_4 that penetrates tissues may well account for about one third of the total activity of the thyroid hormones. After binding to the nuclear receptor, T_3 and T_4 probably act to increase mRNA synthesis by stimulating an RNA polymerase; the result is an increase in *protein synthesis* in the cell.

Nevertheless, a nuclear mode of action does not exclude direct stimulation of protein synthesis in mitochondria or polysomes, and there is some evidence that thyroid hormones act at these sites as well. Thyroid hormones, can, for example, bind directly to a mitochondrial-binding protein in the liver and kidney, a step that may also be of great importance in affecting oxygen consumption since most oxygen transfer occurs in the mitochondrion. Nor does a general enhancement of protein synthesis explain why increased protein breakdown also occurs, resulting in increased turnover of many cell proteins; nor does it explain the specific effects of thyroid hormones on the proper sequential *growth and maturation* of tissues such as the cerebral cortex and bony epiphyses. Still, an effect on synthesis of specific proteins helps to explain increases in several hepatic and adipose tissue enzymes and in cardiac myosin.

How do thyroid hormones increase *oxygen consumption?* This effect was observed in 1895 and is still not completely understood; it is also called the calorigenic or thermogenic effect because oxygen consumption is directly related to heat production. With the realization that the locus of the cell's oxidative reactions was the mitochondrion, the effects of thyroid hormone on the mitochondria were closely studied. Although some data suggest that any action of thyroid hormone on mitochondria is mediated through the cell's nucleus, the idea of at least a partially direct mitochondrial effect seems to have a reasonable foundation. The mitochondrial-binding protein mentioned above, direct binding of thyroid hormone to isolated mitochondria and a rapid (several minutes) increase in oxygen consumption in these isolated mitochondria all support a direct effect, although they do not exclude an additional nuclear-mediated effect. Whether thyroid hormone ultimately effects changes in oxidation by some structural change in the mitochondrion or by some change in the respiratory chain is not clear; whatever the case, there appear to be increases in both stage 3 and stage 4 respiration, i.e., oxygen consumption related to conversion of ADP to ATP and to increased turnover of a high-energy intermediate in oxidative phosphorylation.

Another approach to the effects of thyroid hormone on oxygen consumption has been to examine the mechanism by which sodium is actively pumped out of cells. This is an appropriate site to study because in most tissues the energy needed for active transport of sodium is a major fraction of the total energy supply and because the biochemical equivalent of the sodium pump, the activity of the sodium-potassium (Na-K)-dependent ATPase, is measurable. Thyroid hormone does indeed increase the activity of this enzyme system in liver, kidney and muscle. Whether this effect is mediated by the cell nucleus and subsequent protein synthesis is not known. It is also not clear whether the increased sodium pumping and increased mitochondrial ATP synthesis are causally related or simply parallel effects of thyroid hormone. It should be noted that thyroid hormone does *not* cause increased oxygen consumption in some tissues, including brain, testis and spleen. Some support for the sodium pump as a site of action for thyroid hormone is seen in the lack of its effect on brain Na-K ATPase, an expected result if thyroid hormone-dependent oxygen consumption and the sodium pump are related.

The *cerebral effects* of thyroid hormone are poorly understood. The brain has cytosol and nuclear-binding proteins for thyroid hormones but not mitochondrial-binding proteins. Although thyroid hormone does not stimulate oxygen consumption, it does stimulate protein synthesis. How this relates to the improved alertness and mental acuity in properly treated hypothyroid patients is simply not known.

Thyroid hormones also have actions that can be understood as generally *increasing turnover* of a process, such as cholesterol or bone mineral turnover, or as *permissive*, such as allowing catecholamines to act more effectively on increasing glycogenolysis, lipolysis and cardiac rate. These actions may be mediated by effects on protein synthesis but little is known about them.

CONTROL OF SECRETION

Thyrotropin (TSH) is the principal stimulator of thyroid hormone secretion; without it, the patient becomes hypothyroid. As with ACTH and the adrenal cortex, TSH binds to a receptor site on the plasma membrane of the thyroid cell and activates adenyl cyclase, causing a rise in intracellular cAMP. A cAMP-activated protein kinase then presumably stimulates resorption of follicular colloid, proteolysis of the resorbed thyroglobulin by lysosomal proteases and release of T_4 and T_3. However, linkage of the active protein kinase to these actions has not been directly shown. Since cAMP levels fall despite continued secretion of T_3 and T_4 after exposure to TSH, presumably there is stimulation of cellular events other than activation of protein kinase that causes continued resorption and proteolysis of follicular colloid. The precise role of the sympathetic nerves supplying the thyroid or of thyroid prostaglandins and cGMP in stimulating thyroid hormone secretion is not clear; these may be intermediate or minor regulatory factors.

The TSH stimulates not only the release of preformed thyroid hormone but also the synthesis of new hormone and the growth of the entire gland. The effects on synthesis include increases in iodide pump activity, in peroxidase activity and organification and in coupling of iodotyrosines to form T_3 and T_4.

These effects of TSH are modulated by the amount of circulating iodide; a higher iodide level in the blood (and hence for the moment in the thyroid) decreases the effect of TSH on both the iodide pump and synthesis of new hormone. Further, large amounts of iodide can block the action of TSH on the release of thyroid hormone, an action sometimes used in treating hyperthyroid patients.

TSH also causes increased protein synthesis in the thyroid, an effect mediated by cAMP that probably occurs at the level of translation rather than by new RNA synthesis. How many of the other actions of TSH could be secondary to its effect on protein synthesis? Certainly an increase in cell size could be related to protein synthesis, as could TSH-induced increased glucose oxidation. However, increased protein synthesis would explain neither an increase in cell number, which must be DNA-related, nor the effects of TSH on thyroid hormone synthesis or secretion since they are not blocked by puromycin, a known inhibitor of protein synthesis.

How is *TSH secretion* controlled? On the surface, the relationship between TSH and the thyroid hormones appears to be a straightforward negative feedback control system: TSH stimulates thyroid hormone secretion and thyroid hormone inhibits TSH secretion. However, nothing is as simple as it seems at first. TSH secretion is actually stimulated by a hypothalamic hormone, TRH. Secretion of this hypothalamic tripeptide is in turn stimulated by noradrenergic neurons and perhaps inhibited by serotoninergic neurons in the hypothalamus. Furthermore, there is a slight diurnal variation in TSH secretion, with plasma levels higher at night than during the day. Thus there are clear effects of the CNS on TSH secretion, and thyroid hormone could act at any level—suprahypothalamic, hypothalamic or pituitary—in controlling TSH secretion. To date the major observed inhibitory effect of thyroid hormone on TSH secretion is on the pituitary; however, thyroid hormone does not simply block TSH secretion but, rather, inhibits the ability of TRH to stimulate TSH secretion (Fig 10–18). The amount of TSH actually secreted is thus a result of the balance between TRH and thyroid hormone, as seen by the pituitary thyrotrope cell.

Inhibition of the action of TRH is, of course, another tissue effect of thyroid hormone, and the mechanism is similar to the other actions of thyroid hormone. Thyroid hormones, particularly T_3, bind to pituitary cytosol and nuclear-binding proteins, stimulate RNA and protein synthesis and then inhibit the action of TRH. Whether the presence of specific binding proteins for thyroid hormones in the brain indicates negative feedback control sites in the brain is not clear; some evidence actually suggests the opposite, i.e., thyroid hormone stimulates synthesis of hypothalamic TRH rather than inhibiting it. An unexpected finding in this regard is the presence of TRH in small but definite amounts throughout the CNS. No one knows what it is doing there but it may act as a neurotransmitter.

There may also be regulators of TSH secretion other than thyroid hormone, just as ACTH is regulated by more than cortisol. Cold, for example, stimulates TSH secretion in rats and newborn humans almost certainly by acting via the brain to increase TRH secretion. Since the hypothalamus also has a tempera-

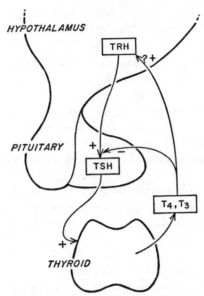

Fig 10–18.–Control of TSH secretion. Thyroid hormones block TRH stimulation of TSH release and may possibly stimulate TRH secretion at the same time.

ture control center, it is tempting to speculate that both body temperature and thyroid hormone control TRH and TSH secretion, the resulting changes in thyroid hormone secretion helping to control body temperature by their thermogenic effect.

THYROID DISEASE

When TSH and thyroid hormone secretion are in balance, the patient is euthyroid and the thyroid gland is barely palpable, if at all. If there is some defect in the thyroid gland and thyroid hormone secretion falls, there follows at least a temporary increase in TSH secretion and the TSH attempts to drive the thyroid gland harder. If this compensatory mechanism is successful, the thyroid enlarges and thyroid hormone secretion rises to normal. The patient now has a *simple* or *euthyroid goiter.* For reasons not entirely clear, serum TSH levels are often not elevated above the normal range in this condition; perhaps they are only mildly elevated or the thyroid has become more sensitive to TSH. Furthermore, the response of the thyroid to exogenous TSH may not differ from normal probably because the range of normal is so wide. So both endogenous and exogenous stimulation tests are not very helpful, and the diagnosis depends on clinical judgment and a normal T_4 level. Iodine deficiency is often suspected in goitrous patients but is a rare cause of goiter in the United States; in fact, the average U. S. diet contains a fairly high level of iodine.

However, if the thyroid does not compensate properly, clinical *hypothyroidism* results, with all the symptoms noted earlier. If the thyroid tissue has grown but simply cannot make enough thyroid hormone, the patient has goitrous hypothyroidism. In most patients, however, the thyroid tissue is damaged and cannot grow; the patient has nongoitrous hypothyroidism. In either case, T_4 and T_3 levels are low and TSH levels are high, which is an abnormal endogenous stimulation test. Injection of TSH into the patient fails to cause the usual rise in thyroid function, an abnormal exogenous stimulation test. This form of hypothyroidism, i.e., in which the cause is a defect in the thyroid gland, is called *primary hypothyroidism*. The treatment is thyroid hormone. Treatment of simple goiter is the same unless some other cause is suspected for the large thyroid, such as carcinoma, and even then thyroid hormone is useful.

If the T_4 and T_3 levels are low and the TSH level is *not* elevated or is even low, the patient has *secondary hypothyroidism* or hypothyroidism owing to TSH deficiency. This could be due to lesions of the pituitary, such as a destructive pituitary tumor, or of the hypothalamus with a deficiency of TRH (hypothalamic hypothyroidism). Since TRH is difficult to measure, we can distinguish between pituitary and hypothalamic hypothyroidism only by giving exogenous TRH. Failure of serum TSH to rise after TRH suggests a pituitary lesion, whereas a rise in TSH in such a patient points to a hypothalamic defect; the results of the test, however, are not always clear cut. Treatment is the same as for other types of hypothyroidism: thyroid hormone. TSH and TRH are not useful modes of treatment because of scarcity and the need for injections.

If thyroid hormone secretion rises and remains elevated, clinical *hyperthyroidism* results. The patient loses weight because of net loss of protein and fat owing to high turnover and increased catabolism. There is a preference for cold weather because of increased oxygen consumption and heat production. Sweating and trembling occur because of increased sympathetic nervous activity. Hyperthyroidism is rarely caused by excessive TSH secretion and, for practical purposes, is always due to a thyroid defect. Serum T_4 and T_3 levels are high, although sometimes only the T_3 level is elevated (T_3 toxicosis). Serum TSH falls to undetectable levels. An exogenous suppression test, needed only in equivocal cases, can be done by giving thyroid hormone and observing the effect on the 24-hour radioiodine uptake by the thyroid; if it does not fall below 25–30%, the patient is probably hyperthyroid. Most patients with hyperthyroidism also have a goiter, since the gland is usually diffusely hyperplastic, suggesting an exogenous stimulus. As noted, the stimulus is not TSH but appears to be immunologic and is related to abnormal infiltration of lymphocytes in the gland.

The treatment is to remove the excessive thyroid hormone—not always easy to do. Subtotal thyroidectomy or radioiodine can remove or destroy enough of the gland to render the patient euthyroid but hypothyroidism often develops later. An antithyroid drug such as propylthiouracil blocks the synthesis of thyroid hormone and renders almost all patients euthyroid, but the disease may relapse when the drug is stopped; nevertheless, this therapy works in many patients and is the preferred first treatment by most physicians.

HORMONAL CONTROL OF CALCIUM METABOLISM

By the late 19th century, surgeons knew that surgery on the neck sometimes caused tetany, a spastic twitching of muscle owing to increased neuromuscular irritability. Although the parathyroid glands had first been described in the Indian rhinoceros almost half a century before, few knew of the description (not surprising) and no one connected these structures with tetany. The experimental physiologists, Gley, Vassale and Generali, then showed that it was the removal of the parathyroid glands that caused the tetany. Still, it was not until 1909 that McCallum and Voegtlin concluded that a low serum calcium (hypocalcemia) was the direct cause of tetany and that the parathyroid glands did something to keep the serum calcium up to normal. *Parathyroid hormone* (PTH) was later extracted from the glands and shown to be the substance that restored serum calcium to normal.

In the normal person the level of serum calcium varies little from its average of 9–10 mg/dl (4.5–5 mEq/L); this is particularly true for that fraction (about 45%) that is ionized, i.e., not complexed or bound to serum proteins. A proper level of serum *ionized calcium* is critical for many physiologic processes such as muscle contraction, blood coagulation, nerve conduction and adenyl cyclase activity. So it is not surprising that serum calcium is controlled by more than one hormone. The control system includes, in addition to PTH, *vitamin D*, which can raise serum calcium, and *calcitonin*, which can lower it.

More than 99% of the body's calcium (1–1.5 kg) is in bone, which is therefore a major site of action for the serum calcium-regulating system. In addition, the kidney and gastrointestinal tract play major roles in calcium regulation. However, much of the calcium resorbed from bone is not used to maintain serum calcium but rather occurs as part of the process of *bone remodeling*. The bony skeleton is the structural support of terrestrial vertebrates and its strength depends on crystallized calcium salts. To do its job, bone must grow when the body grows and change its shape (albeit slowly) when weight-bearing and persistent muscle contraction demand it. This remodeling is a result of resorption of old bone and formation of new. The regulating system for bone remodeling is not as well understood as that for maintaining serum calcium but the same three hormones are involved. Thus the two regulating systems are closely interrelated, although they do not always function in parallel; bone remodeling may increase or decrease without any change in serum calcium.

Synthesis, Secretion and Metabolism of the Hormones

PARATHYROID HORMONE.—The PTH is an 84 amino acid single chain peptide. It is not made as such directly from amino acids; the original product of ribosomal protein synthesis is a larger peptide of about 115 amino acids (pre-pro-PTH), which is then converted to a smaller 90 amino acid peptide (pro-PTH). Six amino acids are then cleaved from pro-PTH to form PTH, which is either directly secreted or stored in small granules. Release of the stored PTH may involve microtubular contraction. Some PTH is destroyed in the para-thyroid glands without being secreted. When PTH is released into the blood, it is rapidly broken down to smaller fragments by tissues such as kidney and liver. A major fragment consists of the first 33 or 34 amino acids (from the N-terminal end); this fragment is biologically active. Other fragments lacking the first one or two amino acids or consisting only of the other end of the mole-cule are not active. Thus the active circulating hormone consists of molecules of several different sizes; furthermore, radioimmunoassays for PTH may de-tect biologically inactive as well as active fragments, depending on the anti-body used.

VITAMIN D.—This is not made in a recognized gland. Ultraviolet radiation in sunlight converts 7-dehydrocholesterol in the skin to vitamin D_3, or chole-calciferol (Fig 10–19). That is why the thick London fog caused rickets (until coal fires were abolished) and why sunlight cures it. Some vitamin D_2, or er-gocalciferol, is consumed in the diet, although the vitamin D that is always added to pasteurized milk in the United States may be either vitamin D_2 or D_3. Both are of equal biologic activity in man but neither is active without being converted to other metabolites. After entry into the blood, where the vitamin is bound to a specific plasma protein, 25-hydroxylation takes place in the liver. The 25-hydroxycholecalciferol or $25[OH]D_3$ is released back into the blood and circulates at a level of 20 ng/ml. Then 1-hydroxylation occurs in the mitochondria of the kidney. The 1-hydroxylation step is stimulated by PTH, perhaps as a result of lowering renal intracellular phosphate concentra-tion, and may be inhibited by calcitonin. The resulting 1,25-dihydroxy-cholecalciferol or $1,25[OH]_2 D_3$ is the principal biologically active form of the vitamin, although there is only about 20–40 pg/ml in blood. The kidney can also hydroxylate $25[OH]D_3$ at the 24 and 26 positions. The biologic signifi-cance of these metabolites is not clear, although 24-hydroxylation may specifi-

Fig 10–19.—Formation of biologically active form of vitamin D_3, $1,25(OH)_2D_3$. Note structural similarity to steroid hormones.

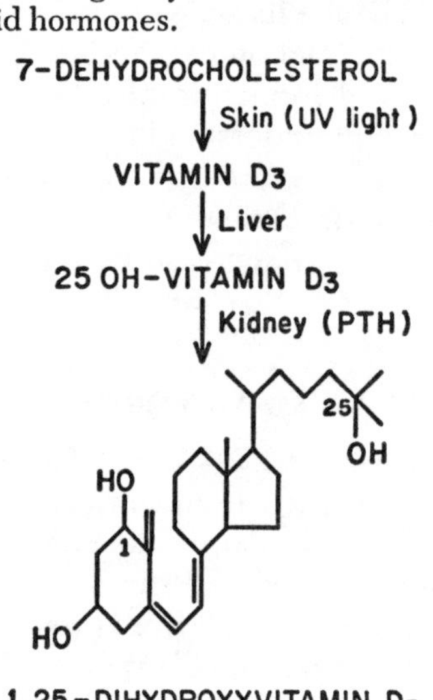

7–DEHYDROCHOLESTEROL

↓ Skin (UV light)

VITAMIN D3

↓ Liver

25 OH–VITAMIN D3

↓ Kidney (PTH)

1, 25–DIHYDROXYVITAMIN D3

cally enhance the vitamin's activity on bone, and 26-hydroxylation its activity on the gut; $1,24,25[OH]_3D_3$ may be an important metabolite since it is about as active as $1,25[OH]_2D_3$. Vitamin D is, then, a hormone and the skin (or the liver or the kidney) is an endocrine gland.

CALCITONIN. — Sometimes called thyrocalcitonin because in mammals most of it is found in the thyroid gland, calcitonin is a 32 amino acid peptide. It is made by the thyroid C cells, also called parafollicular cells, since most of them are found next to but not in the thyroid follicles. Like PTH, calcitonin is first made as a larger molecular weight peptide that is fragmented to form the secreted calcitonin. Once it is secreted, little is known of its metabolism. Measurement in plasma is difficult; recent assays indicate that normal human levels are about 10–30 pg/ml.

ACTIONS OF THE HORMONES

Both physiologic functions of these hormones—*regulation of serum calcium* and *regulation of bone remodeling*—involve direct actions on bone, so some understanding of bone metabolism is needed. However, total calcium balance involves intake and excretion of both *calcium and phosphate*, and the three hormones affect these processes as well.

INTAKE AND ABSORPTION OF CALCIUM AND PHOSPHATE. — Dietary calcium amounts to about 1,000 mg per day, although it is only about 400 mg per day if no dairy products are eaten. Secreted gastrointestinal fluids add another 200 mg to the luminal contents. Of the 1,200 mg presented to the intestinal mucosa per day, perhaps 400 mg are absorbed, most by active transport; the rest is excreted in the feces. Thus about 30–40% of the calcium presented to the mucosa is absorbed, although *net* absorption is only about 15–20% of dietary intake. Absorption is decreased by large amounts of dietary fat, oxalate or phosphate, all of which tend to bind calcium in the intestinal lumen. Absorption also decreases with age but increases when the body needs calcium, as when the diet is deficient in calcium; a patient recovering from rickets or a large fracture may absorb nearly all the calcium ingested. Dietary phosphorus is about 700–1,000 mg per day, and about 70% is absorbed.

Vitamin D is the main stimulator of calcium absorption from the gut; however, as noted, it must first be hydroxylated by the liver and kidney to $1,25-[OH]_2D$. Circulating $1,25[OH]_2D$ acts on the intestinal mucosal cell much as do the steroid hormones it so closely resembles. After binding to a cytosol-binding protein within the cell, it goes to the cell nucleus, where it binds to chromatin. Several hours later, there is a stimulation of RNA (which may not be essential) and protein synthesis, with an increase in the amount of a specific calcium-binding protein (CaBP; molecular weight about 8,000–9,000 in mammals) and in the activity of a microvillous calcium-dependent ATPase. The CaBP and ATPase appear to act at the luminal surface of the cell to bind calcium in the gut lumen and facilitate its entry into the cell. The calcium concentration in the cell rises sharply (up to 5 mM) and in solution this would be toxic; some transport mechanism that keeps calcium out of solution must exist in the cell. The best candidate at present is the intestinal cell mitochondrion but no one really knows. At the serosal border of the cell, the calcium is somehow released from its carrier and actively transported out of the cell, a process that is also enhanced by vitamin D. Since entry of sodium into the cell is coupled to calcium's exit, the Na-K ATPase so familiar in active transport mechanisms may be involved in delivery of calcium out of the cell into the blood stream. Stimulation of calcium transport across the gut by vitamin D is independent of phosphate absorption, but vitamin D may also stimulate part of phosphate transport as well. This action on phosphate transport may be more important than we now think, since a recent study suggests that 1,25-[OH]$_2$D may act primarily on phosphate and that the effect on calcium transport may be secondary.

The *PTH* also stimulates absorption of dietary calcium but not directly. Instead it acts on the kidney to enhance conversion of 25[OH]D to 1,25[OH]$_2$D, which in turn increases calcium absorption. *Calcitonin* in some circumstances can decrease calcium absorption but the evidence is equivocal at best in normal man; it may, however, decrease phosphate absorption.

CALCIUM BALANCE.—Exactly what happens to absorbed calcium is not clear because determining its disposition into bone, urine or gut is difficult. Clearly if the body is in calcium balance, net absorption from the gut and urinary losses should be about equal, as should bone formation and resorption (bone turnover). In normal man, the 100–200 mg of net calcium absorption by the gut approximates the amount in the urine; it is still not clear, however, what method is best for measuring net absorption from the gut. Even less clear is the proper method for assessing bone turnover quantitatively. Several grams of calcium turn over in bone each day, but most of this seems to be calcium "exchange" wherein serum calcium simply exchanges with calcium atoms already in the bone crystal without true bone remodeling. Actual bone turnover, or the sum of resorption and formation of bone by bone cells, probably amounts to only 600–800 mg of calcium per day.

BONE REMODELING.—In adults, bone formation seems to occur only when there has been bone resorption, a phenomenon that makes teleologic sense, since the goal is to remodel the bone in such a way that bone mass stays constant but bone structure is remodeled to suit the needs of the body. Nevertheless, just how bone formation is linked to bone resorption is a mystery. Part of the answer may lie in sequential generation of the different bone cells. Precursor cells may develop into osteoclasts, multinucleated cells that actively resorb bone. Some of the osteoclasts probably then change to osteoblasts, bone-forming cells that make and secrete the bone matrix, or osteoid, made up mainly of collagen and mucopolysaccharides. Next the osteoblasts induce linear calcification along the interface of the new osteoid and existing bone. Finally the osteoid is completely calcified and a small number of osteoblasts become surrounded by new bone and are now called osteocytes. The osteocytes sit in tiny lacunae and are probably connected to each other by cytoplasmic processes; thus much of bone can be viewed as an osteocytic network. The osteocyte can play little role in the remodeling of bone surfaces but may be the most important bone cell for the moment-to-moment regulation of serum calcium. Bone resorption can, therefore, be caused by osteoclasts during remodeling and by osteocytes when raising the serum calcium; bone formation results from osteoblastic activity during remodeling and can also be seen near osteocytes under some conditions.

Bone resorption during remodeling is stimulated by PTH, but only if the serum calcium is not too low. The PTH increases the number and activity of osteoclasts and decreases their conversion to osteoblasts. Vitamin D may be synergistic here but the data are not clear. Decreased calcitonin could also result in bone resorption, but this is speculative since the normal role of calcitonin in bone turnover is not known. There are other factors that can stimulate bone resorption such as the prostaglandins and a substance from activated lymphocytes called osteoclast-activating factor (OAF), but they are of uncertain physiologic importance.

Proper *bone formation* requires both calcium and phosphate; without these no calcified bone will form, although some osteoid may be laid down. Clearly vitamin D is needed; its absence causes poor bone calcification and results in clinical rickets or osteomalacia ("soft bones"). Just what it does is not clear; it would make good sense if vitamin D had a direct action on the osteoblast but the evidence is weak. It may be that vitamin D enhances bone calcification solely by providing proper amounts of calcium and phosphate in the extracellular fluid by increasing both absorption from the gut and resorption from the bone. The PTH may also support osteoblast function. Calcitonin, when injected, decreases osteoclast activity and enhances the osteoclasts' conversion to osteoblasts; it ought to be important in bone formation but proof is still lacking.

MAINTENANCE OF SERUM CALCIUM. — In maintaining a normal serum calcium, bone calcium functions as a major storage depot to be called upon whenever dietary calcium intake falls short. Furthermore, since absorption of calcium from the gut varies, depending on when the last meal was eaten, keeping the serum calcium within the narrow normal range requires a fairly fine-tuned regulatory system that prevents the level of calcium from going too high as well as too low. The fine-tuning occurs in bone; regulation of calcium transport in gut and kidney is probably too slow to correct a sudden change in serum calcium. Although some calcium enters the blood owing to the action of osteoclasts in remodeling, it is likely (though not completely proved) that most regulation of serum calcium is due to changes in the rate of bone resorption by the osteocyte, or *osteocytic osteolysis*. PTH is the major stimulus to rapid mobilization of calcium from bone but it does not act well unless *vitamin D* is present. Whether or not this is a specific synergy between PTH and vitamin D acting on bone cells is unclear. Vitamin D may mobilize some calcium directly and permit PTH to mobilize even more, simply by providing calcium to PTH-sensitive cells, since PTH requires a reasonable amount of calcium to act at all. In any case, vitamin D acts as an amplifier of the action of PTH on bone. As in the gut, $1,25[OH]_2D$ is probably the important vitamin D metabolite; nevertheless, its action on bone may differ in some respects from that on the gut. For example, RNA synthesis seems required for $1,25[OH]_2D$ to act on bone, whereas its necessity is questionable in the gut. Despite reservations about physiologic relevance, *calcitonin* may be an important regulator of the serum calcium level by inhibiting calcium resorption from bone whenever serum calcium rises above normal.

How do these hormones act on bone cells? Bone cells are difficult to isolate and study because of bone's rigid structure so there is a dearth of precise information. The *PTH* is the best studied and, as indicated above, needs a certain level of extracellular calcium to act. Parathyroid hormone binds to a receptor on the plasma membrane of the bone cell and activates adenyl cyclase, thus causing a rise in intracellular cAMP and presumably activation of a protein kinase. At the same time, PTH causes calcium to enter the cell. The raised intracellular calcium concentration may be further elevated by a cAMP-induced mobilization of calcium from the mitochondria. The remaining intracellular events leading to increased osteoclast differentiation and activity and increased calcium resorption are unknown. The *1,25[OH]$_2$D* binds to a cytosol-binding protein and then, as in the gut, binds to nuclear chromatin. What happens next is not clear but bone calcium is mobilized and, depending on the cellular ionic and hormonal environment, is delivered to the blood stream and/or laid down in newly formed osteoid. *Calcitonin* acts like PTH in that it activates bone adenyl cyclase and raises the cAMP content of the cell, although through a different receptor site from PTH. Somehow it then acts as a noncompetitive inhibitor of PTH on bone resorption, perhaps by decreasing intracellular calcium concentration. Its inhibition of PTH action is further enhanced by phosphate, and calcitonin may actually increase phosphate uptake by bone cells. Thus both calcitonin and phosphate can be viewed as modulators of PTH and vitamin D on bone resorption.

REGULATION OF RENAL HANDLING OF CALCIUM AND PHOSPHATE. — Renal excretion of calcium and phosphate is clearly important in homeostasis of serum calcium and bone mineral. The PTH, vitamin D and calcitonin are all involved. The *PTH* has long been known to increase phosphate clearance. It first binds to renal tubular cell receptors, then increases adenyl cyclase activity and cAMP, which results in decreased tubular resorption of phosphate (TRP) and decreased renal intracellular phosphate concentration. Since the actions of PTH on bone and indirectly on gut tend to raise the serum phosphate level, PTH increases the total amount of phosphate in the urine. Parathyroid hormone also enhances the net resorption rate of calcium by the kidney; nevertheless, since the other actions of PTH tend to increase serum calcium, total urinary calcium may rise rather than fall due to the increased filtered load of calcium. How cAMP causes these effects on calcium and phosphate is unknown; one hypothesis is that both effects may be secondary to a primary action of PTH on sodium flux. *Calcitonin* also binds to tubular cell receptors and enhances cAMP production, perhaps in some of the same cells responsive to PTH, although the adenyl cyclase systems probably differ for the two hormones. Like PTH, calcitonin is phosphaturic but, in contrast to PTH, it causes calcium loss into the urine rather than increasing net calcium resorption. *Vitamin D* enhances resorption of calcium, apparently by stimulating formation of CaBP similar to that found in the intestine. It also stimulates phosphate resorption; PTH seems necessary for this effect. Paradoxically, vitamin D may enhance the phosphaturic effect of PTH, although the effect is not clear cut. The overall effect of these three hormones on urinary calcium and phosphate is then a result of many actions on the bone and gut as well as on the kidney; the interaction between PTH and vitamin D at the cellular level is of prime importance but is not well understood.

OTHER ACTIONS OF CALCIUM-REGULATING HORMONES. — Other actions of these hormones are of clinical interest, although they are not well enough defined to fit into a physiologic scheme. For example, *PTH* can increase gluconeogenesis in liver and kidney, apparently mediated by cAMP. It also can increase renal bicarbonate clearance and decrease renal urate clearance, perhaps explaining a tendency toward systemic acidosis, hyperuricemia and gout in patients with excessive PTH. *Calcitonin* can increase lipolysis in adipose tissue and can affect the rate of mitosis of lymphoblasts. The presence of cytosol-binding proteins for 25[OH]D and *1,25[OH]$_2$D* in other tissues such as muscle, brain and pancreas suggests that vitamin D may act on many tissues not usually considered "targets." Such actions and receptor sites may only represent pharmacologic effects of the hormones or evolutionarily atavistic and irrelevant receptors, but more likely their existence indicates that we simply do not know enough about their actions.

CONTROL OF SYNTHESIS AND SECRETION OF THE HORMONES. — The main regulator of *PTH secretion* is the level of serum-ionized calcium; a fall in calcium stimulates PTH secretion and a rise in calcium inhibits it. Since PTH is stored in the parathyroid glands in only small amounts, we would expect changes in serum calcium to affect both synthesis and secretion of PTH in parallel. By and large this is so, but the details are more complex. Rapid PTH secretion may be modulated by cAMP in the parathyroid gland; a higher serum calcium would inhibit adenyl cyclase, decrease cAMP and thus decrease PTH secretion. Since cAMP seems not to affect PTH synthesis, presumably the PTH it releases comes from the storage granules, and suppression of PTH synthesis by calcium occurs by some other mechanism than by decreasing cAMP. A higher calcium level also decreases PTH secretion by actually increasing destruction of PTH within the gland. The 1,25[OH]$_2$D also inhibits PTH secretion, probably by stimulating synthesis of a CaBP similar to that in the gut and kidney. The CaBP could thus enhance uptake of calcium by the parathyroid cell and transport it to wherever calcium shuts off PTH synthesis and/or secretion; this probably is not a rapid action. Finally, calcium does not quickly regulate the conversion of pro-PTH to PTH but may do so over several days' time; for example, 1 or 2 weeks of dietary calcium restriction increases the rate of this conversion in rat parathyroid glands. Thus PTH raises the serum calcium and increases 1,25[OH]$_2$D production; in turn, both calcium and 1,25[OH]$_2$D inhibit PTH secretion (Fig 10–20).

Calcitonin secretion is regulated by serum calcium in the opposite way from PTH secretion; a rise in calcium stimulates calcitonin secretion and a fall probably inhibits it. As with PTH, cAMP may well mediate rapid calcitonin secretion, but there is controversy over whether cAMP stimulates synthesis as well. Other possible stimuli to calcitonin secretion include phosphate and the gastrointestinal hormones, gastrin and pancreozymin; their role remains unclear.

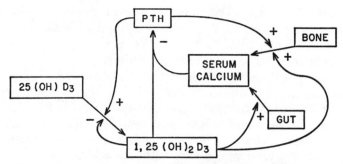

Fig 10–20.—Summary of principal components of control of serum calcium.

Secretion *of vitamin D₃* by the skin is not controlled at all except by the amount of sunlight to which the skin is exposed and perhaps by the darkness of the skin; the intake of dietary vitamin D_2 or D_3 is also under no metabolic control. Production of 25[OH]D by the liver is, however, partially controlled by the product; 25[OH]D inhibits hepatic conversion of vitamin D to 25[OH]D. Renal conversion of 25(OH)D to 1,25[OH]₂D is enhanced by PTH and may be decreased by calcitonin. Thus not only the actions but the control mechanisms of these three hormones are interrelated.

In this discussion so far, no mention has been made of magnesium. In many ways its metabolism resembles that of calcium, although magnesium is more of an intracellular cation. Much is in the bone (about half the body's content), about half the circulating magnesium is bound to serum protein and a low serum level may precipitate tetany. The PTH may enhance renal resorption of magnesium, again a similarity to calcium. Furthermore, the serum magnesium level, normally about 2–2.5 mg/dl, may play a minor role in regulation of PTH secretion similar to the major role played by calcium; a fall in magnesium enhances PTH secretion and a rise inhibits it. Confusingly, a very low serum magnesium seems to inhibit PTH secretion by interfering with its synthesis. Despite these similarities to calcium metabolism, magnesium metabolism and its hormonal interrelationships are not as well studied and cannot be well integrated into the schema presented for calcium.

DISORDERS OF CALCIUM METABOLISM

HYPOCALCEMIA.—A decrease in ionized calcium causes tetany if severe enough but can be much more insidious. The patient may have seizures or a psychiatric abnormality, and hypocalcemia may not even be suspected. Usually the total serum calcium rather than the ionized calcium is measured; this is sufficient, provided the level of serum albumin is normal. Hypocalcemia may be caused by too little PTH or vitamin D or too much calcitonin.

Lack of PTH or *hypoparathyroidism* is not uncommon. The hypoparathyroid patient has too low a level of serum PTH relative to the low serum calcium (endogenous stimulation test). Rarely, a hypocalcemic patient has a normal PTH response to the low calcium, but the PTH fails to act owing to some defect in the target tissues such as deficient cAMP production (pseudohypoparathyroidism); **in** this case we confirm the diagnosis by giving PTH and showing a lack of response, i.e., failure of urinary cAMP to rise (exogenous stimulation test). Treatment of hypoparathyroidism might be the use of PTH, but human PTH is not available; instead, we give oral calcium and large doses of vitamin D.

Hypocalcemia may be due to *vitamin D deficiency,* which, if severe enough, produces rickets in children and osteomalacia in adults, as mentioned earlier. Either lack of sunlight or a deficient diet or both may be the cause. Because vitamin D is added to milk, its deficiency is now more often due to malabsorption of vitamin D from the gut, e.g., after gastrectomy, or to abnormal metabolism of vitamin D so that little $1,25[OH]_2D$ is synthesized, e.g., in severe renal disease or after treatment with diphenylhydantoin, an anticonvulsant drug. Diagnosis would be much easier if measurements of $1,25[OH]_2D$ were readily available; they are not, but may be in the next few years. Treatment of this condition is vitamin D given by mouth.

Excessive calcitonin secretion is only rarely a cause of hypocalcemia; patients usually have a normal serum calcium. It occurs when thyroid C cells become hyperplastic or form a tumor, medullary carcinoma of the thyroid. These patients have few symptoms or signs, so a high level of serum calcitonin serves as a marker for the disease even when a tumor is not clinically evident. Since the disease is often hereditary, family members who are apparently well can be tested. Thus early surgery, the recommended treatment, is possible because of the calcitonin assay.

HYPERCALCEMIA. — This has many causes, among them excessive PTH secretion and excessive vitamin D ingestion. Deficient calcitonin as a cause has not been found. Several types of cancer can cause hypercalcemia by directly invading the bone and perhaps stimulating the release of prostaglandins or OAF; some cancers, however, actually make PTH or something similar to it. If the hypercalcemia is due to taking too much vitamin D, the patient should simply stop. The trick here is to ask the right question. If no other cause is obvious, hypercalcemia is usually caused by excessive PTH secretion by the parathyroid glands, or *hyperparathyroidism.* Most often due to a benign parathyroid tumor, this sometimes can be a result of parathyroid hyperplasia. The patient has a high serum PTH relative to serum calcium (endogenous suppression test) and, when he is given calcium, the PTH level does not fall as much as it should (exogenous suppression test). Treatment is surgical removal of the abnormal parathyroid tissue. In milder cases we may give an oral phosphate solution, which does not correct the high PTH but can return the high calcium and low phosphate to normal.

Many patients with abnormal calcium homeostasis have normal serum calcium but *abnormal bone.* Generally the patient shows thin bone on a routine x-ray or suffers a vertebral fracture after only minimal stress. Patients with decreased bone mass usually are classed together and diagnosed as having *osteoporosis* (porous bone); more likely they suffer from several different diseases. For example, hyperparathyroidism, osteomalacia and invasion of bone by some tumors can all appear as osteoporosis. *Osteopenia* (thin bone) is probably a better term for anyone with decreased bone mineral, whereas osteoporosis could be reserved for patients without a known cause of bone thinning. Defined in this way, osteoporosis may be a specific disorder of bone tissue. It often occurs in the elderly, particularly in postmenopausal women. In part, this is because bone mass decreases with age in everyone, but why it is worse in these patients is unknown. In any case, the patient with osteoporosis has a relative excess of bone resorption compared to bone formation. Loss of gonadal hormones is a factor but is not the entire explanation. Since the cause in most of these patients is not clear, there is no specific test, and treatment is empiric and often unsatisfactory. Until more is known about the disease, reasonable therapies such as increased dietary calcium, moderate doses of vitamin D and estrogen/androgen treatment (if acceptable) are best. Agents such as calcitonin or fluoride (which increases bone formation) may be useful but are still experimental. Ultimately, prevention is the answer since even with effective therapy it takes years to replace the lost bone calcium.

HORMONES AND REPRODUCTION

The individual can survive without reproducing but the species cannot. Every aspect of reproduction is controlled by one or another hormone and the entire process, from the development of germ cells to the free-living new individual, is regulated by a complex interaction of the brain, the pituitary, the gonads—testes and ovaries—and, during pregnancy, the placenta. Each step in the sequence must be made correctly for the next to occur and, in contrast to other regulatory systems, there are no back-up systems to take over if the main system fails. Redundancy lies in the species, not in the individual; if one individual cannot reproduce, others will. The sequence needed for reproduction in an individual includes secretion of fetal hormones; pubertal development; secretion of male hormones and spermatogenesis in men; cyclic hormone secretion and ovulation in women and finally coitus, fertilization, implantation, pregnancy and lactation.

WHAT HORMONES ARE NEEDED?—The pituitary hormones are *follicle-stimulating hormone* (FSH), *luteinizing hormone* (LH) and *prolactin* (PRL). The FSH and LH are gonadotropins, i.e., they stimulate or direct the gonads, and in human beings appear to be made in the same basophilic cell. They are so named because, in females, FSH causes growth of the ovarian follicle containing the ovum and LH converts the follicle into the corpus luteum (luteinization). The FSH may also stimulate follicular hormone secretion, mostly estrogens. LH not only causes luteinization of the follicle but also brings about ovulation in the first place and stimulates hormone secretion by the corpus luteum and the follicle. Both FSH and LH are gonadotropins in males but obviously have different actions, since males have neither follicles nor corpora lutea. The names, however, have stuck because the molecular structures are the same. In males, FSH stimulates spermatogenesis and LH stimulates secretion of hormones, mostly testosterone, by the testicular interstitial (or Leydig) cells. The FSH and LH are, like TSH, glycoprotein hormones containing an α and a β subunit. Both subunits are needed for activity, but specificity of action is due to the β subunit. The α subunits of human FSH and LH are identical, contain about 20% carbohydrate and have a peptide chain of 89 amino acids. The β subunits of human FSH and LH differ in amino acid sequence, although the peptide portions of both have about 115 amino acids. In contrast to many peptide hormones, these glycoprotein hormones tend to have plasma half-lives of many minutes or hours rather than just a few minutes; probably the carbohydrate portion of the molecule, which contains considerable sialic acid, delays degradation.

Prolactin is a major hormone in breast development and milk secretion. Sometimes called luteotropin because it supports the corpus luteum in rats, prolactin is the better name since there is no strong evidence that it is luteotropic in man or many other species; in fact, in man "luteotropin" might be better applied to LH. Prolactin is a peptide hormone containing 191 amino acids; a larger molecular weight form of PRL also circulates in the blood and may conceivably be a precursor of PRL much as pro-PTH is the precursor or prohormone of PTH.

The *gonadal hormones,* like the adrenocortical hormones, are steroids and are made from cholesterol (see Fig 10–2). The principal ones in women are the estrogen, *17β-estradiol* (also called E_2 because it has two hydroxyl groups), which is made by both the follicle and corpus luteum, and the progestin, *progesterone* (Fig 10–21), which is made almost entirely by the corpus luteum. An *estrogen* is a substance that induces an overt increase in sexual activity (estrus or "heat") in animals. Overt estrus does not occur in women, but the corresponding time in the menstrual cycle is halfway between two menses (*not* during menses) when there is a rise in E_2 and ovulation occurs. A *progestin* is anything that supports pregnancy. These are the original definitions of estrogens and progestins; currently, estrogens and progestins are defined either by specific assays or by their stimulation of development of female genitalia such as the uterus. In men, the main gonadal hormone is the androgen *testosterone.* An androgen is anything that induces a male phenotype or masculine behavior. Androgens also have anabolic activity, i.e., they enhance protein synthesis and muscle mass. Most of the estradiol and testosterone in the blood is bound to a specific estradiol-testosterone-binding globulin (E_2TBG). The gonads make many other steroids, some of which may be biologically active, but those mentioned are the most important.

Fig 10–21.—Structures of ovarian hormones. Most estrone is formed by peripheral conversion (not in ovary) from estradiol, but some is directly secreted.

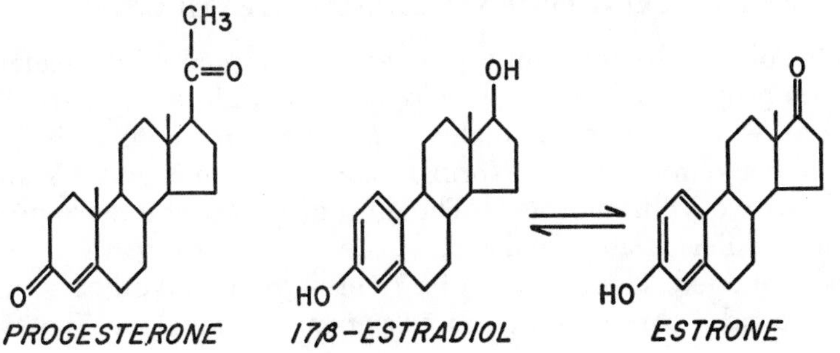

PROGESTERONE   **17β–ESTRADIOL**   **ESTRONE**

Some knowledge of the biosynthesis and metabolism of gonadal hormones is important because various enzymatic defects in their synthesis can cause disease and their metabolism outside the gonads is often a major factor in their action. For example, E_2 can be converted to estrone (E_1) by several tissues, and E_1 can be converted back to E_2 (see Fig 10–21); estrone is a weaker estrogen than E_2 and so the conversion causes a net decrease in the total estrogen action in the body. In many tissues, testosterone must be converted to dihydrotestosterone (DHT) before it can act on them. Furthermore, testosterone can be converted to E_2, which accounts for much of the circulating E_2 in men; since E_2 may be needed for normal sexual behavior in both men and women, this conversion is of obvious importance.

Any list of the *placental hormones* is bound to be incomplete because it seems more are still being discovered. Most are of clear importance in the maintenance and completion of pregnancy and in the preparation of the breast for lactation. They include estradiol and progesterone; the glycoprotein hormone, *human chorionic gonadotropin* (hCG); and the peptide, *human placental lactogen* (hPL or choriomammotropin). The hCG, which is LH-like in its structure and action, helps to maintain the corpus luteum beyond its usual life span and stimulates the fetal testis to make testosterone. The hCG is composed of α and β subunits; the α subunit is identical to that of FSH and LH, whereas the β subunit resembles that of LH but has a longer peptide chain (about 140 amino acids) and contains more carbohydrate. The principal action of hPL is probably breast development during pregnancy, although it can also enhance the action of growth hormone and may have important effects on glucose and fatty acid metabolism. It is a peptide containing 191 amino acids, most of which are the same as in growth hormone itself.

The *control mechanisms* for secretion of gonadotropins and gonadal hormones are discussed below; the precise details are complex and not yet well understood. Nevertheless, in general, there is a basic negative feedback system so that estradiol, with some assistance from progesterone when present, tends to inhibit FSH and LH secretion, and testosterone tends to inhibit LH secretion. Complexities arise because (1) in women there are periodic changes in hormone levels associated with the menstrual cycle, and some aspects of the control system show positive rather than negative feedback control, and (2) in men the testes secrete E_2 as well as testosterone, and control of FSH secretion is not at all clear.

FETAL GONADS AND SEXUAL DEVELOPMENT

The fetus obviously does not reproduce, but normal adult sexual function depends on proper fetal sexual development; fetal hormones are clearly important for this, particularly in males. *Genetic sex* in mammals is determined by the X and Y chromosomes. Normal males have an X and a Y and normal females have two X chromosomes. The presence of a Y chromosome somehow causes the undifferentiated fetal gonad to become a testis. An autosomal factor may also be needed. The absence of a Y chromosome together with the presence of a second X chromosome causes an ovary to develop. Whatever else is present, too many or too few X chromosomes lead to abnormal fetal gonadal development. *Somatic sex*, or a recognizably male or female body habitus, is determined by the presence or absence of testes or ovaries and their hormones. Of great importance is the tendency to develop a female body habitus in the absence of any gonad; thus, a male phenotype can occur only in the presence of a functioning testis.

During development of a male fetus, the fetal testis is stimulated by placental hCG to secrete testosterone. There is a reasonable correlation between maternal serum hCG and fetal serum testosterone, both reaching a peak at 3 to 4 months of pregnancy. The fetal testosterone does several critical things: (1) the testosterone, or possibly the related steroids, Δ^5-androstenediol or DHT, stimulates development of the Wolffian duct into the male internal genitalia (epididymis, vas deferens, seminal vesicles, prostate) and induces them to respond to androgen later in life; (2) fetal testosterone or, more likely, DHT, causes development of male external genitalia, i.e., formation of a normal penis and scrotum; and (3) testosterone has some effect on the brain so that masculine sexual behavior and the noncyclic male pattern of gonadotropin secretion will occur after puberty when serum testosterone reaches adult levels. The testis also secretes something that causes regression of the Müllerian duct, which would otherwise develop into the fallopian tube, uterus and part of the vagina. The responsible agent is *not* testosterone; it appears to be a water-soluble substance of large molecular weight ($>15,000$) made only by fetal or neonatal seminiferous tubules. The testes may also make something that actively suppresses ovarian development.

Less is known about ovarian function in the fetal female. Although much E_2 is present in the serum of female fetuses during the first trimester of pregnancy, male fetuses have almost as much, and the E_2 is more likely to be related to placental than fetal gonadal secretion. Female fetuses, however, have higher levels of FSH than males; this may be related to fetal ovarian development. In general, genetic sex in women determines the presence of the ovary, somatic sex seems largely due to the absence of a testis and adult sexual function is determined both by the absence of testicular secretions during fetal life and by normal ovarian secretion in adulthood.

At birth the serum testosterone level is higher in boys than in girls, although it is much lower than at 4 months of pregnancy. At several months of age, serum testosterone falls in both boys and girls and remains at a low level until puberty. Interestingly, serum FSH and LH actually rise after birth in boys and girls; as in utero, the FSH rise is higher in girls. Since FSH and LH fall to low levels at several months of age, or at about the same time as the testosterone, it is tempting to relate the two; however, neither the relationship between these hormones nor their functional significance in the newborn infant is clear.

PUBERTY

The infant, recognizably a boy or a girl and primed to function as an adult man or woman, goes through childhood with low but detectable levels of gonadotropins and gonadal secretions. At about age 10 or 11 there appear the first signs of puberty, the process by which children are transformed into sexually mature adults. The process takes several years to complete, often lasting until age 16 to 18. On the average it begins about a year earlier in girls than in boys but is of variable onset in both sexes, sometimes not beginning until the mid-teens ("constitutional delayed puberty").

STAGES OF PUBERTY.—Clinically, we can divide puberty into several stages based on the various secondary sex characteristics (the primary sex characteristic is the production of gametes). In girls the stages are (1) prepubertal or no change from childhood; (2) early breast budding and a few pubic hairs; (3) a clear increase in breast and labial size; (4) well-developed breasts, external genitalia and pubic and axillary hair but without menses and (5) adult appearance including menses. In boys the parallel stages are (1) prepubertal; (2) increase in testis size and some pubic hair; (3) further testis and penile growth, with the appearance of some axillary and facial hair; (4) increased pubic, facial and axillary hair with vocal changes and acne and (5) fully adult appearance with normal adult-sized testes (4.5–5 cm in length). In both boys and girls, several more years after puberty is finished are needed for complete maturation. For example, menstrual cycles and ovulation may be quite erratic for several years before becoming more regular. In both sexes there is a growth spurt induced by gonadal hormones, but the rapid increase in height ends when the same gonadal hormones cause fusion of the epiphyses of the long bones.

HORMONAL AND GONADAL CHANGES DURING PUBERTY.—The physical changes of puberty are directly due to an increase in secretion of gonadotropins and gonadal hormones. In both sexes there is first an increase in FSH and LH secretion; the rise in FSH appears to be steeper than the rise in LH. This is followed by ovarian follicular development and increased E_2 secretion in girls and by an increase in testicular size and testosterone secretion in boys. In both sexes, the rise in LH occurs principally at night and is induced by sleep, indicating that the LH rise is dependent on the hypothalamus. Along with the higher LH level, LH secretion now occurs in bursts with a peak every hour or two. In boys the sleep-induced rise in LH correlates well with the rise in serum testosterone. The E_2 and testosterone are directly responsible for most of the physical changes in girls and boys, respectively; their levels can be directly correlated with the several stages of puberty. Along with the increase in E_2 and testosterone, there also is an increase in secretion of the adrenal steroids, dehydroepiandrosterone (DHEA) and dehydroepiandrosterone sulfate (DHEA-S) in both boys and girls. Since these steroids can be eventually converted to small amounts of testosterone, they may contribute to some of the pubertal changes, particularly the growth of pubic and axillary hair. Why these adrenal steroids increase is not known.

Thus, throughout puberty there is a progressive increase in the levels of FSH and LH. In girls the gonadotropins stimulate the ovaries to produce a parallel increase in E_2 secretion and to release ova, whereas in boys the testes are stimulated to secrete gradually increasing amounts of testosterone and to make spermatozoa.

What causes the hormonal changes of puberty? The question remains controversial. The best hypothesis is that there is an increased secretion of GnRH by the hypothalamus, which results in higher gonadotropin and gonadal hormone secretion. During childhood there seems to be effective negative feedback control of FSH and LH by the small amounts of circulating E_2 and testosterone in girls and boys. The low levels of gonadotropins and gonadal hormones keep each other in balance presumably because only small amounts of E_2 or testosterone are needed to suppress the secretion and/or action of GnRH. For some reason during puberty it becomes more difficult for E_2 or testosterone to suppress GnRH, and the resulting higher GnRH secretion stimulates in turn FSH, LH and E_2 or testosterone secretion. In essence the set-point for gonadal hormone inhibition of GnRH/gonadotropin secretion is set at a higher level. This hypothesis requires proof; for example, measurements of GnRH in blood throughout puberty are needed and this has not yet been done. Furthermore, the critical question of why the hypothalamus becomes less sensitive to gonadal hormones is unanswered.

Other hypotheses are possible. Perhaps the change in hypothalamic sensitivity is unrelated to gonadal hormones but simply occurs at a certain age or weight. Or perhaps the initial change in sensitivity occurs in the pituitary or in the gonad. If the pituitary were to become more sensitive, it would secrete more FSH and LH in response to the same amount of GnRH. If the gonad were to do the same, the gonad would secrete more E_2 or testosterone in response to the same amount of FSH and/or LH; for example, FSH is known to increase the Leydig cell's sensitivity to LH. With the latter hypothesis, we must still postulate a hypothalamic component. There would have to be a positive feedback effect, the initial increase in gonadal hormones stimulating either the hypothalamus or pituitary rather than inhibiting them. There is some support for this since immature female rats given small doses of estrogen show a rise in LH.

THE TESTIS AND REPRODUCTION IN MEN

The adult testis has two functions: (1) to make active spermatozoa (spermatogenesis) and (2) to secrete hormones that affect the entire body as well as sexual function and ultimately make it possible for the spermatozoa to fertilize an ovum.

TESTICULAR HORMONES

The principal testicular hormone is *testosterone*, secreted by the Leydig or interstitial cells at a rate of about 7 mg per day (Fig 10–22). The testis also secretes small amounts of testosterone precursors, DHEA and Δ^4-androstenedione; a testosterone metabolite, DHT; and two estrogens, E_2 and E_1. The DHT and E_2 are physiologically active as an androgen and estrogen, respectively, but the other secreted steroids have little biologic activity. Nevertheless, DHEA and androstenedione (which are also secreted by the adrenal cortex) can be converted to testosterone by other tissues so they may indirectly

Fig 10–22.—Structure of testosterone.

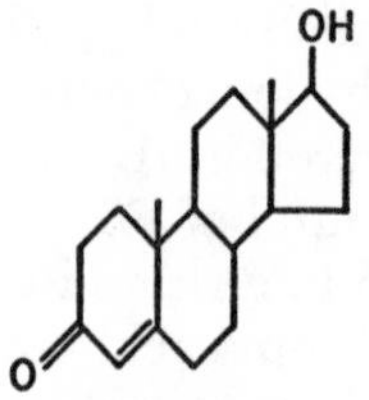

TESTOSTERONE

contribute a small amount of androgen activity. Furthermore, testosterone it-self is converted by other tissues to DHT and to E_1 and E_2; in fact, these tes-tosterone conversions account for most of the circulating levels of DHT, E_1 and E_2 in the blood. Finally, the conversion of testosterone to DHT (and pos-sibly to the even more reduced 5α-androstane-3β,17β-diol) in certain andro-gen-sensitive tissues is an important step in testosterone's action (see below). Very little is known about how or even if these nongonadal conversions are regulated, and yet any such regulation would be of obvious importance in understanding how androgens work. Average serum levels of these steroids in adult men are

$$\text{Testosterone} = 600 \text{ ng/dl (slightly lower in evening)}$$
$$\text{DHT} = 50 \text{ ng/dl}$$
$$E_2 = 3 \text{ ng/dl}$$
$$E_1 = 4 \text{ ng/dl}$$

ACTIONS OF TESTOSTERONE.—Some of the actions of testosterone have been known for thousands of years. Castrated men usually lose their sexual desire. If castration is done in childhood, the body is less muscular, the geni-talia smaller and the arms and legs longer because of late fusion of the epi-physes. The voice remains high pitched because the larynx does not grow in size; during the Renaissance, boys were intentionally castrated (the "castra-ti") so that puberty would not ruin their singing voices. The effects of testos-terone on the genitalia, on other male secondary sex characteristics and on sexual desire or libido (a CNS effect) are considered *androgenic;* those on muscle and on protein synthesis in general are called *anabolic*. The distinc-tion is somewhat arbitrary since there is increased protein synthesis in the genitalia as well, but it is useful because certain synthetic steroids have mod-erate anabolic but little androgenic activity, an important factor if these com-pounds are to be used in women or children.

HOW DOES TESTOSTERONE ACT?—The best studies have been done with *prostatic tissue,* but the same mechanisms appear to hold for the seminal vesi-cles and epididymis. The prostate and seminal vesicles are dependent on tes-tosterone for their growth and function; they secrete most of the seminal fluid needed to carry the spermatozoa.

Testosterone is taken up by the prostate and rapidly converted by a micro-somal 5α-reductase to DHT. This, like other steroid hormones, then binds to one or more cytosol-binding proteins (see Fig 10–6). The DHT-protein com-plex enters the cell nucleus and binds to a nuclear receptor and then to chromatin (? to an acidic protein). This is followed by increased RNA polymerase activity, increased template activity of nuclear chromatin and increased synthesis of mRNA and rRNA, which leads to increased synthesis of various proteins by the ribosomes and to cell growth. Some of the DHT is converted further to 5α-androstane-3β,17β-diol (3β-diol); there is some speculation that DHT may stimulate principally prostatic cell growth and that 3β-diol may stimulate cell function, i.e., the prostatic secretions that contribute to seminal fluid. Roughly similar events occur in *muscle* except that testosterone itself appears to bind to a cytosol protein and to act without being converted to DHT. In some tissues, there may be no need for cytosol binding of the androgen, which then binds directly to the nuclear receptor site (rat dorsal prostate). Finally, not all of testosterone's action on the prostate need be mediated via the cell nucleus; the DHT-cytosol-binding protein complex may directly enhance ribosomal protein synthesis.

Presumably the same sort of mechanism operates in the *brain* and *hypothalamus* to cause changes in behavior and in gonadotropin secretion. Testosterone and/or DHT is taken up by several areas of the brain including the hypothalamus, in which there is a cytosol-binding protein. Aggressive behavior in animals is clearly induced by testosterone; whether a higher level of serum testosterone correlates with more aggressive behavior in men is uncertain. However, male sexual behavior is unquestionably induced by testosterone. Whether testosterone itself is the active steroid is another question. In castrated male rats, DHT is effective in stimulating the prostate but is a poor stimulator of sexual behavior. The current hypothesis is that testosterone must be converted to estradiol, i.e., aromatized, in the brain for normal sexual behavior. In castrated monkeys, however, DHT is effective; a reasonable guess would be that DHT might be effective in men as well.

Nevertheless testosterone alone in the adult is not sufficient for completely normal adult male sexual behavior. Based largely on studies in rats but with some support from work in primates, it seems that fetal testosterone secretion is necessary to "set" the brain to respond properly to testosterone as an adult. And, of course, normal male sexual behavior requires the presence of a female. This seemingly facetious statement has a solid experimental basis. In rats and monkeys, the presence of a sexually active female is needed for normal sexual behavior in the male. An estrogen-deficient female will not do. The reason is that estrogens induce production in the female vagina of volatile short chain fatty acids, which are smelled by the male. These fatty acids are called *pheromones* because they are hormone-like, i.e., they are chemical messengers that stimulate and are carried from one individual to another of the same species. The vaginal pheromones produce overt male sexual behavior in properly primed male rats and monkeys. They exist in women but it is not clear what role, if any, they play in human sexual behavior.

Androgens are also active in stimulating *granulocyte* and *red cell* production, the latter in part by increasing erythrogenin secretion by the kidney. This action may be related to the known androgen-binding protein in the kidney and is clinically useful in certain patients with deficient red cell production.

WHAT STIMULATES TESTOSTERONE SECRETION?—The *LH* in adults and *hCG* in the fetus are the direct stimulators of testosterone secretion. They bind to a specific receptor on the plasma membrane of the Leydig cell and activate an adenyl cyclase, resulting in an increased cAMP. By steps as yet unknown, the cAMP then enhances the mitochondrial side chain cleavage of cholesterol much as it does in the adrenal cortex. The resulting Δ^5-pregnenolone is converted to testosterone by several microsomal reactions. Prolactin may support the action of LH on the Leydig cell but the evidence to date is only suggestive.

SPERMATOGENESIS AND SPERM MATURATION

The bulk of the testis is made up of the seminiferous tubules where *spermatogenesis* occurs (Fig 10–23). The germ cells lining the basal area of the tubules go through many changes to become active motile spermatozoa, which are then released into the lumen of the tubule. During the entire process, the developing germ cells are in close contact with the Sertoli cells, which are located so that they are interposed between the germ cells and the blood and interstitial fluid. The exact function of the Sertoli cell is not clear but it seems to support spermatogenesis in some way. The germ cells develop in clusters so that an entire group of spermatozoa are formed at the same time; probably intercellular bridges between the germ cells are responsible for this coordination.

The early undifferentiated *spermatogonia* (type A) first become more differentiated to type B spermatogonia. Proliferation of spermatogonia leads to formation of *spermatocytes,* a midstage. The primary spermatocytes divide fur-

Fig 10–23.—Sequence of spermatogenesis.

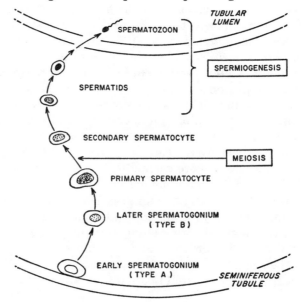

TUBULAR LUMEN

SPERMATOZOON

SPERMIOGENESIS

SPERMATIDS

SECONDARY SPERMATOCYTE

MEIOSIS

PRIMARY SPERMATOCYTE

LATER SPERMATOGONIUM (TYPE B)

EARLY SPERMATOGONIUM (TYPE A)

SEMINIFEROUS TUBULE

ther, followed by reduction division or *meiosis*, forming the secondary spermatocytes, half of which contain an X chromosome and half a Y chromosome. Further proliferation and some nuclear condensation then take place and the cells are then called *spermatids*. Up to this point all the cells are round. The spermatids are gradually transformed into spermatozoa after going through several more stages including the formation of dense chromatin, an acrosome and a tail; this transformation is called *spermiogenesis*. The mature spermatozoon is then released into the central lumen of the tubule; until its release, it had constantly remained in contact with a Sertoli cell. The entire sequence so far takes 1½ to 2 months in normal men.

The formed spermatozoa are not yet fully functional. They proceed first to the epididymis, where their chromatin becomes fully mature and where they acquire their characteristic motility, and then they go to the vas deferens, where secretions of the prostate and seminal vesicles are added to the seminal fluid. By the time the spermatozoa are fully mature and appear in the ejaculate, 2 more weeks have passed. Thus the entire process takes 2 to 2½ months —a fact of clinical importance since any attempt to modify the spermatogenic process takes 8 to 10 weeks to be reflected in the ejaculate.

Spermatogenesis and the subsequent maturation of the spermatozoa are both under hormonal control. Both gonadotropins, FSH and LH, are necessary. The FSH binds to receptors in the tubular cells, principally if not exclusively to Sertoli cells. In the Sertoli cells, FSH stimulates the production of cAMP, activation of a cAMP-dependent protein kinase, and eventually increased synthesis of an androgen-binding protein (ABP) that binds testosterone and DHT and is different from the cytosol-binding protein that mediates the action of testosterone. New data indicate that FSH also stimulates E_2 synthesis in the Sertoli cell.

What the Sertoli cell does to support spermatogenesis is uncertain, but the events after stimulation by FSH are somehow linked to a subsequent increase in RNA and protein synthesis in the adjacent germ cells. It is also not clear at just what point the sequence of spermatogenesis is enhanced by FSH. Some evidence suggests that FSH helps convert type A spermatogonia to type B spermatogonia and to spermatocytes. However, other evidence points to an enhancement of the metamorphosis of spermatids to spermatozoa.

The LH is necessary for spermatogenesis, not because it acts directly on the tubule but because it stimulates testosterone secretion. The high concentration of testosterone in the fluid bathing the tubules, perhaps aided by synthesis of some testosterone in the tubules themselves, is essential for spermatogenesis. Testosterone (or DHT) binds to a specific cytosol-binding protein in tubular cells, probably the spermatogenic cells. This binding protein is not the ABP from the Sertoli cells. The androgen-binding protein complex then enters the cell nucleus, binds to chromatin and presumably stimulates RNA and protein synthesis. In some species, if spermatogenesis is already going well, a high level of testosterone is all that is needed to keep the process going even without FSH. Furthermore, testosterone alone seems to support specifically several stages of spermatogenesis, particularly meiosis. However, if spermatogenesis has not yet started or if for any reason it has stopped or slowed down, testosterone alone is not enough; FSH is essential to get it going. This may be at least part of the explanation of the link between Sertoli cells and germ cells. The ABP induced by FSH may now be critical. Synthesized in the Sertoli cells and perhaps taken up by the immature or inactive spermatogenic cells, ABP may help concentrate any available androgen in the spermatogenic cells until they begin to proliferate. Further work will tell if this hypothesis is correct.

The ABP also is secreted into the tubular lumen along with the spermatozoa, travels through the tubules and appears to be taken up by epididymal cells where it may serve the same function as in the spermatogenic cells, i.e., increasing the amount of androgen available for binding to the specific intracellular cytosol-binding protein. As are the other internal genitalia, the epididymis is a testosterone-sensitive tissue. Since the spermatozoa mature further and acquire the ability to fertilize ova only after passing through the epididymis, the ABP and/or some androgen-dependent epididymal secretion may be the operative agent. There is, however, some doubt about the role of ABP in man, since it has proved difficult to demonstrate it in the human testis. Finally, an unknown factor in the androgen-dependent secretion of the prostate (though not the seminal vesicle) confers on the spermatozoa the ability to move in one direction rather than randomly; this factor may be related to prostaglandin E. Thus by the time the spermatozoa are ejaculated, they are fully mature. This would not be possible without the proper hormonal milieu at every step of the way.

How is gonadotropin secretion controlled in men?— Since FSH stimulates the seminiferous tubules and LH the Leydig cells, it would be intellectually satisfying if the secretion of FSH were controlled by something from the tubule and that of LH by something from the Leydig cell. To some degree this is the case but there remain many uncertainties. LH is controlled mainly by testosterone but, since the testis secretes E_2 and since E_2 can suppress LH secretion, the actual level of LH probably depends on both gonadal hormones. The control of FSH is much less clear. In part it depends on testosterone and E_2 but, in addition, something in the germinal tissue or seminal fluid does seem to regulate FSH secretion. This substance, *inhibin*, has been sought for years. Recent experiments have defined it somewhat, but it remains nebulous. Inhibin is not a steroid, is not **ABP** and may be a protein with a molecular weight of 15,000–20,000 daltons. Overall gonadal control of both FSH and LH secretion in males seems to be a result of testosterone and E_2 levels and the hypothetical inhibin.

The mechanism by which the gonadal hormones control gonadotropin secretion is also complex. Both FSH and LH are stimulated by the hypothalamic decapeptide, GnRH, also called LH-releasing hormone (LRH) or LH-releasing factor (LRF). The gonadal hormones could inhibit GnRH secretion by the hypothalamus or they could act on the pituitary to modulate GnRH stimulation of FSH and LH secretion. Perhaps both possibilities are true since androgen cytosol receptors exist in both the pituitary and hypothalamus of rats. At the pituitary level, evidence in rats suggests that testosterone may decrease the LH response to GnRH and E_2 the FSH response. At the hypothalamic level, data from rats and men suggest that, whereas testosterone inhibits GnRH secretion, small doses of E_2 may actually increase GnRH secretion. But there are other data showing different results, and the precise relationships of GnRH, gonadotropins and gonadal hormones in men remain undefined.

Finally, higher CNS centers can affect hypothalamic GnRH secretion. For example, the pineal gland, the amygdaloid nucleus or even a female phero-mone sensed by the CNS all may affect GnRH secretion. The control of male gonadotropin secretion is obviously complex, though not as complex as in women. Many questions remain to be answered.

THE OVARY AND REPRODUCTION IN WOMEN

THE MENSTRUAL CYCLE: HORMONAL, OVARIAN AND UTERINE

For reasons not yet clear, the hypothalamic-pituitary-ovarian axis functions cyclically in normal women after puberty. Too much androgen will, however, abolish the cycle. Once established, the monthly cyclic pattern is strikingly regular in most women for 30 years or more. The periodic monthly bleeding, or *menses,* is the sloughing of an endometrium primed for implantation of a fertilized ovum; in a sense, the menses is a failed pregnancy. The menses has long been known to be related to hormonal changes but only in the last few years have hormone assays been good enough to define these changes pre-cisely.

A rise in FSH(Fig 10−24), even before bleeding stops, starts the develop-ment of a small group (10−20) of new follicles, one of which will ovulate. Why only one follicle eventually ovulates in women is not clear; it probably relates to the amount of FSH secreted, since injection of larger amounts of FSH causes multiple ovulation (and is one reason for the recent rash of reports of quintuplets). The inner cell lining of the follicle, composed of *granulosa cells,* surrounds the ovum. Outside these inner cells are the *thecal cells.* The FSH binds to specific receptors on the granulosa cells and does not seem to affect the ovum directly but, rather, through these cells. How FSH then stimulates follicular growth is unknown. With growth of the follicle, several layers of granulosa cells develop and a cavity, the antrum, forms. This contains the fol-licular fluid. The cells lining the follicle are responsible for ovarian hormone secretion. Both granulosa and thecal cells contribute to synthesis of the estro-gens, E_2 and E_1; the final steps in estrogen synthesis take place in the thecal cells. The granulosa cells make progesterone, found in the follicular fluid and secreted in relatively small amounts during follicular development (2 mg per day). The granulosa cells may also serve an as yet undefined supportive func-tion for the developing ovum, much as do the Sertoli cells for spermatogene-sis, an analogy supported by the presence of specific FSH receptors in both cell types.

After the initial rise in FSH that starts the follicle on its way, a further in-crease in FSH seems unnecessary. Estrogen secretion by the developing folli-cle results in a gradual rise in serum E_2, which in turn is probably responsible

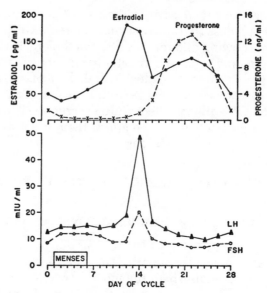

Fig 10–24.—Hormonal changes during menstrual cycle. Ovulation occurs just after peak in FSH and LH; follicular, or proliferative, phase precedes peak; luteal, or secretory, phase follows peak (data representative of several laboratories).

for a slow fall in serum FSH, a negative feedback effect. How does the follicle continue to develop in the face of a falling FSH level? The best explanation is an interesting example of a positive feedback effect at a local level. The initial secretion of E_2 actually enhances the action of FSH on follicular growth and increases the number of granulosa cells. Thus a smaller amount of FSH would still be effective, and E_2 secretion is probably essential to further follicular development.

Whether FSH actually stimulates estrogen secretion is not certain. LH, however, is needed for follicular E_2 secretion. There are specific LH receptors on granulosa cells that are distinct from the FSH receptors; how LH then causes estrogen secretion is unclear, although by analogy with the Leydig cell and corpus luteum, LH probably acts via cAMP with an enhancement of pregnenolone formation. Since serum LH does not rise during follicular growth, how are increasing amounts of E_2 secreted? First, because there are more granulosa cells and, second, because there are more LH receptors in mature follicles, since FSH induces an increase in LH receptors, another example of local positive feedback.

The secreted E_2 and E_1 stimulate the uterus to grow and develop. The uterine endometrium thickens and the endometrial glands elongate. This is the endometrial *proliferative* phase and corresponds to the ovarian *follicular* phase of the menstrual cycle. These phases end with ovulation and take up about the first half of each menstrual cycle.

Toward midcycle, serum E_2 reaches its peak and, at this relatively high level, exerts a *positive* feedback effect on FSH and LH secretion. A day or so after the E_2 peak, there is a sudden burst of LH secretion and a parallel but lesser rise in FSH secretion. Whether or not a rise in progesterone contributes to this effect is controversial. Probably the positive feedback effect of E_2 acts on a different control point from the negative feedback one; the E_2 may increase the pituitary's response to GnRH or increase the secretion of GnRH or both.

The large rise in LH and FSH starts a chain of events that could lead to pregnancy. The wall of the now fully developed follicle becomes thin and the follicle bursts owing to the action of LH with perhaps some help from FSH. The released ovum then travels to the fallopian tube (oviduct) and uterus. Freed from the inhibiting effect of the follicular fluid and perhaps directly stimulated by the high LH, the ovum now undergoes the first meiotic division, i.e., becomes more mature. At the same time, a new endocrine gland, the *corpus luteum* or "yellow body," is formed from the granulosa cells of the collapsed follicle. The LH is responsible both for the process of luteinization in which the granulosa cells change into luteal cells and for the secretion of progesterone by the corpus luteum. The effects of LH on ovulation, ovum maturation, luteinization and progesterone secretion may all be mediated by a rise in cAMP, but this is reasonably certain only for the last two actions. A prostaglandin, perhaps $PGF_{2\alpha}$, may also help mediate the action of LH on progesterone secretion.

The new corpus luteum secretes not only large amounts of progesterone (30 mg per day) but also a good deal of E_2, although the serum E_2 level is somewhat lower than at its preovulatory peak. The levels of FSH and LH fall to the lowest levels of the cycle and remain there as long as the corpus luteum functions. The progesterone and E_2 are probably responsible for the low levels of FSH and LH by a negative feedback effect, but it should be noted that *some* LH is required for normal luteal hormone secretion. Here again is another example of a local positive feedback effect. The corpus luteum has specific E_2 receptors, and E_2 enhances progesterone secretion by the corpus luteum if some LH is present. The probable explanation is that E_2 increases the number of LH receptors in the corpus luteum just as it increases FSH receptors in the follicle; the result is enhancement of progesterone secretion in response to LH.

Progesterone and E_2 change the endometrial pattern so that the glands become tortuous instead of straight, the stromal cells enlarge and contain more glycogen and the endometrial glands secrete fluid containing mucus and glycogen. These changes are due mainly to progesterone, although some estradiol is necessary; they last through the second half of the cycle. The ovarian *luteal* phase thus corresponds to the endometrial *secretory* phase. If pregnancy does not occur, luteal hormone secretion lasts 2 weeks and then falls off. The lack of progesterone (and perhaps estrogen) then causes the endometrium to slough, and the luteal or secretory phase terminates in menstrual bleeding. The fall in progesterone and E_2 then allows the serum levels of FSH and LH to rise again, beginning a new cycle.

Why does the corpus luteum stop functioning? No one really knows. In some nonhuman species the uterus secretes a substance, probably a prostaglandin, that causes the corpus luteum to break down (luteolysis), but this phenomenon almost certainly does not occur in women. Some evidence, as yet scanty, suggests that E_2 or prolactin may be luteolytic in women. Perhaps more gonadotropin is needed to maintain the corpus luteum than is actually present; hCG, an LH-like hormone secreted by the placenta, may cause the corpus luteum to persist during pregnancy, although injections of hCG prolong luteal life only slightly in normal women. Or perhaps a required luteotropic hormone, i.e., a hormone that supports the corpus luteum, disappears. In rats, PRL (prolactin) is a luteotropic hormone and is actually called luteotropin since it is so essential to luteal function. However, in women this seems not to be true since patients with no pituitary function ovulate and have a normal luteal phase when given only FSH and LH without PRL. Although a specific luteolysin or an unrecognized luteotropin could exist in nonpregnant women, for the moment the best guess is that the human corpus luteum has an intrinsic life span in the presence of low normal LH levels and atrophies after 2 weeks unless pregnancy occurs. At that point luteal function is prolonged by placental hormones.

How does a fall in progesterone and E_2 cause menstrual bleeding? Again the answer is not clear. The menstrual flow contains prostaglandins (PGE_2 and $PGF_{2\alpha}$), which might possibly cause endometrial sloughing. In any case they are probably responsible for uterine cramps during menses. Possibly, a release of destructive lysosomal enzymes may produce endometrial lysis.

There are obviously many unanswered questions about the events of the menstrual cycle, despite the availability of new and sensitive hormone assays. The regularity of the cycle, the control of luteal function and the mechanism of menses are not the only problems. The relevance of PRL to any aspect of the cycle in women is unsettled, nor is it known if PRL secretion varies during the cycle. The multiple possible control points (pituitary, hypothalamus or higher brain centers) for gonadotropin secretion make analysis most difficult. The role of the pineal gland, generally inhibitory, or the prostaglandins in regulating gonadotropin secretion is not clear. At least, the new assays have allowed us to ask more difficult questions with some hope of obtaining answers.

NON-UTERINE ACTIONS OF E_2 AND PROGESTERONE

In adult women, as in adult men, the gonadal hormones are responsible for normal sexual development and function. The actions of estrogens and progesterone on female secondary sex characteristics during puberty and on the endometrium during the menstrual cycle have been described. Both hormones also cause more subtle cyclic changes in the mammary epithelium during the menstrual cycle; for example, during the luteal phase, progesterone is probably responsible for increased glycogen deposition and a presecretory state.

Libido or sexual desire in women and estrus or sexual behavior in rats are largely estrogen dependent, although the small amount of active androgen secreted by the ovaries (and adrenal cortex) may contribute. In contrast, in women (and monkeys), progesterone, if anything, decreases libido. Castration in women, however, does not always eliminate sexual desire and activity;

perhaps the adrenal cortical hormones are sufficient in some, or perhaps the CNS, once set, need not have continuous exposure to estrogens or androgens. As noted, estrogens are also responsible for the appearance of vaginal phero- mones, which in monkeys attract the male; whether these pheromones serve the same purpose in human beings is not clear.

Like testosterone and its anabolic effect on muscle protein synthesis, estro- gens have nonsexual effects elsewhere in the body. The E_2 decreases collagen breakdown in the skin and increases the elastin content of the aorta, thus de- creasing its stiffness. It acts on the kidney to enhance sodium retention and perhaps increases renal sensitivity to ADH. It stimulates synthesis of several hepatic proteins, resulting in increased hepatic secretion of triglycerides, re- nin-substrate and several hormone-binding proteins such as TBG, CBG and E_2TBG.

MECHANISM OF ACTION OF E_2 AND PROGESTERONE

How does *estradiol* act? The mechanism is similar to that of other steroid hormones (see Fig 10–6), and E_2 is perhaps the best studied of all. Much of the work has been done with the rat uterus and chick oviduct, but in all like- lihood the results apply to human estrogen-responsive tissues including not only the uterus and fallopian tube (oviduct), but also the breast, hypothal- amus, anterior pituitary and probably any tissue in which E_2 receptors are found. The E_2 enters (? is transported) into the cell and rapidly binds to its specific cytosol-binding protein, causing a change in the molecular shape of the binding protein that allows the E_2-protein complex to enter the cell nucle- us. The complex then binds to nuclear chromatin; the binding site probably includes both DNA and a non-histone (acidic) protein. Shortly thereafter, there is an increase in RNA polymerase activity and increased synthesis of specific mRNAs and later of rRNA and tRNA. The increased RNA synthesis then causes a rise in both specific and general protein synthesis. The rate- limiting step in all of this is synthesis of the specific mRNAs. Synthesis of cer- tain estrogen-specified proteins is presumably responsible for the growth and differentiation of estrogen-responsive cells. In addition, E_2 causes the cells to increase DNA synthesis and eventually divide, thus inducing tissue growth as well.

Although seemingly straightforward, this is a most complex mechanism and some aspects remain controversial, such as whether or not increased protein synthesis must occur before increased RNA synthesis, or whether or not cAMP is somehow involved in the action of E_2 on RNA synthesis. Exactly how the RNA polymerase activities are increased is not known. Finally, some of E_2's actions do not seem mediated by nuclear RNA synthesis; these include transformation of the cytosol E_2 receptor itself, a very early effect of the E_2- receptor complex on increasing nuclear permeability to RNA precursors and another early effect of the complex on increasing the rate of ribosomal peptide elongation. Enhanced nuclear RNA synthesis, although perhaps the most prominent, is not the whole explanation of the actions of E_2.

Progesterone has a similar mechanism of action but with some notable dif-

ferences. The specific cytosol-binding protein for progesterone is different
from that for E_2 and is itself increased by E_2; thus E_2 actually enhances progesterone's actions. The receptor portion of the progesterone-receptor complex appears to have two subunits, one of which binds to DNA and the other
to a nuclear acidic protein. The RNA polymerase activity is stimulated, as
with E_2, but only after an initial progesterone-induced fall, at least in some
estrogen-primed tissues. Although progesterone induces synthesis of specific
mRNAs that code for specific proteins, total cell RNA synthesis is not increased. And progesterone, instead of enhancing the cell proliferation induced by estrogen, actually blocks further proliferation. Overall, progesterone's mechanism of action is similar to that for E_2 but its thrust is to cause further differentiation of cells already primed with E_2; thus the proliferative
endometrium is converted to the glycogen-laden secretory endometrium and
is now suitable for the implantation of a blastocyst.

CONTROL OF GONADOTROPIN SECRETION IN WOMEN

In women, the gonadotropin control system is even more complicated than
in men. Not only does it produce a monthly cycle, but one of the glands, the
corpus luteum, disappears for half of each cycle. Studies in other species may
not be relevant to the human being since there-are undoubtedly many differences between species. However, some kinds of information, e.g., the results
of hormone implants into the hypothalamus, are simply not otherwise available. The most sensible approach is to base an explanation of gonadotropin
control on hormonal fluctuations of the menstrual cycle and seek supportive
data from well-designed human studies and presumably relevant animal data.

At the beginning of the menstrual cycle, during the menses, very low levels
of E_2 and progesterone precede a slight but significant rise in FSH and, to a
lesser degree, LH (see Fig 10–24). High but still physiologic amounts of E_2
and progesterone, as in the luteal phase, are associated with low (though still
necessary) levels of FSH and LH. Thus tonic — or more or less steady — secretion of gonadotropins is controlled by E_2 and progesterone in a classic negative feedback system. Of course, FSH and LH secretion is not really constant
during the follicular or luteal phases but occurs in short bursts every hour or
two; the average of these bursts is the tonic level of secretion. Is there selective control of each gonadotropin by ovarian hormones? To some degree there
probably is. The rising E_2 level in the follicular phase is associated with a fall
in FSH but not LH, although larger amounts of E_2 suppress both gonadotropins; thus, E_2 regulates tonic FSH secretion more easily than LH secretion.
This is not actually a true negative feedback loop since FSH probably does
not directly stimulate E_2 secretion. There may be an ovarian follicular inhibin, like that in the testis, that functions in a true negative feedback fashion to
control FSH secretion. Physiologic amounts of E_2 alone are not sufficient to
control both FSH and LH, since the LH level does not fall in the follicular
phase. The LH, as well as FSH, falls to a low level in the luteal phase when
progesterone is secreted in addition to E_2. Thus tonic secretion of FSH is
primarily related to changes in E_2, but more complete physiologic regulation

of tonic secretion of both FSH and LH requires the synergistic effects of E_2 and progesterone.

On the other hand, the *cyclic* ovulatory surge of LH and FSH that occurs in midcycle is the result of a *positive* feedback effect of E_2 and perhaps progesterone. How do the same gonadal hormones produce feedback effects that are sometimes negative and sometimes positive? A complete answer does not yet exist but it involves the brain, the pituitary and both ovarian hormones.

It is clear that control of gonadotropin secretion involves the brain as well as the anterior pituitary. In some animals, areas of the limbic system (amygdala and hippocampus) as well as the hypothalamus are involved. In primates, however, normal cycling may require only the hypothalamus. Whatever brain impulses are involved, release of GnRH from the hypothalamus stimulates both FSH and LH secretion. The level of GnRH secretion probably directly determines tonic secretion of FSH and LH; whether a burst of GnRH secretion causes the ovulatory surge of FSH and LH in women is not known.

Modest levels of E_2 and progesterone could therefore act on the hypothalamus to inhibit tonic GnRH secretion and thus inhibit FSH and LH secretion. The hypothalamus has specific cytosol receptors for E_2 and, as in the uterus, E_2 enhances progesterone uptake by the hypothalamic neurons; a hypothalamic site of action for E_2 (or an E_2 metabolite such as 2-hydroxyestradiol) and progesterone in suppressing tonic gonadotropin secretion is therefore plausible. The gonadal hormones could act directly on the peptidergic hypothalamic neurons that secrete GnRH or on the monoaminergic neurons that impinge on the peptidergic neurons. In rats (and probably monkeys), norepinephrine stimulates GnRH secretion, and dopamine may stimulate or inhibit it. The E_2 and progesterone could act at this monoamine level as well as on the GnRH neurons; in fact, E_2 does inhibit the synthesis of norepinephrine in the hypothalamus. Other neurons having serotonin or acetylcholine as the neurotransmitter may also be stimulatory for GnRH secretion, as may hypothalamic PGE_2. These neurons may also be potential sites for negative feedback regulation. Finally, tonic gonadotropin secretion could be regulated at the anterior pituitary itself since there are E_2 cytosol receptors here as well. A reasonable case can be made for regulation of tonic FSH and LH secretion by gonadal hormones acting at both the pituitary and hypothalamic levels, although precise details need further study.

Where in this hierarchy of aminergic neurons, GnRH-secreting neurons and anterior pituitary gonadotropes is the switch that responds positively to E_2 and progesterone and turns on the ovulatory surge of LH and FSH? In part, the switch lies in the anterior pituitary. Levels of E_2 equal to those seen in the late follicular phase — or perhaps a slight fall from these high levels — enhance the response of serum FSH and LH to a given dose of GnRH in both women and rats. But there is probably a hypothalamic component to the switch as well, since in rats (1) ovulation can be induced by direct implantation of E_2 into the amygdala, (2) the anterior hypothalamus seems specifically involved in cyclic gonadotropin secretion and (3) E_2 under appropriate circumstances increases secretion of both GnRH and the gonadotropins at the same time.

Since GnRH neurons originate in several hypothalamic areas including the preoptic, supraoptic and anterior hypothalamus, and probably in other areas throughout the brain, we may speculate that moderate amounts of E_2 inhibit some of these neurons and eventually inhibit gonadotropin secretion, whereas higher physiologic amounts of E_2 may provoke other neurons to secrete a burst of GnRH. The role of *progesterone* in inducing the ovulatory peak of LH and FSH is not clear. When given with E_2, it appears to enhance the positive feedback effect of E_2 in women. However, it is not known whether it acts on the pituitary or hypothalamus and there may not be a real rise in serum progesterone before ovulation. In rats, at least, progesterone enhances both LH and FSH secretion when implanted into the hypothalamus and may actually cause release of GnRH from a cell-free hypothalamic preparation without mediation of the cell nucleus. Overall the positive feedback effect of the ovary on the midcycle surge of LH and FSH secretion is not yet well explained but involves at least E_2 and perhaps progesterone and is probably mediated at both pituitary and hypothalamic levels.

A final point: once the initial rise of FSH and LH occurs during the menses after destruction of the corpus luteum from the previous cycle, the remaining events of the new cycle are in large part determined by the ovary. Follicular E_2 secretion is the principal determinant of the ovulatory gonadotropin peak that induces ovulation and corpus luteum formation. The luteal hormones, E_2 and progesterone, suppress FSH and LH to levels sufficient to maintain the corpus luteum for its 2-week life span but are not enough to prolong luteal function beyond that. Lysis of the corpus luteum and the associated fall in E_2 and progesterone lead to the next rise in FSH and LH and a new cycle. That the ovary should determine the timing of events in the cycle makes teleologic sense since the ovulatory gonadotropin peak should not occur before the follicle is mature, i.e., with well-developed granulosa cells and capable of forming the corpus luteum needed for implantation should the ovum be fertilized.

PREGNANCY

We might think, with completed ovulation in the woman and fully developed spermatozoa in the man, that successful pregnancy would be a simple matter. After ejaculation, the selected spermatozoon simply swims upstream, merges with the ovum and events take their course until a new child is born. The current concern about a rapidly increasing population tells us that the process is all too often successful. On the other hand, many patients are infertile and frustrated in their attempts to have children. In fact, almost every step in a successful pregnancy is under hormonal control, and a thorough understanding of these endocrine changes can not only help the infertile patient but can offer a rational basis for some aspects of contraception. In sequence, the events are *coitus, fertilization, implantation, maintenance of pregnancy* to its completion and *birth* or parturition at the correct time.

Successful *coitus* is the result of normal sexual behavior in normal men and women. Normally developed external genitalia are obviously mandatory and, as noted before, the correct gonadal hormone, testosterone or DHT in men and E_2 in women, is needed both for the development of external genitalia in utero and for their growth and function in the adult. Normal sexual behavior depends on both the interest (or libido) of one person and the attractiveness (whatever the criteria) of a sexual partner. Both are, as noted, hormonally determined by the correct gonadal hormone acting on the brain: E_2 in women and testosterone, perhaps metabolized to E_2, in men. For optimal adult sexual behavior, the gonadal hormones should be present in utero as well as in the adult. Sexual activity in men, although highly variable from individual to individual, does not seem to change much until after the age of 60 or so and this correlates fairly well with the fact that serum testosterone levels are fairly stable until that age, when they begin to fall. In women, however, serum E_2 levels rise and fall monthly, and some data (though not all) suggest increased sexual activity around the time of the highest E_2 levels. Nevertheless, it is intriguing that a small percentage of castrate men and women have reasonable sexual activity, indicating that the brain, once set, need not have continued hormonal stimulation.

FERTILIZATION.—Fertilization or the merging of two gametes is not simply a matter of adding a spermatozoon to an ovum. They must meet in the right place at the right time and do the right things.

As ejaculated in the seminal fluid, spermatozoa are not able to fertilize the egg. However, after spending a while in the uterus or oviduct they are able to fertilize. Somehow the female genital tract conditions the surface of the spermatozoon, perhaps by removing an inhibitor to fertilization in the seminal fluid. The conditioning process is called *capacitation* and is estrogen dependent. The inhibitor in the seminal fluid is called *decapacitation factor (DF)*; DF is not yet well defined but may be several small peptides with molecular weights of about 1,000. There is some controversy whether human spermatozoa need to undergo capacitation, since most of the studies have been done in rabbits, but probably they do and in women the process seems to take several hours.

Capacitated sperm can then fertilize the ovum; actual fertilization occurs in the fallopian tube. Even when capacitated, however, the spermatozoon's entry into the ovum is not easy. The ovum does not lie free in the oviduct but is surrounded by a mass of granulosa cells, including a tightly bound corona radiata just next to the ovum and a looser accumulation of cells outside the corona called the cumulus oophorus (Fig 10–25). Furthermore the ovum itself is covered by a clear material, the zona pellucida. The spermatozoon must penetrate all of these as well as the outer surface of the ovum, the vitelline (or egg) membrane. It does so by releasing several enzymes, most of which are contained in the acrosome, a lysosome-like structure in the head of the spermatozoon. Previous capacitation is necessary for the action and release of these enzymes. Acrosomal enzyme release occurs when the spermatozoon's plasma membrane fuses with the underlying acrosomal membrane and the combined membranes gradually disintegrate. This process, called the acrosome reaction, occurs when the spermatozoon is close to the ovum and is stimulated by the granulosa cells surrounding the ovum. Pituitary gonadotropins added directly to these granulosa cells enhance the acrosome reaction so it is likely

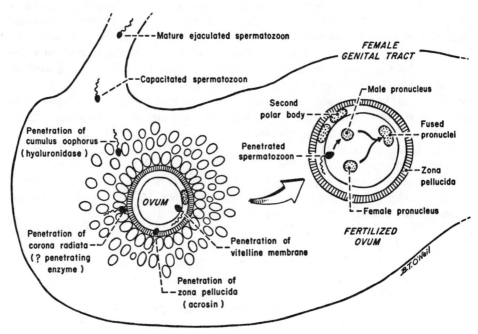

Fig 10–25.— Fertilization: spermatozoon penetrating and fusing with ovum.

that the granulosa cells secrete a hormone ($?E_2$) that mediates the reaction.

The spermatozoon seems to pass through the cumulus oophorus by the action of a species-specific hyaluronidase from the spermatozoon's surface and acrosome. On reaching the corona radiata, the acrosome reaction is usually well developed, and spermatozoon penetration between the coronal cells is probably effected by a separate corona-penetrating enzyme (CPE), which is specifically inhibited by the seminal fluid DF. Now the spermatozoon is at the zona pellucida and penetrates this layer by releasing a proteolytic enzyme, *acrosin* (MW = 30,000 in man). The spermatozoon finally penetrates the vitelline membrane of the ovum (?enzymatically induced) and causes the ovum to release some heat-labile substance (?an acrosin inhibitor) that causes the zona pellucida to become impermeable to other sperm. This block to polyspermy is of clear importance if the fertilized ovum is to have a diploid number of chromosomes. The entry of the spermatozoon into the egg triggers the final step in meiosis, and the second polar body is cast off by the ovum. The dense sperm chromatin becomes dispersed in the ovum and forms the male pronucleus, the counterpart of the female pronucleus already in the ovum. Proof of fertilization can therefore be the presence of at least two polar bodies just outside the ovum, or of pieces of the spermatozoon's tail in the ovum, or of two pronuclei.

The fertilized ovum now travels down the oviduct to the uterus, dividing as it goes, and eventually forms a *blastocyst* with a central cavity. The trip to the uterus takes about a week. Delayed delivery of the fertilized ovum from the oviduct to the uterus is needed to allow for blastocyst development and is regulated by hormonal effects on the oviduct's smooth muscle; E_2 slows transit, perhaps assisted by a relaxing follicular kinin, and progesterone speeds it up.

IMPLANTATION. — Implantation of the blastocyst into the endometrium is nicely timed so that it occurs when luteal E_2 and progesterone secretion are at their peak and when these hormones have produced appropriate development of the secretory endometrium. In fact, a secretory endometrium is essential to implantation and, if the endometrium is out of phase, implantation will not occur. Furthermore, the need for E_2 and progesterone in implantation may be due not only to their effects on the uterus but also to direct effects on the blastocyst. In at least some species, the estrogen-primed and progesterone-stimulated endometrium secretes a substance, blastokinin, into the uterine cavity. Blastokinin binds both E_2 and progesterone and may carry these hormones to the blastocyst, which in turn is known to bind at least E_2. Both blastokinin and E_2 stimulate RNA and protein synthesis in the blastocyst so blastokinin may well be a mediator of hormonal stimulation of the blastocyst. It is also possible that the blastocyst makes its own E_2 and progesterone. Precisely how E_2 and progesterone enhance the blastocyst's ability to implant is not known. However, at the time of implantation the blastocyst appears to be making both a protease and hCG (human chorionic gonadotropin); the protease may assist blastocyst invasion of the endometrium, and hCG keeps the corpus luteum functioning so that the secretory endometrium does not slough and carry the blastocyst with it.

MAINTENANCE OF PREGNANCY. — Maintaining the fetus to its normal term is critical for survival of the species; if the fetus does not have time to develop, it will die. A number of hormonal changes during pregnancy keep things going for the proper period; the hormones include *pituitary, gonadal* and *placental.*

After implantation, some of the blastocyst cells become the invading trophoblasts, whereas the others develop into the fetus. The trophoblast cells become a major portion of the placenta. The placenta is responsible for most (though not all) of the hormonal changes in pregnancy, some of which can be detected in the maternal circulation as early as 9–10 days after ovulation — only 2–3 days after implantation. In maternal serum, hCG rises to a peak value of 50–150 international units (IU)/ml at 8 to 10 weeks of the 40-week pregnancy, falls to less than half that a few weeks later and rises again, but not as markedly, toward the end of pregnancy (Fig 10–26). There is no good explanation for this pattern of hCG since the other hormones made by the placenta generally show a steady increase throughout pregnancy. By the end of pregnancy the serum levels reached for the other placental hormones in the mother are placental lactogen (hPL): 5–10 μg/ml; estrone (E_1): 10 ng/ml; estrone sulfate (E_1-S): 20 ng/ml or more; estradiol (E_2): 20–25 ng/ml; estriol (E_3, or $16\alpha OH$-E_2): 10–15 ng/ml; estetrol (E_4, or $15\alpha OH$-E_3): 1–2 ng/ml; and progesterone: 150–200 ng/ml. Along with the rise in placental hormones is a parallel increase in maternal serum prolactin, (PRL) secreted by the mother's pituitary, possibly in response to high levels of circulating estrogen acting either on the hypothalamus or directly on the pituitary. The maternal serum PRL reach-

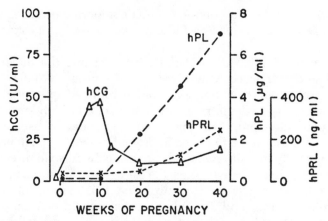

Fig 10–26.—Maternal serum levels of hCG, placental lactogen (hPL) and prolactin (hPRL) during pregnancy. Note that there is 30–40 times more hPL than hPRL (data representative of several laboratories).

es a peak of 200 ng/ml at term. The rise in PRL secretion correlates nicely with a 50% increase in pituitary size during pregnancy, most of which is due to PRL-secreting cells. Perhaps more intriguing is the very large amount of PRL in the amniotic fluid. This PRL is identical to maternal pituitary PRL and yet seems to be made by the chorion or fetus and does not enter the maternal circulation. Its function is unknown but it may have something to do with maintenance of sodium and water balance in utero.

The *hCG* has several actions. It stimulates fetal testicular testosterone secretion (in males) and perhaps fetal adrenal DHEA (dehydroepiandrosterone) secretion. It also stimulates the maternal corpus luteum to keep secreting progesterone. Luteal progesterone secretion is essential for the continuation of pregnancy for the first 7 or 8 weeks; without it, the uterus becomes more contractile, and abortion occurs. Beyond this time, the placenta takes over the bulk of progesterone secretion, although some is still secreted by the corpus luteum until the end of pregnancy. The hCG has also been reported to decrease lymphocyte activation, which may be of critical importance in the mechanism by which the mother is immunologically tolerant of the fetus.

The *hPL* is secreted almost entirely into the maternal circulation and probably acts mostly on the breast, along with prolactin, E_2 and progesterone, to develop and prepare the breast for lactation. The hPL may also act to shunt glucose away from maternal metabolism toward the fetus by blunting the action of insulin in the mother.

The many *estrogens* secreted by the placenta are not made by the placenta de novo but are derived from precursors secreted by both the maternal and fetal adrenal cortex. The principal precursors are DHEA and DHEA-S (DHEA-sulfate), mostly the latter, the bulk of which comes from the fetus. Since the placenta cannot make DHEA or DHEA-S and the fetal adrenal cannot make estrogens, the estrogen rise in maternal blood depends on both the fetus and the placenta; hence the term, *fetoplacental unit*. The placenta converts fetal adrenal DHEA-S and DHEA to E_1 and E_2 by several enzymatic steps. To make E_3 and E_4, however, appropriate precursors must be hydroxylated in the fetal liver in addition to the placental steps. E_2 and to a lesser degree E_1 are responsible for uterine and mammary growth and stimulation of maternal PRL secretion. These estrogens also cause several metabolic changes mediated by the maternal liver such as increases in hormone-binding proteins and triglycerides. The role of the other estrogens, E_3 and E_4, is unknown even though 90% of the total estrogen secreted in pregnancy is E_3 (serum E_3 is lower than serum E_2 because E_3 is metabolized much faster). However, E_3 and E_4 are indicators of fetal health since their formation depends on reasonably normal function of both the fetal adrenals and liver.

Placental *progesterone* also is not synthesized by the placenta de novo but is derived from maternal circulating cholesterol. The progesterone is an important precursor for fetal adrenal glucocorticoid and mineralocorticoid synthesis. In the mother, progesterone maintains the secretory endometrium and decidual tissue, keeps the uterus from contracting and aids in breast development. Progesterone, like E_2, also has metabolic affects on the mother; it causes sodium loss by the kidney, which results in a rise in maternal aldosterone secretion.

Little is known about the regulation of placental hormone secretion. The hCG seems not to have major effects on placental cholesterol metabolism or estrogen synthesis, although it may enhance the conversion of testosterone to E_2 (via cAMP). There may be some end-product inhibition of the conversion of DHEA-S to DHEA and this may be an important regulatory step. Overall, though, it seems that, with the exception of precursor supply, there may not be much regulation and that placental hormone secretion may be related more to placental mass than anything else.

Other maternal and fetal hormones and metabolites also change in pregnancy. Although some of the placental hormones are secreted into the fetus, including hCG, progesterone and about 10% of the estrogens, surprisingly few maternal hormones cross the placenta into the fetal circulation. None of the peptide hormones (e.g., LH, ACTH or TSH from the maternal pituitary or PTH, insulin or glucagon) passes across. Only small amounts of thyroid hormones cross into the fetus, with slightly more maternal T_3 than T_4 appearing in the fetal blood stream. Interestingly, fetal serum T_3 is definitely lower than in the mother but reverse T_3 ($3,3',5'-T_3$), a relatively inactive analogue, is higher in the fetus. Maternal cortisol might cross but the placenta metabolizes most of it to the biologically inactive cortisone. Overall, the secretion of the various fetal hormones is relatively independent of the mother (except for the feto-placental unit noted above) and is dependent on fetal development.

For fuel and calcium, however, the fetus is totally dependent on the mother. Glucose and FFA, the two principal sources of fuel, and calcium cross the placenta into the fetus. Glucose uptake and utilization by the fetus is in fact a drain on maternal glucose supplies. As a result, pregnant women tend to have a lower fasting serum glucose than normal because, in addition to the fetal glucose drain, production of glucose by the maternal liver—the main source of glucose in the fasting state—is decreased. On the other hand, after eating, maternal serum glucose levels rise as expected but stay up longer than usual, the "diabetogenic" effect of pregnancy. This is probably in part owing to hPL and may be a mechanism for transporting more glucose to the fetus before it is removed from the blood stream by the mother.

The fetus is also a drain on maternal calcium; by the time the infant is born, its skeleton contains about 25 gm of calcium. The placenta seems to transport calcium against a concentration gradient so that the maternal serum calcium falls slightly and the fetal serum calcium is slightly higher than the maternal level. As a result, maternal PTH levels go up somewhat and, in addition, more calcium is absorbed from the gut. Furthermore, the mother must be the source of any vitamin D necessary for fetal skeletal development, since the fetus receives little sunlight. Indeed, 25(OH)D does cross the placenta. Once in the fetal circulation, calcium's entry into the skeleton is probably aided by the higher calcitonin levels in the fetus.

BIRTH OR PARTURITION

Just what physiologic change induces the uterine contractions ("labor") that expel the fetus remains a controversial topic. Clearly the event is nicely timed in the vast majority of pregnancies since a normally developed child is usually delivered. Changes in maternal hormones might be the cause. For example, an increase in secretion of oxytocin, a known uterine stimulant, from the neurohypophysis has been proposed as an initiator of labor. Or perhaps prostaglandins, particularly $PGF_{2\alpha}$, which are also known uterine stimulants, are involved. Unfortunately, it has been most difficult to show any rise in serum oxytocin just before delivery in women. In sheep and goats, there is increased oxytocin but only after labor has begun; the rise in oxytocin may help to ensure proper delivery but it is not the initiating event. Prostaglandins almost certainly play an important role in women; $PGF_{2\alpha}$ is found in amniotic fluid, and both its level and that of PGE in the maternal blood rise during labor. Again, however, changes in prostaglandins do not seem to be primary.

Since estrogens enhance uterine contractility and progesterone decreases it, investigators have tried hard to show a rise in maternal estrogen and a fall in maternal progesterone just prior to onset of labor. Although by no means universally agreed upon, exactly these changes in E_2 and progesterone have been found in sheep and (in one study) in women. Of course, this would not really be a change in maternal hormone secretion but rather in placental hormone secretion. Perhaps birth occurs when the placenta has reached a certain age and becomes "senescent"; the problem with this hypothesis is that it is difficult to explain the rise in E_2 on the basis of placental failure.

Another hypothesis is that the fetus itself initiates labor when it reaches the proper stage of development. This makes teleologic sense because birth should not occur until the fetus is ready. Best studied in sheep, the postulate is that the fetal brain somehow recognizes that the time is ripe and stimulates increased secretion of fetal ACTH and then cortisol. The cortisol reduces placental progesterone production and increases its E_2 secretion. The E_2 may possibly increase maternal oxytocin secretion and placento-uterine synthesis of $PGF_{2\alpha}$. Labor then ensues owing to the contractions induced by oxytocin and $FGF_{2\alpha}$ in a uterus made more sensitive by the fall in progesterone and rise in E_2. A major problem with this postulate is that it is not completely proved in any one species; furthermore, there is undoubtedly much variation between species in its exact details. The mechanism in human beings is thus still poorly defined. It is known that human parturition requires an intact fetal brain, that the fetal cortisol level is relatively high at birth (although ACTH may not be high before birth) and that the pregnant uterus has oxytocin receptors. Combined with the maternal hormone changes mentioned above, the postulated scheme may have some applicability to human beings but only as a guide to further investigation.

After birth or post partum, several hormonal changes occur in the mother. Since the placenta is gone, its hormones disappear. The hCG disappears quickly and is gone in several days; the maternal corpus luteum finally stops functioning. The hPL and steroid hormones disappear as well. Maternal pituitary FSH and LH secretion are suppressed during pregnancy not only because of the elevated E_2 and progesterone but also because of the high PRL level; they rise to normal basal levels by 2–4 weeks post partum. The ovary seems slightly more sluggish; E_2 production, ovulation and menses return only after several more weeks have passed and may not return for 2–3 months. The high level of maternal prolactin secretion during pregnancy falls somewhat but stays above normal if lactation begins. The high PRL level continues to suppress at least the ovulatory peak of FSH and LH secretion; if the PRL level remains high for several months owing to continued lactation, return of ovulation is delayed. However, the high degree of individual variation in suppressing ovulation with lactation limits the use of lactation as an effective contraceptive for more than 3–4 months.

In most species—and probably to a large degree in women—there are distinct changes in behavior after giving birth. Described overall as *maternal behavior,* the changes differ widely among species but are aimed at nurturing and protecting the relatively defenseless newborn infant. Even in rats, the best studied animal, the hormonal factors determining maternal behavior are not clear, but some combination of E_2, progesterone and PRL seems responsible. Perhaps the E_2 and progesterone during late pregnancy, plus the continued increased secretion of PRL after birth, prime the CNS to respond to the newborn infant, particularly if nursing, with the various patterns of maternal behavior.

LACTATION

Except in man, lactation is essential for the survival of the newborn. Even in human infants, however, nursing at the breast may be better than any substitute. For the infant to nurse successfully, there must be normal *breast development, milk synthesis and secretion* and *milk ejection.*

Breast development includes not only the normal development of the female breast but particularly the changes that occur during pregnancy. Before pregnancy the usual amounts of E_2 and progesterone are responsible for some degree of ductal and alveolar development, presumably via the same biochemical mechanisms as in any tissue responsive to these hormones. The role of PRL is unclear but it may be permissive for the action of E_2. During pregnancy major changes occur. The alveolar cells divide and differentiate so that they are primed to secrete milk. The high levels of E_2, progesterone, PRL and hPL are primarily responsible. Both PRL and hPL bind to specific receptors on the mammary cell membrane, and both E_2 and progesterone bind to cytosol receptors. The differentiated breast tissue is now capable of making lactose (the predominant milk sugar), casein (a protein specific to milk) and fatty acids.

Nevertheless, lactation or *milk secretion* does not usually occur during pregnancy, although it can be induced by nursing another infant, because actual milk secretion into the mammary alveoli is held in check by the high levels of E_2 and progesterone. After birth, E_2 and progesterone fall to low levels and the effects of PRL are no longer inhibited. Already primed synthetic processes now begin to work. Casein is synthesized. The two subunits of the enzyme lactose synthetase combine, activating the enzyme to make lactose. Activation of mammary lipoprotein lipase converts circulating triglycerides to fatty acids, which enter the mammary cell. These fatty acids and those synthesized in the cell by the activated fatty acid synthetase complex are resynthesized into triglycerides. The triglycerides are enclosed in an envelope probably composed of the cell's plasma membrane (or ? part of the Golgi apparatus), forming fat globules. All the components of milk are then released into the mammary alveoli. Fat globules are largely responsible for the "milky" turbidity of milk.

Milk ejection is the last step in getting the milk to the child. Suckling by the infant is an integral component of normal lactation. Without suckling, milk production decreases and soon stops altogether. What does suckling do? It stimulates sensory neurons near the nipple, and the impulses travel through the intercostal nerves and spinal cord to the hypothalamus. There they probably act on the paraventricular nucleus and directly stimulate oxytocin secretion from the posterior pituitary. Oxytocin circulates in the blood and stimulates specific oxytocin receptors in the myoepithelial cells surrounding the mammary alveoli. Contraction of these myoepithelial cells decreases the alveolar volume and forces milk out of the breast. Once nursing is well established, just the sight of the infant by the mother can induce milk ejection, presumably through oxytocin secretion.

For *maintenance of lactation* essentially the same hormones are needed as for its initiation. Prolactin is all important, as are a little estrogen and normal amounts of T_4, hydrocortisone and calcium—hence normal amounts of TSH, ACTH and PTH. Proper nutrition and a stable mental state are needed; a poor diet or a distraught mother diminishes milk secretion.

The same suckling stimulus that stimulates oxytocin secretion causes the release of PRL. For the first 2 or 3 months after birth, suckling by the child induces bursts of PRL secretion during nursing. This effect is mediated by the hypothalamus. Suckling may stimulate the release of a prolactin-releasing factor (PRF), perhaps mediated by hypothalamic serotonin-secreting or norepinephrine-secreting neurons. Alternatively, it may block the release of a pro-lactin-inhibiting factor (PIF), an effect probably mediated by hypothalamic dopamine-secreting neurons. Either mechanism would result in increased PRL secretion by the pituitary. Thus suckling itself maintains milk secretion and milk ejection by directly stimulating secretion of both PRL and oxytocin.

After the first 2 or 3 months of nursing, however, the basal PRL level is no different from that of the non-nursing mother, and suckling no longer causes a rise in PRL. A high PRL level no longer seems necessary for milk production; milk secretion apparently persists owing mostly to the developed breast tissue and the nursing infant. When nursing stops, lactation stops. The absence of suckling results in no further oxytocin secretion, and the failure to remove milk from the alveoli of the breast probably acts locally to inhibit further milk secretion.

HORMONAL ASPECTS OF REPRODUCTIVE DISORDERS

With the number of hormones and control sites involved in both men and women, it is not surprising that many different endocrine disorders can occur. Any one of the hormones discussed could be missing, there could be too much of others and sometimes the response to the hormone is abnormal.

DISORDERS IN MEN

There can be a deficiency of GnRH, of FSH and/or LH or of testosterone or its tissue response; the result of any of these deficiencies is some degree of *hypogonadism*, i.e., deficient testosterone secretion or poor spermatogenesis or both. The patient reports weakness, poor sexual function or infertility. A hypothalamic or pituitary lesion can cause hypogonadism owing to low gonadotropin secretion; this is *secondary hypogonadism,* so called because the problem is not in the testis but is secondary to a defect in the hypothalamus or pituitary. Usually both FSH and LH are missing (or at least not as high as they should be) and so both spermatogenesis and testosterone secretion are impaired. Rarely, the gonadotropin deficiency is limited to LH; even so, spermatogenesis is still not normal because it requires normal amounts of testosterone. For an *exogenous stimulation test* in suspected FSH and LH deficiency, we give GnRH or the anti-estrogen, clomiphene, thought to act via the hypothalamus to increase GnRH secretion. A deficient response of gonadotropins to GnRH suggests a pituitary lesion, whereas a deficient response to clomiphene suggests a hypothalamic one; unfortunately, the tests overlap to some degree and we cannot use them to locate a lesion (anatomical or functional) with great confidence.

More common is *primary hypogonadism,* in which the defect lies in the testis. The problem may be due to abnormal chromosomes, such as in Klinefelter's syndrome, in which there is an extra X chromosome (XXY instead of XY); this phenotypic male has small testes, almost totally defective spermatogenesis and often (though not always) poor testosterone production. Or there may be an enzymatic block in testosterone synthesis or even absent or misplaced testes or Leydig cells. In all of these instances both testosterone secretion and spermatogenesis are likely to be defective because the latter depends on the former. Often, however, patients may have normal testosterone secretion and there is a defect only in spermatogenesis; the sole complaint is infertility. Unfortunately, though this form of hypogonadism is fairly common, in most patients who have few spermatozoa (oligospermia) or none at all (azoospermia), either no cause is found or a known cause, e.g., mumps orchitis, has resulted in irreversible damage.

Poor spermatogenesis, whatever the origin, is diagnosed when the sperm count (number of spermatozoa in the ejaculate) is low (less than about 20 million/ml); if the prime defect is in the testis, serum FSH is often elevated. Sometimes a biopsy of the testis is needed to demonstrate poor spermatogenesis if there is little or no ejaculate. Rarely, the sperm count is normal and the infertility is due to an intrinsic defect in the spermatozoa, such as lack of the acrosome or excessive prostaglandin F in the seminal fluid.

When *Leydig cell damage* or dysfunction occurs in primary hypogonadism, the serum testosterone is low or at least responds poorly to stimuli such as hCG (exogenous stimulation test). Both FSH and LH are usually elevated, although several measurements may be needed because both are secreted in bursts and a single measurement may miss a high point (endogenous stimulation test). The LH is high because of testosterone (and possibly E_2) deficiency; FSH is high in part because spermatogenesis is likely to be defective owing to testosterone deficiency and in part because of low testosterone and E_2.

Rarely, *defects* occur *in testosterone's action.* When testosterone is totally incapable of acting, the patient has testicular feminization. The patient, a genetic male with XY sex chromosomes, has a woman's body habitus with normal breast development but no uterus. There are no ovaries, but there are two testes that secrete normal male amounts of testosterone and E_2. The defect usually lies in tissues normally responsive to testosterone; they do not have the proper cytosol receptor for testosterone or DHT and so the androgens cannot act. The E_2, though scanty, is unimpeded by androgen and thus causes feminization. Partial degrees of this disorder also exist. In these cases, the genitalia are only partly masculinized, and various anomalies are present at birth. In the partial disorder there may be either some degree of deficiency of the androgen cytosol receptor or relatively poor conversion of testosterone to DHT in the responding tissues.

Finally, both functions of the testis slowly decrease with *age* owing to some intrinsic testicular defect; as a result, half of men over age 70 have elevated serum FSH and LH. The diminished spermatogenesis is not usually a problem, but decreasing testosterone secretion may be. Whereas a lower testosterone may prevent or delay the appearance of obstructive prostatic hypertrophy, the lack of testosterone might lead to osteoporosis and certainly can cause decreased sexual function.

Treatment of most primary and secondary hypogonadism is testosterone, preferably given as a long-acting injectable preparation. Recent work suggests that testosterone itself may be effective by mouth if enough is given. Not much can be done about defective spermatogenesis unless it responds to testosterone, which is not likely since the high testicular levels of testosterone cannot be reached by giving testosterone to the patient. Occasionally a patient may respond to hCG or clomiphene even when FSH, LH and testosterone levels are normal. The patient with secondary hypogónadism who is infertile and strongly desires children may have a spermatogenic response to gonadotropin treatment. However, since gonadotropins must be given by injection, the treatment is long and expensive and the response is uncertain. When endorgan insensitivity is the problem, treatment is complex and is primarily surgical and psychosocial; the main problem is assignment of a gender, usually female, with corrective surgery to accomplish that end. Since the testes in testicular feminization tend to develop carcinoma, they are usually removed at some point; appropriate estrogen therapy is then necessary.

Excessive gonadal function is rare and appears mainly as *precocious puberty* in a young boy. Often signifying a serious lesion, it may be due to an adrenal or testicular tumor making excess androgen or to some anomaly affecting the hypothalamus (such as a pineal tumor) causing premature pubertal levels of gonadotropins.

Sometimes men (or boys) develop breast enlargement, or *gynecomastia*. Often due to use of various drugs, it may rarely be caused by an estrogen- or gonadotropin-secreting tumor of the adrenal cortex or testis. Excessive PRL is almost never a cause. In many instances a cause is never found, although a fair case can often be made for an imbalance between estrogen and androgen, i.e., a high estrogen/androgen ratio, at the level of the breast tissue itself. Treatment consists of removing the cause, if known, and in removing the excessive breast tissue if cosmetically disturbing.

DISORDERS IN WOMEN

The same general types of hormonal deficiencies occur in women as in men. There may be deficient GnRH, FSH and/or LH, or E_2 or progesterone. Partial deficiencies are more apparent in women than in men because ovulation and the menstrual cycle are sensitive to slight deviations from normal. The result is often amenorrhea (absent menses) or changes in the menstrual cycle that are easily noticed. An example is the amenorrhea that occurs with malnutrition or a stress such as leaving home for the first time; the amenorrhea is almost certainly due to a functional hypothalamic defect that affects the ovulatory peak of gonadotropins but not their basal secretion.

As in men, *hypogonadism* owing to hypothalamic or pituitary defects is termed secondary hypogonadism and that due to ovarian disease is primary hypogonadism. It should be noted, however, that to the gynecologist a patient who has never menstruated has primary amenorrhea and one who has menstruated but then stopped has secondary amenorrhea; these are descriptive terms that have no necessary relationship to primary and secondary hypogonadism.

In moderate to severe degrees of hypogonadism there is not only amenorrhea but also infertility and evidence of decreased estrogen secretion, such as decreased libido and thin vaginal mucosa. Low or relatively low serum FSH and LH along with low E_2 levels define secondary hypogonadism. However, the defect may be subtle; a slight deficiency of FSH, for example, may result in the usual menstrual cycle except that it is shorter than normal because the corpus luteum lasts only a week or so instead of 2 weeks (the short luteal phase). Estradiol levels are normal and there is an ovulatory peak of FSH and LH. The mild FSH deficiency is difficult to distinguish from normal; the best test is probably the serum progesterone level, which is lower than expected during the luteal phase. When gonadotropin levels are high in the face of low E_2 levels, the patient has primary hypogonadism. This is exactly what happens in the menopause, but it also occurs in any ovarian disease in which the follicles fail to function properly. Nevertheless, because we do not fully understand the details of the normal menstrual cycle, we are not able to define completely all the possible hormonal abnormalities. The more subtle ones require many measurements of several hormones over a period of time and even then the disorder may remain elusive.

Treatment of overt estrogen deficiency is to replace the estrogen; this is commonly done in menopausal women, although there is some controversy over how long treatment should be continued. Progesterone deficiency, of course, always occurs in the absence of ovulation but is not treated with progesterone except possibly in the instance of the short luteal phase when associated with infertility; bromergocryptine, an inhibitor of PRL secretion, may be helpful (? is prolactin luteolytic). When hypogonadism is secondary and infertility is the main complaint, human gonadotropins or clomiphene may be used; this treatment is not always predictable and may result in birth of quintuplets.

Hirsutism or the more severe *virilism* in women is sometimes due to drugs but often to no known cause; in many instances there is too much androgen production. The excessive androgen is usually testosterone—though in some cases it seems to be DHT or the testosterone precursor androstenedione; the source may be either the adrenal cortex, the ovary or both. Occasionally a tumor is responsible, but more often hyperplastic tissue is found. When the adrenal is the source, the increased testosterone production may be associated with Cushing's disease, with any of several enzymatic defects usually seen in childhood or with nothing in particular. When the ovary is at fault, hirsutism can occur with hyperplasia of thecal cells (hyperthecosis) or with polycystic ovaries (Stein-Leventhal syndrome); enzymatic blocks may be involved here as well. Sometimes, however, the distinction between an ovarian and adrenal cause of increased testosterone secretion is not easy to make because both glands may secrete excessive testosterone at the same time.

Treatment of hirsutism is local and palliative, i.e., depilatories or electrolysis, unless definite proof of increased androgen production can be shown. This may require measurement of DHT, androstenedione, free (unbound) testosterone or the testosterone production rate in addition to total serum testosterone. If proved, treatment can be directed at the adrenal (glucocorticoids for enzymatic defects or as empirical treatment for unknown defects; surgery or pituitary ablation for Cushing's disease) or at the ovary (usually an estrogen/progestin combination such as an oral contraceptive, or surgery).

In addition to disorders of gonadotropin or steroid secretion, women may also have *abnormal PRL secretion*. Too little PRL can occur as an isolated defect as well as in generalized hypopituitarism; the only problem is inability to nurse the infant after birth. Excessive PRL may cause no symptoms at all or may cause excessive milk secretion, or *galactorrhea*. The more severe degrees are associated with amenorrhea and low gonadotropin secretion; most of these instances occur either as postpartum lactation that has lasted too long or with a pituitary tumor. Milder galactorrhea can occur with normal menses. In contrast to more severely affected patients, these patients usually do not have elevated PRL levels and the cause of the problem is unknown.

Treatment of galactorrhea depends on the cause: if drug-induced (often the case), the drug should be removed; if post partum, sometimes simply waiting is enough; if due to a pituitary tumor, surgery may be necessary. However, nonsurgical treatment is now possible. Inducing ovulation with clomiphene sometimes works in postpartum galactorrhea; presumably, stimulation of GnRH secretion is associated with PIF (PRL-inhibiting factor) secretion. Other agents such as L-dopa or bromergocryptine seem to specifically suppress excessive secretion of PRL even when caused by a pituitary tumor; they may stimulate PIF secretion or act directly on the anterior pituitary.

GROWTH AND GROWTH HORMONES

Growth is an indefinite concept: in man, it usually refers to height; in rats it may mean weight or length; in cell cultures it may refer to total protein, DNA or mitoses; and in populations it refers to numbers of individuals. Fortunately, in a single individual, height (or length), total protein and DNA are crudely correlated; thus a taller person generally has more total protein and more cells. Most of the variation in human adult height is due to differences in genetic background or nutrition; tall parents tend to have tall children and poorly fed or emotionally deprived children do not grow as fast as they should. Clearly, however, a normal complement of several hormones is necessary for optimal growth; abnormal amounts of any of these can lead to short stature. Too little thyroxine (hypothyroidism) or insulin (diabetes mellitus) results in poor growth, as does too much hydrocortisone (Cushing's syndrome) or excess gonadal hormones (precocious puberty). Excessive height can result from a partial deficiency of testosterone. Nevertheless even if all these hormones are secreted in normal amounts, the child (and the adult) will be too short if *growth hormone*, also called *somatotropin*, is missing.

GROWTH HORMONE AND ITS ACTIONS

Secreted from acidophil cells of the anterior pituitary, hGH (human growth hormone) is a polypeptide of 191 amino acids. Many of the amino acids are in the same loci as those in hPL (human placental lactogen), but hPL has little or no growth activity in man. Growth hormone (GH) was first isolated and named by noting which pituitary extracts caused hypophysectomized rats to resume normal growth and retain nitrogen, an index of net protein synthesis.

The same molecule, however, when injected into rats or human beings, has additional effects such as increasing plasma FFA (free fatty acids) and inducing insulin resistance and thus a tendency toward a higher plasma glucose level. Growth hormone thus may be poorly named since it has effects other than simply stimulating growth.

How does hGH act? In contrast to most other peptide hormones, cAMP does not seem to be the second messenger for hGH except possibly in thymocytes and the renal medulla. Little is known about its cellular mechanism except that it first binds to specific receptors on the cell's plasma membrane. The growth function of GH is reflected in increased DNA, RNA and protein synthesis in tissues, particularly muscle, and in increased lengthening of long bones by stimulation of cartilaginous growth in the bony epiphyses. For example, GH enhances muscle size in vivo (provided insulin is present) and can increase amino acid uptake and incorporation into protein in muscle in vitro. However, the in vitro effects occur only at high levels of GH that are not found in vivo and may not be physiologically relevant.

How then does GH work? The effects on bone and muscle growth are probably not due to a direct action of GH. More likely they are caused by *somatomedin* or "sulfation factor," so called because it mediates some of the actions of GH and because it stimulates sulfate incorporation into cartilage. Growth hormone stimulates secretion of somatomedin by the liver and possibly by the kidney. Somatomedin is probably identical to NSILA-S (nonsuppressible insulin-like activity, soluble), which has a molecular weight of 6,000–8,000 daltons and has insulin-like actions in certain bioassays but is not insulin and probably has no insulin-like activity in vivo. Thus GH certainly has direct actions on some tissues — it acts directly on the liver to produce somatomedin itself — but it stimulates growth of cartilage and probably muscle only indirectly via somatomedin. Since the optimal growth effect of GH seems to require the presence of insulin, insulin must somehow interact with GH and/or somatomedin, but the cellular mechanism is unknown.

The mechanisms of the metabolic actions of GH, i.e., those on FFA and glucose, are less easy to define in vivo. Part of the problem is that it is not clear whether GH has any role in lipolysis or regulating glucose utilization at the GH levels normally found in serum. For example, there is no question that injection of GH causes a rise in plasma FFA, but FFA are still mobilized in the absence of a rise in endogenous GH. Perhaps a better viewpoint is that the presence of some GH may be permissive for optimal lipolysis, but a rise in GH is not needed for reasonably normal lipolysis to occur. There also may be a role for GH in long-term regulation of adipose tissue mass; it may keep down total body fat since in its absence patients tend to become obese. Similarly, there is no question that a gross excess of GH can inhibit glucose utilization by insulin-sensitive tissues. Physiologic amounts of GH may do this in times of glucose need, i.e., when glucose must be shunted to the brain, a glucose-dependent but not insulin-sensitive tissue. However, GH may also do exactly the opposite: normal amounts of GH seem to enhance the growth and function of the pancreatic beta cells, thereby enhancing glucose use via more efficient insulin secretion. With this degree of confusion regarding the significance of the metabolic effects of GH, it is not surprising that almost nothing is known of the mechanisms of action.

OTHER GROWTH FACTORS

It has long been known that cells in tissue culture grow better when the incubation medium includes serum. Why this is so is not as clear. The effect is probably not due to GH but to some other growth factor or probably factors. Recently there have been advances in defining these factors, all of which probably deserve to be called hormones. It is beginning to seem that there are too many of them to integrate into a coherent scheme. Somatomedin is itself one of these and there are now at least two or three substances that can be called somatomedin, known as somatomedins A, B and C. In addition, serum contains *multiplication-stimulating activity (MSA)*, which comes from the liver and may be the same as somatomedin C; *epidermal growth factor (EGF)*, which stimulates the growth of epithelial cells and fibroblasts; and *nerve growth factor (NGF)*, which maintains cultures of nerve cells in vitro but also stimulates fibroblast growth. Other growth factors can be extracted from the anterior pituitary or brain tissue. Most of this work is in its infancy, and further information is needed before we can begin to put it all together.

CONTROL OF GROWTH HORMONE SECRETION

Because the radioimmunoassay for GH was one of the first developed, much is known about its secretion, which is largely under hypothalamic control. For example, exercise, sleep and many types of stress cause release of GH; GHRF (a specific hypothalamic GH-releasing factor) has not yet been chemically defined, but it almost certainly exists and probably stimulates GH secretion via an increase in pituitary cAMP (or possibly cGMP). A hypothalamic factor that inhibits GH secretion has been chemically defined. Called *somatostatin*, it is a 14 amino acid peptide and is found not only in the hypothalamus but also in other areas of the brain, in the intestine and in the pancreatic islets. Furthermore, it also inhibits secretion of prolactin, TSH, insulin and glucagon. Somatostatin's distribution and multiple actions make it difficult to define its precise physiologic role; presumably the amount of GH actually secreted is determined by the net balance between somatostatin and GHRF.

What controls the secretion of GHRF and somatostatin? Since neither can be reliably assayed in blood, the question cannot be answered. Nevertheless, a number of agents stimulate or suppress GH secretion in man and many of these probably act via the hypothalamus. Some of these agents are (or stimulate the production of) neurotransmitters, which in turn presumably act on the peptidergic neurons that secrete GHRF or somatostatin. For example, apomorphine stimulates dopaminergic receptors and leads to a rise in serum GH; thus dopamine may stimulate GHRF secretion or decrease somatostatin secretion. Dopamine is not the only neurotransmitter involved in GH secretion, however, since there is evidence that norepinephrine (via alpha adrenergic receptors) and serotonin can also enhance GH secretion. Most of these effects, in rats at least, are probably mediated by the ventromedial nucleus of the hypothalamus. Just how these monoamines are linked to GHRF and somatostatin secretion is unknown. Furthermore, since control of GH secretion is different in rats and human beings—stress stimulates GH in man but inhibits it in the rat—there may be sufficient species differences in monoaminergic control of GHRF and somatostatin secretion that each species may have to be studied individually.

Some of the agents affecting GH secretion are related to fuel metabolism. A fall in blood glucose (hypoglycemia), a fall in FFA or a rise in amino acids all stimulate GH secretion; of these, hypoglycemia is the most potent stimulus. Hyperglycemia or a rise in FFA suppresses GH secretion. Hypoglycemia seems to act via the ventromedial (VM) hypothalamic nucleus but the locus of the other effects is not clear, nor is it known whether these metabolic agents act via one or another of the neurotransmitters or by directly stimulating the release of GHRF or somatostatin. The fact that these metabolic agents affect GH secretion makes us suspect that GH has an important role in controlling glucose, FFA and amino acid metabolism, despite the reservations about its metabolic actions noted above.

Abnormal Growth Hormone Secretion

As we might guess, most short children do not have GH deficiency and most basketball players do not have excessive GH. But some do. Growth hormone deficiency may be an indicator of a deficiency of several pituitary hormones or it may occur as the only deficiency; in the latter case, the lesion is likely to be hypothalamic rather than pituitary. Excessive GH can result in a very tall person ("pituitary giant") but only if the bony epiphyses have not fused. After puberty, too much GH does not increase height but causes overgrowth of endochondral bone and soft tissues, resulting in the clinical syndrome of *acromegaly*. So if suspicion of GH deficiency or GH excess exists, the serum GH should be measured. To get the best information we do a suppression test when suspecting excessive GH and a stimulation test when suspecting deficient GH. Oral glucose is the standard suppression test. Any of several stimulation tests may be used: insulin-induced hypoglycemia; L-dopa, which results in monoaminergic stimulation; or glucagon, which acts partly by inducing a rise and subsequent fall in serum glucose. Treatment of GH deficiency is injection of hGH; the supply is limited and must be restricted to patients with proved deficiency who have good potential for further growth. Treatment of acromegaly is usually destruction of the anterior pituitary either by surgery or radiation. Medical therapy with agents that suppress GH secretion, e.g., chlorpromazine or ergocryptine, is possible but not yet of proved reliability.

HORMONAL CONTROL OF FUEL METABOLISM

We are, of course, what we eat. But we do not eat continuously. In order to stay alive, we must store metabolic fuel so that there is a reserve to be called upon when not eating. For example, fully half the calories expended each day are needed to keep the body's metabolic processes going even at complete rest; these processes demand an energy source whether or not we are eating. If the next meal is not in sight, more energy must be spent to get it, which usually means some degree of muscular exercise; the stored reserves of fuel are then even more important.

The proper intake and storage of fuel, the metabolic switch from the fed to the fasting state and the release of stored fuel are finely regulated mechanisms under hormonal control. Sometimes the mechanisms go awry and too much fuel is stored (obesity) or fuel is released too quickly (acute diabetic ketoacidosis); in a patient with another common syndrome, the obese adult-onset diabetic, there seems to be a combination of excessive long-term storage and inefficient short-term storage of fuel.

What are the body's fuel stores? Most is in the form of fat (triglyceride), 98% of which is stored in adipose tissue and 2% in muscle (Table 10–3). A small but critical amount of fuel is stored as glycogen in liver and muscle. The body's protein usually is not considered a fuel depot since it is in the form of working muscles and collagenous structures, but protein is an important indirect fuel source during fasting.

What are the body's fuels? During and shortly after eating—the fed state—they are whatever is in the diet. The carbohydrate, protein and fat in the usual meal appear in the blood stream as glucose, amino acids and triglycerides or FFA. When not eating—the fasting state—the principal fuels mobilized from stores into the blood are *glucose* from liver (and kidney), *FFA* from adipose tissue, and the so-called *ketoacids*, acetoacetate and β-hydroxybutyrate, generated from FFA by the liver. Some of the fuel used during fasting comes from local stores rather than from the blood stream, e.g., muscle triglycerides and glycogen are used in the muscle itself.

What regulates the fuel supply and what are the signals used? Several hormones are involved; the main ones are *insulin* from the beta cells and *glucagon* from the alpha cells of the pancreatic islets. Cortisol, GH and the thyroid hormones are also important for normal control of fuel supply but are more supportive or permissive than directly regulatory.

The *nervous system* is involved as well. Both the parasympathetic and sympathetic system affect insulin secretion, for example. The sympathetic

TABLE 10–3.—BODY FUELS*

FUEL	DAILY INTAKE		BODY STORES	
	GM	K-CAL	GM	K-CAL
Carbohydrate	300	1,200	400	1,600†
Fat	120	1,100	14,000	130,000‡
Protein	75	300	13,000	50,000

*Amounts are approximate and given in grams and kilocalories.

†Almost all stored as glycogen. Most of the glycogen (~80%) is in muscle, but the 50–75 gm in liver is the main emergency supply for blood glucose.

‡Almost all stored as triglyceride: 98% of triglyceride in adipose tissue, 2% in muscle, 0.1% in liver.

nervous system, via *norepinephrine* and *epinephrine* from the sympathetic nerves and adrenal medulla, is particularly important in mobilizing fuel stores during times of stress. The CNS is equally important, since it is the locus for control of *food intake*. The hypothalamus—perhaps including neurons entering and leaving it—has centers that both enhance food intake (perceived as hunger or increased appetite) and suppress it (perceived as satiety).

The known signals sensed by these hormonal and neural regulating mechanisms are probably the plasma levels of the fuels themselves, particularly glucose, amino acids, FFA and perhaps glycerol. However, afferent neural signals clearly contribute, although their relative importance is hard to measure. For example, with proper conditioning, the suggestion of a meal can cause insulin secretion without food actually being eaten, and afferent vagal impulses originating in the stomach may help regulate satiety. Furthermore, there may well be other signals as yet unknown that are part of this complex system; the unknown signal by which the total mass of adipose tissue appears to regulate appetite is a case in point.

INSULIN: ACTIONS AND REGULATION OF SECRETION

Insulin is a 51 amino acid peptide with two chains connected by disulfide bridges. It was "discovered" in 1921 by Banting and Best. Other investigators knew before then that something in the pancreas lowered the blood level of glucose but Banting and Best were the first to prepare a nontoxic active pancreatic extract and successfully treat diabetic patients. Later workers crystallized insulin (Kimball, Murlin, Abel), determined its amino acid sequence (Sanger) and synthesized it (Katsoyannis). Insulin is formed in the beta cell as a cleavage product of a larger peptide, *proinsulin* (see Fig 10–3), and stored as crystals in tiny intracellular vesicles until secreted into the blood. The other cleavage product is the "connecting" peptide (or C-peptide), which, though biologically inactive, is secreted at the same time as insulin. The normal level of serum insulin in the fasting state is 10–20 μU/ml. The amount of insulin secreted actually varies from moment to moment because of dietary effects but, in man, the total daily secretion is about 25–50 units per day. (Insulin was originally measured in arbitrary units, and common usage retains this tradition. Pure insulin is now known to have about 25 units/mg; thus 25–50 units equals about 1–2 mg and a serum level of 10 μU/ml is about 400 pg/ml.) About half the insulin secreted into the portal vessels never gets past the liver; thus, the rest of the body is exposed to 12–25 units per day. The kidney also metabolizes much insulin—roughly 20% of the daily secretion.

ACTION OF INSULIN

Everyone knows that insulin lowers the blood glucose level and, in fact, insulin is the main regulator of carbohydrate metabolism. After binding to receptors on the cell membrane, insulin rapidly increases the velocity at which glucose enters many tissues, particularly muscle and adipose tissue, most likely by affecting the cell membrane so that a glucose carrier moves faster (Fig 10–27). Glucose utilization by tissues responsive to insulin is thus increased. Insulin may have a further specific stimulating effect on glycolysis by increasing phosphofructokinase activity. Some of the glucose taken up by tissues is stored as glycogen; insulin enhances this process by specifically increasing the activity of glycogen synthetase I (largely in liver, muscle and adipose tissue). Although insulin binds to plasma membrane receptors in the liver, as in muscle and adipose tissue, it increases hepatic glucose uptake more slowly than in the other tissues; it is thought to do so by increasing intracellular glucokinase activity rather than by speeding the membrane transport of glucose, but further work on this point is needed. Insulin also decreases hepatic glucose output both by decreasing gluconeogenesis (owing to the presentation of less gluconeogenic substrates to the liver and to a lower level of hepatic phosphoenolpyruvate carboxykinase [PEPCK]) and by decreasing glycogenolysis (owing to a lower level of active phosphorylase). The end result of insulin's action on carbohydrate metabolism is the use of glucose as a fuel and the storage of some glucose as glycogen.

However, insulin also has major effects on fat and protein metabolism. In the liver (and in adipose tissue in some species, though not in man), insulin simulates lipogenesis, the process by which glucose or other substrates are converted into fatty acids. It probably does so by increasing the activities of citrate cleavage enzyme and acetyl-CoA carboxylase, the enzymes responsible for the formation of acetyl-CoA and malonyl-CoA in the hepatic cytosol. The fatty acids so formed are esterified into triglycerides. Insulin also stimulates the release of triglycerides from the liver into the blood as very low density lipoproteins (VLDL). After the triglycerides are in the blood, insulin enhances their deposition in adipose tissue. Plasma triglycerides, whether carried as VLDL or as chylomicra derived from the diet, cannot directly enter adipose tissue. They must first be broken down to fatty acids near the surface

Fig 10–27.—Insulin action on glucose and amino acids in muscle or adipose tissue after binding to its receptor.

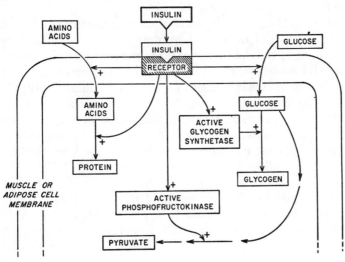

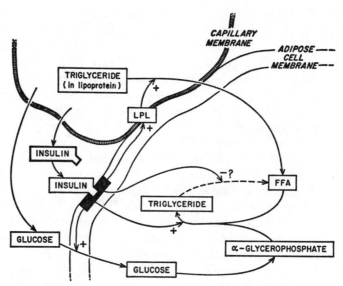

Fig 10–28.—Insulin action on lipid metabolism in adipose tissue. *LPL* = lipoprotein lipase; ? = effect not proved.

of the adipose cell by the enzyme, lipoprotein lipase (LPL) (Fig 10–28). The fatty acids then enter the cell and are esterified back into triglycerides. Insulin enhances LPL activity and thus helps move triglycerides from the blood into the adipose tissue stores. In the adipose cell, insulin stimulates the re-esterification of fatty acids into triglycerides and may inhibit lipolysis, thus decreasing the release of FFA and glycerol. Insulin also stimulates the activity of an LPL found in the liver; this enzyme may also play a role in clearing plasma triglycerides.

In muscle (and some other tissues), insulin decreases the release of amino acids and may increase their uptake. Insulin also stimulates the incorporation of amino acids into protein (see Fig 10–27).

Thus insulin has effects on all metabolic fuels; the net result of its action is an increase in stored fuel and in body protein and it is therefore considered an anabolic hormone. It remains unclear how insulin exerts these effects after it binds to the cell membrane. Much effort has been spent trying to demonstrate a fall in cellular adenyl cyclase activity or in intracellular cAMP; the effort has only occasionally been rewarding. For example, some experiments have shown a fall in hepatic cAMP after insulin, and in adipose tissue insulin blunts the expected rise in cAMP after a known stimulus to cAMP production. In many other experiments, however, insulin caused no change in cAMP. Nevertheless, some effect of insulin on suppressing cAMP or its action is probably important. Recent experiments have shown that insulin enhances the phosphorylation of certain proteins, at least in adipose tissue, and that insulin may somehow block activation of protein kinase by cAMP; perhaps the insulin-induced phosphorylated protein is involved in the protein kinase blockade. If so, there need be no change in cAMP levels in order for insulin to inhibit a cAMP-induced mechanism. Or perhaps an increased cGMP is the intracellular mediator of insulin's action. Time will tell.

Of interest is the fact that insulin has no effect on glucose uptake by certain tissues for which glucose is a required fuel. The most prominent of these tissues is the brain, which burns only glucose except during prolonged fasting when it can use ketoacids. With the exception of certain critical areas in the hypothalamus, the entire CNS is unaffected by insulin. Similarly, the renal medulla and formed elements of the blood fail to respond to insulin and yet require glucose.

Finally, the action of insulin in responsive tissues may be affected by other hormones or by insulin itself. Growth hormone, for example, can decrease insulin's ability to enhance amino acid uptake by thymocytes and may do the same thing to glucose uptake by muscle. Prolactin may do the opposite to amino acid uptake. Insulin itself, when it acts on cells, appears to inhibit further binding of insulin molecules to receptor sites on the same cells. This effect called "negative cooperativity," is discussed further in the subsection on obesity. Should this phenomenon, a form of local negative feedback, be more widespread and not peculiar to insulin, it would be of great importance in understanding hormone actions in general.

REGULATION OF INSULIN SECRETION

Since insulin lowers blood glucose level, an elevated blood level of glucose should stimulate insulin secretion if it were controlled by a simple negative feedback loop. In fact, this does occur and the blood level of glucose is the principal physiologic controller of insulin secretion. A rise in plasma glucose above a critical level of about 80 mg/dl stimulates insulin secretion, whereas a fall in glucose shuts it off. However, the very complexity of insulin's actions and of the entire fuel control system might make us suspect that the regulating mechanism in reality is more complex than a simple closed-loop glucose-insulin control. The suspicion would be correct. Modifiers of the basic glucose-insulin system include (1) gastrointestinal factors, (2) factors that alter the blood glucose, thus indirectly affecting insulin secretion, (3) neural effects and (4) other agents directly affecting the beta cell (Fig 10–29).

Oral glucose stimulates a greater insulin response than does the same amount of glucose given intravenously, so something in the gut responds to the glucose and stimulates insulin secretion. Many of the gut hormones have been proposed as the gastrointestinal insulin-releasing factor because they stimulate insulin secretion when injected; these include gastrin, secretin, cholecystokinin (CCK), gastric inhibitory peptide (GIP), a glucagon-like intestinal peptide (GLI) and a separate (though not yet purified) peptide thought to be specific for insulin secretion called insulin-releasing polypeptide (IRP). Which (if any) of these is the truly physiologic one is not yet clear. Whichever it is, it probably acts by sensitizing the beta cell so that a given amount of glucose causes a greater insulin response than without the gut hormone.

Once the glucose has been absorbed, the actual blood level of glucose is determined not only by the glucose load but also by other factors that affect glucose entry into or exit from the blood. Anything that increases entry of glucose into the blood from gluconeogenesis or stored glycogen (glucagon, hydrocortisone, increased sympathetic impulses to the liver) or decreases exit of glucose from blood into tissues (GH, FFA) results in a higher blood level of glucose after a glucose load. The higher blood level of glucose causes a greater insulin secretion. Conversely, anything that decreases glucose entry into the blood (fasting or absence of hormones that increase entry) or increases its exit (exercise) results in lower insulin secretion.

Although not the primary stimulus and not recognized for a long time, there is no question that neural impulses affect insulin secretion. The pancreatic islets are supplied by both adrenergic and cholinergic nerve fibers. Efferent vagal impulses stimulate insulin secretion (if glucose is present) via acetylcholine. Sympathetic nerve impulses both stimulate (beta adrenergic) and suppress (alpha adrenergic) insulin secretion; the net effect of sympathetic stimulation is usually suppression. In addition, epinephrine released from the adrenal medulla, acting as an alpha adrenergic stimulus, also inhibits insulin secretion. It is possible that serotoninergic nerve fibers in the pancreas act to block insulin's secretion.

How much these neural impulses regulate glucose-induced insulin secretion is hard to say in quantitative terms. In the basal state without stress or external stimuli, there probably is some tonic neural control of insulin secretion, but overall the total effect seems minor. Stress, however, sets off sympathetic impulses, and conditioned response to eating (sham eating or smelling food without eating) triggers vagal impulses. The best evidence is that these responses are mediated via the hypothalamus. Stimulation of the ventromedial (VM) hypothalamus blunts insulin secretion both by direct sympathetic impulses to the pancreas and by adrenal epinephrine secretion. Stimulation of the ventrolateral (VL) hypothalamus, the area responsible for eating behavior, increases insulin secretion at least in part via the vagus and possibly by causing secretion of an unknown hypothalamic hormone. Although the hypothalamus mainly acts by integrating sensory neural inputs, some evi-

Fig 10–29.—Glucose is main stimulus to insulin secretion; other factors modulate insulin secretion. ? = probable or possible effects.

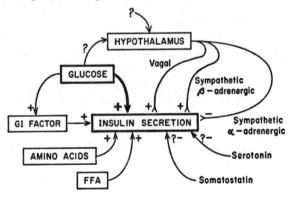

dence suggests direct perception by the hypothalamus of the blood glucose and/or insulin level. A high glucose (or insulin) level in the blood may, via the hypothalamus, trigger the vagus and enhance the pancreatic insulin response to glucose, a reinforcing effect reminiscent of oral glucose and gut hormones. Thus the hypothalamus and the autonomic nervous system can play a major role in regulating insulin secretion in the nonbasal state, and neural control of insulin may itself be regulated by the blood level of glucose or insulin.

Finally, other hormones and metabolic fuels can change insulin secretion, although glucose still dominates. Amino acids are a potent stimulus, particularly in infants, and triglycerides and FFA can also act as mild stimuli if glucose is present; these effects are relevant after eating. Glucagon, and perhaps GH, also directly stimulates insulin secretion under some circumstances, although these effects and the mild stimulation by acetoacetate may only be important during fasting or over the long term. Somatostatin—surprisingly found in one type of pancreatic islet delta cell—blocks insulin secretion, as does epinephrine, as noted earlier. Whether somatostatin restrains insulin secretion under physiologic circumstances is not yet clear.

How do these stimuli act on the beta cell? Glucose causes insulin to be secreted from the granular stores with a rapid early burst of secretion and a more prolonged second phase. Glucose also increases insulin synthesis, but almost all the insulin secreted immediately after a glucose load comes from the granules. The storage granules actually move to the surface of the beta cell and, after fusion of the granular membrane with the cell membrane, the granule is released into the extracellular space; this process is called emiocytosis (see Fig 10–4). The movement of the granules from within the cell to the cell surface is aided by a system of microtubules (20 nm in diameter), whereas the exit of the granules from the cell tends to be inhibited by a system of microfilaments (6 nm in diameter) just inside the cell surface.

How glucose triggers this release mechanism is a matter of controversy. Some evidence suggests that glucose may be detected by a specific receptor at or near the cell surface. The alpha anomer of D-glucose is a better insulin secretogogue than is the beta anomer, indicating that the structure of the glucose molecule itself rather than that of a glucose metabolite is important. On the other hand, other data indicate that metabolism of glucose may actually be required for optimal insulin secretion. D-Glyceraldehyde directly causes insulin release, and total blockade of glucose metabolism clearly decreases insulin secretion.

The problem of whether glucose itself or a glucose metabolite is the stimulus is complicated by the fact that both the intracellular calcium and cAMP concentrations probably act as second messengers for glucose and other stimuli inside the beta cell. How it all fits together is uncertain. The rapidity of the early phase of insulin secretion, which occurs a minute or two after glucose stimulation, suggests a membrane effect of some sort; glucose (or a metabolite) may raise the intracellular calcium by transiently blocking calcium efflux from the cell, and the high intracellular calcium then may directly cause contraction of the microtubules. There seems to be no increase in islet cAMP content owing to glucose during this early phase, although the beta cell has an adenyl cyclase in its cell membrane. However, the *presence* of some cAMP may be needed for proper calcium translocation and/or the later phase of sustained insulin secretion. And later on there may also be a stimulation of adenyl cyclase by glucose or its metabolites, which then maintains the calcium-stimulated insulin release—and synthesis of new insulin—as long as glucose is present. Thus glucose may cause early insulin secretion by its effects on calcium and may maintain more prolonged insulin secretion and synthesis by its effects on both calcium and cAMP.

Other agents acting directly on the beta cell may or may not act via cAMP. Glucagon and beta adrenergic stimuli raise the islet cAMP and calcium content, and insulin secretion follows more or less in parallel. There may be specific receptors for amino acids but the amino acids do not elevate islet cAMP. Alpha adrenergic stimuli decrease cAMP and calcium in the islet cell and thus decrease insulin secretion. Clearly there are opportunities for interactions within the cell between these agents and glucose; the actual amount of insulin secreted is probably a result of these interactions as they affect calcium, cAMP and the microtubules and microfilaments.

GLUCAGON: ACTIONS AND REGULATION OF SECRETION

In the 1920s it was observed that pancreatic extracts, thought to contain only insulin, produced a transient rise in blood glucose. A second hormone of the pancreatic islet was therefore postulated and named *glucagon*. Later, glucagon was isolated, chemically defined and synthesized; it is a 29 amino acid peptide with a molecular weight of about 3,500 and is secreted by the alpha cell of the pancreatic islet. In some species, e.g., the dog, pancreatic glucagon is also found in the gut, but in most species the GLI (glucagon-like immunoreactivity) found in the gut is due to an agent with a larger molecular weight, which has little or none of glucagon's physiologic actions. In the alpha cell, glucagon is formed from a larger proglucagon containing at least 37 amino acids and is stored in granular form. The normal fasting plasma glucagon level is about 50–150 pg/ml. As expected, the portal vein contains about twice as much glucagon as peripheral veins, indicating substantial removal by the liver; the kidney may also be involved in clearance of glucagon since plasma glucagon rises in renal failure.

ACTIONS OF GLUCAGON

Glucagon is best viewed as a metabolic counterweight to insulin. Whenever blood glucose level falls below normal, glucagon tends to keep it up; whenever blood glucose level rises above normal, insulin drives it back down again. Nevertheless, it is important to note that, in man, the predominant hormone of the two is insulin and that glucagon is a modulator. Human beings who have had pancreatectomy have an absolute need for insulin but get along well enough without glucagon.

Glucagon increases the blood glucose level by increasing hepatic glucose output; it does this by stimulating both glycogen breakdown (glycogenolysis) and synthesis of new glucose (gluconeogenesis) within a few minutes. Glucagon is a potent stimulator of glycogenolysis. One molecule of glucagon can cause the release of about 3×10^6 molecules of glucose from glycogen. How is this multiplier effect brought about? By an endocrine cascade that begins with the binding of glucagon to specific receptors on the hepatic cell membrane. These receptors differ from the adrenergic receptors, but subsequent events are the same as with catecholamine-induced glycogenolysis (see Fig 10–14). First, there is activation of adenyl cyclase. The resulting increase in cAMP then activates a protein kinase, which in turn converts a phosphorylase kinase to its active form. The phosphorylase kinase then converts inactive phosphorylase to active phosphorylase and glycogen breakdown follows. Interestingly, the protein kinase is probably the same enzyme that phosphorylates the active form of glycogen synthetase, causing it to become inactive. Thus glucagon not only enhances glycogenolysis but at the same time also shunts glucose away from glycogen synthesis.

Glucagon increases hepatic glucose synthesis—or gluconeogenesis—as well. Again mediated by cAMP, glucagon causes increased conversion of amino acids, such as alanine, and of lactate and pyruvate to glucose. How does this happen? The link with cAMP is not as clear as with glycogenolysis but glucagon increases the activity (and often synthesis) of hepatic enzymes responsible for converting several amino acids to glucose precursors and for converting pyruvate (via oxaloacetate) to phosphoenolpyruvate (PEP) and thence to glucose. Glucagon can also increase protein breakdown and decrease protein synthesis in the liver, thus supplying more amino acids for gluconeogenesis. Because fatty acids enhance gluconeogenesis and because glucagon may stimulate lipolysis (in adipose tissue and/or liver), it is possible that glucagon acts as well by increasing hepatic FFA content. However, this is unlikely and the effects of glucagon and FFA on gluconeogenesis are probably separate and additive.

Glucagon has some interesting effects on hepatic lipid metabolism other than its possible effect on lipolysis (which may not occur at physiologic glucagon concentrations). It decreases triglyceride release from the liver, the opposite of insulin action. It also switches the metabolism of hepatic FFA from triglyceride synthesis to ketoacid formation. Glucagon (probably via cAMP) does this by increasing the activity of mitochondrial carnitine acyltransferase II, the enzyme responsible for generating acyl CoA inside the mitochondria. This is an important control point and so glucagon may be essential for ketoacid synthesis and, in severe diabetics, for the development of ketoacidosis.

There are several extrahepatic actions of glucagon, such as increasing cardiac contractility, lowering serum calcium and stimulating GH secretion, as well as its effect on increasing insulin secretion and adipose tissue lipolysis. To date it is uncertain whether any of these occur at levels of glucagon found in the blood.

Most, and perhaps all, of glucagon's actions are opposed or modulated by insulin, provided enough insulin is present. There is some disagreement on this point, since several experiments indicate that glucagon's action may not always be blocked by insulin, e.g., its action on glucose production by canine liver. It now appears that the ratio between insulin and glucagon may be the determining factor in in vitro perfusion experiments, and the disagreement may simply be owing to differences in this ratio. In vivo, however, the insulin/glucagon ratio may not be as important as the absolute level of circulating insulin. Glucagon raises the blood glucose level when insulin is at or below its normal fasting level but, when insulin is clearly higher, as after a glucose load, physiologic levels of glucagon seem to have little or no effect. How does insulin counteract glucagon if insulin does not seem to act by directly affecting hepatic cAMP production, whereas glucagon does? The best answer is that insulin probably restricts the rise in cAMP induced by glucagon, although the mechanism is obscure, or possibly blunts activation of the protein kinase. Thus neither insulin nor glucagon can be thought of as acting alone; both must be taken into account, keeping in mind insulin's dominant position in mammals relative to glucagon.

REGULATION OF GLUCAGON SECRETION

As with insulin, the basic regulator of glucagon secretion is the level of blood glucose: a fall in blood glucose level stimulates glucagon secretion, whereas a rise in blood glucose level suppresses it. The glucagon response to induced hypoglycemia resembles the insulin response to glucose: there is a brisk secretion of glucagon in the first few minutes, followed by a second, more prolonged phase lasting as long as hypoglycemia persists. The physiologic moderate fall in blood glucose level during fasting is probably responsible for the modest elevation in plasma glucagon.

As with insulin, however, things are not that simple and glucagon secretion is modulated by factors other than glucose. The fall in glucagon that occurs after a rise in blood glucose is in part due to a direct suppressant effect of glucose on the alpha cell but is probably assisted by the local rise in insulin from the neighboring beta cells, since insulin is known to decrease glucagon secretion (Fig 10–30). Somatostatin blocks secretion of glucagon as well as of insulin; many of the delta cells containing somatostatin are near the alpha cells, but whether or not somatostatin regulates glucagon physiologically is not known. Addition of amino acids or a protein meal, in contrast to the opposite effects of glucose on insulin and glucagon secretion, causes a rise in both glucagon and insulin secretion. The rise in glucagon after adding amino acids — and glucagon's subsequent effect on hepatic glucose output — acts to prevent the fall in blood glucose that would otherwise result from the concomitant rise in insulin. Further, the glucagon response to protein taken orally is partly mediated by a gastrointestinal hormone, similar to the oral glucose-insulin relationship. The nature of this hormone is not clear, although both gastrin and cholecystokinin have been proposed. The FFA also are a modest stimulus for glucagon secretion as well as for insulin secretion.

Neural impulses also affect glucagon secretion. Sympathetic nerve stimulation increases glucagon secretion by activating beta adrenergic receptors in the alpha cells of the islets. This is the opposite of the effect on insulin secretion in which sympathetic stimulation exerts a predominantly alpha adrener-

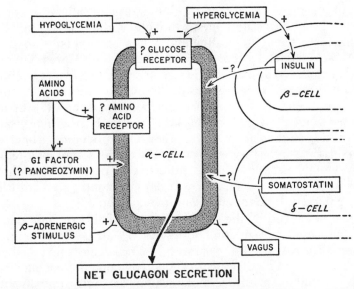

Fig 10–30. — Factors affecting glucagon secretion. Intracellular events are unclear, although cAMP may be a mediator of some stimuli. The existence of specific glucose and amino acid receptors is not proved, nor is role of neighboring islet cells (see text).

gic effect. Vagal stimulation (cholinergic) reduces glucagon secretion, again the opposite of the effect on insulin; there is, however, some suggestion that the vagus may mediate part of the glucagon response to hypoglycemia. Thus the rise in glucagon after hypoglycemia may result not only from direct stimulation of the alpha cell but also from sympathetic and parasympathetic stimulation, presumably mediated via the hypothalamus. The sympathetic neural effect fits with the rise in glucagon seen with many types of stress such as exercise, trauma, burns, infection, painful stimuli or even loud noise. The increased glucagon may then provide the fuel needed during the time of stress. Nothing is known of the role of the hypothalamus in integrating the alpha cell's response to these stresses.

How does the alpha cell respond to these stimuli? There are probably receptors for both glucose and amino acids, as there are in the beta cell. The cAMP may mediate increases in glucagon secretion since the alpha cell membrane has an adenyl cyclase; however, suppression of glucagon secretion by glucose is probably not mediated by cAMP. Microtubules and microfilaments may play some role in release of the glucagon-containing granules but, if anything, both structures seem to inhibit granular release. Clearly, not enough is known about intracellular events in the alpha cell, perhaps because they are outnumbered eightfold by the beta cells in the mammalian pancreas.

EATING: STARTING AND STOPPING

We eat when we are hungry and stop eating when we are satisfied. In a behavioral sense, hunger or the urge to eat is synonymous with eating, provided there is food. Satiety or the sensation of having eaten enough is the same as not eating in the presence of food. Both hunger and satiety, or appetite, are hypothalamic functions. The general thought is that the ventrolateral (VL) hypothalamic area stimulates eating and, when enough is eaten, the ventromedial (VM) area somehow senses it and shuts off the VL area. There remains a good deal of controversy over (1) the signals perceived by the hypothalamus, (2) whether hunger and satiety are controlled by specific nuclei in the VL and VM areas or by neurons passing through these areas, and (3) the actual relationship between the two hypothalamic nuclei. Furthermore, satiety and hunger as usually studied are relatively short-term phenomena operating over a few hours. Eating behavior is also subject to some sort of long-term control since body weight usually changes little over a period of months or even years; how food intake is regulated over days to months so that little weight is normally gained or lost is another controversial topic.

The signals perceived by the hypothalamus probably include one or more of the metabolic fuels or regulating hormones as well as neural impulses from the periphery (Fig 10–31). Much of the evidence suggests that the hypothalamus can sense a rise in blood glucose. The hypothalamic sensing mechanism is probably related to insulin-stimulated glucose entry into hypothalamic cells, but the hepatic portal bed also seems able to detect a rise in glucose and to transmit this information to the hypothalamus via the vagus. Possibly the rise in insulin itself is what is sensed. With a rise in glucose, the hypothalamus then induces satiety and stops eating behavior. Conversely, a fall in glucose and/or insulin should induce hunger and increased eating. Hypoglycemia does cause hunger. Yet after a normal meal, hunger and increased eating do not usually occur until a few hours after serum glucose and insulin have returned to normal. Hence there is some question that glucose is the only regulator ("glucostatic" theory of appetite regulation). Possibly a slight fall in blood glucose within the normal range is an effective stimulus. Other evidence suggests short-term control of eating by the gastrointestinal tract itself:

Fig 10–31.—General scheme of factors regulating eating. Inappropriately increased energy stores (largely as adipose triglyceride) might result from several defects, including (1) excessive response of stores to storing stimuli, (2) poor response of stores to mobilizing stimuli, (3) poor generation of inhibitory signals to hypothalamus and (4) poor perception of these signals by hypothalamus.

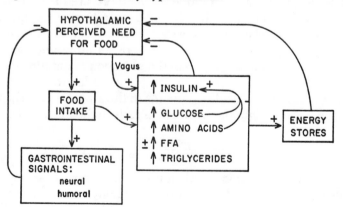

eating can be inhibited both by vagal afferent impulses stimulated by gastric distension and by CCK (cholecystokinin) released by the small intestine (CCK is identical to pancreozymin). Furthermore, exercise, which causes a fall in both blood glucose and insulin levels, also decreases eating; here, perhaps the hypothalamus is responding to beta adrenergic stimuli, which are known to decrease eating.

A major current hypothesis is that these short-term inhibitors of eating may act by triggering the VM hypothalamus to actively inhibit the VL hypothalamus and thus decrease eating. However, although the VM area is related to satiety and the VL area to eating, there is some doubt whether the two areas are directly connected, as required by the hypothesis. Short-term eating behavior may be regulated by the VL area with the stimuli — e.g., glucose, insulin, cholecystokinin, vagal afferents — sensed and acted upon in the VL area itself. An alternate hypothesis proposes that the VM area is involved in long-term rather than short-term control of eating.

The relative constancy of body weight and adipose tissue over reasonably long periods of time — several months or a few years — requires rather fine control of caloric intake. A positive error of only 50 calories per day in food intake, less than 2% of the total, would result in a weight gain of almost 30 lb in 5 years. Few people gain this much, and yet no one is conscious of a control mechanism as precise as this. The details of long-term regulation of eating are hazy at best. One possibility is that the VM area may integrate the signals and fine tune the activity of the VL hypothalamus. The signals may include an integrated glucose and/or insulin level but probably also include some factor that reflects total body adipose tissue, perhaps the blood level of glycerol. If such a signal from adipose tissue exists, adipose tissue mass would regulate eating; thus adipose tissue mass would itself be more or less constant. This is the lipostatic theory and is the long-term counterpart of the short-term glucostatic theory. In reality, however, we suspect that such an important and complicated system is regulated by multiple factors. One factor, glucose, may be predominant at certain times, such as after meals, depending on the composition of the meal. At other times, i.e., more than 2 hours after eating, other signals may be more important, such as FFA or glycerol released from adipose tissue.

REGULATION OF METABOLIC FUEL SUPPLY DURING EATING, FASTING AND EXERCISE

This subsection will synthesize some of the preceding material and relate it to the changes actually observed in man during daily life.

EATING. — The usual mixed meal contains carbohydrate, protein and fat. These are processed by the gut and appear in the blood stream as glucose, amino acids, triglycerides and FFA. Glucose alone causes a rise in serum insulin (directly and via a gastrointestinal factor), a fall in serum glucagon (a fall enhanced by the increased insulin) and a fall in serum GH. Amino acids alone cause a similar rise in insulin but, in contrast to glucose, also cause a rise in both glucagon and GH. Triglycerides and FFA have a slight effect in increasing insulin and decreasing GH but the effects are minor and difficult to show in man. In fact, if the fat content of the meal is not high, there may not be a rise in FFA or even in triglycerides after eating. Since the predominant hormonally effective fuel after a mixed meal is glucose, the net hormonal change is a rise in insulin and a fall in both glucagon and GH.

What then happens to the metabolic substrates and the urge to eat?

With *glucose* there are the following:

1. Increased total body uptake and oxidation owing to increased insulin.

2. Decreased hepatic output with decreased glycogenolysis and gluconeogenesis owing to more insulin and perhaps to a resulting decreased cAMP (the decreased glucagon probably has little effect here).

3. Increased storage as glycogen in liver, kidney, and muscle owing to increased insulin.

4. Increased storage as triglyceride via glycerophosphate synthesis in adipose tissue and liver and via fatty acid synthesis in liver owing mainly to increased insulin.

With *amino acids* there are the following:

1. Increased uptake by muscle (of at least some amino acids) and synthesis of new protein owing to increased insulin aided by the presence of some GH (or somatomedin).

2. Decreased release by muscle, particularly of alanine, owing to unknown mechanisms (possibly insulin).

3. Decreased uptake by liver, which in part causes the decreased gluconeogenesis; the decreased uptake is caused by increased insulin and perhaps decreased glucagon.

With *FFA and triglycerides* there are the following:

1. Decreased release of FFA from adipose tissue, which results from (a) decreased lipolysis of stored triglycerides perhaps owing to increased insulin (and possibly decreased glucagon and growth hormone) (see Fig 10–28); there is also an associated fall in the active form of triglyceride lipase; and (b) increased re-esterification of FFA into stored triglycerides owing to increased insulin and the associated increased conversion of glucose and glycerol to glycerophosphate.

2. Decreased oxidation of FFA in general, owing to decreased lipolysis and decreased FFA oxidation specifically by the liver perhaps owing to lower glucagon and to lower growth hormone.

3. Increased storage of FFA in adipose tissue and ?liver owing to increased insulin and the rise in re-esterification noted above.

4. Increased storage of circulating triglyceride in adipose tissue and liver owing to increased insulin, which induces a rise in the active lipoprotein lipase needed for triglyceride entry into cells (as noted before, triglycerides cannot enter these tissues directly but must first be broken down to FFA, which enter the cell and are re-esterified as above).

Eating behavior is decreased owing probably to increased glucose (and probably insulin) acting on the hypothalamus.

Overall, eating is a state in which there is abundant glucose and a high insulin level. The result is a replenishment of carbohydrate and fat stores and a shunting of amino acids away from gluconeogenesis and toward the synthesis of new protein.

FASTING. — Beginning a few hours after eating, the body switches to endogenous stores for its fuel supply. The longest time most people fast is overnight, i.e., about 12–15 hours, at which time liver glycogen is a major source of glucose. Most of the calories needed come from FFA or their metabolites but a good supply of glucose is critical during fasting because without it the brain cannot function. It should be noted that the brain needs glucose but does not need insulin to burn it (except for the hypothalamus, which is insulin sensitive). Eventually the brain learns to use ketoacids for some of its energy needs, but defense of blood glucose level is essential for survival. Thus the metabolic fuels during fasting are glucose, FFA and the ketoacids, acetoacetate and β-hydroxybutyrate (which is not really a ketoacid), formed from FFA by the hepatic mitochondria.

After overnight fasting there is a modest fall in blood glucose, a rise in FFA and ketoacids and an increase in release of amino acids into the blood, particularly of alanine and glutamine. The concomitant hormonal changes are a fall in insulin and a rise in glucagon largely owing to a lower blood glucose. There is also a probable rise in sympathetic nervous system activity and an erratic rise in GH. Levels of cortisol and thyroid hormone are not changed but normal amounts of these hormones are necessary for mobilization of stored fuels.

The *blood glucose* level may fall to 50–70 mg/dl but no further, even with prolonged fasting (except that in some women it may fall as low as 35 mg/dl). It is defended by the following:

1. Increased glycogenolysis in the liver owing to the increased glucagon and probably the increased sympathetic nerve activity (muscle glycogen is broken down and used in the muscle).

2. Decreased glycogen synthesis from glucose in the liver owing to the fall in insulin and the rise in glucagon (muscle glycogen synthesis is also decreased).

3. Increased gluconeogenesis in the liver probably owing to the fall in insulin and the rise in glucagon. (The main substrates are amino acids and lactate from muscle proteolysis and glycolysis; it is uncertain whether the hormones control these processes. The rise in FFA also contributes somehow to increased gluconeogenesis, perhaps by increasing needed NADH and/or by increasing mitochondrial acetyl CoA, which activates pyruvate carboxylase.)

4. Decreased glycolysis in the liver possibly owing to decreased insulin.

As the blood glucose falls, the *blood FFA* rises and plateaus at a level of 1,500–2,000 μEq/L; the ketoacids rise in parallel. These effects result from the following:

1. Increased lipolysis in adipose tissue owing to decreased insulin and increased sympathetic neural activity (beta adrenergic). (Although glucagon and GH can be lipolytic, they have no proved role in fasting lipolysis; lipotropin is another hormone with no proved function.) Muscle and liver triglycerides can be hydrolyzed and used by those tissues.

2. Decreased re-esterification of FFA in adipose tissue owing to decreased insulin.

3. Increased ketogenesis (production of ketoacids in the liver) owing to an increased load of plasma FFA presented to the liver and to increased glucagon in the presence of a low insulin level.

The blood levels of many *amino acids,* such as alanine, fall during fasting, although others such as valine and leucine rise. Regardless of changes in individual amino acids, total muscle release of amino acids increases; the increase is largely accounted for by alanine and glutamine. Yet alanine and glutamine make up only a small percentage of the total amino acids in muscle protein. How does all of this occur? It results from (1) increased proteolysis in muscle due to unknown causes, perhaps related to decreased insulin; (2) increased transfer of muscle amino groups to pyruvate (derived from muscle glycolysis), forming extra alanine, and to glutamic acid, forming extra glutamine due to unknown causes; (3) increased uptake of alanine and other gluconeogenic amino acids by the liver probably due to the fall in insulin and rise in glucagon.

Thus, early in fasting (overnight up to a day or two) the brain and a few other glucose-dependent tissues, such as blood cells and renal medulla, receive the 150 gm or so of glucose they need every day, largely from hepatic glycogen stores and indirectly from muscle protein and glucose metabolism (Fig 10–32). If the glucose is not completely oxidized but is metabolized only to pyruvate and lactate, these 3-carbon compounds return to the liver and are recycled into glucose. Perhaps 75 gm of protein is broken down during each day of fasting, much of which winds up as glucose and, incidentally, urea and ammonium. About 150 gm per day of fat (triglycerides) is mobilized as FFA, of which about 60 gm is converted to ketoacids. These are used almost entirely as fuel, although there may be an excess of FFA, which can be converted back to triglycerides in liver or muscle. Clinical ketoacidosis does not normal-

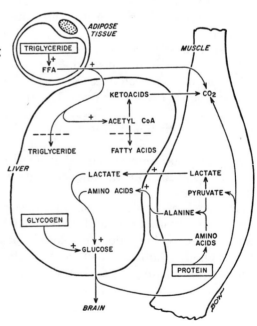

Fig 10–32.—Sources of circulating fuel in fasting state. + = process enhanced by fall in insulin and/or rise in glucagon. FFA also enhance gluconeogenesis.

ly happen because the small amount of insulin present exerts sufficient restraint on lipolysis to prevent massive release of FFA.

The brain thus gets its needed glucose, whereas the rest of the body burns mostly FFA and ketoacids. The brain can burn ketoacids (though not FFA) and will do so readily if fasting is prolonged for several days, provided some glucose is available. What stops other body tissues from using the glucose the brain needs? The answer is unsettled but important factors may be (1) decreased insulin, which impairs glucose uptake by muscle but not by brain since the brain does not need insulin for glucose uptake in the first place; (2) increased FFA, which is probably preferentially used as fuel by muscle instead of glucose and which may directly decrease muscle glucose uptake; and (3) increased GH, which might decrease glucose uptake by muscle.

Thus the brain and muscles are supplied with proper fuel, which permits the next meal to be sought and caught. Eating restores fuel stores and muscle protein, and the cycle begins anew.

EXERCISE. — The next meal, for mammals, rarely comes without effort. Work is usually needed. Work or muscular exercise requires fuel, which must be supplied rapidly. Even if food has recently been eaten, the body must depend on its stored fuels. We might guess, correctly, that fuel stored in the muscle itself would be used first. What happens in the contracting muscle? The local changes are probably not hormonally induced but are due to the contraction itself. They include (1) increased glycogenolysis and ultimately increased production and release of pyruvate and lactate (a good supply of muscle glycogen is essential for prolonged exercise), (2) increased alanine release probably due both to proteolysis and to increased conversion of pyruvate to alanine and (3) increased lipolysis of muscle triglycerides and thus increased FFA oxidation (a relatively minor source of energy during exercise).

However, although essential both for early, sudden muscle contraction (a few minutes) and for prolonged, sustained exercise, muscle fuel stores alone are not enough. The contracting muscle must also take up glucose and FFA from the blood. The supply of these fuels is under neural and hormonal control. The following changes occur: (1) increased sympathetic nervous activity; (2) decreased serum insulin due largely to alpha adrenergic sympathetic stimuli; (3) increased serum glucagon due to sympathetic stimuli, possibly to a rise in serum alanine and later to a fall in blood glucose and (4) increased serum GH, which is said to be a response to stress but the exact cause is not known.

During exercise the blood glucose level rises or at least stays the same and does not fall unless the exercise is prolonged for several hours. Since glucose uptake by the exercising muscle is greatly increased, there must be a rise in glucose entry into the blood. Plasma FFA also rise after an initial transient fall, again indicating increased entry into the blood from storage sites. Increased hepatic glycogenolysis is responsible for most of the increased supply of blood glucose. Glycogenolysis is probably stimulated by increased sympathetic activity and is supported by the rise in glucagon and the fall in insulin when they occur. The exact details are not entirely clear. There is also increased gluconeogenesis in the liver (owing to increased sympathetic activity, rise in glucagon and fall in insulin) and in the kidney (owing to mild lactic acidosis). Gluconeogenesis contributes little of the glucose needed immediately during short-term exercise but, after an hour or so of exercise, hepatic gluconeogenesis contributes half the glucose released into the blood. Adipose tissue lipolysis is also enhanced by increased sympathetic nervous activity, and perhaps by the fall in insulin, and is responsible for the rise in FFA.

Both the glucose and FFA from the blood, important though they are in brief exercise, become even more important during prolonged exercise. Some muscle glycogen is needed in prolonged exercise, but most of the muscle's stored fuels are depleted by then and the principal energy sources become blood glucose and FFA. Curiously, the high glucose uptake by muscle occurs in the face of a fall in insulin levels; the glucose uptake is due to some exercise-induced change in glucose transport but requires a small amount of insulin to occur. As exercise continues, blood FFA become the more important of the external fuels and account for twice as much muscle oxygen consumption (60% of the total oxygen consumed) as does blood glucose (30% of the total).

Thus the energy supply for exercise is initially the exercising muscle itself but, after a few minutes, other sources of energy are needed. The liver and adipose tissue supply the required glucose and FFA. They do so in response to the neural and hormonal changes of exercise, i.e., increased sympathetic nervous activity, a rise in glucagon and a fall in insulin. If all goes well, the next meal is by now at hand.

DISORDERS OF FUEL METABOLISM

DIABETES MELLITUS. — Diabetes mellitus ("sweet flowing through" or excessive sweet-tasting urine) is an ancient disease. It was probably first described in written records in an Egyptian papyrus dating back to 1500 B.C. A clear distinction from other causes of large urine volume, such as diabetes insipidus (or "tasteless" urine), was not made until the 17th century when Thomas Willis noted and described the sweet taste of the urine. Later it was shown that the sweetness was due to glucose and that this was secondary to a high blood glucose level. Naturally this finding led to more esthetically acceptable modes of diagnosis; a high blood level of glucose is now diagnostic of diabetes mellitus. By the 19th century attention had focused on the pancreas as the source of the disease. Insulin deficiency was proposed as the cause, and in 1921 for the first time an active pancreatic extract was prepared that contained enough insulin to treat patients effectively.

Diabetes mellitus is a common disease and affects about 2% of the population in the United States, although in certain groups, such as Pima Indians, as many as 50–60% may have it. Most patients have a mild form of the disease and only a few have the severe, ketoacidotic type. In fact, about half of those who have diabetes mellitus do not know it. The typical diabetic patient has an elevated fasting blood level of glucose but, in the milder form of the disease, the blood level of glucose may be high only after a glucose load. The *glucose tolerance test* is designed to detect this. Glucose, 100 gm, is given orally, and the rise and subsequent fall in blood glucose level is measured at hourly intervals. If the blood level of glucose rises above defined normal limits during the test, diabetes mellitus can be diagnosed, even though the fasting glucose level is normal. Empirically, the blood glucose value at 2 hours is the most useful; in normal persons it is below 120–130 mg/dl (below 140–150 mg/dl if serum is assayed instead of blood). The glucose tolerance test is really a *stimulation test* for an insulin-deficient state since it assesses glucose disappearance from the blood, which in turn is a result of insulin secretion and action.

As we might expect, there is total insulin deficiency in severe diabetes mellitus and only partial deficiency in the milder types. The cause is unknown. Clearly, genetic factors are involved since many members of the same family may have the disease and there is a higher incidence in children of diabetic patients. However, environmental factors such as viral infections may also be important. Probably multiple genetic factors are involved in the tendency to become diabetic, and the actual expression of the disease may depend on what happens later in life. The exact defect in the beta cell is unclear. Total damage to the beta cells owing to inflammation (e.g., infection or pancreatitis) would, of course, cause total insulin deficiency, but many patients with diabetes mellitus have reasonable numbers of beta cells. These latter patients may have total or partial insulin deficiency.

What is wrong then with the beta cell? Possibilities include a defective signal from the gut when glucose is eaten (probably not true), a defective glucoreceptor on the beta cell (a good possibility), a problem within the beta cell in transmission of the glucose-induced signal to the insulin-secreting apparatus (not proved) or a functional defect in any of the several steps involved in insulin synthesis and secretion (not proved).

Despite our ignorance, the range of possible causes is exciting and clearly suggests that diabetes mellitus may in fact be several different diseases, all of which happen to be characterized by hyperglycemia and insulin deficiency. Since the severe type often (though by no means always) occurs in younger patients, it is also called the juvenile-onset type; the more common milder type almost always occurs later in life, and hence is called adult-onset.

The metabolic defects in the juvenile diabetic patient almost totally lacking in insulin are almost all predictable (and are discussed below), provided insulin's multiple effects are kept in mind. Perhaps not as predictable is the elevated glucagon level. In part, we might expect a high glucagon level because, in the normal person, insulin probably acts as a physiologic restraint on glucagon secretion. However, it is also possible that there is a separate defect in human diabetes causing abnormally high glucagon secretion from the alpha cell as well as poor insulin secretion by the beta cell; the point is controversial and not settled. In any case, glucagon levels are higher than normal, particularly when the high blood glucose level, which should suppress glucagon, is taken into account. Since glucagon generally acts on metabolic fuels in an opposite direction from insulin, the metabolic abnormalities are therefore worse than we might expect from insulin deficiency alone.

The blood glucose level is high not only because of defective glucose entry into tissues but also because the liver puts out more glucose. Gluconeogenesis from several substrates is increased, as is the supply of amino acids from muscle. Glycogenolysis may also be increased, but the total amount of glycogen may not fall because the high blood glucose level alone may maintain a reasonable rate of glycogen synthesis. Other hormones such as cortisol and GH may be elevated in the acutely ill patient, thus increasing gluconeogenesis even more. Furthermore, with eating, a higher percentage of glucose from the diet gets past the liver into the systemic circulation because the usual 85% or so of ingested glucose is not taken up by the liver. It should be noted, however, that, although the blood sugar level is high and glucose exit from the blood slower than normal for a given level of blood glucose, the total amount of glucose oxidized per day is normal in most diabetics, provided at least a little insulin is present. Glucose entry into metabolizing tissues simply occurs at a higher blood glucose level.

A high blood glucose level alone has some untoward results. When the blood glucose level is clearly elevated, i.e., consistently above 200 mg/dl or so, more intracellular glucose is metabolized to sorbitol and then to fructose — steps catalyzed by aldose reductase and sorbitol dehydrogenase, respectively. Little sorbitol is normally found in cells but, with increased conversion of glucose to sorbitol, the sorbitol accumulates because its metabolism to fructose is relatively slow. A high level of intracellular sorbitol may cause cell damage by swelling due to increased intracellular osmolality or perhaps by directly interfering with normal biochemical processes. Other unknown osmoles can also contribute to cellular edema when blood level of glucose is high. These shifts in carbohydrate and fluid metabolism may be responsible for certain complications seen in diabetes mellitus, such as cataracts, peripheral neuropathy and possibly CNS damage, and may perhaps contribute to the atherosclerosis that occurs so frequently in large blood vessels of the diabetic patient.

Whether a high blood glucose level causes the small vessel disease that is also so common in diabetic patients is a matter of great controversy. These capillary and arteriolar lesions are characterized by a thickening of the endothelial basement membrane and are a principal cause of disability and death in anyone with long-standing diabetes mellitus. They ultimately result in major retinal and renal disease and contribute to the peripheral neuropathy and gangrenous feet seen in many patients. Diabetes mellitus is now one of the major causes of new cases of blindness in the United States and is a common cause of renal failure requiring hemodialysis and renal transplantation.

Many workers think that the elevated blood glucose level causes these lesions. Why? Because when lesions occur, they appear only after 10–15 years of diabetes mellitus and because experimental hyperglycemia in rats can lead to biochemical and anatomical changes in renal glomeruli. However, others feel that the basement membrane defect is a separate part of the overall syndrome, perhaps also genetically determined. The microvascular and carbohydrate defects thus could coexist; the one might make the other worse but neither need be the cause of the other. This viewpoint is supported both by the existence of renal or retinal disease or neuropathy in some patients before there is abnormal glucose tolerance and by the presence of abnormally thickened capillary basement membranes in nondiabetic relatives of patients with diabetes mellitus. Some scattered evidence suggests that pituitary GH may be causally involved in the microvascular disease but again there is disagreement. Resolution of this controversy is of obvious importance in preventing these irreversible complications of diabetes mellitus.

The blood glucose level may be fairly high (200–300 mg/dl) in the juvenile diabetic patient but he may feel reasonably well because the amount of insulin given as treatment, although not quite enough to keep the blood glucose normal, is enough to control lipid and protein metabolism. It should be noted, however, that insulin injections do not mimic normal pancreatic insulin secretion, which fluctuates with changes in the blood level of glucose (and other regulatory factors) from moment to moment; hence the current interest in an electronic implantable artificial pancreas and in beta cell transplantation. These approaches to therapy are, however, several years away at best, and false hopes should not be raised over their ultimate usefulness. The goal of truly bringing the blood glucose level to normal still remains. If this is done, we shall be in a position to know whether or not a high blood glucose causes diabetic microvascular disease.

Should all insulin be gone, as seen in undiagnosed juvenile diabetes, the patient would become ketoacidotic. This is a direct result of the release of vast amounts of FFA from adipose tissue owing to unrestrained lipolysis and deficient re-esterification of FFA. The amount of ketoacids, acetoacetate and β-hydroxybutyrate, produced by the hepatic mitochondria is roughly proportional to the presented load of FFA. Ketogenesis is increased even more, however, by the presence of a high glucagon level unopposed by insulin and by an increased amount of GH, both of which are found in the acutely ill ketoacidotic patient. How glucagon affects a mitochondrial process is unclear; perhaps it is mediated via cAMP. Clinical acidosis results from overproduction of these ketoacids and is further worsened by decreased ketoacid removal from the blood, a process that is normally speeded by insulin. Coma then results if the acidosis is allowed to persist.

Thus, in the diabetic patient lacking insulin, a higher fraction of the FFA presented to the liver becomes ketoacids, and a lower fraction becomes triglycerides. Nevertheless, total triglyceride synthesis may actually be increased because the influx of FFA is so great. The liver normally would attach much of the new triglyceride to an apolipoprotein and secrete it into the blood as VLDL (very low density lipoprotein). However, there appears to be a relative defect in apolipoprotein synthesis so that much triglyceride is retained in the liver. The triglyceride that is secreted, along with any chylomicra formed from ingested fat, encounters further difficulties. It cannot easily leave the blood because the activity of the adipose tissue enzyme needed to remove triglyceride, lipoprotein lipase, is deficient owing to the lack of insulin. Thus hypertriglyceridemia results; occasionally this is so marked that the serum is milky or lipemic. Protein metabolism is also adversely affected in the complete absence of insulin. Amino acids do not enter muscle easily and muscle protein breaks down faster. Protein catabolism thus parallels adipose tissue fat catabolism.

Overall, the body is breaking down its stored fuels so fast and in such a way that, instead of meeting the body's needs, the body is being destroyed. Catabolism continues unabated and clinically the patient becomes somnolent and comatose and dies unless proper treatment is started. Insulin is specific therapy and reverses all these defects. It does not work instantaneously, however, in a patient in an advanced state of ketoacidotic coma. Additional therapy with intravenous saline, potassium and phosphate is essential, since these elements are lost in large amounts while the illness is developing. Massive glucosuria carries with it sodium and chloride; vomiting, diarrhea, and somnolence may compound the fluid and electrolyte loss; and muscle breakdown, insulin deficiency and secondary aldosteronism due to volume contraction induce and worsen losses of potassium and phosphate. Careful attention to all the metabolic abnormalities is necessary for successful treatment.

The patient with adult-onset diabetes rarely becomes ketoacidotic because the insulin deficiency is only partial. In fact, this type of diabetes mellitus may be an entirely different disease from the juvenile type according to genetic analysis of families of diabetic patients. The principal presentation is thus hyperglycemia alone, with its attendant problems as discussed above, but without ketoacidosis. The patient with adult-onset diabetes may have fairly large amounts of circulating insulin. How can this be? Many of these patients are obese, and obesity by some unknown mechanism causes resistance to insulin's action. A normal person who becomes obese must therefore secrete extra insulin in order to maintain normal glucose disappearance from the blood (or normal glucose tolerance). The obese diabetic patient is unable to provide the extra insulin and so is diabetic even though there is enough insulin for a nonobese person; the patient is therefore relatively insulin deficient. The lesson is clear. Obesity is bad for patients with diabetes mellitus; weight reduction always helps and may result in return of normal glucose tolerance. Why these patients become obese in the first place is no more clear than why anybody becomes obese. If weight reduction does not work in the diabetic patient with symptoms or, more commonly, cannot be achieved, the blood glucose level must be lowered to some reasonable level though not necessarily to normal. Insulin is effective treatment, though it may not affect the vascular complications that occur as well in this type of diabetes mellitus. Whether or not oral hypoglycemic agents, such as tolbutamide, should be used is another current controversy; a personal opinion is that they are clinically useful in selected patients.

HYPOGLYCEMIA.—Since a normal glucose level is essential for normal brain function, hypoglycemia is a dangerous state. It reflects a major disorder of fuel metabolism because it means that several defense mechanisms designed to maintain normoglycemia have been overridden. What we call a normal blood glucose level is a result of glucose influx and efflux. Glucose influx into the blood must keep up with efflux if the blood glucose level is not to fall. Normally several mechanisms do this including a rise in glucagon and/or a fall in insulin secretion; increased sympathetic activity, causing hepatic glycogenolysis; induction of hunger, causing increased eating; and maintenance of normal levels of cortisol, GH and thyroid hormones. This degree of redundancy in the means of maintaining the blood level of glucose indicates how important the process is. Failure or deviation of one or more of these mechanisms can result in hypoglycemia unless the other mechanisms can compensate.

Hypoglycemia is commonly overdiagnosed. Proof of hypoglycemia requires demonstration of a low blood glucose level *and* symptoms consistent with hypoglycemia occurring at the time of low glucose. Both requirements are necessary because the range of normal glucose values is so wide that there is much overlap between normal persons and patients who are truly hypoglycemic.

Clinically, hypoglycemia can occur within a few hours after eating (reactive hypoglycemia) or in the fasting state, although the distinction is often blurred. *Reactive hypoglycemia* is usually attributed to excessive insulin secretion in response to a meal, thus causing a low blood glucose 1–2 hours after eating. In fact, with the exception of patients who have had partial gastrectomy, insulin excess is not commonly found. If there is no demonstrable defect in the mechanisms noted above, the cause remains unknown. Avoidance of large carbohydrate meals and sometimes use of anticholinergic or beta adrenergic blocking drugs may help. These measures may work in part by blunting the insulin response to meals; other unknown mechanisms must be involved because frequently the insulin response is not changed but the patient is better. *Fasting hypoglycemia* may be due to an insulin-secreting tumor of the pancreas (insulinoma), to islet-cell hyperplasia involving beta cells or to any defect in the glucose-maintaining regulatory system. Insulinoma is actually a rare tumor and can be diagnosed by showing an inappropriately high insulin level in the presence of hypoglycemia (which should normally decrease insulin secretion) (*endogenous suppression test*). We can also perform an *exogenous suppression test* by giving porcine insulin and inducing hypoglycemia. Failure of endogenous C-peptide levels to fall indicates excessive insulin secretion. Unusual causes of fasting hypoglycemia include glucagon deficiency, GH deficiency or poor hepatic glucose output. The most common cause of hypoglycemia is overtreatment of a diabetic with insulin, the *"insulin reaction."* Treatment of most patients with fasting hypoglycemia consists of replacement of any missing hormones and surgical removal of excess insulin-secreting tissue.

OBESITY. – Obesity is common in our society and increases mortality when it is severe enough. It aggravates other diseases, such as diabetes mellitus or hypertension, and increases the risk of any surgical procedure. It is associated with more heart disease than are other diseases, although the higher dietary cholesterol may be the more direct causal factor. Fat may or may not be beautiful, but it is not healthy.

Obesity is an increase in total body fat. Since the range of body weight in the total population is a continuum, there is no clear break-point above which a person is obese and below which a person is not. There are fairly elaborate ways of specifically measuring body fat, but these are too cumbersome for clinical use or even for large-scale studies. Although not as precise, simpler methods are more practical. Standard height-weight tables are probably the most widely used aid; arbitrarily, a patient who weighs more than 25% above ideal weight (corrected for height, sex and size of body frame) can be considered clearly obese. Skin-fold thickness is also useful since the thickness of subcutaneous adipose tissue is roughly proportional to total body fat. Simply looking at the patient turns out to be about as reasonable a diagnostic method as any for the practiced observer.

Almost all of the extra fat is in adipose tissue in the form of triglyceride. Every extra pound of adipose tissue represents about 3,600 extra calories. Each adipose cell is larger than normal and more stuffed with triglyceride. However, many (perhaps most) obese patients have a larger number of adipose cells as well. Whereas a normal person has about 30×10^9 adipose cells, the obese patient may have three times as many, particularly when the obesity has begun in childhood or in the teens. Weight reduction in such a patient does not decrease the number of cells, although the cells are smaller and have less triglyceride. If more calories are eaten, the cells are still there, ready to become large again and they do. Therein lies part of the difficulty in maintaining weight loss once it is achieved.

A large fat cell usually is also an insulin-resistant cell, i.e., it takes more insulin to have the same effect. Does this mean that all obese patients have a high blood glucose level or diabetes mellitus? No, because if their pancreatic islets are normal, they will secrete more insulin, thus maintaining normal glucose tolerance. In fact, nondiabetic obese patients have a normal blood glucose level but a higher than normal serum insulin level; the serum insulin correlates fairly well with the size of the fat cells. What causes the insulin resistance? Herein lies a major controversy. Some investigators believe that large fat cells have fewer insulin receptors per cell. Other data suggest that the receptors are normal in number but fail to transmit their signal to the cell's interior. In either case obesity and the large fat cells would cause the insulin resistance.

Another point of view is that large fat cells are not inherently insulin resistant at all. How then does the obese patient have a normal glucose level but a high insulin level in the blood? The explanation is that the large fat cells are insulin resistant but are made that way by the diet of the obese patient. If the high caloric intake that is necessary if a person is to become obese includes an excess of carbohydrate, the high carbohydrate intake will cause both an increase in insulin secretion and in fat stores. Increased insulin can decrease the total number of apparent insulin receptors on the fat cell membrane. Functionally this is termed "negative cooperativity." How does this happen? When insulin binds to one receptor, there is a receptor-receptor interaction (or cooperation), mediated by an intracellular process inside the cell's membrane, such that insulin's binding to other receptors is decreased; functionally the result is fewer insulin-binding sites than expected. Thus we would see insulin resistance and large fat cells. The answers are not all in, but the concept of negative cooperativity is an important addition to our discussion of both normal physiology and disease.

In obesity, decreased insulin-binding sites are also found in other cells such as lymphocytes and hepatic cells. Furthermore, with weight loss, the insulin resistance tends to disappear and the number of insulin-binding sites rises toward normal. Thus not only is obesity associated with insulin resistance, whatever the mechanism, but the effect is systemic and not limited to the adipose cell. Although we do not know how it all happens, there is no question that the added strain on the pancreatic beta cell may push the patient with a diabetic tendency into overt clinical diabetes mellitus. Other unexplained hormonal changes owing to obesity include decreased secretion of GH and increased metabolic clearance of cortisol.

If the hormonal abnormalities disappear when obesity is corrected, what then causes the obesity in the first place? For a time attention focused on possible primary abnormalities in adipose tissue itself. However, with the exception of a few patients in whom adipose tissue has difficulty breaking down triglycerides, this does not appear to be the answer to obesity in human beings.

What about the calories eaten? Clearly, a person cannot become obese while undernourished. Calories do count and obesity is the result of eating too much. But overeating, though necessary, is not the whole answer. *Why* do obese patients overeat? Normal nonobese human volunteers who intentionally eat extra food for several months become obese but then, if allowed to eat whatever they wish, spontaneously return to their previous weight. How do they do it? They simply decrease their food intake until the extra fat is mobilized. Somehow the larger mass of adipose tissue decreases their appetitie until the adipose tissue mass is normal again.

Therefore, to become obese spontaneously and remain so probably involves some defect, however slight, in the appetite control mechanism. The defect may be in the generation of a signal from the adipose tissue (see Fig 10–31), in the sensing of some signal by the hypothalamus, or in hypothalamic function. Since we do not understand normal appetite control, it is not possible to define defective control more precisely. Hypotheses such as decreased glucose receptors in the hypothalamus are plausible but not proved. A genetic component may underlie all of this in some patients. Obese parents tend to have obese children; we might say that this could be entirely environmental, but then it is difficult to explain why other children in the same family are thin.

Thus, overeating is the cause of obesity, which in turn results in a number of metabolic defects. The cause of the overeating itself is unknown but may be a functional hypothalamic defect with a disturbance in appetite control. The only effective treatment is to eat less. Many patients lose weight but treatment is not often really effective because they usually regain the weight shortly thereafter. Various types of behavior modification programs seem to offer the best approach to date but this does not mean obesity is solely or even primarily a behavioral problem. Several diets relatively low in carbohydrate and relatively high in protein and/or fat are often useful, perhaps because the high fat intake effectively induces satiety and suppresses hunger. But whatever the approach, the number of calories eaten must be reduced. It is little solace to the patient to look upon a thin physician and wonder how it is done, and certainly an obese physician does not inspire great confidence. It is indeed hard to eat less and keep on eating less, but it is the only way.

APPENDIX: ACRONYMS AND ABBREVIATIONS*

ABP:	androgen-binding protein
ACTH:	adrenocorticotropic hormone (=corticotropin)
ADH:	antidiuretic hormone (=vasopressin)
AT-I:	angiotensin I
AT-II:	angiotensin II
CaBP:	calcium-binding protein
cAMP:	cyclic adenosin monophosphate
CBG:	corticosteroid-binding globulin (=transcortin)
CCK:	cholecystokinin (=pancreozymin)
cGMP:	cyclic adenosine monophosphate
CNS:	central nervous system
CPE:	corona-penetrating enzyme
CRF:	corticotropin-releasing factor
CT:	calcitonin
D:	vitamin D. Also vitamin D_2, vitamin D_3. See also 25OH-D, 1,25$(OH)_2$D.
DF:	decapacitation factor
DHEA:	dehydroepiandrosterone
DHEA-S:	DHEA-sulfate
DHT:	5α-dihydrotestosterone
DIT:	diiodotyrosine
E_1:	estrone
E_2:	17β—estradiol
E_3	estriol
E_4:	estetrol
EGF:	epidermal growth factor
E_2TBG:	estradiol-testosterone binding globulin
FFA:	free fatty acids
FSH:	follicle-stimulating hormone
GH:	growth hormone (also hGH) (=somatotropin)
GHRF:	growth hormone-releasing factor
GIP:	gastric inhibitory peptide
GLI:	glucagon-like immunoreactivity
GnRH:	gonadotropin-releasing hormone (=LRH)
hCG:	human chorionic gonadotropin
hCS:	human chorionic somatomammotropin (=hPL)
hGH:	human growth hormone
hPL:	human placental lactogen (=hCS)
IRP:	insulin-releasing polypeptide
IU:	international units
LH:	luteinizing hormone

*Attempts are now underway to eliminate the use of abbreviations and to give each hormone a name. In many instances the proposed name may not last and, in any case, common usage of many of the abbreviations requires that they be known—hence this list, which includes some (but not all) of the proposed new names.

LPL:	lipoprotein lipase
LRF:	see LRH
LRH:	luteinizing hormone-releasing hormone (=GnRH)
MIT:	monoiodotyrosine
MSA:	multiplication-stimulating activity
MSH:	melanocyte-stimulating hormone
NGF:	nerve growth factor
OAF:	osteoclast-activating factor
25-OH D_3:	25-hydroxyvitamin D_3
1,25(OH)$_2$ D_3:	1,25-dihydroxyvitamin D_3
PCZ:	pancreozymin (=cholecystokinin)
PEPCK:	phosphoenolpyruvate carboxykinase
PG:	prostaglandin
PGE:	prostaglandin E
PGF:	prostaglandin F
PIF:	prolactin-inhibiting factor
PRF:	prolactin-releasing factor
PRL:	prolactin
PTH:	parathyroid hormone
Reverse T_3:	3,3',5'-triiodothyronine
T_3:	triiodothyronine
T_4:	thyroxine
TBG:	thyroxine-binding globulin (also binds T_3)
TGB:	thyroglobulin
TRH:	thyrotropin-releasing hormone
TSH:	thyroid-stimulating hormone (=thyrotropin)
VL:	ventrolateral
VLDL:	very low density lipoprotein
VM:	ventromedial
VMA:	vanillylmandelic acid

BIBLIOGRAPHY

Axelrod, J.: The pineal gland, a neurochemical transducer, Science 184:1341, 1974.

Axelrod, J., and Weinshilboum, R.: Catecholamines, N. Engl. J. Med. 287:237, 1972.

Baulieu, E. E.: Steroid receptors and hormone receptivity, J.A.M.A. 234:404, 1975.

Bedford, J. M.: Maturation of the fertilizing ability of mammalian spermatozoa in the male and female reproductive tract, Biol. Reprod. 11:346, 1974.

Bitensky, M. W., and Gorman, R. E.: Chemical mediation of hormone action, Ann. Rev. Med. 23:263, 1972.

Bray, G. A., Davidson, M. B., and Drenick, E. J.: Obesity: A serious symptom, Ann. Intern. Med. 77:797, 1972.

Chan, L., and O'Malley, B. W.: Mechanism of action of the sex steroid hormones, N. Engl. J. Med. 294:1322, 1976.

Chopra, I. J.: An assessment of daily production and significance of thyroidal secretion of 3,3'5'-triiodothyronine (reverse T_3) in man, J. Clin. Invest. 58:32, 1976.

Cuatrecasas, P.: The insulin receptor, Diabetes (suppl. 2) 21:396, 1972.

Daughaday, W. H., Herington, A. C., and Phillips, L. S.: The regulation of growth by endocrines, Ann. Rev. Physiol. 37:211, 1975.

Davis, J. O.: The control of renin release, Am. J. Med. 55:333, 1973.

DeLuca, H. F.: Vitamin D endocrinology, Ann. Intern. Med. 85:367, 1976.

Diczfalusy, E.: Endocrine functions of the human fetus and placenta, Am. J. Obstet. Gynecol. 119:419, 1974.

Edelman, I. S.: Thyroid thermogenesis, N. Engl. J. Med. 290:1303, 1974.

Federman, D. D.: Genetic control of sexual difference, Prog. Med. Genet. 9:215, 1973.

Felig, P., and Wahren, J.: Fuel homeostasis in exercise, N. Engl. J. Med. 293:1078, 1975.

Fournier, P. J. R., Desjardins, P. D., and Friesen, H. G.: Current understanding of human prolactin physiology and its diagnostic and therapeutic applications: A review, Am. J. Obstet. Gynecol. 118:337, 1974.

Franchimont, P., Chari, S., and Demoulin, A.: Hypothalamus-pituitary, testis interaction, J. Reprod. Fertil. 44:335, 1975.

Frohman, L. A., and Stachura, M. E.: Neuropharmacologic control of neuroendocrine function in man, Metabolism 24:211, 1975.

Genuth, S. M., Castro, J. H., and Vertes, V.: Weight reduction in obesity by outpatient semi-starvation, J.A.M.A., 230:987, 1974.

Gerich, J. E., Charles, M. A., and Grodsky, G. M.: Regulation of pancreatic insulin and glucagon secretion, Ann. Rev. Physiol. 38:353, 1976.

Gill, G. N.: Mechanism of ACTH action, Metabolism 21:571, 1972.

Goldstein, A.: Opioid peptides (endorphins) in pituitary and brain, Science 193:1081, 1976.

Halmi, N. S.: The current status of human pituitary cytophysiology, New Zealand Med. J. 80:551, 1974.

Havel, R. J.: Caloric homeostasis and disorders of fuel transport, N. Engl. J. Med. 287:1186, 1972.

Hers, H. G.: The control of glycogen metabolism in the liver, Ann. Rev. Biochem. 45:167, 1976.

Higgins, C. B., and Braunwald, E.: The prostaglandins, Am. J. Med. 53:92, 1972.

Himms-Hagen, J.: Cellular thermogenesis, Ann. Rev. Physiol. 38:315, 1976.

Ingbar, S. H., and Braverman, L. E.: Active form of thyroid hormone, Ann. Rev. Med. 26:443, 1975.

Kahn, C. R.: Membrane receptors for hormones and neurotransmitters, J. Cell Biol. 70:261, 1976.

Luttge, W. C.: The role of gonadal hormones in the sexual behavior of the rhesus monkey and human: a literature survey, Arch. Sex. Behav. 1:61, 1971.

Martin, J. B.: Neural regulation of growth hormone secretion, N. Engl. J. Med. 288:1384, 1973.

Morris, D. J., and Davis, R. C.: Aldosterone: Current concepts, Metabolism 23:473, 1974.

Nelson, D. N.: Regulation of glucocorticoid release, Am. J. Med. 53:590, 1972.

Oldham, S. B., Fischer, J. A., Capen, C. C., et al.: Dynamics of parathyroid hormone secretion in vitro, Am. J. Med. 50:650, 1971.

Parfitt, A. M.: The actions of parathyroid hormone on bone: relation to bone remodeling and turnover, calcium homeostasis and metabolic bone disease, Metabolism 25:809, 1976.

Queener, S. R., and Bell, N. H.: Calcitonin: a general survey, Metabolism 24:555, 1975.

Rasmussen, H., and Bordier, P.: The cellular basis of metabolic bone disease, N. Engl. J. Med. 289:25, 1973.

Reichlin, S., Saperstein, R., and Jackson, I. M. D., et al.: Hypothalamic hormones, Ann. Rev. Physiol. 38:389, 1976.

Robertson, R. P., and Porte, D., Jr.: The glucose receptor, a defective mechanism in diabetes mellitus distinct from the beta adrenergic receptor, J. Clin. Invest. 52:870, 1973.

Ross, G. T., Cargille, C. M., Lipsett, M. B., et al.: Pituitary and gonadal hormones in women during spontaneous and induced ovulatory cycles, Recent Prog. Horm. Res. 26:1, 1970.

Ruderman, N. H.: Muscle amino acid metabolism and gluconeogenesis, Ann. Rev. Med. 26:245, 1975.

Sawin, C. T., and Hershman, J. M.: Clinical use of thyrotropin-releasing hormone, Pharmacol. Ther. C. 1:351, 1976.

Soll, A. H., Kahn, C. R., Neville, D. M., Jr., et al.: Insulin receptor deficiency in genetic and acquired obesity, J. Clin. Invest. 56:769, 1975.

Steinberger, E.: Hormonal control of mammalian spermatogenesis, Physiol. Rev. 51:1, 1971.

Sterling, K., and Milch, P. O.: Thyroid hormone binding by a component of mitochondrial membrane, Proc. Natl. Acad. Sci. USA 72:3225, 1975.

Thompson, E. B., and Lippman, M. E.: Mechanisms of action of glucocorticoids, Metabolism 23:159, 1974.

Unger, R. H.: Diabetes and the alpha cell, Diabetes 25:136, 1976.

Wallach, E. E.: Physiology of menstruation, Clin. Obstet. Gynecol. 13:366, 1970.

Yen, S. S. C.: Gonadotropin-releasing hormone, Ann. Rev. Med. 26:403, 1975.

Index